# Advanced Digital Audio

# Advanced Digital Audio

Ken C. Pohlmann
*Editor-in-Chief*

***SAMS***
*A Division of Macmillan Computer Publishing*
*11711 North College, Carmel, Indiana 46032 USA*

©1991 by SAMS

FIRST EDITION
FIRST PRINTING—1991

All rights reserved. No part of this book shall be reproduced, stored in a retrieval system, or transmitted by any means, electronic, mechanical, photocopying, recording, or otherwise, without written permission from the publisher. No patent liability is assumed with respect to the use of the information contained herein. Although every precaution has been taken in the preparation of this book, the publisher and author assume no responsibility for errors or omissions. Neither is any liability assumed for damages resulting from the use of the information contained herein. For information, address SAMS, 11711 N. College Ave., Carmel, IN 46032.

International Standard Book Number: 0-672-22768-1
Library of Congress Catalog Card Number: 90-60171

Publisher: *Richard K. Swadley*
Acquisitions Editor: *Charlie Dresser*
Illustrators: *Don Clemons & T.R. Emrick*
Cover Design: *George Harris*
Production: *Martin Coleman, Bob LaRoche, Howard Pierce, & Tad Ringo*
Indexer: *Sharon Hilgenberg*
Composition: *Shepard Poorman Communications Corporation*

*Printed in the United States of America*

# Contents

**Preface** *xv*

**Trademark Acknowledgments** *xvii*

## 1. Human Auditory Capabilities *1*
*Daniel C. Mikat*

| | |
|---|---|
| Introduction | 1 |
| Auditory Physiology | 1 |
| Psychoacoustical Foundations | 6 |
| Spatial Audiology | 8 |
| Other Psychoacoustical Phenomena | 9 |
| Auditory Limitations in Digital Audio Design | 10 |
| Noise and Distortion | 14 |
| Error Correction | 17 |
| Conclusion | 18 |

## 2. Pulse Modulation and Sampling Systems *21*
*Ken C. Pohlmann*

| | |
|---|---|
| Introduction | 21 |
| Pulse Modulation | 21 |
|     Pulse Code Modulation | 22 |

| | |
|---|---|
| Theoretical Performance of Pulse Code | 23 |
| Practical Limitations of Pulse Code | 25 |
| Bit Coding | 27 |
| Sampling Spectra | 29 |
| The Sampling Theorem | 33 |
| Low-Pass Filtering | 37 |
| Sample-and-Hold Circuits | 39 |
| Oversampling | 42 |
| Quantization | 44 |
| Dither | 47 |
| Conclusion | 51 |

## 3. Multibit Conversion     55
*Ken C. Pohlmann*

| | |
|---|---|
| Introduction | 55 |
| Digital-to-Analog Conversion | 55 |
| Conversion Nonlinearity | 56 |
| Zero-Cross Distortion | 60 |
| Eighteen-Bit D/A Conversion Architectures | 63 |
|     Bit-Switching D/A Converter | 64 |
|     Bipolar D/A Converter | 66 |
|     Linear 18-Bit D/A Converter | 67 |
|     Multiple-Bias D/A Converter | 68 |
|     ROM-Compensated D/A Converter | 71 |
|     Dual D/A Converter with Bias | 73 |
| Analog-to-Digital Conversion | 76 |
| Dual-Channel A/D Converters | 77 |
| Conclusion | 82 |

## 4. Laser Technology     85
*Brent A. Karley*

| | |
|---|---|
| Introduction | 85 |
| A Brief History | 85 |
| Laser Physics | 87 |
|     Quantum Mechanics | 88 |
| Laser Elements | 88 |
| Laser Theory | 89 |
|     Spontaneous Emission | 89 |
|     Induced Transitions | 90 |

|   |   |
|---|---|
| The Lineshape | 91 |
| Amplification | 92 |
| Gain Saturation | 93 |
| Oscillation | 95 |
| Laser Characteristics | 99 |
|     Coherence | 99 |
|     Monochromatic Light | 99 |
|     Beam Directionality | 100 |
|     Beam Polarization | 100 |
|     Mode Structure | 101 |
|     Power | 101 |
|     Efficiency | 102 |
|     Lifetime | 102 |
|     Continous-Wave Lasers | 103 |
|     Pulsed Lasers | 103 |
|     Cooling Requirements | 104 |
| The Laser Diode | 104 |
|     Operation | 104 |
|     Structure | 107 |
|     Laser Diode Modulation | 110 |
|     Manufacturing | 111 |
|     Maintenance and Reliability | 111 |
|     Driver Circuitry for the Semiconductor Laser | 113 |
|     Audio Applications | 114 |

## 5. Fiber Optics      *119*
*Brent A. Karley*

|   |   |
|---|---|
| Introduction | 119 |
| Advantages | 119 |
| Disadvantages | 120 |
| The Principles of Fiber Optics | 120 |
|     The Optical Spectrum | 121 |
|     Diffraction | 122 |
|     Reflection and Refraction | 122 |
|     Total Internal Reflection | 124 |
|     The Numerical Aperture | 125 |
|     Dispersion | 126 |
|     Absorption | 127 |
|     Scattering | 127 |
|     Polarization | 127 |
| Optical Fibers and Cables | 128 |
|     Optical Fibers | 128 |
|     Fiber-Optic Cable Design | 130 |

|  |  |
|---|---|
| The Manufacture of Fiber-Optic Cable | 133 |
| Common Failures of Optical Fibers | 136 |
| Maintenance | 137 |
| Transmitters | 137 |
| The Light-Emitting Diode | 137 |
| The Laser Diode | 141 |
| Transmitter Couplers | 142 |
| The Photodetector | 142 |
| Cooling Systems | 142 |
| Receivers | 143 |
| The PIN Photodetector | 143 |
| The Avalanche Photodiode | 144 |
| Fiber-Optic Switches | 144 |
| Mechanical Switches | 144 |
| Electro-Optic Switches | 144 |
| Optical Couplers | 147 |
| **Multiplexers** and Demultiplexers | 149 |
| Wavelength Dispersive Devices | 149 |
| Wavelength Selective Devices | 151 |
| Optical Splicing | 152 |
| Mechanical Splices | 152 |
| Fused Splices | 155 |
| Connectors | 156 |
| Butt-type **Connectors** | 157 |
| Lens-type **Connectors** | 158 |
| Fiber-Optic Audio Applications | 158 |
| A Design Example | 159 |
| Conclusion | 164 |

## 6. Optical Disc Technology *167*
*Brent A. Karley*

|  |  |
|---|---|
| Introduction | 167 |
| Recording Systems | 167 |
| The Compact Disc | 168 |
| The Compact Disc Message | 168 |
| The Compact Disc Medium | 169 |
| Data Format | 169 |
| Structure | 171 |
| Compact Disc Encoding | 172 |
| Sampling | 172 |
| CIRC Encoding | 173 |
| Subcode Word | 174 |
| EFM Modulation | 178 |

|  |  |
|---|---|
| Synchronization Word | 179 |
| The Compact Disc Player | 180 |
| The Optical System | 180 |
|     The Pickup | 180 |
|     Focusing | 183 |
|     Tracking | 184 |
|     Single-Beam Pickups | 184 |
|     Pickup Control | 186 |
| The Electrical System | 186 |
|     Decoding | 186 |
|     The Digital Filter | 187 |
|     The D/A Converter | 188 |
|     The Low-Pass Filter | 188 |
| Optical Recording Systems | 188 |
| Write-Once Systems | 189 |
|     Optical Recording Methods | 189 |
|     CD-WO Discs | 190 |
|     CD-WO Disc Structure | 192 |
|     CD-WO Applications | 192 |
|     Other Write-Once Systems | 193 |
|     Structures | 195 |
|     Audio Applications | 195 |
| Erasable Optical Systems | 196 |
|     Magneto-Optical Recording | 197 |
|     Dye Polymer Recording | 200 |
|     Phase Change Recording | 200 |
|     Disc Drives | 201 |
|     Erasable Optical System Applications | 201 |
| Conclusion | 201 |

## 7. Digital Audio for Video and Film    *203*
*Michael Shawn Ballman*

|  |  |
|---|---|
| Introduction | 203 |
| Digital Audio for Video | 203 |
|     D1—Component Digital Video | 203 |
|     D2—Composite Digital Video | 214 |
|     8 mm Video | 221 |
|     1-Inch Type C Video | 228 |
| Digital Audio for Film | 231 |
|     Design Issues for Digital Sound on Film | 231 |
|     Fluorescent Layer System | 234 |
|     Imaging Dye Technique | 234 |

## 8. Data Compression    241
*Michael Shawn Ballman*

| | |
|---|---|
| Introduction | 241 |
| A Short History of Data Compression | 241 |
| The Measurement of Data Compression | 242 |
| Data Compression in Theory | 242 |
| Information Theory and Data Compression | 243 |
| Distortion Rate Theory | 245 |
| Data Compression in Practice | 246 |
|     Noiseless versus Noisy Coding | 246 |
|     Block versus Nonblock Compression | 247 |
|     Quantization | 247 |
|     Prediction versus Interpolation | 248 |
|     Adaptation | 248 |
| Data Compression Techniques | 249 |
|     Huffman Coding | 249 |
|     Arithmetic Coding | 251 |
|     Ziv-Lempel Coding | 253 |
|     Differential Pulse Code Modulation | 254 |
|     Delta Modulation | 255 |
|     Adaptive Delta Modulation | 256 |
|     Adaptive Delta Pulse Code Modulation | 256 |
|     Transformations | 257 |
| Digital Audio Data Compression Systems | 258 |
|     Design Considerations | 259 |
|     The DAT 16/12, BBC 14/10, and Video 8 10/8 Schemes | 259 |
|     The CD-I and DVI Systems | 260 |
|     A 4:1 Compression System | 261 |
|     The Eureka Systems | 262 |
|     DCC (Digital Compact Cassette) and the PASC (Precision Adaptive Subband Coding) System | 263 |
|     Conclusion | 265 |

## 9. Digital Audio Satellite Broadcasting    267
*Brent A. Karley*

| | |
|---|---|
| Introduction | 267 |
| The Digital Studio | 269 |
|     Baseband Signals | 269 |
| Digital Transmission Processing | 271 |
|     Multiplexing | 272 |
|     Modulation | 272 |

Satellite Transmitters     276
Satellite Design     278
    Station-Keeping Subsystem     279
    Power Systems     279
    Antennas     280
    Transponders     280
Satellite Receivers     281
Direct Broadcast Satellites     283
    DBS in Japan     284
    DBS in Australia     286
    DBS in Europe     286
Terrestrial DAB Systems     288
    Cable DAB Systems     289
    CATV Audio Tuners     290

## 10. Digital Signal Processing: Theory     *293*
*Jayant Datta*

Digital Signals and Systems     293
    The Impulse Response     297
    Convolution     298
    Spectra     301
Transforms     302
    The Laplace Transform     302
    The Fourier Transform     304
    The Discrete Fourier Transform     304
    The Fast Fourier Transform     306
    The $z$-Transform     307
    The Region of Convergence     310
    The Unit Circle and Digital Frequencies     312
Digital Filters     313
    Pole and Zero Patterns and the System Response     315
    Stability     316
    DSP Components     319
    Filter Classification     319
    Comparison of IIR and FIR Filters     324
    Examples of Digital Filters     324
    Errors     329
    Types of Digital Filter Realizations     331
Digital Filter Design Techniques     333
    General Analog-to-Digital Transformations     334
    The Design of FIR Filters with Arbitrary Responses     336
    Windowing     339
    Computer-Aided Design of FIR Filters     340

|  |  |
|---|---|
| Frequency Sampling | 342 |
| Equiripple FIR Filters | 343 |
| Conclusion | 343 |

## 11. Digital Signal Processing: Applications     347
*Jayant Datta*

|  |  |
|---|---|
| Introduction | 347 |
| Digital Soundfield Processing | 348 |
|    Soundfields | 348 |
|    Soundfield Processing in the Home | 350 |
|    Soundfield Processing in the Car | 355 |
| Other DSP Applications in Consumer Audio | 359 |
| Loudspeaker Crossover Networks | 362 |
| Digital Mixing Consoles | 365 |
| The Tapeless Studio | 368 |
| Early Restoration Processes | 369 |
| A Restoration Workstation | 370 |
| Conclusion | 371 |

## 12. Low-Bit Conversion and Noise Shaping     375
*Ken C. Pohlmann*

|  |  |
|---|---|
| Introduction | 375 |
| Low-Bit Conversion | 375 |
| Sigma-Delta Modulation | 377 |
| Noise Shaping | 380 |
| Second-Order Noise Shaping in D/A Conversion | 390 |
| Third-Order Noise Shaping in D/A Conversion | 394 |
| Quasi–Fourth-Order Noise Shaping in D/A Conversion | 401 |
| Low-Bit A/D Conversion | 403 |
| A Low-Bit A/D Converter Chip | 413 |
| A Multilevel Quantizer A/D Converter | 414 |
| Low-Bit A/D-D/A Converter | 417 |
| Conclusion | 419 |

## 13. Digital Signal Processing: Architectures     423
*Matt Booty*

|  |  |
|---|---|
| Introduction | 423 |
| Basic Architecture | 426 |

| | |
|---|---|
| DSP Chip Architectures | 431 |
|    Motorola DSP56001 | 431 |
|    Texas Instuments TMS320 | 435 |
|    AT&T WE-DSPl6A and WE-DSP32C | 437 |
|    Motorola DSP96002 | 441 |

## 14. Digital Signal Processing: Programming and Interfacing     445
*Matt Booty*

| | |
|---|---|
| Introduction | 445 |
| Programming Models | 446 |
| Instruction Sets | 452 |
| Implementing Algorithms with Software | 470 |
| Interfacing | 476 |
| Assemblers, Compilers, and Development Packages | 481 |

## 15. New Markets in Digital Audio     491
*Daniel C. Mikat*

| | |
|---|---|
| Introduction | 491 |
| The Third Wave | 492 |
| A Perfect Audio System | 494 |
| New Digital Audio Markets | 497 |
|    High-Tech Education | 498 |
|       Biomedical Applications | 499 |
| Conclusion | 501 |

**Index**     *503*

# Preface

Imagine, if you will, an inverted triangle with THE BASICS at the bottom, AD-VANCED THEORY in one upper corner, and THE REAL WORLD in the opposite upper corner. The middle of the figure is empty. Unfortunately, this represents the state of affairs in many technical fields. Following successful introduction, it is often unnaturally difficult for the novice to progress to a more sophisticated level, either to pursue research and advanced study or to put either basics or advanced theory into practice. A kind of Escher-like recursion can develop, similar to that in the adage "you can't get a job without experience, and you can't get experience with a job." Sadly, too many students and professionals fail to overcome the problem.

In the spirit of overcoming, it is the aim of this book to help fill the void in the center of the figure, in the technology of digital audio. This book takes a renaissance approach to digital audio, attempting to tie basic topics to more advanced ones, and the theoretical to the practical—a unique and ambitious undertaking.

There are now many books which describe the basic theory of digital audio; this author's *Principles of Digital Audio* is one example. Such books open the door of the art and science of digital audio to anyone willing to expend just a little effort.

Periodicals such as the *Journal of the Audio Engineering Society* address the needs of academics and advanced practitioners alike. Their pages contain the ongoing record of the progress of the evolution of digital audio and provide a forum vital for the exchange of ideas.

Commercial industry publishes a variety of information on the products of digital audio. Specification sheets, schematic diagrams, instruction sets, applications notes, and other documents are important tools for the practicing engineer.

To successfully link the points of this triangle, to allow students to successfully progress from the basics, and practitioners to move between theory and practice, a very specific menu of information was identified for this book: (1) A broad range of topics which encompass the spectrum of digital audio studies, ranging from fiber optics to data compression, from low-bit conversion to DSP interfacing. (2) A selection of topics that is timely and in particular emphasizes areas that stand on the horizon as key technologies. (3) Capstone discussions of digital audio theory in which the most important elements of theory, especially those which find direct applications, are pinpointed. (4) Content that addresses practical, real-world concerns of the engineer and is thus very pragmatic in nature. (5) Specific applications examples, often employing commercial devices, which both illustrate the theory in use and demonstrate how implementation is carried out. (6) A bibliography for each chapter which helps point the way to more advanced reading, especially in theoretical areas.

To carry out this agenda I enlisted the aid of five talented author-engineers. After considerable discussion, we settled on a list of topics that we considered extremely relevant and timely. Individual topics, usually areas of personal specialization, were assigned to each author. Following a lengthy editing and review process, the result is now in your hands. To the best of our ability, *Advanced Digital Audio* comprehensively presents those topics most currently of interest and presents them in a coordinated manner, with a special emphasis on newly developed theories and techniques. Above all, these are practical discussions.

This book is primarily intended for the intermediate student and practicing audio engineer. Although theoretical basics are summarized, discussions are generally sophisticated in that they presume an engineering background. Problematically, the book tackles a range of topics that is extremely wide. However, that spectrum's breadth is an important aspect of today's digital audio field, and one aim of the book is specifically to provide a concise summary of the many technologies that pertain to digital audio systems.

Every chapter of this book addresses a topic that has an impact on the development of future digital audio technology: human auditory capabilities, pulse code modulation and sampling systems, multibit conversion, laser technology, fiber optics, optical recording, digital audio for film and video, data compression, digital audio satellite broadcasting, DSP theory, DSP applications, DSP low-bit conversion and noise shaping, DSP hardware architectures, DSP programming and interfacing, as well as a discussion of new commercial markets in digital audio. This is a list of "hot buttons" that will dictate the development of digital audio over the next decade and beyond. Each chapter concludes with a list of references.

On behalf of the authors of this book, I hope that its text provides a path that is both intellectually and commercially useful to you. I hope that its panorama between theory and practice serves to unite these sometimes disparate fields. I hope that its diversity of topics helps persuade you that digital audio has quickly grown into a discipline that is integral in our technological society.

*Ken C. Pohlmann*

# Trademark Acknowledgments

All terms mentioned in this book that are known to be trademarks or service marks are listed below. In addition, terms suspected of being trademarks or service marks have been appropriately capitalized. SAMS cannot attest to the accuracy of this information. Use of a term in this book should not be regarded as affecting the validity of any trademark or service mark.

ADS is a trademark of Motorola, Inc.
AIB is a trademark of Texas Instruments, Inc.
Apple is a trademark of Apple Computer Corporation.
Apt X 100 is a trademark of Audio Processing Technology, Ltd.
Betamax is a trademark of Sony Corporation.
Cinedigital/Fluorescentsound is a trademark of Peter Custer.
Cinema Digital Sound is a trademark of Optical Radiation Corporation and Eastman Kodak Corporation.
DFDP is a trademark of Atlanta Signal Processors, Inc.
Digital Filter Design Package (DFDP) is a trademark of Atlanta Signal Processors, Inc.
DSP16 is a trademark of AT&T.
DSP32 is a trademark of AT&T.
DSP56001 is a trademark of Motorola, Inc.
DSP96002 is a trademark of Motorola, Inc.
DVR-100 is a trademark of Sony Corporation.
Flying Erase is a trademark of Sony Corporation.
IBM is a trademark of IBM Corporation.
Intel is a trademark of Intel Corporation.
Intertools is a trademark of Intermetics, Inc.
Kevlar is a trademark of DuPont Company.
Macintosh is a trademark of Apple Computer Corporation.
Media Engine is a trademark of Motorola, Inc.
MS-DOS is a trademark of Microsoft, Inc.

OnCE is a trademark of Motorola, Inc.
Sun is a trademark of Sum Microsystems, Inc.
SWDS is a trademark of Texas Instruments, Inc.
TMS is a trademark of Texas Instruments, Inc.
TMS320xx is a trademark of Texas Instruments, Inc.
UNIX is a trademark of Bell Laboratories.
VAX is a trademark of Digital Equipment Corporation.
Video 8 is a trademark of Sony Corporation.
WE is a trademark of AT&T.
XDS is a trademark of Texas Instruments, Inc.

Chapter 1

# Human Auditory Capabilities
Daniel C. Mikat

## Introduction

The principles of audiology and the study of psychoacoustical perception are important in understanding the limitations of our hearing mechanism. Frequency range, threshold of hearing (minimal sound level perceivable), maximum loudness, distortion, envelope and localization all guide engineers in the design of digital audio systems since it is the listener's perception which is the most important factor in the design of an audio system. The motive behind the study of the limitations of the human auditory system is not insight into how much distortion or noise is audible per se. More appropriately, our understanding of audiology leads research and study in engineering which will give improvements in sound production quality. Indeed, this dual enterprise has already resulted in digital audio devices in which the quality of the device surpasses the limitations of the most critical listener. To that end, in this chapter, basic auditory physiology is explained, principles of psychoacoustics are presented, and auditory capabilities with respect to frequency, loudness, localization, and distortion are discussed as they relate to the digital audio system.

## Auditory Physiology

The auditory mechanism can be described in three parts, each of which performs functions mandatory for normal hearing ability. Any deficiency in the chain will result in either altered perception or no perception at all. The outer ear, middle ear, and inner ear are shown in Figure 1. The outer and middle ear are separated by the tympanic membrane or eardrum. The middle and inner ear are joined at the oval window and round window.

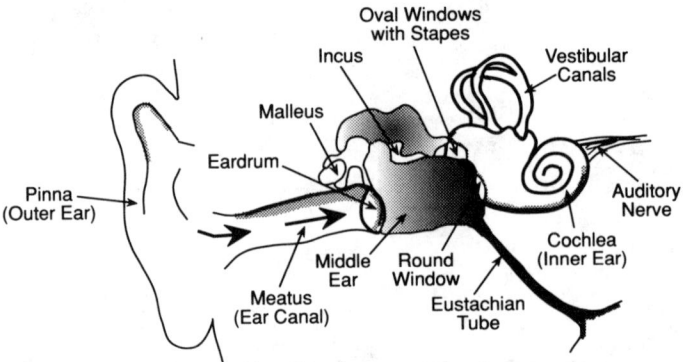

**Figure 1.** The human auditory mechanism.

The visible ear or pinna serves little useful purpose in our discussion other than in the collection of sound. It has been surmised by some that it functions in some higher-order localization functions. This is evidenced in animals, such as dogs or foxes, which have rotatable pinnas. When the animal hears a sound and is uncertain of its location, it rotates the outer ear so that sound reaching the ear canal becomes either more or less direct, and the frequency spectrum reaching the eardrum changes, giving the animal further directivity cues. In humans, the same function can be realized by turning the whole head since the outer ear is for the most part immobile. This type of localization skill is attributed to diffraction. It is well established that low-frequency content is more easily diffracted than high frequencies. Through higher-order acoustic processing in the brain, a listener makes the distinction that as the head is rotated, the point at which the high-frequency content is most prevalent is the point where the ear is on-axis with the sound source. This occurs subconsciously and quickly so that good location discrimination is possible even if only one ear is functioning.

The ear canal (meatus) serves several functions. First, it protects the thin eardrum from foreign objects such as fingers, or instruments which may cause damage. The other major function of the ear canal is as a resonator for acoustic frequencies centered at about 3 kHz. This coincides with the region of frequencies which hold a great deal of intelligibility information of human speech.

The eardrum (tympanic membrane) functions in two important ways. First, it transduces the acoustical energy impinging upon it into mechanical energy, which is then transferred further through the tiny bones of the middle ear to the oval window. These bones work as a pivoting lever, so that a compression of acoustic energy at the eardrum will result in a tension of mechanical energy at the oval window. This results in the transmission of signal 180° out of phase with the actual sound, as shown in Figure 2. The Eustachian tube acts as a port for pressure equalization so that the motion of the eardrum is not hindered by a positive or negative pressure inside the middle ear in relation to the atmospheric pressure of the outer ear. It is interesting to note that the reason you may not hear as well when you have a cold or ear infection is because the Eustachian tubes are blocked and the changes in middle ear pressure cannot be released. The second major function of the eardrum is to stop acoustical energy from passing through to the middle ear. If the eardrum is torn or perforated, the in-phase acoustic energy will partially cancel the out-of-phase mechanical energy reaching the oval window, resulting in inefficient transduction.

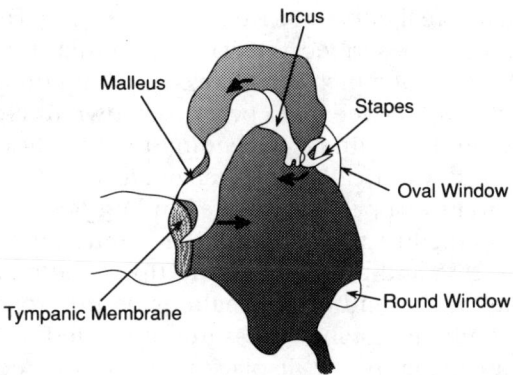

**Figure 2.** The bones of the middle ear. A forward motion of the malleus arm pivots the incus and stapes to move in the opposite direction. This results in motion of the oval window 180° out of phase with that of the tympanic membrane.

The need for the middle-ear bone transduction becomes obvious when one considers the inefficiency of passing acoustic energy from a gas (air) to a fluid (cochlear fluid). When acoustic energy moves from air into saltwater (which has approximately the same density as fluid in the human cochlea) 99.9% of the energy is reflected, allowing only 0.1% to pass into the water. Once sound pressure enters a fluid, however, the density of the medium allows for very efficient transmission at high speeds. Fluids and solids transmit sound several times more efficiently than gases.

If there were no middle ear or tympanic membrane and the air were directly coupled to both the oval window and the round window, the variations in air pressure would meet both windows concurrently, affecting the fluid pressure slightly on the inside of the cochlea. When the closely packed molecules of a fluid are further pushed together they exert significant resistance to the force exerted on them. If, instead, the oval window were connected to the vibrating arm of the stapes and the round window was allowed to move freely, we would notice a tremendous increase in efficiency. As the stapes moves toward the cochlea, the oval window gives slightly, moving the small quantity of fluid inside the cochlea longitudinally along its length. When the pressure meets the round window, it is released back into the middle ear also aided partially by the rarefaction of air pressure in the middle ear. The rarefaction is caused by the 180° out-of-phase design that the three pivoting bones achieve.

The middle-ear function is a clever means of overcoming an extremely severe efficiency problem.

The middle ear, as mentioned, conducts the mechanical energy of the eardrum to the oval window via three small bones called the malleus, incus, and stapes. Since the area of the eardrum is much larger than that of the oval window, there is a benefit from the areal effect; the displacement of the eardrum results in a much greater force per unit area at the oval window. This improves hearing efficiency by an average of 25 dB.

The areal effect was first investigated by Hermann von Helmholtz in 1863. Helmholtz described the tympanic radial fibers as individual tension units. Each fiber (of which there are several hundred) functions the same as a chain suspended between two posts, as shown in Figure 3. The posts exert outward force on the chain far exceeding the weight of the chain, pulling it taut. If a force acted vertically on the chain, it would be displaced but would be pulled back to its rest position by the binding posts. Each radial fiber of the tympanic membrane functions in a similar manner, with one end anchored to the tympanic ring and the other to the tip of the malleus. The middle of the membrane, at the end of the malleus, is free to move. This force acting on the tympanic membrane (pressure multiplied by membrane area) multiplied by the mean amplitude displacement must be equal to the malleus force multiplied by its amplitude displacement. Since the motion of the malleus is significantly less than the total displacement of the tympanic membrane, its force must be significantly greater as transduced to pressure on the oval window. The areal effect accounts for a 15 to 65 dB increase in sensitivity depending on the nature of signal being transmitted.

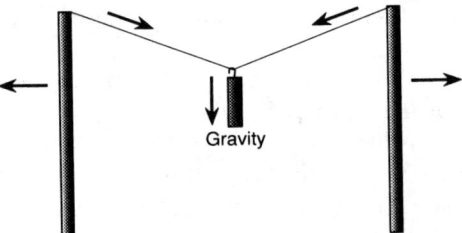

**Figure 3.** Helmholtz's illustration describing the areal effect. The weight suspended by the chain exerts downward force on the chain which is resisted by the binding posts affixed to the ends of the chain.

The middle-ear bones are arranged in such a way as to give a further mechanical advantage, increasing the average sound level by about 2 dB. With increased age, the bones of the middle ear may become calcified or deteriorated. Reconstructive surgery may be performed to remove calcium deposits that inhibit movement or the bones may be replaced with prosthetic ones.

The inner ear, or cochlea, is where the magic of hearing takes place. The fluid-filled cochlea is in the shape of a snail shell which, if unrolled, would extend about an inch and a half. The point of connection of the cochlea to the middle ear is termed the base, while the farthest extended point along the length of the cochlea is the apex. A diagram of a cross section of the cochlear duct is shown in Figure 4. A compression at the oval window travels through the cochlea via the scala vestibuli to its apex, where it returns and travels back through the cochlea via the scala tympani. At the end of the scala tympani is the round window which releases the pressure back into the middle ear 180° out of phase with the pressure at the oval window. A disturbance in pressure at the oval window causes a vibration at some point along the basilar membrane. The basilar membrane holds the organ of Corti, which consists of nerve hair cells and adjoining auditory nerve fibers. The point of disturbance on the basilar membrane is determined by the frequency of the vibration. At this point of vibration, the tectorial membrane moves against the nerve hair cells on the organ of Corti. When these hair cells move, the friction causes an electrical stimulation to be sent through various ganglia, axons, and synapses to the brain where it is processed.

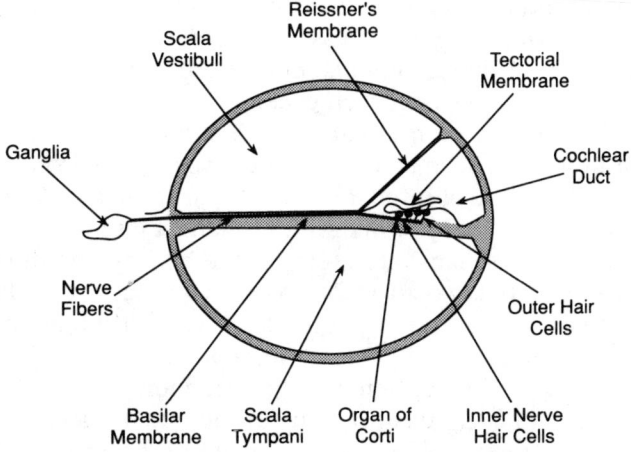

**Figure 4.** Expanded cross section of the inner cochlea.

When a complex acoustical wave stimulates the hearing mechanism, several regions are excited along the basilar membrane. One region is excited for each component of the signal within the audible hearing range. The individual nerve hair cells transmit their data through the auditory nerve to the brain. At the brain, judgments are made regarding the relationships (both in frequency and amplitude) of the frequency component partials present in the acoustic wave. The manner in which the various components are related is delivered to the ear as a ratio of their respective amplitudes. These ratios are deciphered by the brain to yield a perception of harmonic richness, or timbre. In this way, the cochlea functions as a kind of mechanical Fourier transform in which the individual component frequencies of a signal are separated and each stimulates a specific point along the basilar membrane. This is not, however, a perfect transform but instead a transform with leakage on either side of the intended point of excitation. The brain examines both the period of the incoming stimulus and the point of maximum excitation to determine the actual frequency of the waveform component. The surrounding regions also excited are ignored by the central nervous system in the processing of the stimulus.

Each of the three sections of the hearing mechanism are of great importance to normal hearing ability. Any deficiency or dysfunction at any point in the hearing chain will degrade perception in some way. It is clear that for audio engineers an understanding of these principles should be the cornerstone of the effectiveness of a system design.

## Psychoacoustical Foundations

Psychoacoustics is the study of how acoustic transmissions are perceived by a listener. It relates physical quantities, such as frequency and intensity, to perceptual qualities, such as pitch and loudness. Some knowledge of psychoacoustics is necessary in order to understand the manner that music, speech, or environmental cues are processed and how different attributes of sound interact with each other.

Pitch in the musical sense is the perceived correlate of frequency in the physical sense. As the number of cycles per second increases linearly, our perceived sense of pitch increases logarithmically. To illustrate this, consider the musical half-step, which is the smallest interval between two notes in Western music. Musicians often refer to the half-step as an interval of a minor second, for example two adjacent notes on a piano (regardless of whether they are at the bottom or the top of the keyboard). If tuned in an equal tempered scale, each half-step will present the same ratio of frequencies, namely $2^{1/12}$. In other words, the higher note will be a frequency 1.05946 times the frequency of the lower note. For example, the difference in cycles per second between the two adjacent notes A1 and A#1 is given by

$$(110 \text{ Hz} \times 1.05946) - 110 \text{ Hz} = 6.541 \text{ Hz}.$$

The same interval four octaves higher yields a difference of

$$(1760 \text{ Hz} \times 1.05946) - 1760 \text{ Hz} = 104.650 \text{ Hz}.$$

The minor second interval in the low-frequency range presents a frequency difference much less than the frequency difference of the same interval several octaves higher. This logarithmic increase in pitch with a linear increase in frequency establishes musical intervals as being related in ratios of frequency, not differences.

According to pitch-place theory a pure tone will stimulate one small portion along the basilar membrane. A specific point of displacement will always provide the sensation of the same pitch. The area on the membrane farthest from the oval window (i.e., the apex) is responsible for low-frequency perception. As stimulation occurs farther and farther from the apex, the sense of pitch rises until the maximum perceptual frequency is stimulated at the base of the cochlea near the oval window.

Pure tones occur rather infrequently in nature and in music. More common is a complex signal which is composed of several frequency components. The number of frequencies present and their mathematical frequency relationship determine the timbre or tone quality of sound. If the partials are all whole number multiples of the lowest frequency present, they are said to be harmonically related where the lowest frequency is the fundamental frequency and the others are overtones. As noted, when a complex tone reaches the cochlea, several different regions along its length are stimulated. Each stimulated region sends a separate signal to the brain, where higher level processing reveals to the perceiver the manner in which they are related. Although considerable research into this high-level processing is under way, the actual processes involved are as yet unknown.

Intensity is a physical property which determines the perceived loudness of sound. Although intensity is the major factor in loudness, other factors such as timbre, pitch, and durational envelope also contribute to it. A violinist can give the impression of varying loudness without an actual change in intensity merely by changing the manner in which the bow is used, which changes the timbre but not necessarily the intensity. With proficient bowing technique, a change in timbre can be achieved by bowing closer or farther from the bridge, and the slope of the durational envelope can be changed by controlling the speed of the bow. Both duration and timbre changes affect the perceived loudness of the stimulus. Tones with more prominent overtones tend to sound louder. In order to obtain reliable objective judgments of loudness, a minimum duration of up to 500 ms may be necessary depending on the frequency of the stimulus (lower frequencies require a longer duration). A particularly long and steady sound will trick a listener into thinking that its intensity is lower than it is due to the aural phenomenon of auditory habituation, described below.

Intensity is measured in decibels (dB) using the equation

$$x \text{ dB} = 10 \log(I/I_{\text{ref}}),$$

where $x$ is the number of decibels. The decibel is a measure of difference between a measured value and a reference. For purposes of measuring loudness intensity, dB SPL (sound pressure level) is used. Intensity is measured in watts per square meter (w/m$^2$) and is inserted into the above equation as $I$. The term $I_{\text{ref}}$ is the reference intensity for the normal threshold of hearing at 1000 Hz. This threshold is $10^{-12}$ w/m$^2$. The displacement of the eardrum at this power level is about one tenth the diameter of a hydrogen molecule. This demonstrates the incredible sensitivity of the human hearing mechanism. At the nominal threshold of pain (120 dB), the intensity ($I$) is 1.0 w/m$^2$. This intensity will cause immediate temporary threshold shift (TTS). Until a recovery period of perhaps several hours has elapsed, the softest audible sound will have shifted to a louder level. Prolonged exposure to levels even above 80 dB has been shown to cause permanent hearing loss, primarily in the higher frequency range. Due to the logarithmic relationship of loudness to intensity, a 6 dB increase in intensity is perceived as twice as loud, so that the difference of 6 dB between 100 dB and 106 dB seems considerably more significant than the same difference of 6 dB between 30 dB and 36 dB.

Localization is a complex higher-order hearing process to determine the direction from which a sound appears to be coming. Stereo imaging describes the reproduction of audio to achieve localization effects. The sense of localization depends on three factors: time delay, spectral differences, and loudness differences between the two ears. A sound occurring to one's far right will reach the right ear approximately 0.67 ms before reaching the left. The right ear will hear the sound directly, while the left ear is blocked by the head. This will result in a frequency spectrum difference between the ears. The left ear will note an attenuation of higher frequencies, while lower frequencies can diffract around the head, arriving intact. The sound will also be perceived as being louder in the right ear than the left because of the inverse square law. These three cues work together to provide localizing differentiation. If someone is deaf in only one ear, good localization ability can still be attained by moving the head (usually done subconsciously) to hear how the frequency spectrum changes when the sound is perceived at a different angle to the stimulus. The order of importance in the three factors in stereo imaging is not clear. However, it has been shown that if a clear time delay is perceived between the ears, for it to appear in a central location the intensity of the signal to the delayed ear requires an increase of as much as 20 dB. This leads us to believe that the use of time delays in stereo imaging is at least as useful and dramatic (and probably more so) than amplitude panning.

## Spatial Audiology

Spatial processing has been employed in ambience processing devices for many years. Formerly, ambience processing meant the addition of echoes or reverberation to warm the sound or mimic the effect of a certain type of hall. Recently, with efficient and complex digital signal processing, greater versatility has been achieved. Now, digital delay, looping schemes, and cyclic panning can achieve incredible spatial effects.

The brain can be fooled rather easily by feeding complex information to the ears. From birth we are conditioned to perceive location (where the sound is coming from) and localization (where the listener is with respect to the sound source) in a consistent way. For example, a certain kind of ambience having certain delay ratios is generally associated with a certain type and size of room. If that ambience can be mimicked closely enough, the listener will make the necessary associations and be fooled aurally into feeling he or she is in the room.

However, understanding how ambience is perceived is no small endeavor. It involves very complex three-dimensional processes. For example, manufacturers have made complex studies of spatial perception to develop automobile DSP ambience processors which very realistically mimic ambience conditions. A car's passenger compartment can take on many characteristics of the copied environment and seem to have a tremendous fidelity for certain hall settings. The ambience was characterized by making a multitude of recordings in the environments studied. The delay characteristics in all directions were measured and recorded. To be truly effective, 360° sound reproduction is employed with the initial sound arriving from on-axis in front of the listener. After a short delay first-order echoes are heard from the front and immediate sides of the source. After longer delays, more echoes and reverberation fill the rear and front of the soundstage. This surround sound can be extremely effective in fooling the brain regarding the true ambience of the actual environment. Similarly, manufacturers have studied similar effects in an audio/visual environment. The effect can be quite stunning; even if the speakers are close to the television screen, the sound source can appear to move beyond the spatial location of the speakers. The manipulation of auditory perception through spatial processing is very complex but can be quite convincing to the listener. To achieve maximum effectiveness from these smart ambience processors, dry recordings are preferable where as little reverberation as possible is recorded with the program material. In this way, the prerecorded reverberation will not interfere with the added ambience provided by the processor. These and other digital signal processing applications are investigated at some length in Chapter 11.

# Other Psychoacoustical Phenomena

When one is sitting in a concert hall listening to classical music, the extraneous noise occurring throughout the performance may not be particularly obvious. Through selective discrimination, one may focus attention on the program material while perceptually editing out audience noise such as coughing, fidgeting, loud breathing, or background hall noise from air-conditioning and lighting systems. In a recording, however, such distractions become more apparent and even exaggerated. Live recordings are difficult to engineer for this reason and are often deemed unsuitable for release because of obtrusive auditory distractions which are perceptually benign at the concert site. Apparently the visual aid of a live concert gives extraordinary benefit in selective discrimination.

Auditory habituation is a natural phenomenon in which listeners are more aware of changes in program level or noise floor than a static noise floor, even if the static level is greater overall. A noise or distortion that is correlated with a signal may be replaced or masked by an uncorrelated noise floor which is more easily ignored and is perceptually benign. Methods of noise processing are discussed in Chapters 2 and 12.

# Auditory Limitations in Digital Audio Design

The limitations of auditory perception specifically affect the design of a digital audio system. An understanding of the limits of normal human hearing is paramount in determining performance specifications of the system and in finding which aspects in the digital conversion; processing, compression, and storage processes will produce artifacts that will be audible. Fundamental to the design of any digital audio system is an understanding of frequency response and dynamic range requirements.

Through audiometric testing (measuring thresholds of hearing for various frequencies) it has been found that most people cannot hear outside the frequency range from 20 Hz to 20 kHz. Practically, most people cannot hear much above 15 kHz without the sound level being boosted considerably. At Jasna Gora Cathedral in Katowica, Poland, a very elaborate pipe organ can produce tones over the entire audible range (and beyond). Some of the largest pipes can emit frequencies as low as 6 Hz. It was surmised by the organ designers that even though these low-frequency tones are clearly below the normal range of hearing, they can be felt tactilely, enhancing the overall effect of the other tones that are within the audible range. During a blind demonstration with the same passage played several times, with and without the bass subaudible tones, observers agreed that the passages which included the subsonic frequencies seemed broader and more full, thus supporting the designer's theory. In a like manner, perhaps supersonic frequencies higher than 20 kHz can be perceived nonaurally just as subsonic frequencies are.

One's threshold of hearing is raised naturally with increased age and cumulative sound and noise exposure during daily activities. Figure 5 shows how hearing ability at high frequencies in particular decreases dramatically with increased age, especially for men. This degradation is accelerated significantly in people who work in noisy environments or voluntarily expose themselves to loud music, as shown in Figure 6. Many loud workplaces, such as machine shops or construction sites, require hearing-protection equipment to be worn in accordance with OSHA (Occupational Safety and Health Association) regulations. Anyone who has conversed with a long-time employee of a machine shop understands the severity of this functional loss, a loss that will never be mended naturally.

Studies show that teenagers today are experiencing significant hearing loss especially at high frequencies. This is due partially to the influx of portable players which encourage young listeners to play music at high levels. Another significant factor is attendance at concerts where the extreme loudness is thought (and rightly so) to add to the excitement. The Who even went so far as to advertise a concert as "the loudest concert in history." While it is difficult to persuade people to not attend such events, hearing protection can and should be encouraged. It is interesting that even some audio engineers demonstrate disregard for their own auditory health by willfully degrading their auditory acuity, which is the very vehicle of their employment. It is a shame that the

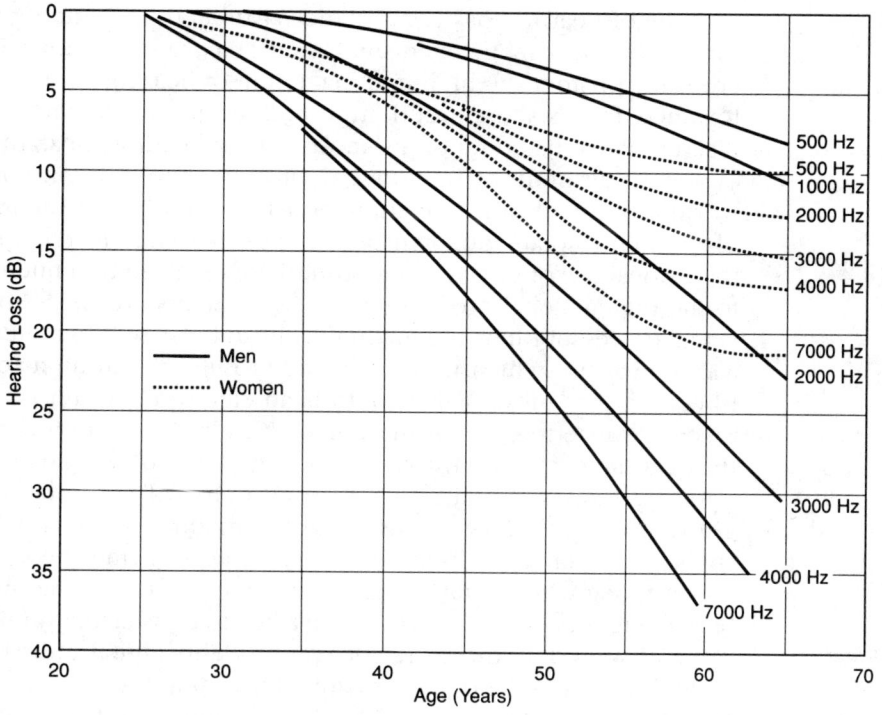

**Figure 5.** Typical hearing loss with increased age.

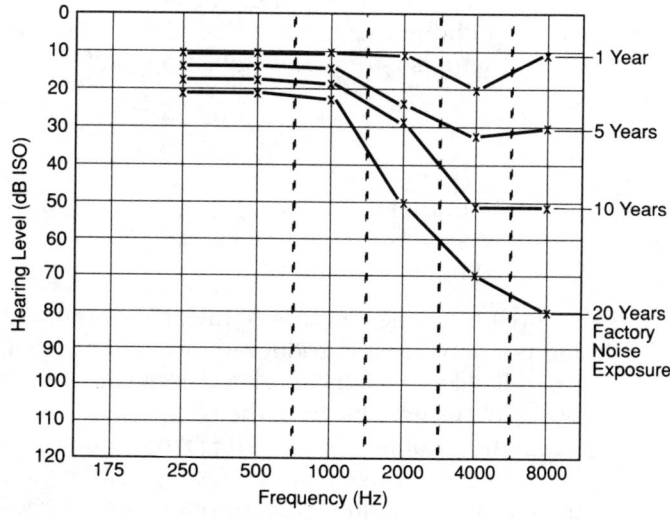

**Figure 6.** Increase in hearing loss as a result of repeated exposure to loud environments.

majority of hearing loss cases could have been prevented given respect and care for the hearing mechanism. Under practical conditions, it would be generous to consider the high-end maximum of hearing ability to be 20 kHz. Allowance for this extended high-end frequency range would allow for discriminative listening of the most healthy hearing apparatus.

As an example of how digital audio technology can correlate to a real world physiological dysfunction, consider the biomedical application of implantation. Implantation can be a useful treatment to restore hearing to the totally deaf. Incoming acoustic stimuli are received by a microphone and sent to the digital signal processing unit worn in a shirt pocket or on a belt. The signal processor filters the incoming signal into a number of discrete bands via a DSP program which mimics cascaded Butterworth high-pass and low-pass filters. Each of these bands controls an electrical signal that is emitted by an electrode inserted within the cochlea itself. With multiple-frequency bands the deaf can hear not only when acoustic energy is being received but its loudness, approximate pitch, and harmonic content. The results are still somewhat crude, but an understanding of environmental cues and intelligibility of speech are attainable. By understanding nerve-firing theory and place-pitch theory along with digital signal processing technology, biomedical engineers have come very close to overcoming a health issue faced by millions of people.

With regards to the design of a high-fidelity digital audio system, the range of hearing guides the determination of the sampling rate and filtering scheme. In order to digitally record the highest frequency in the audible range, a sampling frequency of at least twice that frequency must be employed. This will guarantee two samples for the highest frequency and increasingly more samples as the signal frequency decreases. For a maximum frequency of 20 kHz a sampling frequency of 40 kHz is needed. If a frequency greater than 20 kHz enters the system a phenomenon known as aliasing will occur. With aliasing there are not enough samples available to encode a frequency and the system incorrectly outputs a frequency given by the following equation:

$$f_{out} = f_s - f_{in},$$

where

$f_{out}$ is the output frequency,
$f_s$ is the sampling frequency,
$f_{in}$ is the input frequency.

As the input frequency increases, the aliased output frequency decreases. This aliasing is audible and introduces unwanted components at the output, which are generally harmonically unrelated to the program material. Aliasing is further confusing to a listener because the component frequency causing the alias tone is beyond the audible hearing range. To eliminate the possibility of aliasing, a low-pass filter is employed just below one half of the sampling frequency so that it is impossible for frequencies above that limit to be converted. By placing this filter below half the sampling frequency, aliasing is avoided. For practical low-pass filter implementation, consumer digital audio systems generally employ a sampling frequency of 44.1 kHz with a cutoff frequency at 20 kHz. Alternatively, systems may use oversampling techniques thus allowing a more gradual low-pass filter which produces less phase shift in the audible range.

A digital conversion system can handle frequencies on the low end of the audible spectrum down to DC. If we want to record an organ with subsonic pedal tones we can do so easily as long as the microphone is able to transmit those frequencies. A common criticism of digital audio is that the resulting sound has a harshness to it which is not present in analog recordings. This criticism is invalid at least with respect to frequency response since the audio sampling process can deliver flat frequency response from DC all the way through the human hearing range.

The decibel, as we have discussed, is a representation of the intensity of a sound in relation to the threshold of hearing. This threshold (0 dB at 1 kHz) of course varies from person to person, but the 0 dB standard is defined as the normal threshold for healthy young adults ($10^{-12}$ w/m²). As a stimulus increases in intensity from this threshold, the motion of the eardrum is increased and likewise the disturbance of the tectorial membrane against the nerve hair cells increases to deliver a sensation of increased loudness. The loudness doubles with every 6 dB increase in the stimulus, until approximately 120 dB is reached. At this point the eardrum vibrates at its maximum excursion and will begin to distort the waveform. As intensity is further increased (beyond the threshold of pain), the eardrum may stretch a bit further but will transmit little more perceptual loudness. As this waveform becomes distorted to something resembling a square wave, the actual loudness delivered will begin to decrease as illustrated in Figure 7. In relation to a digital audio system, the maximum signal level corresponds to the exercising of all the quantization levels available in a PCM digital audio system. This level yields the minimum signal distortion throughout the digitization chain. A signal level above this, on the other hand, results in severe clipping distortion.

The minimum sound pressure level available from a PCM digital audio system would be on the order of one quantization level. (Dither can extend a digital audio system's dynamic range, as described below.) This signal will deliver a level below the normal threshold of hearing. The maximum playback level for consumer digital systems is approximately 105 dB SPL as limited by the standard home speaker specifications. The maximum professional level allows for an additional 24 dB SPL output. This professional level is 129 dB SPL. The maximum playback levels vary from device to device, but these approximate values shall serve in our discussion of dynamic range and signal-to-noise ratio.

More important than the actual maximum and minimum levels available on the system is the effect of dynamic range. Dynamic range is the difference in decibels between the maximum SPL of the system and the white noise floor of the output. The minimum white noise floor just noticeable for most listeners is at 4 dB SPL. For the system to be audibly perfect, the dynamic range must be at least 101 dB for consumer systems and 125 dB for professional systems. To achieve dynamic ranges of these magnitudes, 16-bit resolution conversion

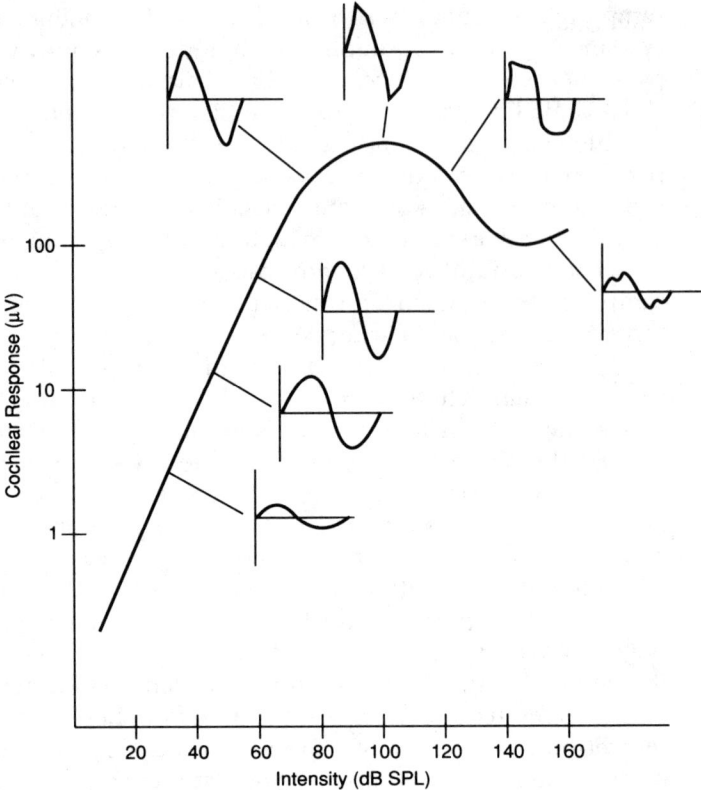

**Figure 7.** Cochlea response with increasing sound intensity. The distortion is due to the physical restraints of the tympanic membrane.

is required for consumer PCM conversion systems and as much as 20-bit resolution conversion for professional use. (In practice, 1-bit converters may be employed to achieve this resolution.) More bits are needed for professional use to reduce quantization error with an increase in dynamic range. Quantization error is discussed below.

# Noise and Distortion

The study of the audibility of noise and distortion is of significant interest. Eliminating noise and distortion in a digital system has been of primary concern since the introduction of the technology. In analog sound systems, noise is in the form of additive white Gaussian noise (AWGN) which is spectrally constant and benign to a listener; this noise floor is considered benign since it is uncorrelated with the audio signal. Distortion in an analog system can often be minimized through stringent design and manufacture of its components. For example, wow and flutter are decreased with speed regulators and phase-locked loops. Noise and distortion in a digital system are of another nature. Namely, they can be correlated with the signal, that is, they change according to the signal behavior. A correlated noise seems to increase perceptually at low signal levels and can be quite audible. There are, however, methods for dealing with this. They often require an investigation of the critical band concept and masking curves, to relate distortion and noise to our auditory mechanism and how they are perceived.

The idea of masking can be easily illustrated by considering a phone conversation in which one converser is speaking in a noisy environment. For the listener to hear the speaker, the signal level must be of a level significant in loudness compared to the environmental noise. The environmental noise is the masker and the masked signal is that from the speaker at the receiving end of the phone line. This condition seems intuitively obvious but becomes complicated when frequency-critical bands are taken into account. Given a pure sinewave tone occurring at some arbitrary frequency (masker), in order for another tone (masked) to become audible in the presence of the first tone, it must attain a certain amplitude. The required amplitude is much greater than that necessary for the tone to be audible without the masker present. If this masking tone is very close in frequency to that of the first tone, a large amplitude is required to overcome the masking effect of the first tone, while if the frequency of the masked tone is far removed from that of the masker only a moderate amplitude may be required. To clarify this, the masking principle may be illustrated using masking curves. Consider the masking curve in Figure 8. When a masking signal of 500 Hz is present at an amplitude of 70 dB, another signal must rise above the curve to be audible. Perhaps more significantly for our discussion, only noise occurring below the curve is inaudible.

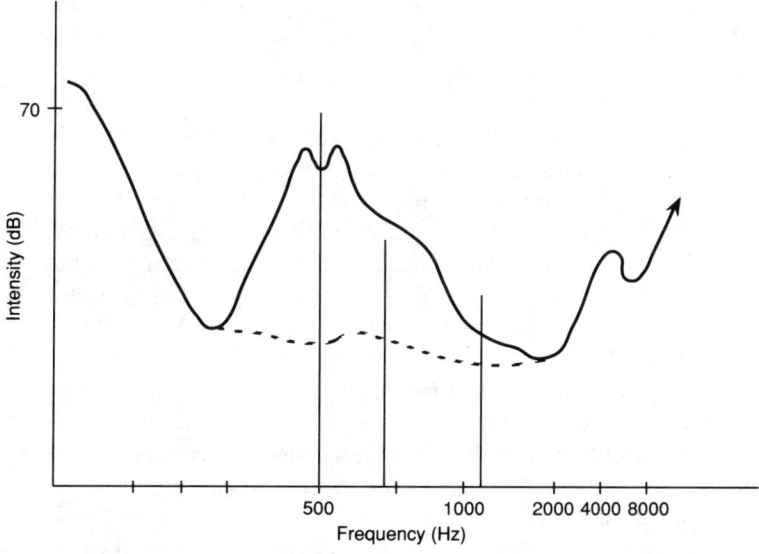

**Figure 8.** Masking resulting from 500 Hz sinewave tone at 70 dB.

The curve's height at a certain frequency represents the level to which a masked signal must be raised in order to become audible in the presence of the masker. This is demonstrated when listening to a noisy sound system. Even with a significant static noise floor, when broad-band, high-level music is played, it is impossible to hear the noise. As the level of the masker is increased, the masking curve is raised in height and broadened in frequency.

Quantization error is inherent in any digital audio system. In a PCM converter it stems from approximating the value of the analog input at a sample time to a predefined quantization level. In ideal converters, the degree of error in approximation can be as much as one half of a quantization level and on the average is one quarter of a quantization level. Since this type of error is constant regardless of the level of the input signal, its audibility increases as the level of the signal decreases. Low-level signals that operate on only a few quantization levels have a greater error relative to that signal.

A floating-point converter attempts to make use of the unused headroom of quiet signal conditions through an automatic gain control. It keeps the program material at the maximum level for digital encoding and decoding. By doing this, the quantization noise will always be small relative to the signal. The difficulty in designing this system is in keeping the gain switching inaudible. The switching of the program material may be inaudible when it is done on a sample-by-sample basis, but the change in quantization noise may be audible. Due to auditory habituation the ear makes note of changes in noise much faster and easier than in a static noise floor even if the static level is higher than that of the changing level. In a floating-point converter, the quantization noise problem is in fact quite the reverse of that of the non–floating-point system. As the signal level decreases, the quantization error will decrease, but as the level increases and the gain element switches, the quantization error changes with respect to the signal level. The theory of the floating-point system attempts to use the notion of auditory masking to cover the quantization noise at all levels, since it should be insignificant in relation to the signal level at all times.

The concept of dither relies on auditory habituation to replace quantization error with a low-level noise floor. The added noise is by nature similar to that which listeners are accustomed to hearing in sound systems and generally is of a lower level. The effect of the dither is to "decorrelate" converter error. A dithered PCM system typically adds white noise to the input analog signal before conversion. This changes the sharply changing low-level signal to a pulse width modulated (PWM) signal, as shown in Figure 9. This increases the resolution of the system beyond that which the least significant bit would normally allow. The addition of the small amount of noise is quite insignificant in comparison with the improvement in sound quality achieved by decreasing the quantization noise. Dither is further discussed in Chapter 2.

Quantization error can be further decreased by using error feedback converters in a technique known as noise shaping. These converters employ sampling rates many times that of normal converters (nominally 44.1 or 48 kHz). Quantization error can be significantly decreased in the audible band while increasing the quantization error in a higher frequency range, beyond the normal

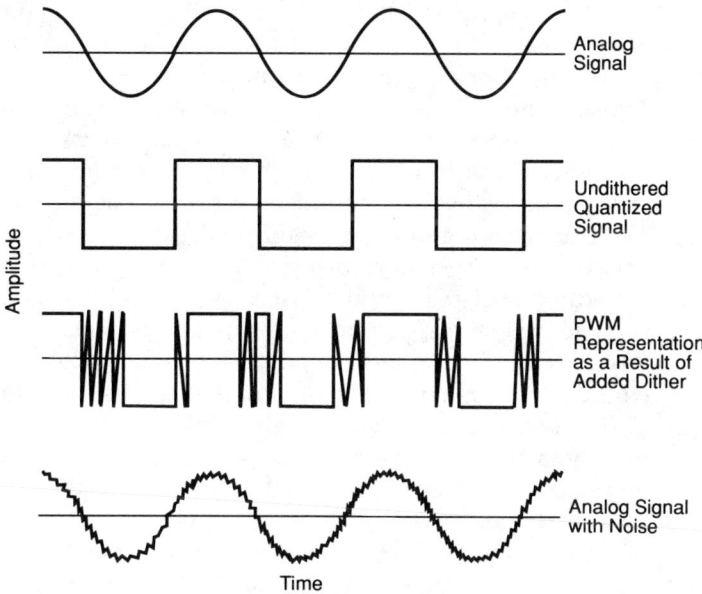

**Figure 9.** The effects of dither on a quantized audio signal.

hearing range. The use of A/D converters with low resolution (1 to 4 bits) and sampling rates as high as 6 MHz reduces quantization error to a low-level noise floor, uncorrelated with the signal. Figure 10 shows an A/D converter which uses error feedback. The low-bit resolution and high sampling rate are converted to the standard employed by the particular medium being used. Low-bit conversion and noise shaping are discussed at some length in Chapter 12.

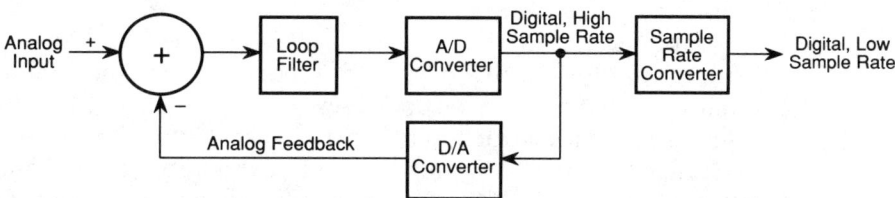

**Figure 10.** Analog-to-digital converter using error feedback.

# Error Correction

The nature of music and sound in general is predictable and quite redundant on the microsecond digital timebase. A single bit stream in the digital domain generally contains words which change only a few quantization levels at a time and contain much the same information as the groups of words in the immediate time frame. In orchestral music, the average note of music is held for a duration of approximately 350 ms while the timbre, pitch, and loudness can be easily represented in 10 to 15 ms. For pop or rock music the average note is somewhat shorter due to the further complications of percussive transients. The average note is about 200 ms, about 185 ms of which are largely redundant. Of course, exact repetition is not particularly common because of changing timbre and loudness within each note, but the understanding of basic redundancy is of great benefit in error concealment.

The handling of huge quantities of data, as with that contained in a compact disc, will by nature bring about errors. Nonuniformities on the disc surface can be misread by the laser; for example, a scratch in the substrate or a particle of dust can cover thousands of bits of data. The detection of errors is often accomplished rather easily by checking the data against previously encoded error detection information that is recorded with the program material. The error correction data contains redundancy in the form of parity, to check for errors and correct invalid data. Random errors involving only a relatively few data words can be completely corrected; there is absolutely no audible artifact. A burst error involving thousands of words may require concealment.

Interpolation is often employed to conceal data based on the surrounding data. Zero-order interpolation holds the last valid word until the next valid one. First-order interpolation calculates the mean value for a word as the average of the words on either side of it. Higher-order interpolation for consecutively invalid words requires increased hardware and design considerations.

Another method of concealing blocks of missing or invalid data is an automatic mute function. This gradually attenuates the last block of valid data to zero to avoid an audible click, then holds a zero value until valid data resumes. The level is restored to the true value, again gradually to avoid an audible click. Muting is far preferable to the audibly distracting output which occurs with invalid data. It has been argued by Doi [1983] that smoothly contoured muting cannot be perceived.

An option to muting, which can be employed for large blocks of invalid data, would be the use of pattern recognition circuitry. As noted, music is highly redundant. If a block of 100 ms is missing and a 10 ms pattern had been established in the prior 100 ms, that pattern could be repeated until new valid data is received. This would be much less audible than a mute of more than 30 ms. The drawback with this approach is the requirement for real-time analysis of patterns. The hardware implementation of such a concealment scheme would be rather extreme and possibly not cost-effective for the number of errors of this magnitude which occur naturally in digital data media. The use of redundancy in coding digital data is discussed further in Chapters 2 and 8.

## Conclusion

In the world of digital audio, the available technology can directly challenge the limits of the human auditory mechanism. Because of this, it is indeed important to understand these limits in order to design digital systems that produce few artifacts while delivering high-fidelity reproduction throughout the audible range. In the chapters that follow, topics involved in the design and manufacture of high-fidelity audio systems are described. These topics include conversion methods, data transmission and broadcast, laser, fiber-optic, and optical-disk technologies, data compression, and DSP theory and practice. As digital audio technology grows to maturity and becomes more closely allied to its sister fields, it will surely perform at the limits of the human auditory system and in turn stimulate a better understanding of the way we hear.

# References

Doi, T. T., "Error Correction for Digital Audio Recordings," *Digital Audio Collected Papers*, Audio Engineering Society, 1983.

Fielder, L. D., "Human Auditory Capabilities and their Consequences in Digital Audio Converter Design," *Audio in Digital Times: 7th AES Conference Proceedings*, May, 1989.

Freeland, A. P., *Deafness: The Facts*, Oxford University Press, 1989.

Gilchrist, N. H. C., "Digital Audio Impairments and Measurements," *Digital Audio Collected Papers*, Audio Engineering Society, 1983.

Hill, F. J., and G. R. Peterson, *Digital Systems*, 3rd edition, John Wiley & Sons, 1987.

Humes, L. E., "Psychoacoustic Foundations of Clinical Audiology," *Handbook of Clinical Audiology*, 3rd edition, J. Katz, editor, Williams & Wilkins Co., 1985.

Katz, J., "Clinical Audiology," *Handbook of Clinical Audiology*, 3rd edition, J. Katz, editor, Williams & Wilkins Co., 1985.

Lipshitz, S. P., "Are the Purists Wrong?" *Journal of the Audio Engineering Society*, vol. 34, no. 9, September, 1986.

Melnick, W., "Industrial Hearing Conservation," *Handbook of Clinical Audiology*, 3rd edition, J. Katz, editor, Williams & Wilkins Co., 1985.

Miller, J., "Theories of Speech Perception as Guides to Neural Mechanisms," *Cochlear Prosthesis: An International Symposium*, C. W. Parkins and S. W. Anderson, editors, New York Academy of Sciences, 1983.

Miller, J., "Digital Implant Awakens Sound in Deaf Ears," *New Scientist*, vol. 121, February 11, 1989.

Radocy, R. E., and J. D. Boyle, *Psychological Foundations of Musical Behavior*, 2nd edition, Charles C. Thomas, 1988.

Stevens, S. S., Davis, H., and M. Lurie, "The Localization of Pitch Perception of the Basilar Membrane," *Journal of General Psychology*, vol. 13, 1935.

Trahiotis, C., "Progress and Pitfalls Associated with Scientific Measures of Auditory Acuity," *Digital Audio Collected Papers*, Audio Engineering Society, 1983.

Van Den Honert, C., "Reproducing Auditory Nerve Temporal Patterns with Sharply Resonant Filters," *Cochlear Implants*, J. Miller and F. Speelman, editors, Springer-Verlag New York, 1990.

Wever, E. G., and M. Lawrence, *Physiological Acoustics*, Princeton University Press, 1954.

Zimmerman, R., "Neurologic Considerations for Audiologists," *Handbook of Clinical Audiology*, 3rd edition, J. Katz, editor, Williams & Wilkins Co., 1985.

*Chapter 2*

# Pulse Modulation and Sampling Systems

Ken C. Pohlmann

## Introduction

There are many ways in which information can be conveyed by a pulse modulation system. Pulse amplitude modulation, pulse position modulation, and pulse width modulation all offer various advantages. However, following its invention in 1937, pulse code modulation was regarded as an almost ideal modulation code for conversion, storage, and transmission of digital signals, with moderate bandwidth requirements and low coding errors. As with any pulse coding method, there are many benchmarks, such as efficiency and bit error probability, which permit evaluation of the format's suitability in an application. The nature of any sampling system is perhaps best described by its frequency domain properties. Fourier analysis clearly shows the way in which a rectangular time pulse modulates the amplitude of a carrier frequency. For example, it can be seen that transmission of narrow pulses necessitates use of a channel with a high bandwidth. Any sampling system is bound by the sampling theorem, which defines the relationship between message and sampling signal rates. In addition, the theorem dictates that the message be bandlimited. Quantization introduces an approximation error which can be predicted with great accuracy, and minimized with techniques such as dither.

## Pulse Modulation

In many applications, including the transmission or recording of speech and music, analog information may be converted to a digital format using one of many types of pulse modulation. For example, a pulse's width or variable duration, or variable position in time may represent information about the signal; pulse width modulation (PWM) is an example of the former, and pulse position modulation (PPM) is an example of the latter. Pulse width modulation and pulse position modulation samples may be formed by comparing analog samples of the input signal with a sawtooth waveform. The time between the sampling signal and the intersection of the analog signal with the sawtooth determines the width of a pulse or its position. A signal may also be conveyed through an uncoded pulse modulation system; pulse amplitude modulation (PAM) is an example of this method. Samples are transmitted via pulses with amplitudes that represent the amplitude of the signal at sample time. Although PWM, PPM, and PAM may be used in the context of conversion, they are not directly suitable for transmission or recording because they are overly affected by noise in the transmission channel and in the storage media.

## Pulse Code Modulation

The most commonly used modulation method is pulse code modulation (PCM). This method was devised in 1937 by Sir Alec Reeves while working as an engineer in the International Telephone and Telegraph Company (ITT) laboratories in France. In PCM, the input signal must undergo sampling, quantization, and coding. By coding the measured analog amplitude values of sampled information into a series of pulses, binary numbers may be used to represent the information. At the receiver the pulse code can be used to reconstruct an analog waveform. The binary words which represent the amplitude of the signal are directly coded into PCM waveforms as shown in Figure 1.

Given a converter with $k$-bit wordlength, $2^k$ unique code words are created to represent $2^k$ amplitude values, called quanta. For example, a 16-bit system would encode 65,536 amplitude quanta. In the simplest incarnation, binary 0000 0000 0000 0000 would represent decimal 0, and binary 1111 1111 1111 1111 would represent 65,535. In practice, however, that might not be the most efficient mapping of the audio waveform. Thus a different arrangement of the PCM data might be employed. Two examples of alternative binary coding are ($a$) sign and magnitude binary and ($b$) two's complement binary notation. In sign and magnitude notation the absolute value of samples are expressed in binary code; however, their sign is expressed in the leftmost bit. In two's complement notation two ascending binary counts are used, the leftmost bit again representing sign. More specifically, negative numbers are formed by taking the complement of the positive equivalent and adding 1; for example, the two's complement of 01000 is 10111 + 00001 = 11000. Humans appreciate two's complement because the left digit always denotes the sign of the number; digital circuits appreciate it because subtraction can be performed with an addition operation, thus simplifying calculations.

With methods such as PWM, PPM, and PAM, only one pulse is needed to represent the amplitude value whereas in PCM several pulses per sample are required. As a result, PCM may require a higher-bandwidth channel. However, PCM forms a very robust signal in that only the presence or absence of a pulse is necessary to read the signal. In addition, a PCM signal can be regenerated without loss. Therefore the quality of a PCM transmission depends on the qual-

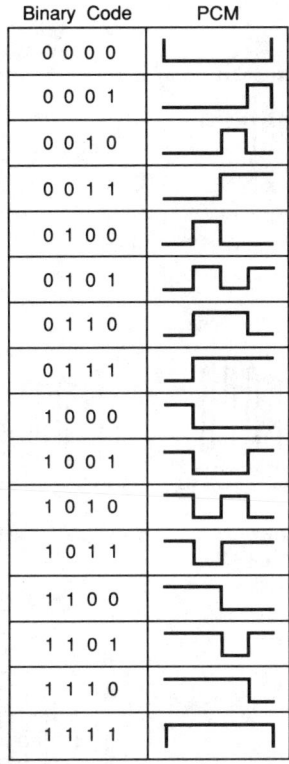

**Figure 1.** Binary words are coded directly into the PCM waveform.

ity of the sampling, quantizing, and coding processes, not the quality of the channel itself. In addition, depending on the sampling rate and the capacity of the channel, several PCM signals may be simultaneously conveyed by using time division multiplexing. This greatly expedites use of PCM. Several types of modulation are reviewed in Figure 2.

## Theoretical Performance of Pulse Code

A pulse code's efficiency can be measured through comparison of the information capacity of the message versus the capacity of the encoded signal itself. As Shannon [1948] has pointed out, the information capacity of the message characterizes the entropy of the signal. The coding or format efficiency may be defined as

$$\eta = C_m/C_c,$$

where $C_m$ is the information capacity of the signal and $C_c$ is the information capacity of the coded signal.

Substituting expressions for $C_m$ and $C_c$, we obtain

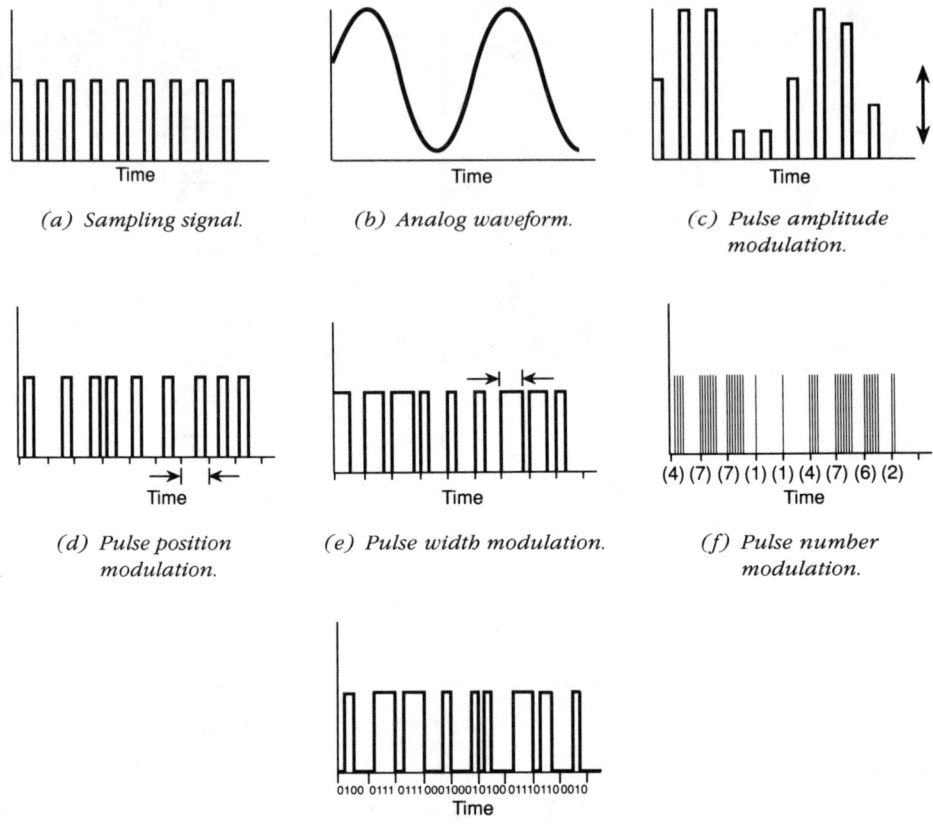

*(a) Sampling signal.*  *(b) Analog waveform.*  *(c) Pulse amplitude modulation.*

*(d) Pulse position modulation.*  *(e) Pulse width modulation.*  *(f) Pulse number modulation.*

*(g) Pulse code modulation.*

**Figure 2.** A variety of modulation methods may be used to represent an analog waveform.

$$\eta = \frac{S_m \sum_{i=1}^{n} P_i \log_2(1/P_i)}{S_c \sum_{i=1}^{n} Q_i \log_2(1/Q_i)},$$

where

$S_m$ is the message symbol rate in symbols per second,
$S_c$ is the coded symbol rate in symbols per second,
$P_i$ is the probability of a logical 1 or 0 message bit occurring,
$Q_i$ is the probability of a coded level occurring,
$i$ is the $i$th message symbol.

Alternatively, coding efficiency may be characterized by the relationship between timing of the minimum duration and the format's clock period:

$$\eta = t_{\min}/T,$$

where $t_{min}$ is the minimum duration timing and $T$ is the clock period.

In codes in which each symbol represents a bit, efficiency can be defined as the ratio of actual information transmitted to the theoretical maximum transmitted rate. In other words, this compares the number of data bits to the maximum number of symbols. Nominally, the highest efficiency of such codes is 1, with efficiency of 1/2 common in codes devised to exclude DC components. However, codes may be written in which several data bits are expressed by one symbol. Since a clock period can convey a number of bits, the coding or format efficiency must be redefined as the number of data bits contained in one symbol compared to a symbol with a maximum number of transitions:

$$\eta = r/s,$$

where $r$ is the number of data bits in a symbol and $s$ is the maximum number of transitions within a clock period.

The reciprocal of this value is known as the format conversion ratio.

A formula proposed by Hartley and Shannon [1949] may be used to study the information transfer of a pulse modulation system. Given a transmitted signal's power and bandwidth, and power of disturbing white noise, the Hartley-Shannon law determines the maximum information that can be accurately conveyed:

$$I = B \log_2(1 + T/N),$$

where

$I$ is the maximum information capacity in bits per second,
$B$ is the bandwidth of the channel in hertz,
$T$ is the power of the transmitted signal,
$N$ is the power of disturbing Gaussian noise.

For example, ideally a PAM system would always transmit an amount of information equal to $I$; information lost will be proportional to the noise in the system.

According to a derivation undertaken by Haykin [1978], the maximum information capacity of a PCM signal in bits per second may be shown to be

$$I_{PCM} = B \log_2(1 + 12T/a^2N),$$

where $a$ is a constant typically equal to 10. Pulse code modulation is a quantized noise-limited modulation method, and, in particular, in a PCM system, power and bandwidth are interchanged logarithmically.

## Practical Limitations of Pulse Code

An efficient pulse code format must restrict DC components which could disrupt timing synchronization, and it must observe the maximum transmission bandwidth available. Data patterns must be analyzed to determine presence of these two limits: DC components and maximum signal rate. Existence of a DC component or noise can seriously compromise the success of the transmission of a signal. Any amplitude variation upward or downward, referred to as baseline wander, changes the reference by which the receiver distinguishes positive or negative values. In AC-coupled receivers, such as in fiber-optic systems, the baseline is established purely by the demodulated data signal itself.

To compound the degradation, the base line wander and other problems can cause a time-axis variation known as jitter to accumulate through a system, leading to random errors. For example, the effects of jitter on the sampling clock of an A/D converter are quite similar to FM modulation; the input frequency acts as the carrier, and clock jitter acts as the modulation frequency. The jitter acts to reduce the amplitude of the input signal and to add sideband components equally spaced at either side of the input frequency, at a distance equal to multiples of the jitter frequency. The effect of sinusoidal sampling clock time jitter on an A/D converter may be described as

$$v(t) = A \cos[\omega_i(t + J \sin \omega_j)],$$

where

$A$ is the amplitude of the input signal,
$\omega_i$ is the frequency of the input signal,
$J$ is the peak amplitude of the jitter,
$\omega_j$ is the jitter frequency.

The effect of jitter thus increases as the input signal frequency increases; specifically, jitter amplitude error increases with input signal slew rate. In practice, the first sideband is potentially audible and can thus decrease the dynamic range of the system. Peak clock jitter of less than 200 ps is often mandated.

Use of crystal-controlled phase-locked loops and other measures can limit jitter to an acceptable level. Master clocks must not be distributed through a studio. Rather, each piece of sampling equipment must contains its own clock, frequency locked to a distributed master clock. A signal can be successfully recovered provided the jitter margin does not exceed

$$\Delta t = \pm \tau/2,$$

where $\tau = 1/T$ and $T$ is the sampling period.

Signal integrity and problems such as DC offset, jitter, peak shift, and noise, which affect signal integrity can be observed by an eye pattern, as shown in Figure 3. This display uses an oscilloscope to overlay a superimposed collection of regenerated data sequences. When the oscilloscope is triggered at the data rate, the dynamic changes in the signal are visible. Amplitude and temporal variations which deteriorate the signal will reduce the eye opening and hence the receiver's performance to the point where pulse shaping can no longer retrieve the signal. The amount of deterioration can be gauged by measuring the extent of amplitude variations and forming an eye opening ratio,

$$E = \frac{a_2}{a_1} = \frac{a_1 - 2\Delta a}{a_1},$$

where

$a_1$ is the outside amplitude,
$a_2$ is the inside amplitude,
$\Delta a$ is the amplitude variation.

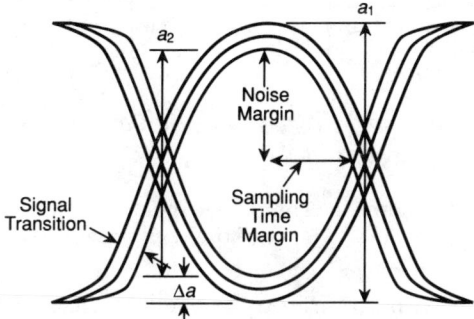

**Figure 3.** Signal integrity can be evaluated through the eye pattern. The eye opening ratio is taken from the ratio of the amplitudes $a_1$ and $a_2$.

Variations are measured from the center of the opening in the eye pattern. The width of the eye gives the percentage of the data period available to ascertain its logical value, and the height shows the maximum difference between these levels during the available time.

## Bit Coding

As we shall see, the analysis of a sampling pulse is relatively easy because of its periodic nature in the time domain; Fourier analysis clearly shows its spectrum. However, a data stream differs in that the data pulses occur aperiodically and in fact can be considered to be random. Analysis may require evaluation of the power spectrum at each frequency, a difficult task. The resulting power spectral density, or power spectrum, $G(f)$, statistically shows the response of the data stream $f(t)$. For example, Figure 4a shows two data streams with identical data content but different pulse types: full binary NRZ and half binary RZ. Figure 4b shows the power spectrum curves for the two pulse types. Signal peak amplitudes are normalized, and the pulses have equal periods. Absolute magnitude is used; thus all frequency components are plotted positively. Viewing the power spectrum, we see that a transmission waveform ideally should have minimal energy at low frequencies to expedite equalization, and minimal energy at high frequencies to reduce crosstalk. A code is optimized through manipulation of pulse shape, number of logical levels, and repetition rate.

Using statistical methods, the worst-case (widest) spectrum of a signal can be determined from its probability distribution. For NRZ, the widest spectrum takes place when data alternates between 1 and 0; for RZ, the worse case is

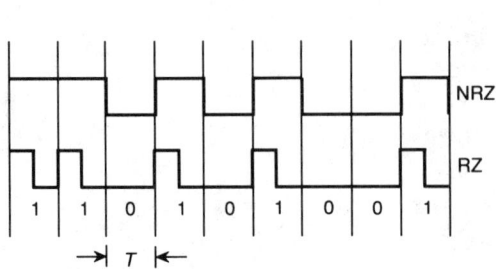 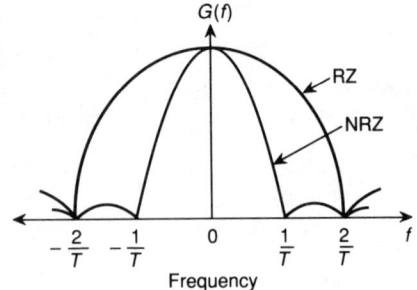

*(a) Two data streams of different type but of identical data.*   *(b) The power spectrum curves for the data streams.*

**Figure 4.** The power spectrum statistically shows the response of the data stream. *(From Morris)*

incurred with continuous 1s. In both cases, when this occurs, the signal level spends its time exactly split, with equal probability, between the positive and negative levels. The average power is thus $f(t)^2$ for both NRZ and RZ. They differ in their pulse durations, however, and because of this NRZ is more efficient in terms of spectrum. The NRZ signal uses the entire bandwidth to convey its information whereas the RZ signal with a pulse duration of $T/2$ uses only half of its bandwidth, wasting half.

Performance may also be analyzed in terms of bit error probability, that is, the receiver's net ability to distinguish between a 1 and 0 value. The error probability is determined as a function of the signal-to-noise (S/N) ratio:

$$P_e = f(S_v/N_{rms}),$$

where $S_v$ is the magnitude of the signal voltage, varying from $+V$ to $-V$, and $N_{rms}$ is the rms noise voltage in the received signal.

In the case of a NRZ signal, a bit error occurs when the noise has a higher (and opposite) magnitude than the signal. The error probability can be written

$$P_e = \text{Prob}(N_{rms} > S_v/2).$$

And the bit error probability is written

$$P_e(\text{NRZ}) = \tfrac{1}{2} \,\text{erfc}(S_v/N_{rms})^{1/2},$$

where erfc is the complementary error function defined as

$$\text{erfc}(u) = \frac{2}{\sqrt{\pi}} \int_u^\infty \exp(-z^2)\, dz.$$

To further minimize decoding errors, formats can be developed in which data is conveyed with data patterns that are as individually unique as possible. For example, in the EFM code devised for the compact disc format, 8-bit symbols are translated into 14-bit symbols, carefully selected for maximum difference between symbols. In this way, invalid data can be more easily recognized. Similarly, a data symbol could be created based on previous adjacent symbols, and the receiver could recognize the symbol and its past history as a unique state. A state pattern diagram is used in which all transitions are defined, based on all possible adjacent symbols.

## Sampling Spectra

Discrete time sampling is founded on the concept of a rectangular impulse of infinitesimal width. Because of its infinite frequency response, and the requirement that the channel have infinite bandwidth, an ideal impulse is physically impossible to create and eludes everyone but mathematicians. In practice we consider a rectangular pulse of fixed amplitude $A$ and finite width $\tau$. Such an impulse is represented in Figure 5.

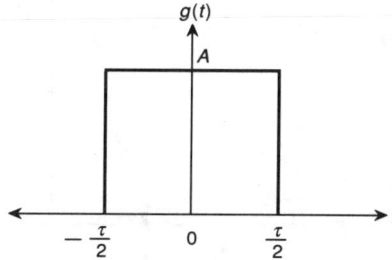

**Figure 5.** A rectangular pulse of fixed amplitude and width: an impulse.

Although data signals are often characterized by their time domain properties, the performance of a transmission channel is usually best described by its frequency domain properties. Specifically, it is important to know the bandwidth required for successful transmission of a single pulse. The Fourier transform may be used to describe the time domain function in the frequency domain. The Fourier transform of a time function $T(t)$ can be found:

$$F(f) = \int_{-\tau/2}^{\tau/2} T(t) e^{-j\omega t}\, dt,$$

where $\omega = 2\pi/\tau$.

Given a single rectangular pulse of duration $\tau$ and amplitude $A$, centered at $t = 0$ (Figure 6),

$$F(f) = \int_{-\tau/2}^{\tau/2} A e^{-j\omega t}\, dt = A\tau\, \frac{\sin(\omega\tau/2)}{(\omega\tau/2)}.$$

If we define $x = \omega\tau/2$, then the transformation of the pulse is shown to be

$$F(f) = A\tau\, \frac{\sin x}{x}.$$

This function is shown in Figure 6. It can be seen that it is composed of a fundamental cosinewave and its harmonics, that its maximum value occurs at $x = 0$, and that it approaches zero as $x$ approaches $\pm\infty$. The width of the center lobe is exactly $2/\tau$, and the frequency response passes through zero at multiples of $1/\tau$. Importantly, it demonstrates the fundamental nature of sampling as a modulation process; the frequency pattern of the function shows that the rectangular time pulse has modulated the amplitude of a carrier frequency. The center frequency may be shifted without altering the shape of the envelope itself. Clearly, this spectrum extends to infinity, and thus ideal transmission of the pulse would require a system with infinite bandwidth. As we shall see, however, only the central lobe is required: A finite bandwidth will suffice.

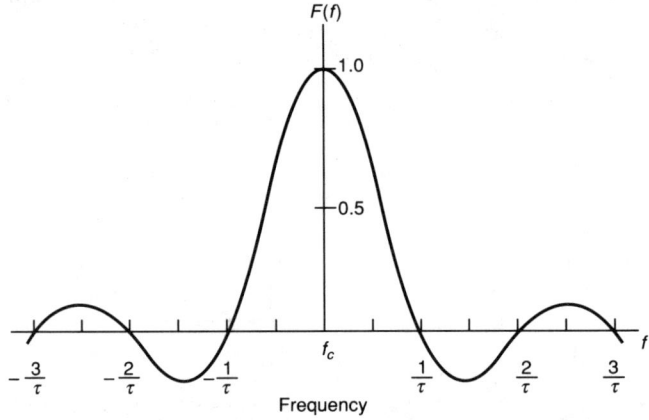

**Figure 6.** A single-pulse impulse response composed of a fundamental cosinewave and its harmonics.

Given an understanding of the properties of a single pulse, it is useful to examine a series of such pulses with a periodic repetition of $T$. This leads to the creation of a practical sampling signal as a periodic series of pulses of fixed amplitude and finite width. It is defined as follows:

$$S(t) = \begin{cases} A & \text{when } -\tau/2 < t < \tau/2, \\ 0 & \text{for the remainder of the period.} \end{cases}$$

where $\tau$ is the width of the pulse, and it is periodically repeated with a period of $T$ seconds. This sampling signal is shown in Figure 7a. Evaluating the complex Fourier coefficient $C_n$, we observe

$$C_n = \frac{1}{T} \int_{-\tau/2}^{\tau/2} A\, e^{-j2\pi nt/T}\, dt = \frac{A}{n\pi} \sin(n\pi\tau/T).$$

To simplify the results, we can employ the sinc function defined as

$$\operatorname{sinc} x = \frac{\sin \pi x}{\pi x}.$$

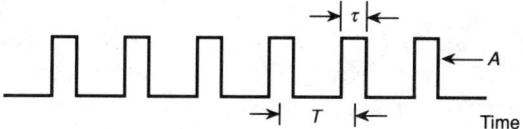

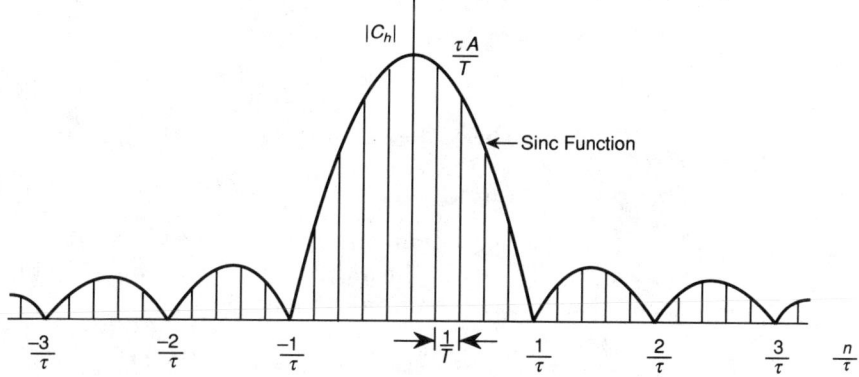

(a) A series of pulses with a periodic repetition rate T.

(b) The frequency spectrum of (a) is defined at discrete values of 1/T along the contour of a sinc function.

**Figure 7.** Illustration of the properties of a series of pulses—a sampling signal.

The sinc function has a maximum value at $x = 0$, and approaches zero as $x$ approaches infinity. It oscillates through positive and negative values passing though zero at integer multiples. Using the sinc function, we may rewrite the expression for $C_n$:

$$C_n = \frac{\tau A}{T} \mathrm{sinc}(n\tau/T).$$

The frequency spectrum of this function is defined at discrete values of $n$; that is, as equally spaced spectral lines with heights corresponding to the discrete frequency components. This spectrum is shown in Figure 7b. The spectral lines are spaced according to the period $T$, and the envelope is determined by the pulse duration $\tau$, duty ratio $\tau/T$, and pulse amplitude $A$ with zero crossings at frequencies that are multiples of $1/\tau$.

The spectral response of a series of sampling pulses thus creates spectral lines with amplitudes that follow the same contour as that of a single pulse. The spectrum bandwidth is not affected by the pulse repetition frequency; rather, the bandwidth is determined by the pulse width $\tau$. The shorter the duration of the pulse, the greater is the frequency spread of the bandwidth. It is thus clear that transmission of narrower pulses requires a channel with higher bandwidth. From a frequency domain standpoint, wider pulses might appear advantageous. However, as viewed in the time domain, narrower pulses permit a greater repetition rate and, for example, permit time multiplexing of channels. In any case, it is not a higher repetition rate that necessitates higher bandwidth but the narrow width of the pulses. Similarly, as we shall see, a condition known as aperture error can be minimized by decreasing the duration of the pulse width.

Using the sampling signal $S(t)$, it is possible to define the nature of the sampled signal $f_s(t)$:

$$f_s(t) = f(t) S(t),$$

where

$f_s(t)$ is the sampled signal,
$f(t)$ is the message signal,
$S(t)$ is the sampling signal.

Moreover, we may obtain an expression for the frequency spectrum of the sampled signal. The multiplication of these two time domain functions may be represented as the convolution of their spectra:

$$F_s(\omega) = \frac{1}{2\pi} F(\omega) * S(\omega),$$

where * denotes the convolution of these frequency domain signals.

Mathematical derivation yields the following equation:

$$F_s(\omega) = Ad \sum_{n=-\infty}^{\infty} \frac{\sin n\pi d}{n\pi d} F(\omega - n\omega_0).$$

More obviously, a graphic representation of this expression is shown in Figure 8. It shows that the spectrum of the message signal contains both positive and negative sidebands centered at the impulses defined by the sampling

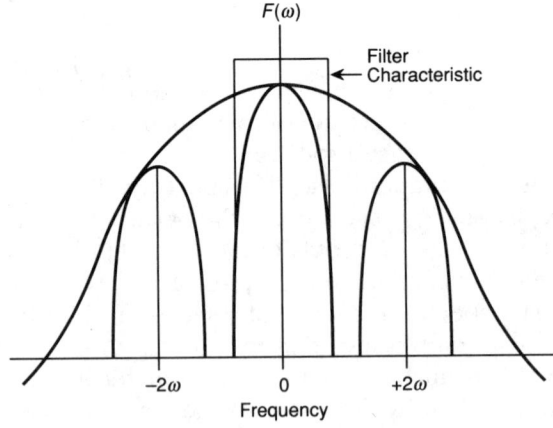

**Figure 8.** The spectrum of the message signal contains positive and negative sidebands centered at the impulse defined by the sampling function.

function, and placed at multiples of the sampling frequency. In addition, their amplitude again clearly follows the $(\sin x)/x$ contour predicted by the Fourier transform of the sampling signal. It is important to note that complete information of the message signal is held in each sideband. Filtering of the sidebands can thus retrieve complete information. Clearly, the bandwidth is finite and has insignificant energy at high frequencies.

A POSTSCRIPT: Before leaving the topic (for the moment) we should note that the $(\sin x)/x$ function occurs repeatedly in sampling theory and in fact is often called the sampling function. In addition, the function is found in other fields; for example, the $(\sin x)/x$ function occurs in optics, where Fraunhoffer lines are analyzed.

ANOTHER POSTSCRIPT: In the case of ideal sampling with a pulse of infinitesimal width and infinite bandwidth, its spectral lines are placed at multiples of the sampling rate, as in natural sampling. The amplitudes of the lines, however, remain constant across the spectrum.

# The Sampling Theorem

Given the fact that any practical channel is bandlimited, it is important to know the maximum transmission rate afforded by a channel. This question was answered by Harry Nyquist, an engineer working in Bell Telephone Laboratories, in his classic paper, "Certain Topics in Telegraph Transmission Theory," published in 1928. He demonstrated that a message of $S$ Hz may be completely characterized by samples taken at less than $S/2$ seconds apart. This sampling condition is now well known as the Nyquist sampling theorem, which, in other words, states that the periodic sampling signal must be at least twice the highest information signal frequency. Looked at in another way, the message signal must be bandlimited to a frequency half that of the sampling frequency. Because messages do not always follow this criterion, it is necessary to low-pass filter them at half the sampling frequency, sometimes known as the Nyquist frequency. Although Nyquist proposed his result in terms of telegraph transmission, the result is equally valid for any kind of digital data transmission including, of course, digital audio.

The basis of sampling theory can be demonstrated with a simple example. Given an arbitrary signal $g(t)$, it may be sampled at a uniform rate $T_s$, yielding a pulse amplitude signal $g_s(t)$ defined at sample times. Sampling theory determines the conditions under which the sampled signal can uniquely define the original signal. The Fourier transform of the sampled signal consists of a series of delta functions spaced $1/T_s$ Hz apart. In particular, the Fourier transform may be expressed as

$$G_s(f) = \frac{1}{T_s} \sum_{n=-\infty}^{\infty} G(f - n/T_s).$$

Thus as a result of time sampling the original signal in the time domain, we obtain a periodic spectrum in the frequency domain with a period equal to the sampling rate. Suppose now that the signal is bandlimited with no components above $S$ Hz, or in other words, the Fourier transform of the signal has the property that $G(f) = 0$ for $|f| > S$. In addition, let the sampling period be $T_s = 1/2S$. Then we may write the spectrum as

$$G(f) = \frac{1}{2S} \sum_{n=-\infty}^{\infty} g(n/2S) e^{-jn\pi f/S}, \qquad -S \leq f \leq S.$$

Because of the relationship of $g(t)$ to $G(f)$ via the inverse Fourier transform, the signal $g(t)$ is uniquely defined by the sample values $g(n/2S)$ for $-\infty \geq n \geq \infty$. Thus the sample values $g(n/2S)$ contain all the information of $g(t)$. Furthermore, this information may be extracted from the samples using reconstruction methods. Using the inverse Fourier transform, we obtain

$$g(t) = \int_{-\infty}^{\infty} G(f) e^{j2\pi ft} \, df.$$

This may be expressed as

$$g(t) = \sum_{n=-\infty}^{\infty} g(n/2S) \frac{1}{2S} \int_{-S}^{S} e^{j2\pi f(t - n/2S)} \, df.$$

Evaluating, we obtain

$$g(t) = \sum_{n=-\infty}^{\infty} g(n/2S) \, \text{sinc}(2St - n).$$

Thus the sinc function may be used as an interpolation function to reconstruct the original signal $g(t)$ from the sample values $g(n/2S)$. Each sample is multiplied by its interpolation function, and added to the functions of all other samples to obtain the signal waveform $g(t)$. Importantly, the sinc function obtained above also represents the response of an ideal low-pass filter of bandwidth $S$, given an input signal of samples $g(n/2S)$. In other words, the original signal may be reconstructed exactly by passing the representing samples through a low-pass filter with a bandwidth $S$. Thus as Nyquist stated, a signal bandlimited to $S$ Hz may be completely represented by specifying values of the signal at a period of $1/2S$ seconds. Further, a signal bandlimited to $S$ Hz may be completely reconstructed from samples taken at a rate of $2S$ Hz. In other words, for a signal bandwidth of $S$, a sampling rate of $2S$ is required. This, of course, is the key component which permits the transformation of analog signals and digital sequences.

Summarizing, the spectrum of the message signal itself is repeated at multiples of the sampling frequency within the envelope of the sampled signal. When proper bandlimiting is provided, the spectrum repeats itself without overlap; if the signal's bandwidth is less than the Nyquist half-sampling frequency, the image sidebands are separated by a guard band, as shown in Figure

9a. The signal can easily be retrieved by filtering out higher-frequency sideband spectra, leaving only the first sideband. As noted, complete information is contained in each sideband, and thus, for example, a negative sideband could theoretically be used.

If the signal's bandwidth is exactly equal to the Nyquist frequency, a condition called critical sampling, it is still possible to retrieve the original signal by filtering. This, however, is the limiting case; it is shown in Figure 9b. If the signal's bandwidth exceeds the Nyquist sampling frequency, spectra will overlap, as shown in Figure 9c. In this case, aliasing occurs, and it is impossible to completely retrieve the original signal. Degradation increases as overlap increases, that is, as the new bandwidth increases. At best, filtering at half the sampling frequency will lose information in the spectrum above half the

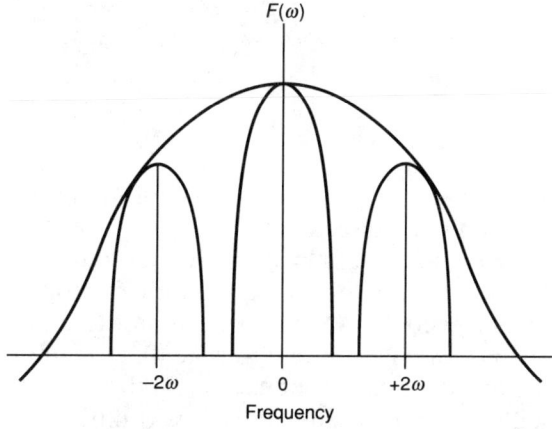

*(a) The signal's bandwidth is less than half the sampling frequency.*

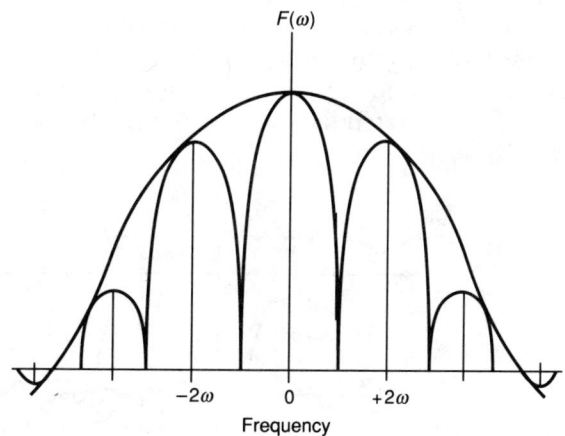

*(b) The signal's bandwidth is equal to half the sampling frequency.*

**Figure 9.** Illustration showing the relationship between signal bandwidth and sampling frequency.

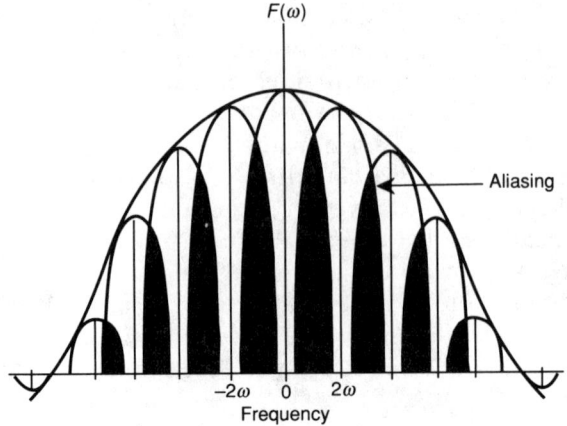

*(c) The signal's bandwidth is greater than half the sampling frequency; aliasing results.*

**Figure 9.** (cont.)

sampling frequency, and yet pass information from the inverted tail of the higher-frequency spectrum.

Information data, sent as pulses of duration $T$, can utilize a signaling rate of $1/T$ through a channel of bandwidth $f = 1/2T$, as shown in Figure 10. Each pulse conveyed through the bandlimited channel will exhibit a pulse response of $(\sin x)/x$. It is important to note that for any given pulse response at a sampling time, the pulse response of all other received signals theoretically passes through zero at that time, and at exact multiples of $T$ seconds. This is because the channel acts as an ideal low-pass filter with ideal cutoff. The condition in which the response of side lobes does not pass through zero at sampling time decreases the system's ability to distinguish and accurately regenerate data; this interference condition is known as intersymbol interference, and can be caused by amplitude distortion and group delay. As noted, an eye pattern composed of regenerated data sequences can be used to evaluate a channel's performance.

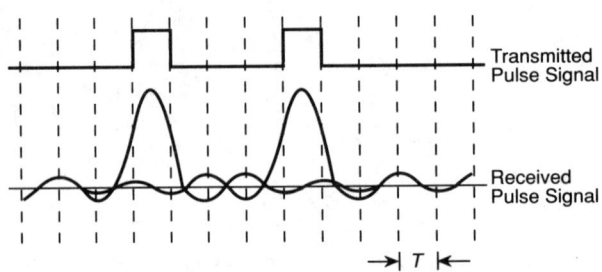

**Figure 10.** Message data sent as pulses of duration $T$ can utilize a signaling rate of $1/T$ through a channel of bandwidth $f = 1/2T$.

## Low-Pass Filtering

As noted, a low-pass filter must precede and follow every digitization system to bandlimit the signal and thus prevent aliasing, and to remove image spectra. In either case, an analog low-pass filter presents the prototypical circuit to accomplish these tasks. Its transfer function may be written as

$$f(s) = \frac{V_{out}}{V_{in}} = \frac{1}{1+S} = \frac{1}{1+j(\omega/\omega_0)},$$

where the angular frequency is $\omega = 2\pi f$ and $\omega_0$ is the angular frequency at the filter's cutoff frequency of $f_c = 1/RC$.

The filter's cutoff frequency can be calculated:

$$f_c = 1/2\pi RC.$$

An ideal low-pass filter perfectly transmits all frequencies inside the passband and infinitely attenuates all frequencies in the stopband. An ideal filter is noncausal; its impulse response shows response prior to the time the unit impulse is applied. More importantly, the impulse response exhibits the sinc $x$ function.

In practice, either analog or digital filters may be used. Although cascaded analog filter stages, resulting in a so-called brickwall characteristic, would bandlimit the signal, other artifacts such as ringing and phase nonlinearity encourage use of digital filters.

Such digital filters accept an input data sequence and output a low-pass filtered data sequence. Generally, given an input $x(n)$, output $y(n)$, and $a$ and $b$, which are constant coefficients describing the filter's response,

$$y(n) = \sum_{i=0}^{M} a_i x(n-i) - \sum_{i=1}^{N} b_i y(n-i).$$

It can be seen that past output samples are used to form present output samples. Such a filter is said to be *recursive*; an example of such a digital filter is shown in Figure 11. When only past and present input samples are used, the equation is simplified to

$$y(n) = \sum_{i=0}^{M} a_i x(n-i).$$

The filter is said to be nonrecursive; an example of a nonrecursive filter is shown in Figure 12. A nonrecursive filter is especially attractive in many digital audio applications because of its phase linearity. The theory underlying the design of digital filters is explored in detail in Chapter 10. Oversampling, a method employed in these filters, is discussed below.

In any case, whatever the type of low-pass filter employed, the output waveform is comprised of the summed response of the individual impulse responses, as shown in Figure 13. Thanks to the miracle of superposition, together these individual responses form the reconstructed waveform.

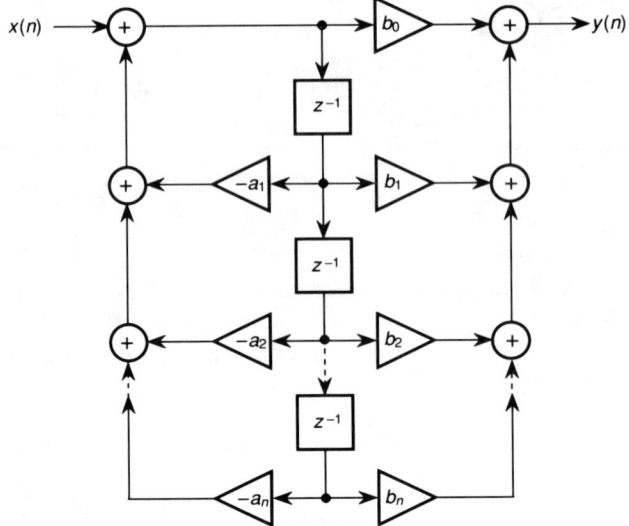

**Figure 11.** An example of an IIR filter using recursive structure.

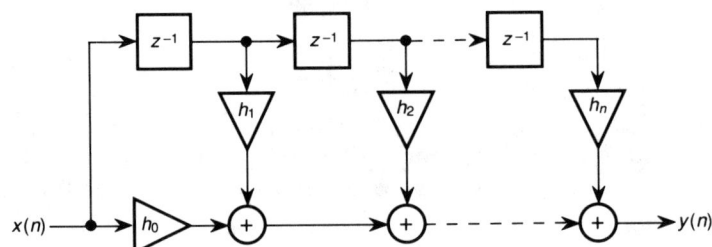

**Figure 12.** An example of an FIR filter using nonrecursive structure.

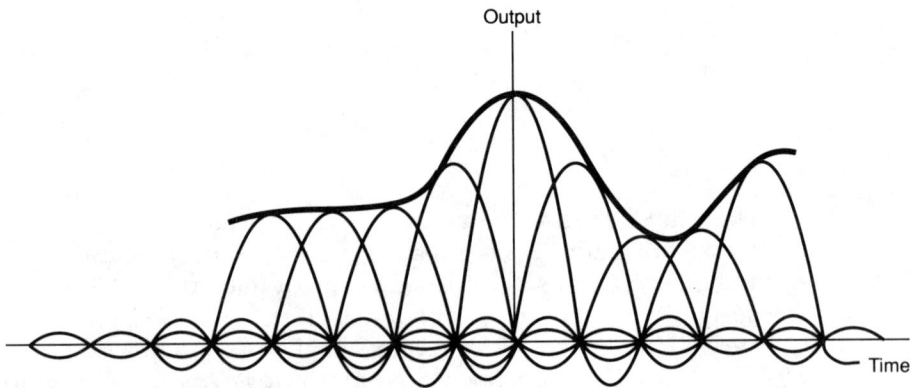

**Figure 13.** The summed response of the individual (low-pass filtered) impulse responses yields the reconstructed analog waveform.

Moreover, this waveform passes exactly through the same points as the original filtered input waveform. On the other hand, we are dealing with systems of finite bandwidth; in other words, in practice the impulse response must be truncated. Whereas the Fourier transform of an infinite (ideal) impulse response would create a rectangular pulse, a finite (real-world) impulse response would create a function exhibiting the Gibbs phenomenon, as shown in Figure 14. This is certainly not ringing as in analog systems but the mark of a finite-bandwidth digital system.

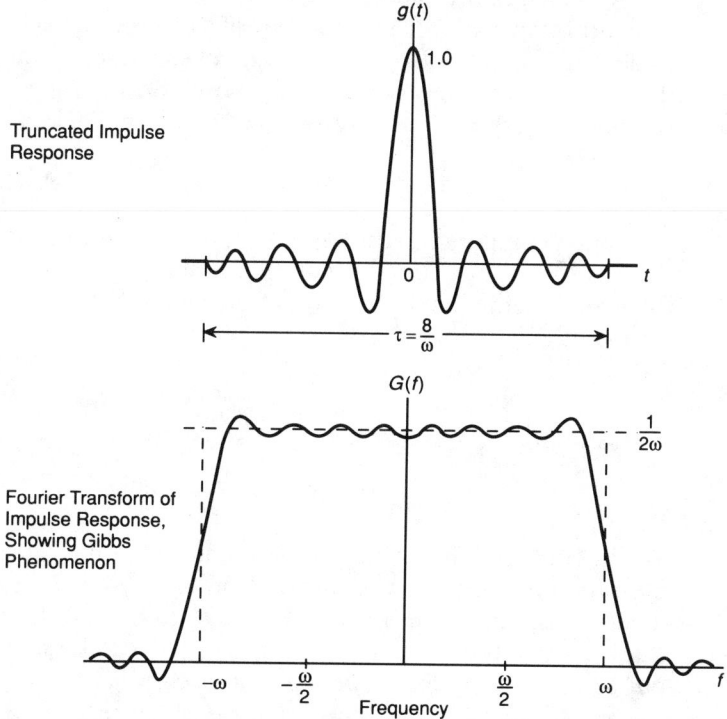

**Figure 14.** Illustration of the Gibbs phenomenon. *(From Haykin)*

# Sample-and-Hold Circuits

Clearly, a sampling system must incorporate input filtering to prevent aliasing, as well as output filtering to remove spurious spectra. A sampling system may also employ two sample-and-hold circuits. An input sample-and-hold circuit may be used to periodically select sample points from the input signal at the sampling period $T$, and hold those analog values for a duration $\tau$. Ideally, the amplitude of the pulse does not change. During this time, an input encoder or quantizer forms a representation of the sample. An output sample-and-hold circuit may also be required, for entirely different reasons; a slight frequency error must be corrected at the system output.

A sample-and-hold circuit, in lengthening pulses from duration $\tau$ to the sampling period $T$, performs the same transform as a low-pass filter. Given an input pulse of duration $\tau$, its transform is equal to

$$H(\omega) = \frac{\tau}{T} \frac{\sin[(\tau/2)\omega]}{(\tau/2)\omega},$$

where $\omega$ is the angular frequency of the input signal.

We observe the $(\sin x)/x$ function once again, which accounts for the filtering that takes place. The contour of this function is the same as that present in defining the sampling spectra; however, the zero points differ. Specifically, the contour of a sample-and-hold circuit is zero at the sampling frequency. At the output of the sample-and-hold circuit, $\tau = T$. Thus

$$H(\omega) = \frac{\sin[(\tau/2)\omega]}{(\tau/2)\omega}.$$

When compared to the input signal bandwidth, as shown in Figure 15, the attenuation of the sample-and-hold contour represents high-frequency loss in the signal. At the maximum input frequency (half the sampling frequency) we observe that

$$\omega = \pi/T.$$

Thus the high-frequency value at half the sampling rate is

$$H(\pi/T) = \frac{\sin(\pi/2)}{\pi/2} \cong 0.64.$$

This is equivalent to a high-frequency attenuation of approximately 4 dB at half the sampling frequency. One solution is equalization; if an equalizer with an inverse response were placed in series, an overall flat response would result. Alternatively, the width of the output pulse can be manipulated using an output sample-and-hold circuit. As noted earlier in our discussion of sampling spectra, the bandwidth of the response is determined by the pulse width $\tau$. The shorter the duration of the pulse, the greater is the frequency spread of the bandwidth. Using the same principle, we observe that aperture error can be minimized by decreasing the duration of the pulse width, as shown in Figure 16. In other words, by decreasing the pulse duration, an output sample-and-hold circuit can be used to more closely approximate the output of an impulse train of an ideal D/A converter. For example, whereas the frequency error at half the sampling frequency is roughly 4 dB when $\tau = T$, it decreases to 0.2 dB when $\tau$ is decreased to $1/4\,T$:

$$H(\pi/T) = \frac{T}{4} = \frac{\sin(\pi/8)}{\pi/8} \cong 0.97.$$

Equalization could easily be used to compensate for this small error.

Further shortening of pulse duration would overly decrease average voltage level and hence diminish the S/N ratio.

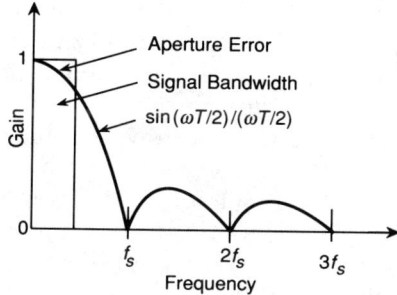

**Figure 15.** The sample-and-hold function creates a high-frequency loss in the signal when compared with the input bandwidth; this is aperture error.

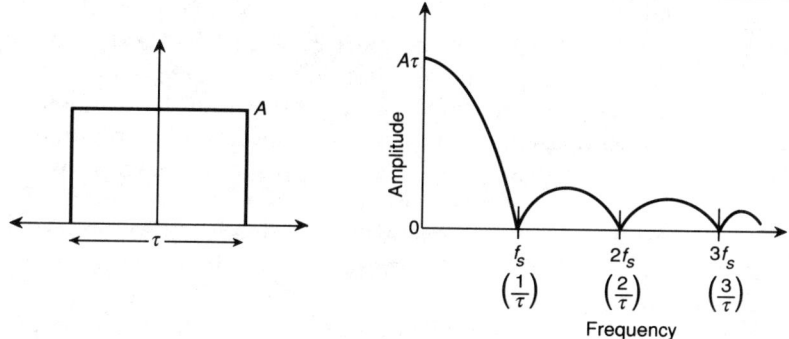

*(a) A longer duration pulse yields rapid high-frequency rolloff.*

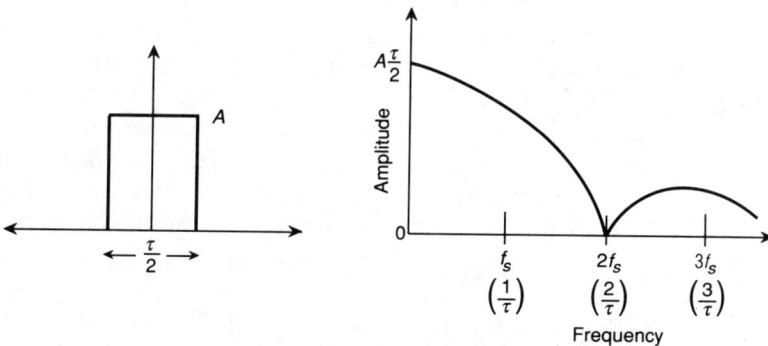

*(b) A shorter duration pulse yields an extended high-frequency response.*

**Figure 16.** Aperture error can be minimized by decreasing the duration of the output pulse width.

# Oversampling

Most contemporary digital audio systems employ a sampling rate of 44.1 or 48 kHz. Some critics have suggested that these sampling rates be doubled. These higher sampling rates would extend frequency response from 20 kHz to 50 kHz. Unfortunately, this extended frequency response is beyond the range of human hearing. The only practical advantage of a higher sampling rate is the decrease in demands on the low-pass filters which precede and follow a digital audio system. The need to sharply limit audio energy at frequencies higher than half the sampling frequency encouraged the use of analog brickwall filters; such filters introduce phase nonlinearities. This problem, however, can be avoided in both A/D and D/A conversion by using oversampling techniques; with these techniques, phase nonlinearities are negligible.

As we have seen, a look at the spectrum of the reconstructed signal reveals the identity of the high-frequency signals, which are images of the original audio signal's spectrum, repeated at multiples of the original sampling frequency. In the case of a 44.1 kHz sampling format, this generates image spectra at 88.2 kHz, 132.3 kHz, 176.4 kHz, 220.5 kHz, etc. As we observed, this is a natural consequence of using sampled waveforms. Although these frequencies are well above the highest frequency audible to humans, they could interfere with downstream electrical equipment. For example, they could affect an analog tape recording by beating against a bias oscillator, or they could affect FM stereo radio transmission. A high-quality digital audio product must suppress these high frequencies. Although this high-frequency information can be removed from the analog waveform using an analog filter, there is a more efficient way. Using a digital filter, the data itself may be processed before it is reconstructed as an analog waveform. With proper processing, the problem of filtering may be largely solved while the data is still in digital form.

In an oversampling filter, audio samples are input and subjected to computation which implements digital filtering of the audio signal. Additional audio samples are generated between the original samples through interpolation, hence the output sampling rate is increased. Intermediate samples are multiplied by fixed coefficients corresponding to their contribution to the overall response of the filter. The output filtered sample is produced by summing together the multiplication products. The spectrum of the signal is changed, with the images appearing at multiples of the new (oversampled) sampling rate. This additional data creates a more linear waveform and shifts unwanted modulation noise to an extreme frequency, where it can be removed without audible effect.

For example, in an eight-times oversampling filter, seven new audio samples are computed for each input sample; an input data rate of 44.1 kHz would be raised to an output rate of 352.8 kHz. Modulation artifacts are shifted to a band centered at 352.8 kHz, where they are easily removed with an analog low-pass filter. The accuracy of such digital filters is precise, yielding passband ripple on the order of 0.00001 dB. In addition, the stopband suppression is greater than 120 dB.

We may define an oversampling ratio as

$$R = f_a/f_s,$$

where $f_a$ is the oversampling frequency and $f_s$ is the input Nyquist sampling frequency.

Oversampling initially requires insertion of $(R-1)$ zero samples per input sample. The samples created by oversampling must be placed symmetrically between the Nyquist input samples. A low-pass filter is used to bandlimit the input data to $f_s/2$, with spectral images at integer multiples of $(R \times f_s)$. Moreover, low-pass filtering creates the intermediate sample values, providing interpolation. Rather than perform multiplications on zero samples, redundancy may be observed to design a transposed FIR filter in which the number of multiplications is reduced by $(R-1)/R$.

Oversampling benefits a digital system's dynamic range by reducing requantization noise. Requantization noise adds $Q^2/12$ noise power to the signal, where $Q$ is 1 LSB; this noise is uniformly distributed over the entire baseband. With oversampling, the spectrum is increased so the noise in the audio band is relatively reduced. For example, a four-times oversampling filter, without any noise shaping, reduces in-band requantization noise by 6 dB. Noise shaping is considered in more detail in Chapter 12.

Consider the eight-times oversampling digital filter in Figure 17 in which seven intermediate samples are generated for each input sample. This filter consists of a shift register of 24 delay elements, each delaying a sample for one sampling period. Each 16-bit sample is multiplied by a 12-bit coefficient. This generates words of 28 bits. The multiplication products are summed eight times during each period and are then output from the filter. Thus eight times as many samples are present after oversampling, with new intermediate values

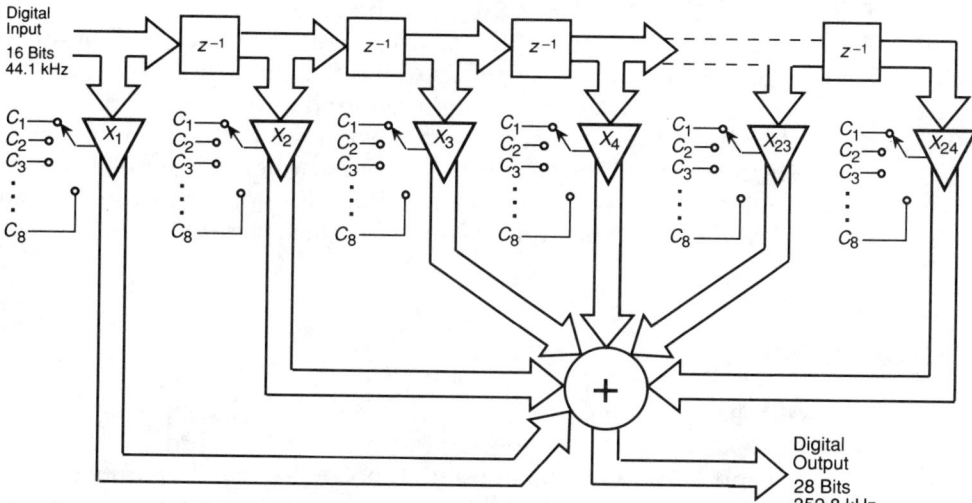

**Figure 17.** An example of an eight-times oversampling filter using a twenty-four element transversal architecture.

calculated by the filter. The sampling frequency is increased to 352.8 kHz. The filter's coefficients produce a transition region between 20 and 24.3 kHz and again around 330 kHz. Figure 18 shows the effect of the filtering. Note that when the effect of the output sample-and-hold circuit is accounted for, the amplitudes of the output spectra are attenuated.

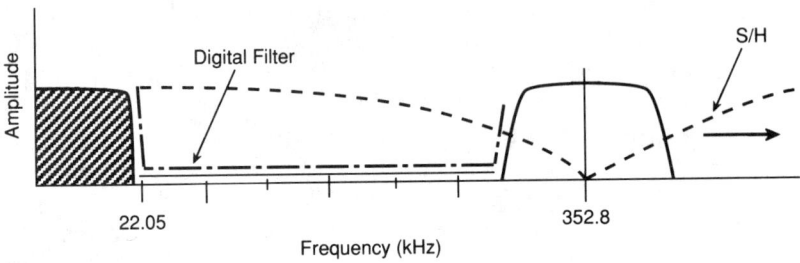

**Figure 18.** Digital filtering extends the placement of the image spectra, suppressing intermediate components. The sample-and-hold function further suppresses high-frequency image spectra.

Following digital filtering, the data is converted to the analog domain with a D/A converter. Because the multiplication in the filter generates words of 28 bits, designers are free to choose words of 16, 18, or 20 bits or more, to apply to 16-, 18-, or 20-bit D/A converters. However, any remaining bands must be completely suppressed by an anti-imaging analog filter following the converter. The oversampling rate determines where the images are placed and hence the kind of post filter needed to remove them. Generally, since the remaining band is so high in frequency, a filter with a gentle response can be used. It is a noncritical design, and its low order guarantees good phase linearity; a simple *RC* low-pass filter may be employed.

A digital filter thus efficiently accomplishes the task of anti-image filtering without resorting to analog brickwall filters. In addition, oversampling may be used to decrease the wordlength required to represent a digital audio signal. Information theory assures us that when the output sampling rate is greater than the input sampling rate, the output wordlength may be less than the input wordlength without loss of source data. This opportunity leads to the concept of low-bit D/A converters, as described in Chapter 12.

# Quantization

With PCM systems, degradation occurs when a sample is measured and assigned an approximate amplitude value. In other words, quantization results in a quantization error, manifested as noise that can be correlated to the signal. This error may be reduced by increasing the number of quantization quanta (ie. longer wordlength); this in turn increases the bandwidth required. The signal-to-quantization noise power ratio increases exponentially with bandwidth. This is an efficient relationship which approaches the theoretical maximum, and is a hallmark of coded systems, such as PCM.

With uniform quantization, an analog signal is mapped into a number of quanta of equal height and placement. The infinite number of points on the analog waveform must be quantized to the finite number of quanta; clearly, this introduces an error. No matter what the medium, the quality of the representation increases as the number of quanta is increased; for example, a high-quality color picture may require 512 color levels, and high-quality music may require 65,536 amplitude levels. On the other hand, only a few levels can still carry reasonable information content; for example, two amplitude levels can (barely) convey intelligible speech.

An analytical means of expressing this quality, in terms of degradation, is the mean square quantizing error $\overline{E^2}(t)$. If $Q$ is the amplitude of a quantum, and $E$ is the difference between the analog signal and the quantized signal, then $E$ must lie between $-Q/2$ and $+Q/2$. Assuming that $E$ possesses equal probability in the range from $-Q/2$ to $+Q/2$,

$$\overline{E^2} = \int_{-\infty}^{\infty} E^2 P(E) \, dE,$$

where $P(E)$ is the probability density function of $E$ such that

$$P(E) = \begin{cases} 1/Q & \text{when } -Q/2 \leq E \leq Q/2, \\ 0 & \text{otherwise.} \end{cases}$$

Thus

$$\overline{E^2} = \int_{-Q/2}^{Q/2} E^2 \frac{1}{Q} \, dE = \frac{1}{Q} \int_{-Q/2}^{Q/2} E^2 \, dE = \frac{Q^2}{12}.$$

Furthermore, if $k$ is the number of bits used to represent the signal,

$$Q = 1/2^{k-1}.$$

Thus,

$$\frac{Q^2}{12} = \frac{2^{-2k}}{3}.$$

This assumes that the transmitted signal is bandlimited and sampling adheres to the Nyquist theorem. Given a noise spectrum of $\pm f_n$, and sampling rate $f_s$, it can be shown that the total quantization noise power is equal to

$$N_Q = \int_{-f_n}^{f_n} N(f) \, df = \frac{Q^2}{12}.$$

The noise power is flat, and the level of the noise power spectral density is

$$N(f) = \frac{Q^2}{12} \cdot \frac{1}{f_s} = \frac{2^{-2k}}{3 f_s}.$$

Additional analysis would show that the spectral content of the quantization error signal lies in the bandwidth of the message signal. The result is only valid given the assumption that the error probability is equal for any amplitude. Furthermore, as noted above, oversampling decreases in-band noise power as a function of the oversampling rate. Specifically, the in-band noise power is equal to

$$N_B = \int_{-f_n}^{f_n} N(f)\ df = \frac{Q^2}{12} \frac{2f_n}{f_s}.$$

Given an expression for noise power, it is relatively simple to derive an expression for S/N ratio for a PCM system. Assuming a sinusoidal message signal with peak amplitude $A$ normalized between $\pm 1$, the average signal power $S$ is

$$S = A^2/2.$$

The S/N ratio power is thus

$$\text{S/N} = \frac{A^2}{2} \frac{12}{Q^2} = \frac{6A^2}{Q^2}.$$

If there are $n$ quantization steps, and the wordlength is $k$ bits in length, then

$$Q = 2/n,$$

and

$$Q = \frac{2}{2^k - 1}.$$

Thus

$$\text{S/N} = \frac{6A^2}{4}(2^k - 1)^2 = \frac{3}{2}A^2(2^k - 1)^2.$$

Written as a decibel expression,

$$(\text{S/N})_{dB} = 10\log_{10}(3/2) + 20\log_{10} A + 20\log_{10}(2^k - 1).$$

Thus, for a sinusoidal signal of maximum allowed input amplitude, this expression can be simplified to

$$(\text{S/N})_{dB} = 6.02k + 1.76.$$

This result, however, assumes that the quantization error is uniformly distributed, and quantization is accurate enough to prevent signal correlation in the error waveform. This is generally true for high-amplitude audio signals but certainly is not the case for low-amplitude signals.

To reduce quantization error, as noted, the quantum size can be made smaller; however, this requires an increase in transmission bandwidth. In some PCM systems, quantization error is minimized by use of nonuniform quantization quantum sizes. Such systems attempt to tailor quantum sizes to best suit the statistical properties of the signal. This is logical because the probability distribution of most message signals is not uniform, as assumed above. For example, speech signals would be best served by an exponential-type quantization distribution; this assumes that small-amplitude signals are more prevalent than large signals. Many quantization levels at low amplitudes, and fewer at high amplitudes, should result in decreased error. Companding, with compression prior to uniform quantization, and expansion following quantization, can be used to achieve this result. Quantization and requantization conversion may be performed with a variety of techniques. Multibit circuits are described in Chapter 3; low-bit techniques are described in Chapter 12.

# Dither

The effects of quantization error may be reduced by increasing wordlength, or by using methods such as companding. A far more efficient technique, however, is the use of dither. Fundamentally, dither is a low-level noise signal added to the message signal prior to quantization. Dither decorrelates and minimizes the effect of quantization and requantization error to the point of elimination, causing a digital system to act more like an analog system in this respect; however, a digital system, properly dithered, far exceeds the signal-to-noise performance of an analog system. On the other hand, an undithered digital system can be inferior to an analog system, particularly under low-level signal conditions. A high-quality digital audio system demands use of dither prior to quantization; in addition, digital computation may require use of dither. For example, gain changing, oversampling interpolation, equalization, and sample rate conversion all require dithering to reduce the effects of quantization and noise modulation as well as to limit cycle oscillations possible in IIR and noise shaping filters. There are many possible dither signals, nonsubtractive as well as subtractive; alternatively, as noted, digital dither may be employed.

Analog dither, applied prior to a linear A/D converter, causes the A/D converter to make additional level transitions which preserve low-level signals through duty cycle modulation—a kind of pulse width modulation. This linearizes the quantization process and helps eliminates error because the average value of the output reflects the input signal. Harmonic distortion and intermodulation products are converted to wide-band noise. A price, however, is paid in a slightly raised noise floor; noise power is $Q^2/4$, where $Q$ is 1 LSB. Because of the ease with which Gaussian noise may be generated in the analog domain, it is often selected for this application. The mathematical expression for a probability density function (PDF) of a zero-mean random noise with rms value of $v_{rms}$ is given by

$$P(v) = \frac{1}{v_{rms}\sqrt{2\pi}} \exp(-v^2/2v_{rms}^2).$$

A Gaussian dither eliminates distortion and reduces noise floor modulation. However, it results in a relatively higher noise floor compared with other kinds of dithering. Other types of dither generally produce better results than Gaussian dither. In some cases, the input analog noise floor itself acts as sufficient dither. For example, a pseudo random noise generator may be used to produce the required dither signal. Lipshitz and Vanderkooy [1989] have shown that dither with a rectangular probability density function (RPDF) eliminates distortion products caused by quantization but does not eliminate noise floor modulation. RPDF dither of $\pm Q/2$ adds $Q^2/12$ noise power. RPDF dither is a uniformly distributed random voltage between $\pm 1/2$ LSB:

$$P(v) = \begin{cases} 1/Q & \text{when } -Q/2 \leq v \leq Q/2, \\ 0 & \text{otherwise,} \end{cases}$$

where $Q$ is one quantizing quantum or LSB step size.

Alternatively, dither with a triangular probability density function (TPDF) can be employed; TPDF can be obtained by summing two random numbers with uniform distribution, that is, by summing two RPDF signals. The total width of the signal is $2Q$. TPDF dither minimizes both distortion and noise floor modulation; the noise floor, however, is somewhat higher than in RPDF dither; TPDF dither adds $Q^2/6$ noise power. RPDF and TPDF dither are easily generated in the digital domain and are preferable to Gaussian dither. Alternatively, a high-frequency dither placed at the Nyquist frequency can be used to alter the spectrum of quantization error to minimize its audibility. An example of the effects of high-frequency triangular dither is shown in Figure 19; a 22 kHz triangular dither of 1/2 LSB peak amplitude is added to a $-90$ dB, 330 Hz sinewave quantized with 16 bits. The dither creates a pulse width modulation signal in which the average value approximates the sinewave input signal. The frequency response plots show that harmonic distortion components are clearly reduced in the dithered signal. In addition, this high-pass dither does not add as much noise to the signal as broadband dither. In this case, however, use of a discrete dither frequency leads to intermodulation products. A wideband dither signal would help alleviate this artifact.

As noted, digital dither is advantageous where signal manipulation takes place. For example, the truncation associated with multiplication can cause objectionable error, especially as it accumulates through multiple calculations. In effect, such rounding adds the same distortion and noise modulation errors as requantization, even if the input digital signal is properly dithered. Infinite impulse response noise shaping filters with digital feedback can exhibit limit cycle oscillations with low-level signals if gain reduction or certain equalization processing (resulting in gain changes) is performed; digital dither can be used to randomize these cycles.

As Vanderkooy and Lipshitz [1987] have shown, truncated or rounded digital words can be redithered with RPDF or TPDF fractional numbers, as shown in Figure 20. An RPDF dither word $D_1$ is added to a digital audio word with integer part $P_i$ and fractional part $P_f$. The carry bit acts as dither for the rounding process in the same way that 1 LSB analog RPDF dither would affect an A/D converter. When a statistically independent RPDF dither $D_2$ is added, TPDF dither results. This TPDF dither noise power is $Q/6$, and rounding noise power is $Q^2/12$, so total noise power is $Q^2/4$. The final sum has integer part $S_i$ and fractional part $S_f$, which become $S$ upon rounding.

In cases of gain fading, TPDF appears to be a better choice than RPDF because of its elimination of noise modulation as well as distortion, at the expense of a slightly higher noise floor. To minimize audibility of this noise penalty, high-pass TPDF dither may be most appropriate in rounding. The TPDF statistics are not changed; however, dither samples are correlated. The power spectrum of the dither signal, normalized for total noise power $Q^2/6$, showing the autocorrelation is

$$f(V) = \int_{-\infty}^{\infty} f(V + v)p(v) \, dV.$$

Average dither noise power is $Q^2/6$, with no noise at 0 Hz and double the average value at the Nyquist frequency—hence the term high-pass dither. This

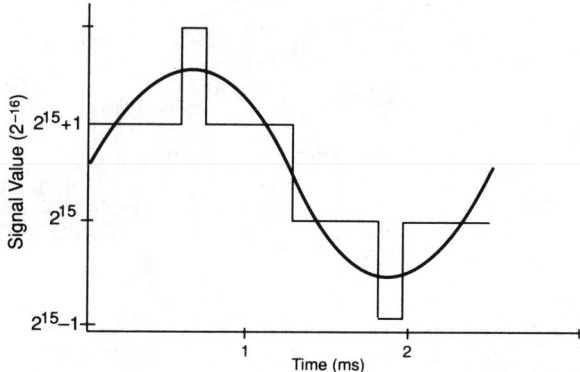

*(a) A −90 dB, 330 Hz sinewave quantized to 16 bits, without dither.*

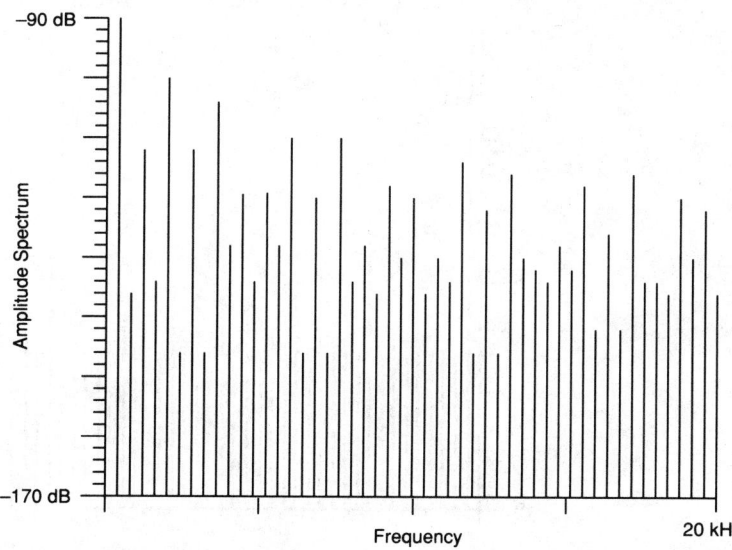

*(b) Frequency spectrum of the undithered quantized sinewave showing harmonic distortion components.*

**Figure 19.** Illustration of the effects of high-frequency triangular dither. *(From Dijkmans and Naus)*

shaping becomes more pronounced with noise increasingly shifted outside the audio band, as the oversampling ratio is increased, for example, by two or four times. These and related topics are discussed in more detail in Chapter 12. As already noted, the audible effect of a noise penalty is lessened when oversampling is employed because in-band noise is relatively decreased proportional to the oversampling rate. Further reduction of the $Q^2/12$ requantization noise power can be achieved through noise shaping circuits, as described in Chapter 12.

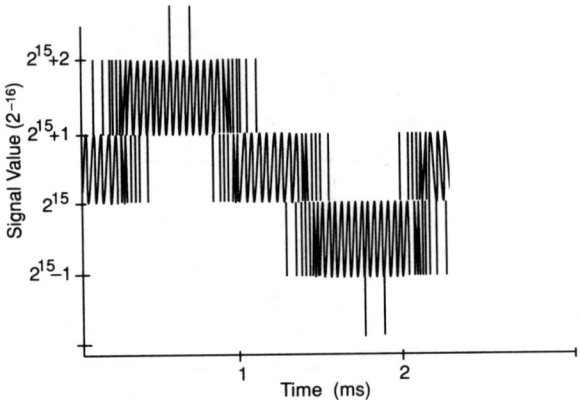

*(c) A 22 kHz triangular dither of 1/2 LSB peak amplitude is added to the −90 dB, 330 Hz sinewave quantized to 16 bits.*

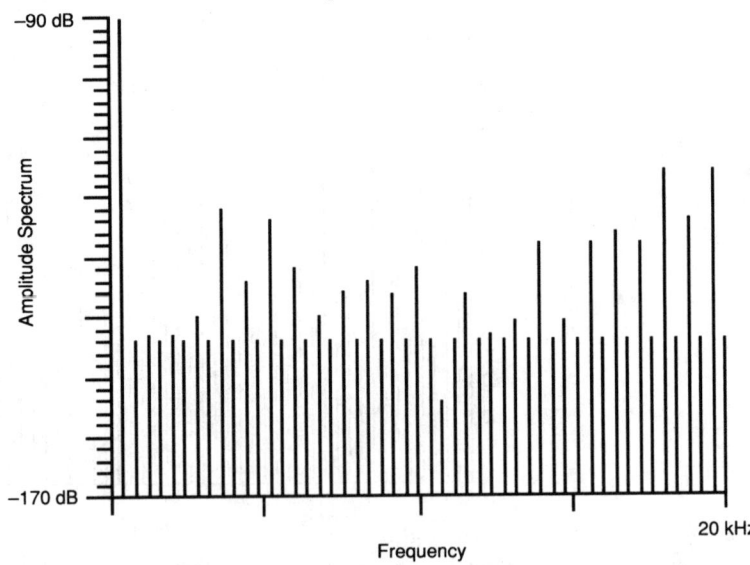

*(d) Frequency spectrum of the dithered quantized sinewave showing reduced distortion components.*

**Figure 19.** (cont.)

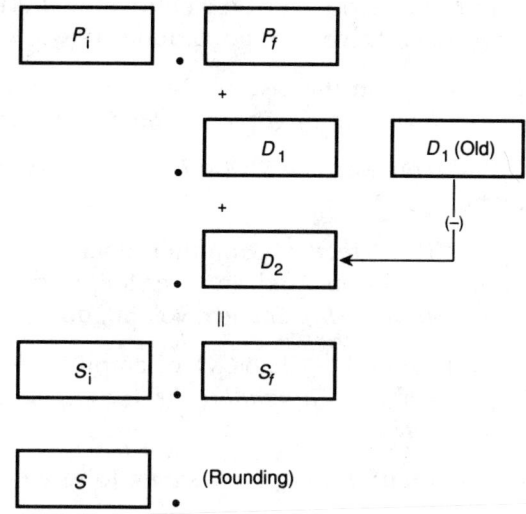

**Figure 20.** An example of the use of digital redithering during truncation or rounding computation. *(From Vanderkooy and Lipshitz)*

# Conclusion

Although complex, the world of digital audio modulation and sampling is based on well-understood principles. Different forms of modulation, although none are perfect, offer a variety of advantages for virtually any application. In practice, a digital audio system may variously employ a PAM signal during encoding, PCM during transmission, and PWM during decoding—each providing a unique asset. Throughout the process, the cornerstone of discrete time sampling yields completely predictable results, thanks to properties of the impulse response, summation, and other fortunate happenstances. Unfortunately, bandwidth demands, quantization, and other negative attributes can affect the success of a system. However, engineers can make use of clever methods, such as dither, to minimize their harm and help ensure the successful operation of the audio digitization system.

# References

AES Recommended Practice for Professional Digital Audio Applications Employing Pulse Code Modulation: Preferred Sampling Frequencies. AES5-1984 (ANSI S4.28-1984), *Journal of the Audio Engineering Society*, vol. 32, no. 10, October, 1984.

Blesser, B. A., and B. N Locanthi, "The Application of Narrow-Band Dither Operating at the Nyquist Frequency in Digital Systems to Provide Improved

Signal-to-Noise Ratio over Conventional Dithering," *Journal of the Audio Engineering Society*, vol. 35, no. 6. June, 1987.

Finger, R. A., "On the Use of Computer Generated Dithered Test Signals," *Journal of the Audio Engineering Society*, vol. 35, no. 6, June, 1987.

Gersho, A., "Quantization," *IEEE Communication Society Magazine*, vol. 15, September, 1977.

Harris, S., "The Effects of Sampling Clock Jitter On Nyquist Sampling Analog-to-Digital Converters, and on Oversampling Delta-Sigma ADCs," *Journal of the Audio Engineering Society*, vol. 38, no. 7/8, July/August, 1990.

Hawksford, M. O. J., "Chaos, Oversampling, and Noise Shaping in Digital-to-Analog Conversion," *Journal of the Audio Engineering Society*, vol. 37, no. 12, December, 1989.

Haykin, S., *Communication Systems*, John Wiley & Sons, New York, 1983.

Jayant, N., and L. Rabiner, "The Application of Dither to the Quantization of Speech Signals," *Bell System Technical Journal*, vol. 51, 1972.

Morris, D. J., *Pulse Code Formats for Fiber Optical Data Communication*, Marcel Dekker, 1983.

Muraoka, T., Yamada, Y., and M. Yamazaki, "Sampling Frequency Considerations in Digital Audio Standards," *Journal of the Audio Engineering Society*, vol. 26, no. 4, April, 1978., Erratum, *ibid.*, vol. 26, no. 7/8, July/August, 1978.

Owen, F. F. E., *PCM and Digital Transmission Systems*, McGraw-Hill Book Co., 1982.

Pohlmann, K. C., *Principles of Digital Audio*, 2nd edition, Howard W. Sams & Co., 1989.

Rossum, D., "Digital Dither in Music and Sound Synthesis," *Audio in Digital Times: 7th AES Conference Proceedings*, May, 1989.

Shannon, C. E., "A Mathematical Theory of Communication," *Bell System Technical Journal*, vol. 27, 1948.

———, "Communication in the Presence of Noise," *Proceedings of the Institute of Radio Engineers*, vol. 37, 1949.

Trischitta, P. R., and E. L. Varma, *Jitter in Digital Transmission Systems*, Artech House, 1989.

Vanderkooy, J., and S. P. Lipshitz, "Digital Dither," *AES Preprint 2412*, December, 1986.

———, "Digital Dither: Signal Processing with Resolution Far Below the Least Significant Bit," *Audio in Digital Times: 7th AES Conference Proceedings*, May, 1989.

———, "Dither in Digital Audio," *Journal of the Audio Engineering Society*, vol. 35, no. 12, December, 1987.

———, "Resolution Below the Least Significant Bit in Digital Audio Systems with Dither," *Journal of the Audio Engineering Society*, vol. 32, no. 3, March, 1984, Erratum, *ibid.*, vol. 32, no. 11, November, 1984.

Widrow, B., "A Study of Rough Amplitude Quantization by Means of Nyquist Sampling Theory," *IRE Transactions on Circuit Theory*, vol. CT-3, December, 1956.

———, "Statistical Analysis of Amplitude-Quantized Sampled-Data Systems," *Transactions of the AIEE (Applications and Industry)*, vol. 79, January, 1961.

Wong, P. W., "Quantization Noise, Fixed-Point Multiplicative Roundoff Noise, and Dithering," *IEEE Transactions*, ASSP, vol. 38, no. 2, February, 1990.

*Chapter 3*

# Multibit Conversion

### Ken C. Pohlmann

## Introduction

Formerly, the design of the input and output stages of a digital audio system was considered straightforward. A brickwall analog input filter and 16-bit A/D converter and a 16-bit D/A converter and brickwall output filter were nominally specified. However, in practice a number of problems appear, including phase distortion, bias-dependent glitches, noise switching, and converter nonlinearity. The audibility of these artifacts soon persuaded designers that the classic method and traditional specifications were not suitable; it is ironic that it is these analog parts that cause the greatest deficiencies in a digital audio system. In particular, multibit converters can degrade the audio signal through nonlinearity, and zero-cross distortion can be manifested as distortion, at low signal levels. To provide adequate linearity, converters can require laser trimming of the converter current summing ladder and careful calibration of MSBs using external trimmer potentiometers. In an effort to overcome limitations in conventional multibit conversion, new multibit converter techniques, such as multiple bias and compensated converters, have been developed.

## Digital-to-Analog Conversion

A digital-to-analog converter must produce an analog voltage corresponding to each digital audio sample. As we observed in Chapter 2, the sample voltages ultimately create an analog waveform, as the impulse response of each sample is summed with the response of all other samples. The analog signal output from the D/A converter is theoretically equivalent to the signal appearing at the output of the sample-and-hold circuit in the recording side of the chain. Specifically, it is a pulse amplitude modulation (PAM) waveform. Whether or not oversampling is employed, an output analog filter is required to remove unwanted supersonic images centered at multiples of the output sampling frequency.

There are several methods available to accomplish multibit D/A conversion. Classically designed D/A converters use $R\text{-}2R$ resistor ladder networks, with a two's complement or binary input code. A reference voltage source is used to create a voltage along the ladder; it is converted to a current by each part of the ladder acting as a current divider. Because of the $R\text{-}2R$ configuration, each ladder current differs by a binary power of two from its successor. These binary-weighted current levels correspond to the power of two weighted values of the bits represented in a digital word. The bits of the input digital word control solid-state switches to access only current values corresponding to a binary 1. For example, a 4-bit word of 1001 would access current from the first and fourth ladder rungs. The currents from the individual ladder rungs are summed and converted to a voltage. The result is a output voltage which corresponds to the input binary word. Because the converter responds to the audio signal around a DC offset of one MSB, large voltages are switched when small voltage signals are required, for example, when the audio signal changes polarity. This presents the worst case of high nonlinearity conditions exactly when high linearity is required. Moreover, every voltage source in a converter adds noise proportional to its weighted value.

Circuit designers have devised D/A converters which differ significantly in design from $R\text{-}2R$ converters. Two examples are integrating and dynamic element matching converters. An integrating D/A converter uses a constant current source and a timing circuit. The length of time that current flows to an integrator circuit is regulated according to the magnitude of the input binary word. The current, and the output voltage derived from it, are proportional to the time duration and hence the magnitude of the input binary word. In practice, because of the speed required for audio conversion, multislope designs must be employed. Although differential linearity of a single ramp converter can be very good, uneven transitions between multiple ramps can cause nonlinearity. In addition, such converters tend to be susceptible to jitter.

Another D/A design called dynamic element matching (DEM) uses the principle of current adding. A series of current sources with binary-weighted values are selectively summed together. As in a $R\text{-}2R$ ladder, the values are selected according to the value of the bit controlling each source. When individual currents are summed and converted to a voltage, conversion is accomplished. These and other conventional converter architectures are discussed in great detail in many of the references for this chapter.

## Conversion Nonlinearity

In theory, digital audio reproduction is relatively simple. For example, 16-bit audio data words can be reproduced through conversion into the analog domain, using a 16-bit D/A converter. However, only a perfect 16-bit converter could reproduce 16-bit audio data with complete accuracy; such perfect converters cannot exist. Analog-to-digital converters experience an identical shortcoming. Thus, in practice, A/D and D/A converters present a weak link in the signal chain.

By definition, quantization levels are spaced 1 LSB apart thus quantization error presents an uncertainty of ±1/2 LSB. In practice, converters impose deviations in addition to the ideal error; resolution is chiefly determined by absolute linearity error and differential linearity error. Absolute linearity error is a deviation from the ideal quantization staircase. Differential linearity error is a relative deviation in the ideal step size of any individual step. Generally the error is uncorrelated with high signal levels but correlated with low signal levels; as a result it is most apparent at low signal levels as distortion. Differential nonlinearity appears as wide and narrow steps and can cause entire sections of the transfer function to be missing. Differential nonlinearity is shown in Figure 1a; an extreme case of nonlinearity called nonmonotonicity is shown in Figure 1b. The effect is usually minor with high-amplitude signals, but the errors can dominate low-level signals. For example, any signal 80 dB below full scale will pass through only six or seven codes in a 16-bit quantizer; if half of those codes were missing, the result would be 14-bit performance. As Dijkmans and Naus [1989] have pointed out, depending on bias, the differential linearity error in a 16-bit D/A converter for a −90 dB sinewave signal can result in generated levels ranging from −85.9 dB to −98.2 dB; this is represented in Figure 2. Since bits and their associated errors switch in and out throughout the transfer curve, their effect is signal dependent. Thus harmonic and intermodulation distortion and noise can vary with input signal conditions. Because this kind of error is clearly correlated with the audio signal, it tends to be quite perceptible.

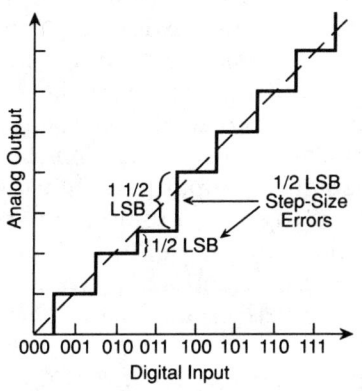
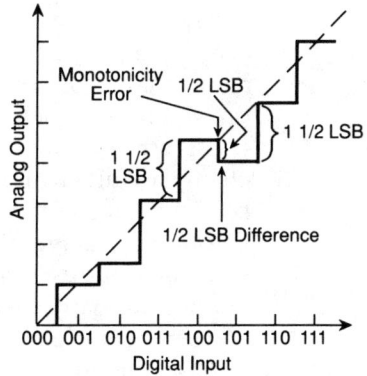

(a) *Differential nonlinearity is any relative deviation from ideal step size.*

(b) *Nonmonotonicity results from a step-size error such that the next higher step does not increase the output level.*

**Figure 1.** Two examples of converter nonlinearity errors.

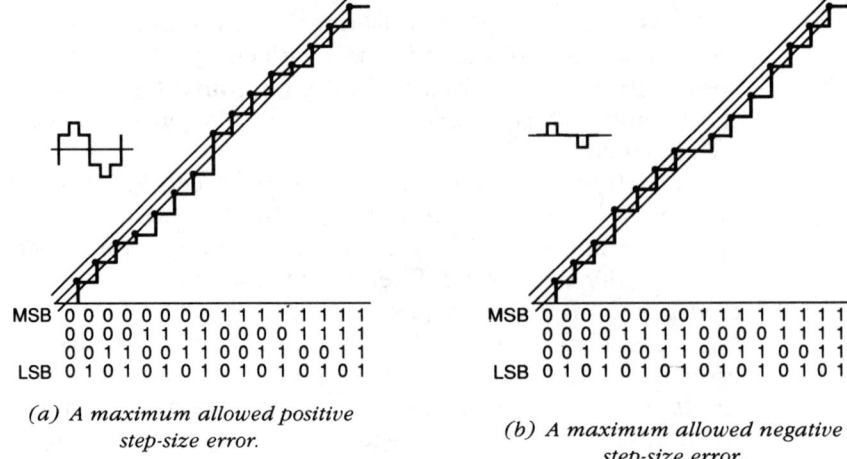

*(a)* A maximum allowed positive step-size error.

*(b)* A maximum allowed negative step-size error.

**Figure 2.** Examples of differential nonlinearity with a −90 dB input signal. *(From Dijkmans)*

Clearly, in practice, 16-bit conversion is insufficient for 16-bit storage or processing. To realize the greater potential of audio fidelity in a digital medium, the signal digitization (and processing) steps must have a dynamic range greater than the audio signal itself. As a result, the engineering methods used to accomplish conversion must be quite sophisticated for digital recording and reproduction.

No matter how many bits are converted, the accuracy of the conversion, and hence the fidelity of the audio signal, hinges on the linearity of the converter. A linearity test measures the converter's ability to record or reproduce various signals, particularly low-amplitude signals, at the proper amplitude. Specifically, linearity refers to the converter's ability to input or output an analog signal which directly conforms to a digital word. For example, when a bit changes from 1 to 0 in a D/A converter, the analog output must decrease exactly by a proportional amount. Any nonlinearity results in distortion in the audio signal. Of course, the amount that the analog voltage changes depends on which bit has changed. Every PCM digital word contains a series of binary digits, arranged in powers of two. The most significant bit (MSB) accounts for a change in fully half of the analog signal's amplitude, while the least significant bit (LSB) accounts for the least change, for example, in an 18-bit word, an amplitude change of less than four parts in a million. Physical realities conspire against this accuracy. As noted, conventional converters exhibit differential nonlinearity due to bit-weighting errors; thermal or physical stress, aging, and temperature variations, are also factors.

To help equipment manufacturers use D/A converters to their best advantage, many D/A chips provide a means to calibrate the converter. Some D/A chips offer calibration of the MSB. This trimming must be performed with great care. Since the MSB is so much larger than the other bits, an MSB error of only 0.01% (one part in 10,000) would completely negate the contribution of the two least significant bits (which account for one part in 21,845 of the total signal amplitude). An error of 0.1% in the MSB would swamp the combined values of the five smallest bits. Other converters may offer calibration of the four most significant bits. It is interesting to note that fully 93% of the total analog output is represented through these four most significant bits. It is these most significant bits that steer the converter's output amplitude; when they are properly calibrated, the entire output signal of a well-designed D/A converter will be more accurate. This accuracy is most important at low levels, usually below −60 dB. Any nonlinearity in D/A conversion will be apparent as a deviation from a nominal amplitude. Moreover, such nonlinearity can easily be heard; for example, when using a test disc containing a dithered fade to silence tone, poor D/A linearity is plainly audible.

By using tests for D/A linearity, the results of poorly calibrated D/A converters can be evaluated. For example, Figure 3a shows a CD player with highly inaccurate conversion. Reproduced signals lower than −60 dB in amplitude begin to show considerable nonlinearity. For example, a −70 dB signal is reproduced at an amplitude of −71 dB, and a −80 dB signal is reproduced at −84 dB. At −90 dB the nonlinearity is even more pronounced. Clearly, this kind of expansion of low-level dynamics would alter low-amplitude information considerably. For example, ambient information would be audibly changed.

Figure 3b shows another CD player with converter nonlinearity. A −70 dB signal is reproduced as −69 dB, a −80 dB signal as −79 dB or −78 dB (depending on channel), and a −90 dB signal as −88 dB or −86 dB (again, depending on channel). This nonlinearity results in compression of low-level information. Again, the affect could be audible. Figure 3c shows a CD player with extreme nonlinearity problems. The output signal is compressed through −80 dB, then suddenly the signal is expanded in a radical manner. A low-level audio waveform would be reproduced in a highly nonlinear fashion. The resulting distortion could greatly detract from the fidelity of music reproduction.

Clearly, the number of bits (e.g., 16 or 18) in a D/A converter is a highly artificial way of determining quality of conversion. In fact, the number of bits employed by the converter is irrelevant compared with the nonlinearities exhibited by the converter. Unless the manufacturer is willing to purchase high-quality converters and adjust them during production, the results can be highly unpredictable and unsatisfactory. On the other hand, when good converters are individually calibrated using the MSB adjustment described above, the results can be impressive. For example, Figure 3d shows the output of a CD player with appropriate nonlinearity compensation. The output is highly linear through −90 dB. In fact, the small error present is barely measurable by the sophisticated test set used to analyze it. It would be safe to say that this nonlinearity would not be audible under most listening conditions.

Careful converter selection and the calibration of individual converters are important aspects of digital audio recording and reproduction. By adjusting the four most significant bits, or the MSB itself, the linearity of PCM converters can be greatly improved. Likewise, the type of converter design and overall

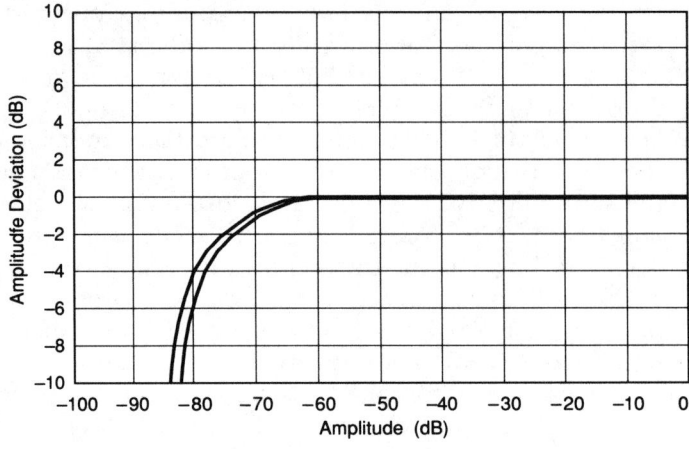

*(a) Data expansion nonlinearity.*

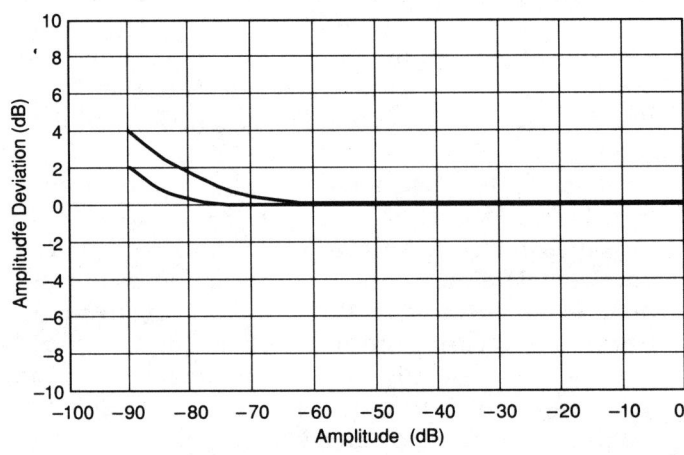

*(b) Data compression nonlinearity.*

**Figure 3.** Examples of D/A converter linearity evaluations.

architecture plays an important role. On one hand, converters with longer wordlengths provide the opportunity for higher conversion accuracy of a 16-bit audio signal. However, the accuracy of any converter, whether 16 or 18 bits, hinges on the precision of its calibration.

## Zero-Cross Distortion

Zero-cross distortion in a converter can occur at the zero-cross point between the positive and negative polarity portions of an analog waveform. When a conventional ladder D/A converter is switched around its MSB, that is, from 0111 1111 1111 1111 to 1000 0000 0000 0000, to reflect a polarity change, large

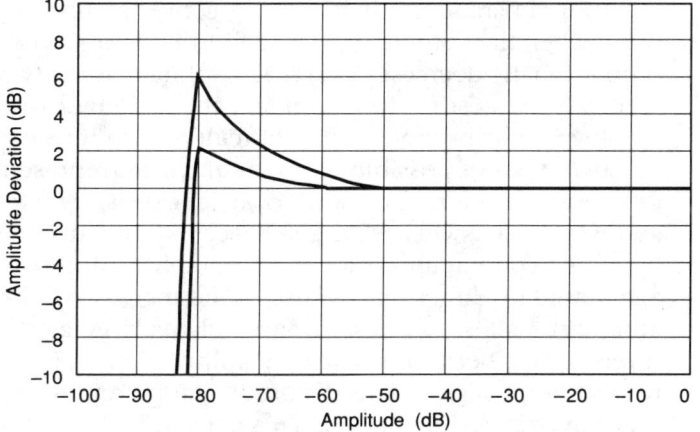

*(c) Data expansion/compression nonlinearity.*

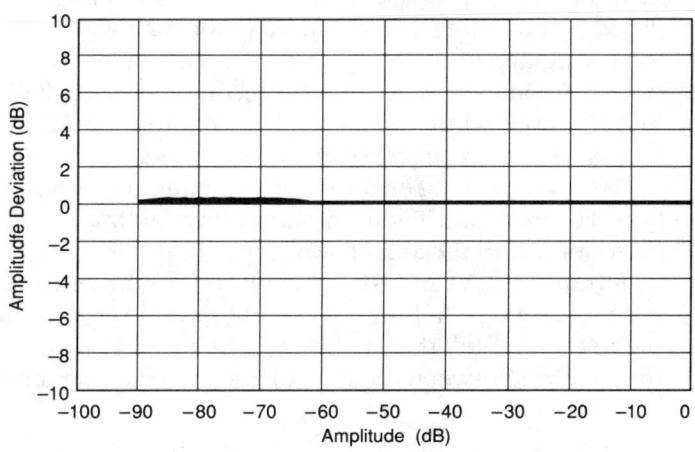

*(d) Relatively accurate linearity.*

**Figure 3.** (cont.)

current fluctuations and variations in bit switching speeds conspire to create considerable differential nonlinearity and glitches. These defects are known collectively as zero-cross distortion. Zero-cross distortion presents an ongoing error. Since musical waveforms continually change between positive and negative polarity, twice each cycle, the zero point is crossed constantly with the MSB continually being turned on and off.

Ideally, when there is no error difference between the smallest and largest resistors in the ladder, there is no error. However, the MSB (largest) resistor generally has an error large compared with the LSB (smallest) resistor; the error can be larger than the value of the LSB resistor itself. This is because the MSB in a 16-bit converter must be represented by a current value with an accuracy greater than 1/65,536 of the LSB. (This demonstrates why adjustment of the MSB is paramount in achieving converter linearity.) Similarly, error can occur when the second, third, and fourth bits are switched; however, the error proportionally decreases as signal level increases. By the same token, error is relatively greatest for low-level signals. As longer-wordlength converters are employed, it is proportionally difficult to match resistor values.

Two types of possible discontinuities are represented in Figure 4. In this example, the internal network of ladder resistors is not accurate to the LSB, and when it is switched, a glitch occurs. This discontinuity contains a high harmonic content and can be quite audible. Moreover, the audibility can be aggravated by dithering because of the increase in the number of transitions around the MSB. The distortion is relatively greater when reproducing low-level signals, because the fixed-amplitude glitch is proportionally large with respect to the signal. This distortion can be removed by careful alignment of the converter; as noted, some D/A converter chips provide for external adjustment of the MSB or several MSBs to allow for this. Calibration, however, can be laborious and is prone to drift over time. Zero-cross distortion can also be alleviated by providing a D/A converter for each waveform polarity; in that way, a total switching of digits never occurs. The digital signal must be split between two D/A converters; this can be accomplished digitally by upstream processing. If properly implemented, this technique can theoretically be glitch-free, at least as far as zero-crossing is concerned.

As noted, timing errors in bit switching can contribute to zero-cross distortion. For example, Figure 5 shows how bit switching timing can result in a discontinuity in the analog waveform. In this example a (9) value is generated by turning the (1) and (8) value bits on. In the subsequent sample a (5) value is generated by combining (1) and (4) value bits. To move from the first sample to the second, the (4) bit must be turned on and the (8) bit turned off. A timing error between them would generate a momentary (1) or (13) value—a

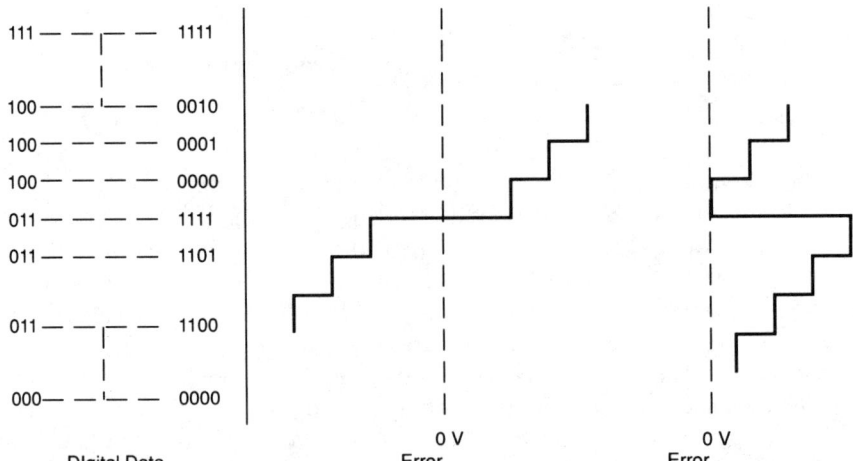

**Figure 4.** Examples of two types of converter discontinuities resulting in zero-cross distortion. *(Courtesy Nakamichi America)*

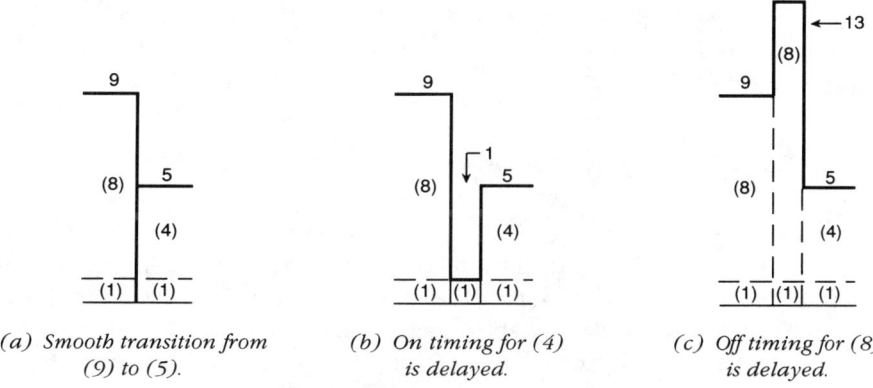

*(a) Smooth transition from (9) to (5).*  *(b) On timing for (4) is delayed.*  *(c) Off timing for (8) is delayed.*

**Figure 5.** Three examples of timing errors resulting in glitches. *(Courtesy Sony Corp.)*

glitch. This error differs from that caused by ladder nonlinearity and can be addressed only by careful design of the converter chip's controller circuitry.

# Eighteen-Bit D/A Conversion Architectures

A converter performing an 18-bit conversion can provide a more accurate reproduction of 16 bits, hence providing measurably lower noise and distortion in the audio signal. Some manufacturers improve performance through hybrid 18-bit methods in which 18 bits of audio data are processed through 16-bit D/A converters specially configured to provide improved performance. Oversampling enables 18-bit conversion when the output of a medium is only 16 bits. For example, when the 44.1 kHz, 16-bit signal from a compact disc is oversampled, both the sampling frequency and number of bits are increased, the former because of oversampling and the latter because of the multiplication in oversampling which must take place. For example, the output of an oversampling filter may be a 176.4 kHz, 28-bit signal. Normally, only the 16 most significant bits are used, for conversion through a 16-bit D/A converter, and the rest are discarded following a rounding-off process (in some designs they are used for noise shaping). An 18-bit architecture uses 18 bits from the output of an oversampling circuit, instead of only 16.

Eighteen-bit D/A conversion technology cannot improve the 16-bit data from the disc; rather, its intent is to make better use of that data. Specifically, 18-bit technology attempts to overcome linearity problems in 16-bit converters which may limit performance. An analogy may be made to oversampling: While the sampling rate is increased, the method doesn't create new information; it merely attempts to make better use of the existing information. It is important that a filter design does not truncate unused bits; instead, least significant bits must be rounded off. When proper oversampling techniques are used, the extra bits convey useful amplitude information, at levels below the most significant sixteen bits. For example, in a filter outputting 18 bits, rounded information is only one sixteenth the magnitude of that rounded in a filter outputting 16 bits.

## Bit-Switching D/A Converter

One bit-switching D/A converter design uses a 16-bit D/A converter configured so that 18 bits from the oversampling filter's output are directed through switches to the inputs of the D/A converter. This circuit is shown in Figure 6a. When the audio signal has a large amplitude (all 16 bits are used to convey the signal) the upper 16 bits are applied to the 16-bit converter. However, when the signal's amplitude is lower (the two upper bits from the oversampling filter are not used to convey the signal) 18 bits are shifted downward, so that the unused bits are ignored, and the 16 lower bits are utilized instead. In other words, until the first 14 bits are occupied, the output is four times its nominal

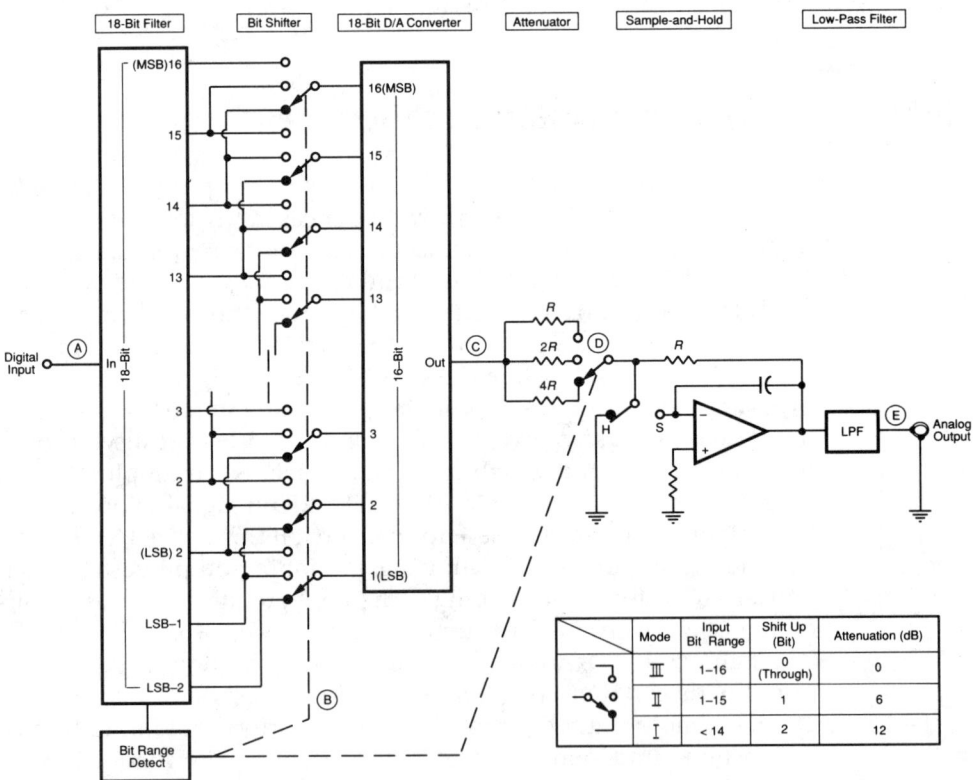

(a) Circuit schematic showing five key points (A–E).

**Figure 6.** A hybrid D/A architecture using bit switching and gain compensation. Eighteen-bit performance is derived from a 16-bit converter.
*(Courtesy Yamaha Electronics)*

amplitude. A one-fourth attenuator is used to compensate for the gain change. When all 14 bits are occupied, the output voltage is maximum. When the fifteenth bit is on, the bits are shifted, and the attenuator is removed. The output increases at normal gain until full 16-bit voltage is reached. In practice, with a music signal, the two upper bits are rarely used, and then only for a brief period. In that respect, the switching system provides dynamic headroom.

Because bit switching changes (through multiplication) the weighted values of the binary word, the gain through the system must be carefully adapted to the switching for any given sample. Specifically, whenever the two lower bits are shifted in, the word's value is effectively multiplied by four, thus the gain of the signal must be reduced by one fourth to compensate. An analog

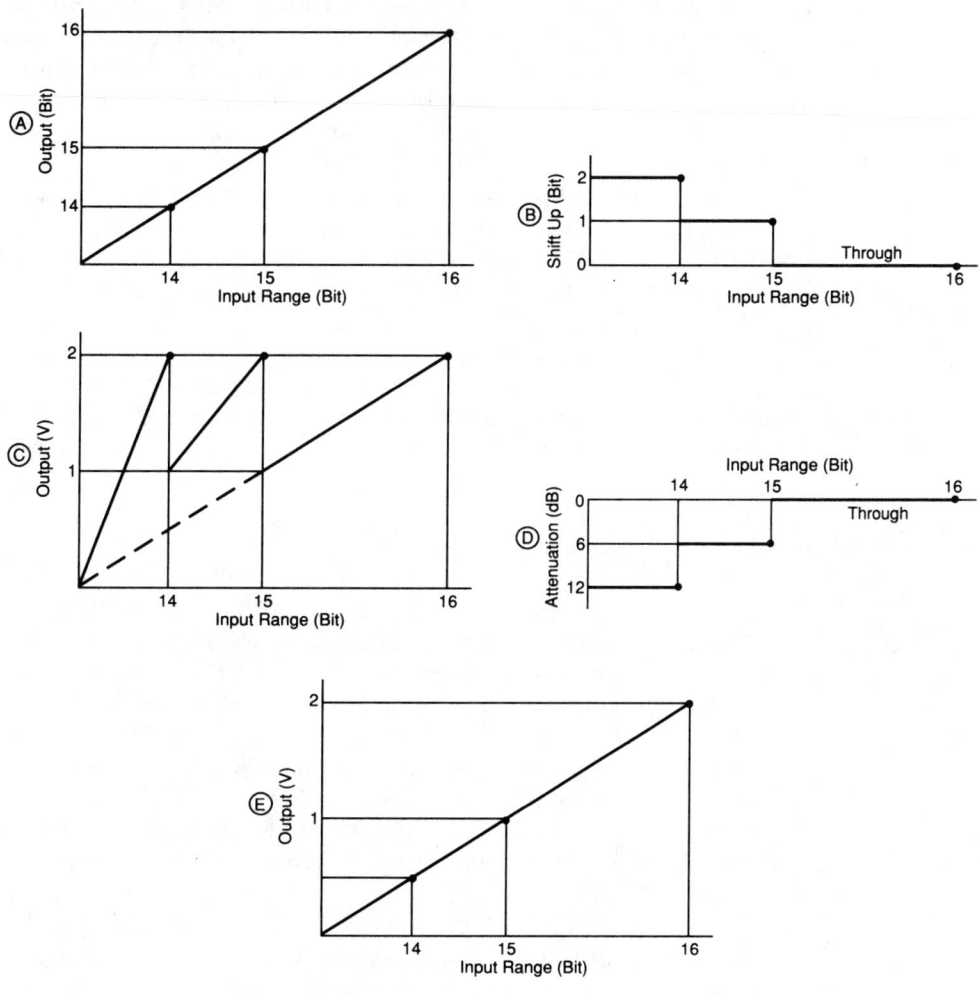

*(b) Signal changes at five key points (A–E).*

**Figure 6.** (cont.)

gain block downstream of the D/A converter performs this function. A one-fourth reduction is required: In the decimal system, when a digit is shifted to the left one place, the result is ten times larger. In the binary system, a shift results in a doubling of value. A shift of two places (two bits) quadruples the value. Thus when the lower bits are shifted up, the amplitude is four times greater, and so the output must be attenuated by a fourth. Operation of the circuit is illustrated in Figure 6b, which shows the changing signal at five key points in the circuit. Through the use of bit switching and gain changing, this method allows a 16-bit converter to effectively process an 18-bit audio signal, theoretically yielding a converter dynamic range of 18 bits.

This adaptive conversion method is a kind of dynamic noise reduction scheme in that the signal is expanded at the D/A converter. As a result, the residual noise of the converter as well as its conversion nonlinearities are proportionally reduced. In this case, an analog output four times larger in magnitude is achieved without increasing the D/A residual noise and conversion error. When the gain is reduced by a fourth, the noise and conversion errors are reduced by a fourth. The result is a theoretical increase of 12 dB in the S/N ratio, and distortion reduction by a fourth.

However, when bits are shifted, it is difficult to immediately and simultaneously shift the gain of the analog output to compensate. The attenuator could introduce a static error, owing to component tolerances. It would not be significant over the first 14 bits, affecting only the attenuation ratio. But when switched on or off, any difference between the attenuator's error and the absolute value of the output would create a glitch in the waveform where the attenuated and nonattenuated signals are joined. The glitch would always be present at $-12$ dB at any signal frequency. Careful manufacturing methods can minimize these artifacts. This technique was devised by Yamaha International.

## Bipolar D/A Converter

In one converter design, four 16-bit D/A converters are configured in a parallel architecture, as shown in Figure 7. In addition, a bit-shifting technique is used. Two D/A converters are used per channel, one for the positive portion and one for the negative portion of the reproduced analog waveform. The use of multiple 16-bit D/A converters aims to improve low-amplitude resolution compared with single D/A converters; in particular, this design attempts to overcome zero-cross distortion.

Eighteen-bit data words are output from a four-times oversampling filter and directed to a distribution processor. The positive or negative portion of the waveform is output to the corresponding converter. If the most significant bit (MSB) of the audio data is high, the word has positive polarity and is applied to the positive-polarity D/A converter. If the MSB is low, the data word is directed to the converter for the negative portion of the waveform. The positive and negative waveforms are summed by a differential push-pull amplifier to produce the output waveform. Because dual D/A converters are used to convert

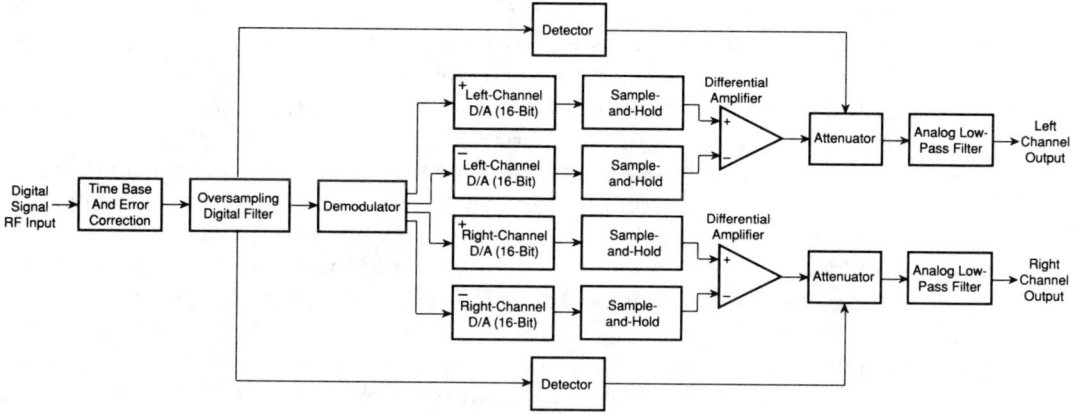

**Figure 7.** A hybrid D/A architecture using four D/A converters in a push-pull mode, with bit-switching and gain compensation.

the bipolar waveform, bits are not switched at polarity changes, and thus zero-cross distortion is theoretically eliminated. In addition, the differential amplifier cancels common-mode noise, additionally improving the S/N ratio.

Furthermore, when the audio signal is less than −12 dB, a detector circuit is used to direct a bit-shifting operation to improve resolution of the low-amplitude converted signal. The audio data word is shifted down by two bits; when the signal is below −12 dB, the second most significant bit ceases to change (it is always low when the MSB is high, and it is high when the MSB is low). The D/A converters thus receive bits 3 through 18. Since the gain of the signal increases when bits are shifted, attenuators must be switched in, to reduce gain proportionally. Attenuators after the D/A converters compensate for the gain change caused by shifting. When the signal is greater than −12 dB, improved low-amplitude resolution is not required, so bit shifting is not engaged, and linear 16-bit conversion takes place.

A VU meter would verify that in most cases an uncompressed music signal is below −12 dB. In other words, the benefits of the system are usually present; normal 16-bit operation only occurs at high levels, of usually brief duration, when the amplitude of the waveform largely masks converter nonlinearities. This technique was devised by Matsushita Electric.

Although hybrid 18-bit conversion methods offer advantages, they may also promote problems, such as switching noise and subtle gain matching errors. Floating conversion systems can introduce errors in the output signal because of the need to compensate for the resulting shifts in amplitude caused by shifting bits.

## Linear 18-Bit D/A Converter

A 16-bit D/A converter must determine which of 65,536 output analog voltages corresponds to the input digital word, nominally within a 20 $\mu$s sampling period. As we have seen, no practical 16-bit converter can perfectly accomplish this. Hybrid 18-bit architectures use modified 16-bit converters in an effort to improve performance. Another approach is linear conversion of a longer data word. Each additional bit reduces error by one-half. A linear 18-bit D/A converter, for example, would have 262,144 levels, exactly four times as many output levels as a 16-bit converter. Any conversion nonlinearity would thus be correspondingly smaller. In other words, an 18-bit converter would give a better 16-bit conversion. In fact, the two extra bits of a linear 18-bit converter would not even have to be connected to yield improved 16-bit performance.

One example of a linear 18-bit D/A converter provides a harmonic distortion specification of 0.0008% for a 1 kHz sinewave. A conventional binary-weighted *R-2R* ladder is used in this converter; however, several special design considerations are employed to attain the desired accuracy. The most significant three bits access seven individual current sources; this reduces thermal errors. Bits 4 through 16 use unit-valued current sources to feed the ladder network. Currents for bits 17 and 18 are derived from unit-valued sources. The relative gain of the three upper bits can be adjusted against the total weight of the 15 lower bits by trimming.

The chip is a 40-pin hybrid with a special divided layout to permit laser trimming of the upper and lower bits. Supporting circuitry is located external to the chip to help maintain the substrate's thermal balance, a critical concern in D/A converter design. For example, even small amounts of heat build-up can affect both the absolute and relative values of the precision resistors in a ladder network, degrading accuracy. The low power consumption of complementary metal-oxide semiconductor (CMOS) technology is attractive in this respect. The PCM 64P D/A converter is manufactured by Burr-Brown Corporation.

The use of linear 18-bit D/A converter chips results in specifications clearly superior to 16-bit designs. As described above, an 18-bit converter has a resolution fourfold greater than a 16-bit converter. In other words, 18-bit conversion of a 16-bit signal theoretically provides a 12 dB increase in S/N ratio. This improvement is clearly measurable. Of course, there is no risk of gain-matching errors as with hybrid 18-bit conversion methods.

## Multiple-Bias D/A Converter

In an effort to further increase S/N ratios and lower distortion in converter design, signal processing techniques can be applied in the digital domain, prior to conversion. One such method uses a multiple-bias converter. This design seeks to eliminate zero-cross distortion with digital data biasing. As noted, zero-cross distortion is particularly critical with low-level signals where masking cannot take place, hence the method distributes the distortion to higher signal levels.

Figure 8 shows a block diagram of a multiple-bias converter. A processor is located between the digital filter and the D/A converter. It adds a negative or positive digital bias to the data thus creating two data streams which are sent to

# Chapter 3: Multibit Conversion

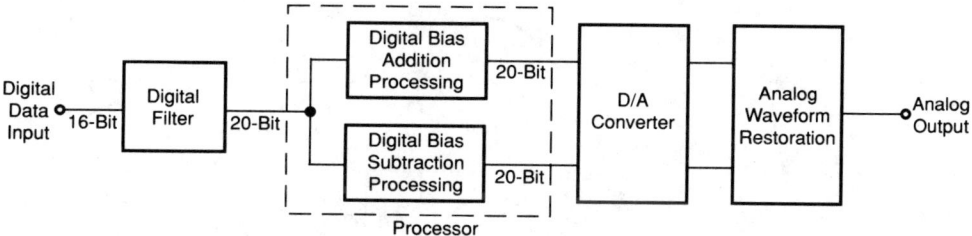

**Figure 8.** Block diagram of a D/A converter architecture, using multiple digital bias. Data biasing minimizes zero-cross distortion.
*(Courtesy Denon America)*

the D/A converters. These DC components are canceled when the two signals are later combined. Figure 9a shows how a low-amplitude audio waveform is shifted, then recombined. (The waveforms consist of digital data but are shown in analog form for illustration). Waveform $A$ (not shifted) would typically exhibit zero-cross distortion. Waveforms $B$ and $C$ are shifted symmetrically, positively and negatively, by an amount of $\pm x$. Because shifted waveforms $B$ and $C$ do not cross the zero line, zero-cross distortion will not arise, as long as peak levels are below the shifted amount $x$. Waveform $D$ shows the recombination of the shifted waveforms, free from zero-cross distortion.

The case when the signal level is higher than the shift amount is shown in Figure 9b. Waveforms $B'$ and $C'$ cross the zero line twice. However, the time is different for $B'$ and $C'$. In the combined output, distortion is present in half of the waveform; thus the magnitude of the distortion at shifted points is half that of a conventional converter. In other words, the effects of zero-cross distortion are distributed and averaged. Moreover, because signal level is high, the distortion is masked. This also applies for distortion of the second, third, and subsequent bits.

If negative and positive bias were applied constantly, a maximum signal level would lead to overload. The signal could be attenuated in the digital filter, but this would decrease the S/N ratio slightly. Instead, the shift is disabled when overload would occur, and the converter operates in a conventional linear fashion. This is easily accomplished because there is no inversion of the MSB at these high amplitudes. In practice, the shift operation is virtually always in operation, being disabled only when the signal level passes to within 0.1 dB of the maximum level. For example, operation with a 20-bit converter is shown in Figure 10. The lower 11 bits, or $\pm 2048$ levels, are shifted. In other words, up to 99.6% of the maximum amplitude level, the converter operates in a shifted mode:

$$\frac{1{,}048{,}576 - 4096}{1{,}048{,}576} \times 100 = 99.6\%.$$

Because two data streams are generated, two D/A converters per channel may be used for maximum S/N ratio performance. Alternatively, one D/A converter may be employed. The performance of this multiple-bias method can be demonstrated in several ways. For example, Figure 11 shows a comparison of

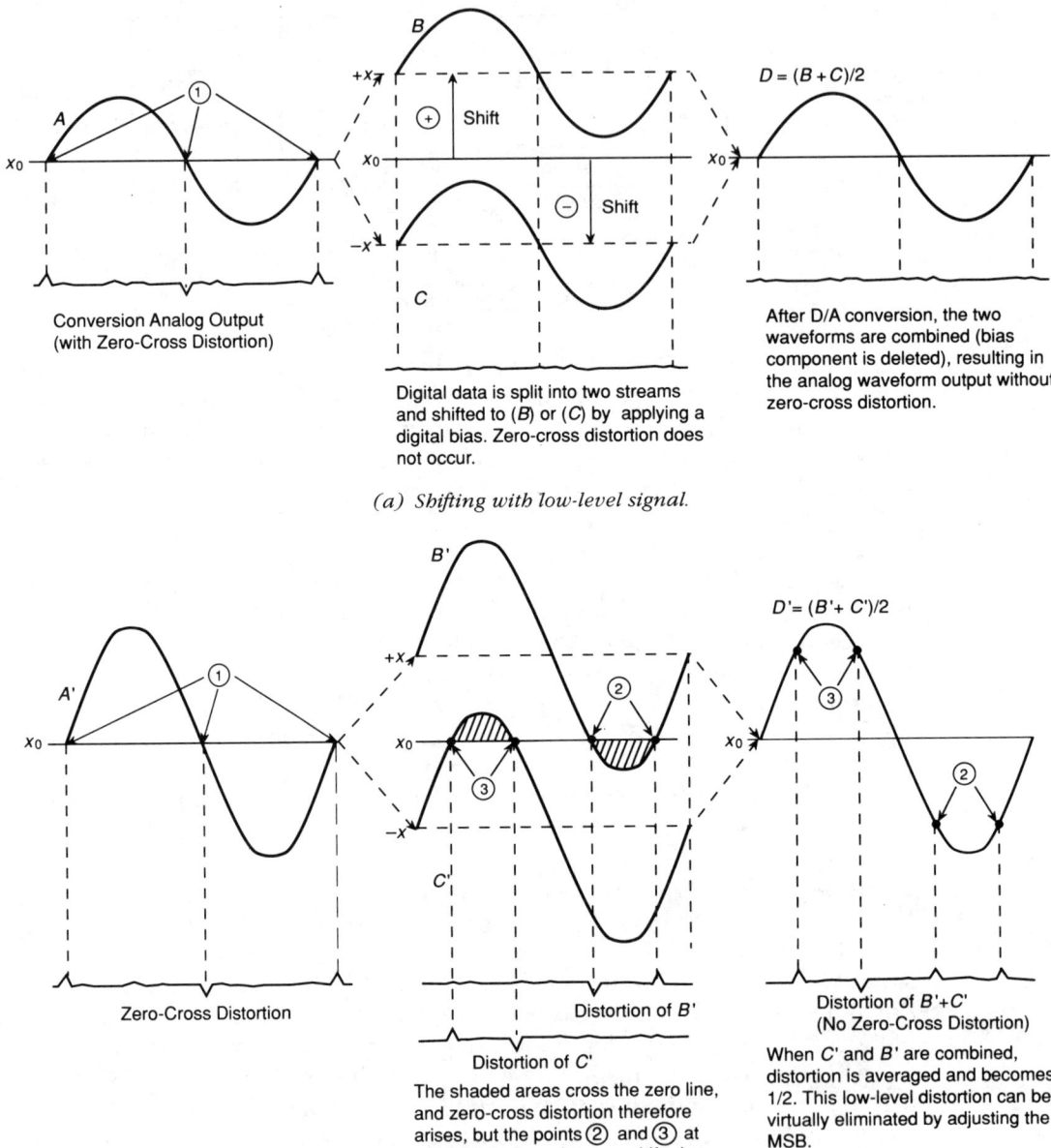

(a) *Shifting with low-level signal.*

(b) *Shifting with high-level signal.*

**Figure 9.** Waveform shifting in a multiple-bias converter. *(Courtesy Denon America)*

the linearity of a conventional D/A converter with zero-cross distortion with the same converter preceded by the multiple-bias processing. This comparison is made before adjustment of the MSB, and the converter has a tolerance of

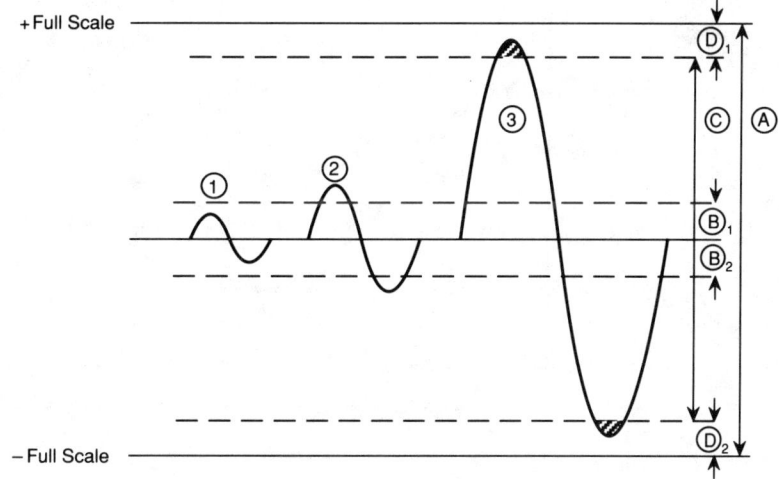

Ⓐ : Full-scale range; in 20-bit system, $2^{20}$ = 1,048,576 steps
Ⓑ$_1$Ⓑ$_2$ : Shift amount; in 20-bit system lower 11 bits, $2^{11}$ = 2048, ±2048 steps
Ⓒ : Range where digital shift is carried out
Ⓓ$_1$Ⓓ$_2$ : Range where digital shift is not carried out: Full-scale −2048 steps
① : Input data level lower than shift amount (2048/1,048,576) × 2 = 1/256(1/256 of full-scale level)
② : Input data level higher than shift amount (higher than1/256 of full-scale level)
③ : Input data peak enters the range where digital shift is not carried out (shaded area in diagram). Here circuit operates as normal converter

**Figure 10.** Operation of multiple-bias converter under varying signal conditions. *(Courtesy Denon America)*

1 LSB for the MSB. The multiple-bias method is also more robust with respect to temperature variations. This technique was devised by Nippon Columbia.

## ROM-Compensated D/A Converter

Much of the error present in a conventional D/A converter is static in nature; bit error distortion is a function of differential nonlinearity in the D/A converter ladder network. In one design, upper and lower bit groups are handled by separate 16-bit D/A converters; the main converter is calibrated using compensation data held in ROM and converted through the sidechain converter. Each primary D/A converter is measured at the factory; for accuracy, high-precision A/D converter measurements may be repeated ten or more times, and averaged error values stored. Error data is stored in ROM chips, and these memories are used in conjunction with the respective D/A converter to provide compensation. Figure 12 shows a block diagram of the converter. The primary D/A converter handles the upper 14 bits, and the sidechain handles the lower 6 bits; likewise there are upper and lower compensation ROMs with 14 and 6 bits, respectively. This data is summed and applied to a common converter. The gain of this converter is set to $1/k$ by data stored in ROM, where

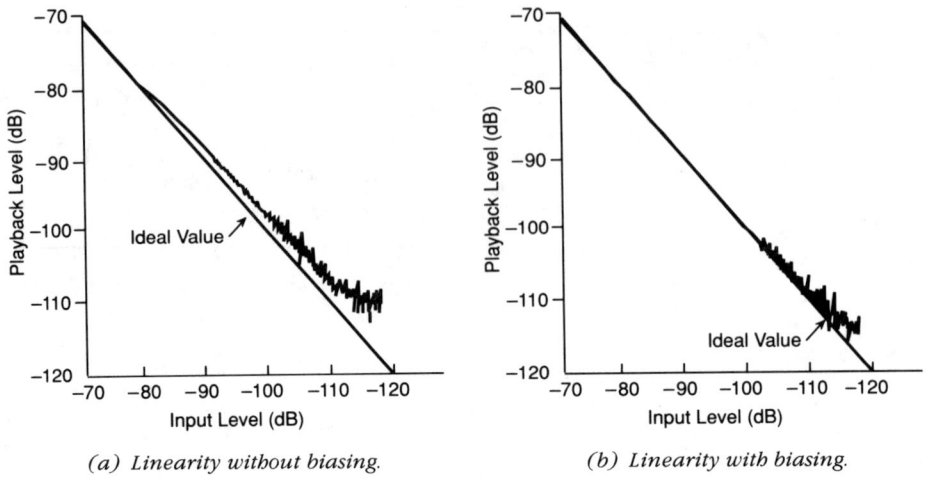

*(a) Linearity without biasing.*   *(b) Linearity with biasing.*

**Figure 11.** Linearity evaluation of multiple-bias converter.
*(Courtesy Denon America)*

$1/k$ is slightly higher than the sum of the lower bit and the compensation output range. Following conversion of compensation data, the latter is summed with the output of the primary converter. A compensation converter is not required for the lower six bits; in practice only the upper nine bits of the sidechain converter are used. This is because additional compensation would not yield meaningful improvement.

Performance results of the compensation method may be seen in Figure 13, showing a comparison of raw 20-bit distortion characteristics versus results after ROM calibration. Distortion is reduced over much of the frequency range. Improvement at high frequencies, however, is partially masked by glitch-induced distortion.

Converter glitches are similar to zero-cross distortion in that they occur at very low levels. Glitches, however, can take place at times other than at the zero-cross point, when the bit values change. Fortunately, glitch points can be

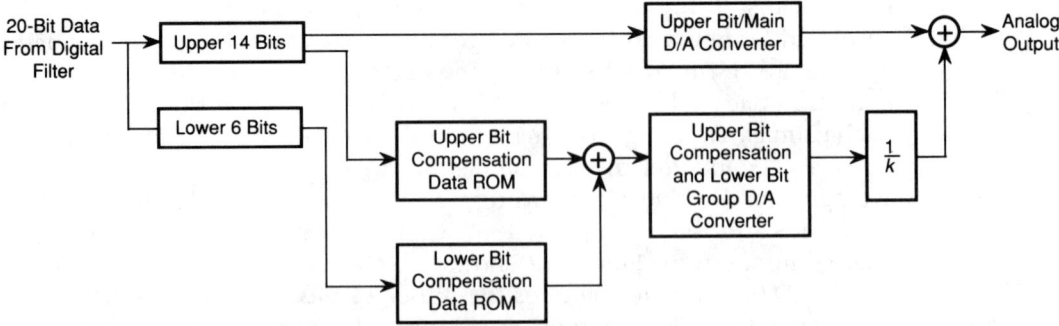

**Figure 12.** Block diagram of a 20-bit, ROM-compensated D/A converter using primary and correction converters. *(Courtesy Nakamichi America)*

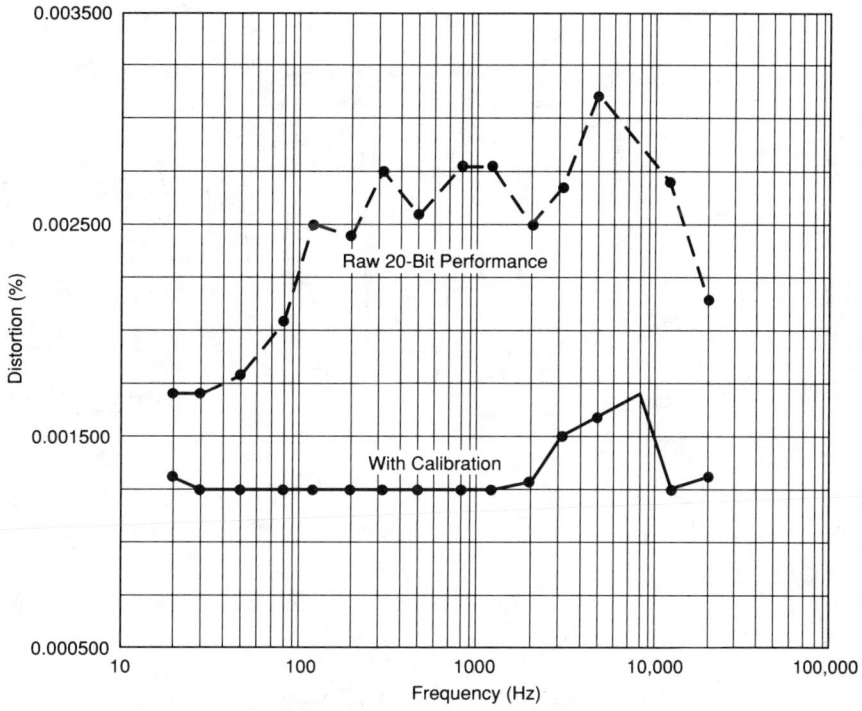

**Figure 13.** Distortion characteristics of a 20-bit, ROM-compensated D/A converter showing performance with and without compensation (0 dB input level) *(Courtesy Nakamichi America)*

predicted. To extend performance, converter glitch-cancellation circuitry may be employed in which timed, inverted pulses are generated at each known glitch point. In this way, glitches are canceled and linearity and distortion specifications are improved. Figure 14 shows the offset characteristics of a 20-bit ROM-compensated converter before and after glitch-cancellation circuitry has been applied. This circuit replaces the conventional sample-and-hold circuit. This technique was devised by Nakamichi.

## Dual D/A Converter with Bias

In one design, two 18-bit D/A converters per channel are configured to create a conversion architecture with 20 bits of linearity. Following oversampling, the 20-bit output from the filter undergoes code conversion to configure the data into two 16-bit streams. One D/A converter is used exclusively for portions of the signal above −24 dB, while another low-amplitude converter is used exclusively for portions below −24 dB, as shown in Figure 15. In other words, when the level of the digital signal is at an equivalent level of −24 dB or less, only one converter is used. At higher signal levels, both D/A converters are active. Data representing signal levels below −24 dB are digitally boosted through

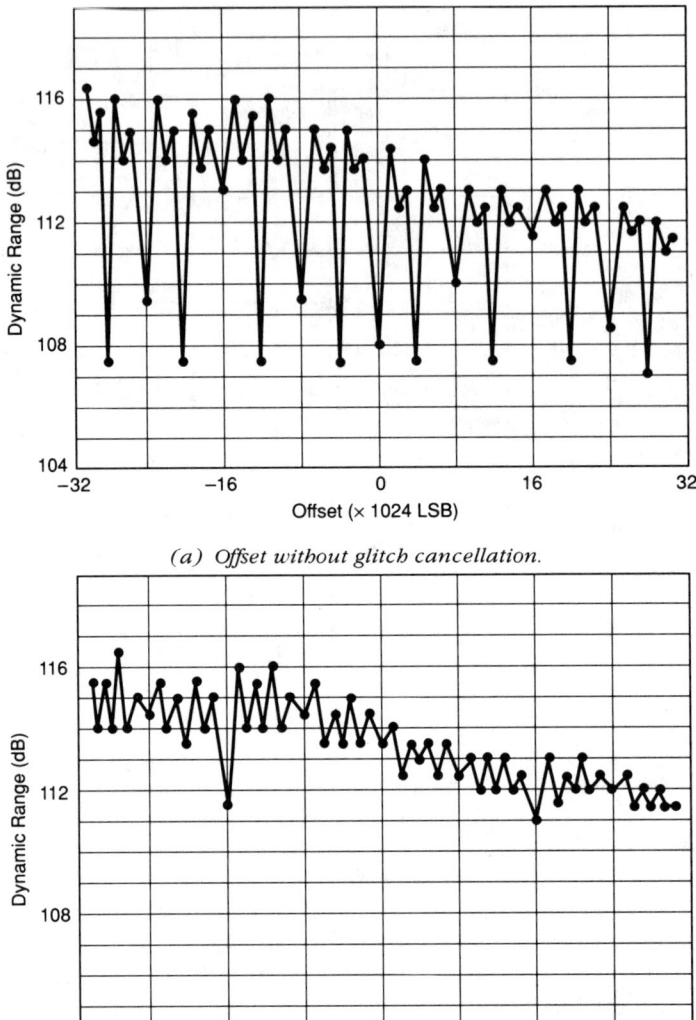

**Figure 14.** Offset characteristic evaluation showing the effect of glitch-cancellation circuitry in a 20-bit, ROM-compensated D/A converter. *(Courtesy Nakamichi America)*

computation by a factor of sixteen (24 dB) before they are processed by the lower-amplitude D/A converter. The output of that converter is then attenuated (in the analog domain) by the same factor of sixteen before being summed with the output of the first converter. In this way, the smallest incremental changes are represented with sixteen times the accuracy (equivalent to four bits) of uncompensated converters, thus improving linearity.

Figure 16 shows the outputs of the two D/A converters, using a sinewave,

Chapter 3: Multibit Conversion

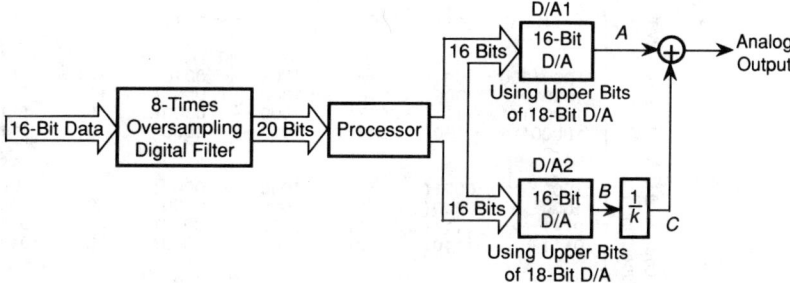

**Figure 15.** Block diagram of a D/A converter design using dual converters to improve linearity. *(Courtesy Nakamichi America)*

at two signal levels: over −24 dB, and less than or equal to −24 dB. In each case, plot 1 shows the output of the first D/A converter, plot 2 is the output of the second D/A converter, plot 3 is the output of the second D/A converter after it is attenuated by 24 dB in the analog domain, and plot 4 (the sum of 1 and 3) is the output waveform. When the signal is below −24 dB the first D/A converter is inactive and all low-level signals are converted through the sec-

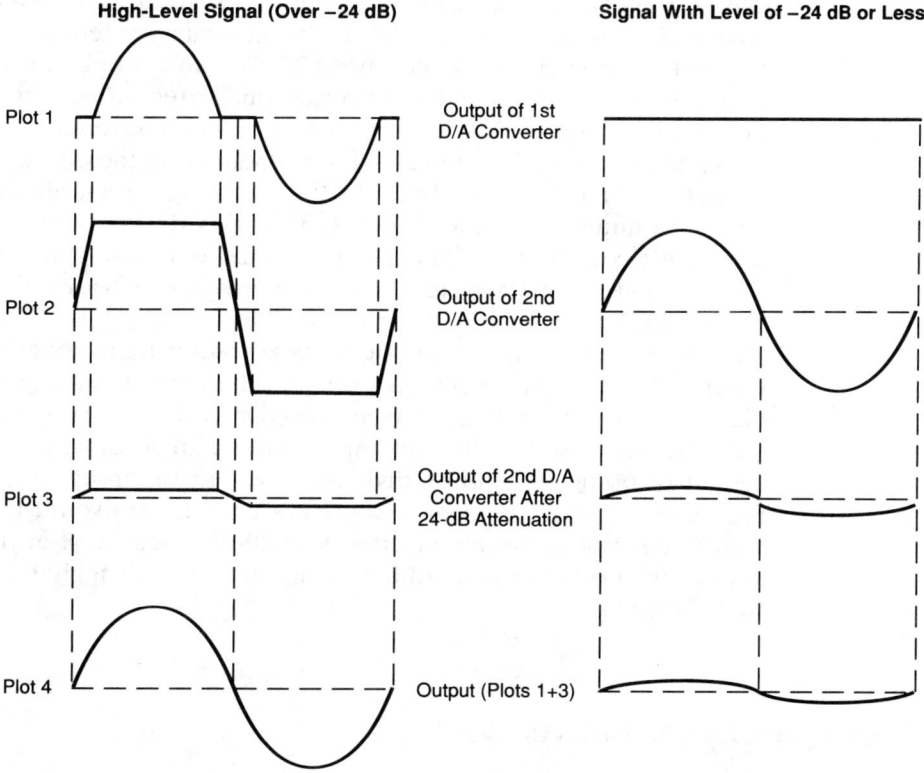

**Figure 16.** Plot of the waveform output of dual D/A converter design showing two signal levels. *(Courtesy Nakamichi America)*

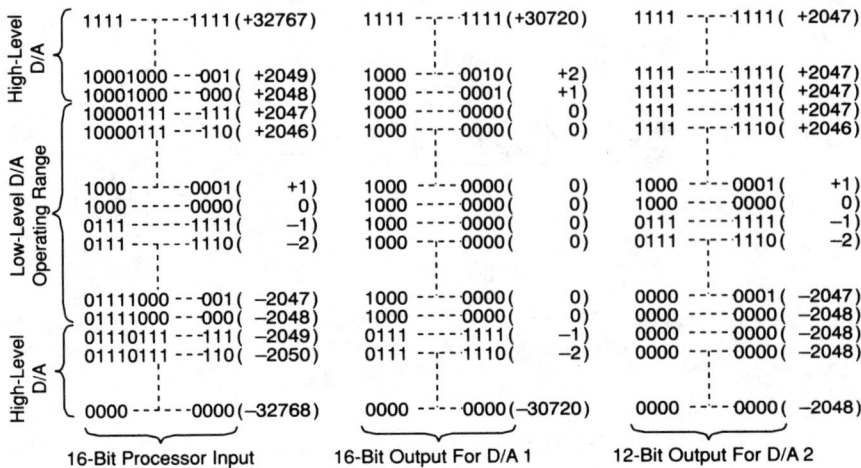

**Figure 17.** Code conversion table for D/A converter using dual converters. A 20-bit data stream from the digital filter is configured as two 16-bit data streams. The four LSBs are omitted for clarity. *(Courtesy Nakamichi America)*

ond D/A converter. The 24 dB boost in the second data stream enables the second converter to represent 1 LSB changes with sixteen times greater accuracy, with low-level signals. Because a 20-bit converter provides sixteen times greater accuracy than a 16-bit converter, this sixteen-times increase in precision in this design yields performance of a 20-bit converter.

Figure 17 shows how the 20-bit data stream from the digital filter is configured as two 16-bit streams. Because the 4 least significant bits are not altered, they are omitted from the illustration. In this design, although 16-bit data streams are converted, 18-bit converters are used. By using the most significant 16 bits of 18-bit converters, resolution is improved over use of 16-bit converters.

It is important to note that in this design the digital data is apportioned to the two converters based on signal level rather than upper and lower bit groups. The highest bit activity occurs at the zero-crossing point but in this design the most significant bits are inactive at the lowest signal levels. This serves to decrease the difficulty in outputting an accurate LSB change at the zero-cross point. In contrast, division of data by bit groups does not improve precision; for example, simply appending a 4-bit converter to a 16-bit converter does not create a converter with 20-bit linearity; if anything, such an attempt would only worsen differential linearity. This technique was devised by Nakamichi.

## Analog-to-Digital Conversion

Analog-to-digital conversion faces the same problems as digital-to-analog conversion, but to a greater degree. For example, filtering is much more critical in

A/D conversion because aliasing components fall in-band in A/D conversion whereas image components fall out-of-band in D/A conversion. Moreover, differential and absolute nonlinearity, zero-cross distortion and other errors are all present in multibit A/D converters. However, multibit A/D converters generally offer fewer options for modification, particularly in the case of digital audio, which demands both high precision and speed. It is true that progress in D/A conversion has led that in A/D conversion; this is ironic because generally all audio that passes through a D/A converter has previously passed through an A/D converter; A/D conversion is thus the limiting factor. In other words, generally it is the recordings themselves that limit fidelity, rather than the hardware used to play them back. This is unfortunate because whereas playback hardware will certainly progress to even higher standards, the fidelity of a recording is generally an unchanging artifact, a testament to the recording technology of its day. For these and other reasons, manufacturers have worked hard to develop high-quality A/D converters.

## Dual-Channel A/D Converters

An example of a dual-channel (stereo), 16-bit A/D converter designed for digital audio applications is shown in Figure 18. The CMOS converter uses a traditional successive approximation algorithm employing a charge redistribution architecture. Instead of an array of binary-weighted resistors, the internal D/A converter uses an array of binary-weighted capacitors. The converter tracks the input analog signal, which is applied across each leg of the D/A capacitor array, thus performing voltage-to-charge conversion, as shown in Figure 19. The conversion command causes charge to be trapped in the capacitor array, and the input is ignored. Thus the array serves as an analog memory, providing the sample-and-hold function and avoiding use of an external circuit.

Conversion is accomplished by manipulating the binary-weighted legs of the array with respect to a reference voltage and ground. The legs of the array share a common node at the input of a comparator. This acts as a binary-weighted capacitive divider. The charge at the comparator's input is fixed; thus the voltage at that point varies proportionally to the capacitance connected to the reference voltage. A successive approximation algorithm is used to find the appropriate proportion of capacitance (that needed to cause an output on the comparator); that binary fraction of capacitance represents the converter's digital output.

Onboard self-calibration circuits help maintain accuracy of the internal D/A converter; this in turn improves low-level linearity. Each binary-weighted bit capacitor is composed of several capacitors which can be addressed to adjust the overall bit weight. During power-up initialization, an onboard microprocessor addresses the subarrays used to apportion the bit weights. Each bit is adjusted to balance the sum of all less significant bits. Beginning with the two LSBs the values are summed to produce a new value for the neighboring bit. A

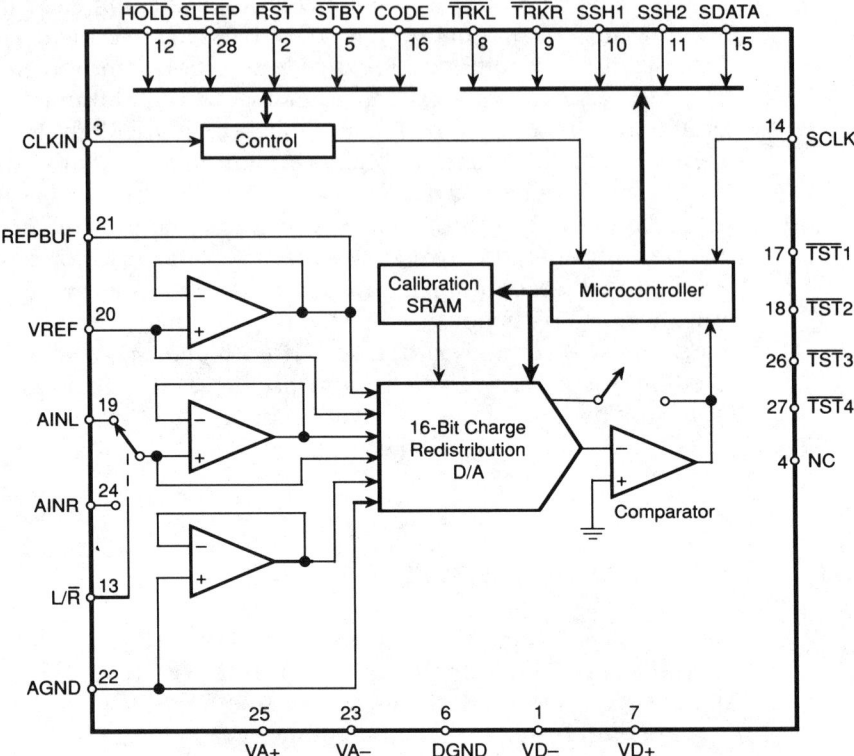

**Figure 18.** An example of an A/D converter chip using an internal array of binary-weighted capacitors. *(Courtesy Crystal Semiconductor Corp.)*

comparison is made, and the value of the new bit is trimmed as needed. The summed reference value is added to the value of the newly calibrated bit to produce the reference for the next neighboring bit. This process is repeated for all remaining bits; calibration is accomplished within 1.4 s. Calibration resolution is 18 bits; differential nonlinearity is ±1/4 LSB; that is, codes range from 3/4 to 5/4 LSB wide.

In stereo mode, sampling rates up to 50 kHz may be processed. However,

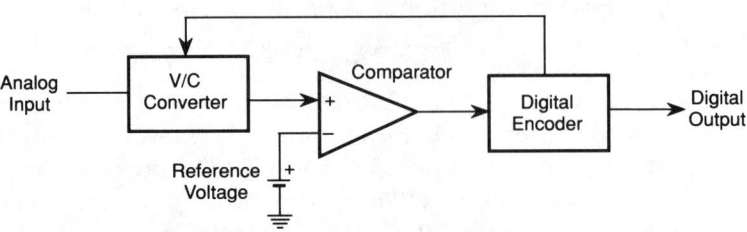

**Figure 19.** Block diagram showing operation of voltage-to-charge conversion. *(Courtesy Nakamichi America)*

the chip may be configured as a monaural converter with rates up to 100 kHz; in the latter case the chip is used in a two-times oversampling mode. At double the nominal sample rate (96 kHz), the anti-aliasing filter's corner frequency may be extended with lower rolloff; this allows for lower passband ripple and greater phase linearity. The converter's S/N ratio is increased in the oversampling mode because quantization noise is spread across the 48 kHz bandwidth halving the audio-band noise, giving a 3 dB improvement in the S/N ratio. During oversampling, digital filtering must be employed after the A/D converter to reject noise out of the audio band. A typical oversampling configuration for two-channel conversion is shown in Figure 20.

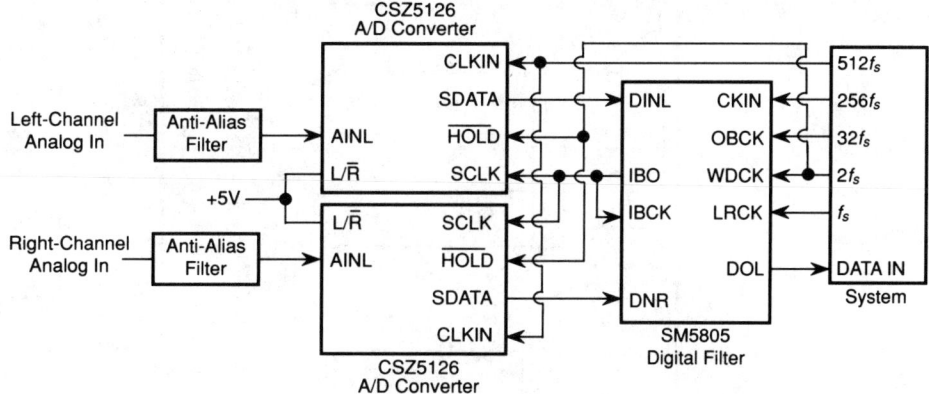

**Figure 20.** An applications circuit showing an oversampling configuration for stereo A/D conversion. *(Courtesy Crystal Semiconductor Corp.)*

A master clock is set at 512 times the per-channel sampling rate in the stereo mode and 256 times in monaural mode; for 48 kHz sampling this is 24.576 MHz and 96 MHz, respectively. All other timing and control inputs may be divided down from the master clock thus yielding a completely synchronous system. Either two's complement or binary coding may be used. The S/N ratio is quoted as 92 dB in stereo mode and 95 dB in monaural mode. This A/D converter was developed by Crystal Semiconductor Corporation.

Another example of a dual-channel A/D converter also uses a successive approximation register with switched capacitor architecture. It provides an inherent co-phase sample-and-hold function for each channel so simultaneous channel sampling can be performed. Fast settling times permit operation at rates of two- or four-times oversampling. A block diagram of the converter is shown in Figure 21; the decimating filter shown is external, and used when oversampling conversion is employed.

A weighted capacitor array is used to form the D/A converter in the successive approximation function. In addition, another switched capacitor converter is used to provide small correction voltages to the latching comparator. The primary capacitor both samples and stores the input signal, thus providing a sample-and-hold function. The weighted capacitors are fabricated with a CMOS process, and thin film resistors help ensure accurate, stable performance.

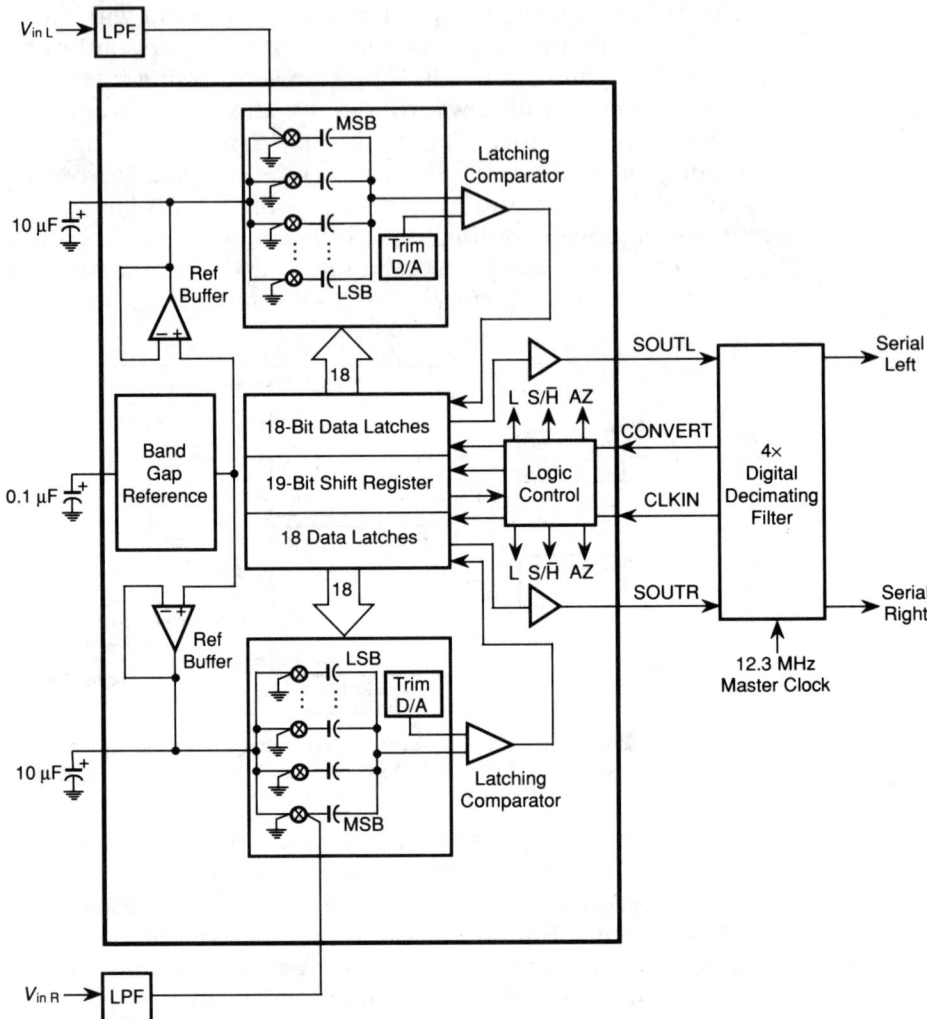

**Figure 21.** Block diagram of a stereo A/D converter chip configured with input low-pass filters and four-times decimating filter.
*(Courtesy Burr-Brown Corp.)*

In most converters, bipolar transistors are employed because of their low-noise properties; binary-weighted current sources are thus constructed. However, the resistors required in the current sources exhibit a Johnson thermal noise that is often unacceptable for the performance levels required for conversion of 16 bits. For example, resistor noise in a 16-bit converter may be 50 µV rms, yielding an S/N ratio of about 90 dB—equivalent to 15 bits.

Alternatively, capacitors are theoretically noiseless (the MOSFET switches needed in the converter have an on-resistance which generates Johnson noise). In this case the noise generated by the capacitor array along with associated components is approximately 15 $\mu$V rms, yielding an S/N ratio of about 97 dB—equivalent to 16 bits or more.

Another noise consideration is noise generated in the silicon chip by currents coupled through parasitic capacitances. For example, transient currents are produced in switching circuits and through ringing from package and wirebond inductances. To minimize this type of noise, digital transients must be isolated from analog circuitry. For example, in this case, separate power and common pins are provided. Similarly, separate power and common pins for the left and right analog channels improve crosstalk specifications. As with any converter, good applications practices are especially important. The user must take care to provide power supply decoupling, to separate digital and analog lines, and to observe good ground-plane technique.

An input sample-and-hold switching delay (aperture delay) of less than 10 ns contributes a small phase shift in the sampled signal. Aperture jitter, a noise-induced random variation in aperture delay, is 50 ps for a full-scale 20 kHz signal, resulting in a variation of 20 $\mu$V peak to peak—an artifact below the converter's noise floor.

A 19-bit shift register, shown in Figure 22, is used to test and control the eighteen data latches present for each channel. The bits of both channels are tested together, beginning with the MSB, and working downward. For each bit a positive pulse from the appropriate shift register is applied to the corresponding data latch and NOR gate. The bit is switched at the beginning of a test interval, and the comparator latch is strobed to provide a logic level and thus determine whether the bit in question should be set or reset. At the same time, the comparator's input is moved closer to zero potential. This process is continued sequentially until all bits are tested, zero potential is reached, and conversion is ended. This algorithm operates synchronously with an external clock to minimize switching noise.

Two serial outputs, one for each channel, are synchronously derived from the respective latched comparator outputs. They deliver two's complement data; as noted, an external decimation filter is required when the converter is used in an oversampling mode. Laser trimming of the conversion ladder contributes to performance levels of 0.0025% THD + N (total harmonic distortion plus noise), an S/N ratio of 92 dB, and channel separation of 100 dB. This A/D converter was developed by Burr-Brown Corporation.

# Conclusion

For all their advantages, conventional multibit A/D and D/A converter techniques pose limitations in high-fidelity digital audio applications. Although these converters can achieve low quantization error, problems exist. For example, the reference voltages supplied to an 18-bit converter must be highly accurate and stable; accuracy to the least significant bit requires precision equivalent to four parts per million. In addition, the converter must be care-

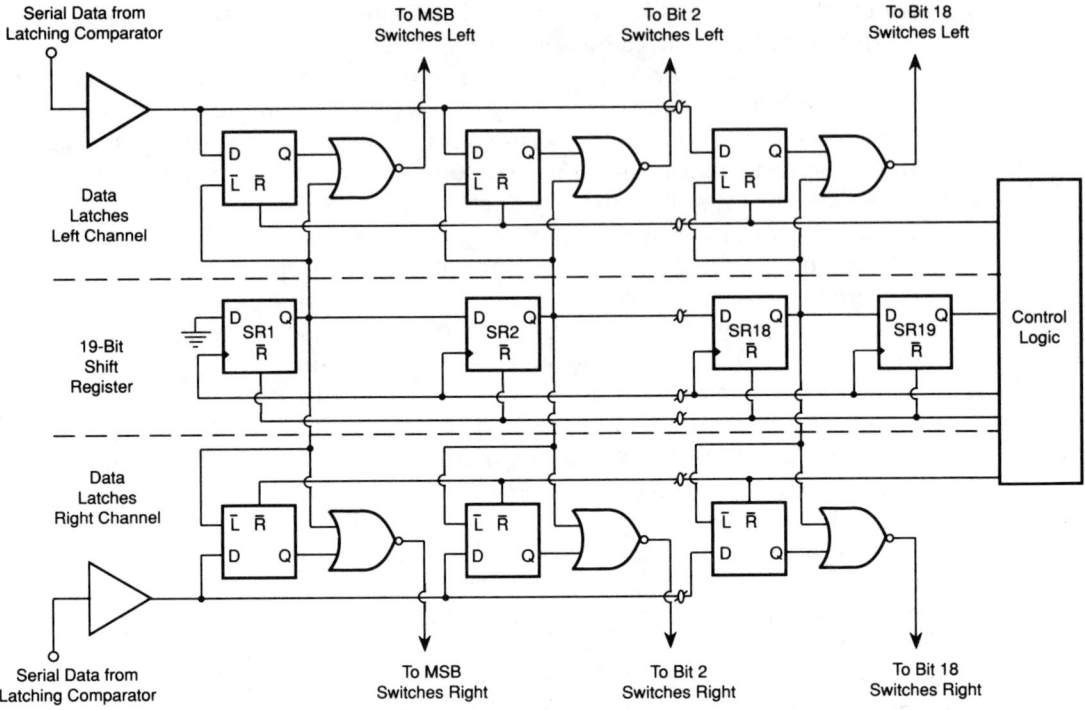

**Figure 22.** Dual successive approximation filter used in A/D converter chip.
*(Courtesy Burr-Brown Corp.)*

fully calibrated, and extreme care must be taken to prevent multiple channels from drifting relatively. In fact, some manufacturers use 18-bit converters in a way which offers performance inferior to that of 16-bit converters. In other words, although it is more precise than a 16-bit converter, an 18-bit converter has all the real-world problems of a 16-bit converter, only more so.

Moreover, these designs generally do not offer the same benefit in A/D conversion. As Adams [1986] has pointed out, an 18-bit successive approximation A/D converter would require a settling time less than 1 $\mu$s, a sample-and-hold circuit with acquisition time of 1 $\mu$s and droop of less than 40 $\mu$V over the sample period, and a comparator with response time of 150 ns with an overdrive of 40 $\mu$V and noise voltage less than 40 $\mu$V—essentially impossible constraints. Filter requirements also present limitations. An oversampling 18-bit A/D converter would still require a high-order analog input filter; only mild oversampling rates are possible because of speed limitations in the D/A converter contained in the A/D converter, and its need to operate at $n$ times the converter's sampling rate, where $n$ is the number of bits undergoing conversion.

The answer lies in alternative conversion methods. Instead of converting parallel 16- or 18-bit digital words, much shorter wordlengths can be converted at faster rates. In the same way that oversampling removes the deficiencies of analog output filtering by performing the operation in the digital domain, low-bit conversion methods remove the deficiencies of multibit conversion. As we shall see in Chapter 12 (following a look at the DSP theory required to achieve these results) this low-bit technique is applicable to both D/A and A/D conversion. Still, giving credit where it is due, conventional multibit converter designs have performed well for many years and have shown themselves to be adaptable to modification and improvement. For many digital audio applications, multibit converters will continue to fulfill the performance criteria.

# References

Adams, R. W., "Design and Implementation of an Audio 18-Bit Analog-to-Digital Converter Using Oversampling Techniques," *Journal of the Audio Engineering Society*, vol. 34, no. 3, March, 1986.

Blesser, B., "Advanced Analog-to-Digital Conversion and Filtering: Data Conversion," *Digital Audio Collected Papers*, Audio Engineering Society, 1983.

Bristow-Johnson, R., "Effect of DAC Deglitching on Frequency Response," *Journal of the Audio Engineering Society*, vol. 36, no. 11, November, 1988.

Burr-Brown Corporation, *PCM58 Data Sheet*, 1988.

Dijkmans, E. C., and P. J. A. Naus, "The Next Step Towards Ideal A/D and D/A Converters," *Audio in Digital Times: AES 7th Conference Proceedings*, May, 1989.

Fielder, L. D., "Evaluation of the Audible Noise and Distortion Produced by Digital Audio Converters," *Journal of the Audio Engineering Society*, vol. 35, no. 7/8, July/August, 1987.

Halbert, J. M., and M. A. Shill, "An 18-Bit Digital-to-Analog Converter for High Performance Digital Audio Applications," *Journal of the Audio Engineering Society*, vol. 36, no. 6, June, 1988.

Lipshitz, S. P., and J. Vanderkooy, "Are D/A Converters Getting Worse?" *AES Preprint 2586*, March, 1988.

Matsushita Electric, "Technical Information: Technics 4-DAC 18-Bit High Resolution System," nd.

Naylor, J. R., "A Dual Monolithic 18-Bit Analog-to-Digital Converter for Digital Audio Applications," *Audio in Digital Times: AES 7th Conference Proceedings*, May, 1989.

Pohlmann, K. C., *Principles of Digital Audio*, 2nd edition, Howard W. Sams & Co., 1989.

Van de Plassche, R. J., and E. C. Dijkmans, "A Monotonic 16-Bit D/A Conversion System for Digital Audio," *Digital Audio Collected Papers*, Audio Engineering Society, 1983.

*Chapter 4*

# Laser Technology
### Brent A. Karley

## Introduction

Digital audio is a continuously growing technology with seemingly boundless applications. Much of this growth is due to the development of other types of technology, such as optical storage systems, fiber optics, and lasers. As digital audio continues to evolve, the audio engineer/technician will also be required to understand an array of new technologies.

The laser provides a significant contribution to the advancement of digital audio. Laser is an acronym for *l*ight *a*mplification by *s*timulated *e*mission of *r*adiation. Although lasers have been available for commercial use for over 25 years, major applications in audio were not forthcoming until the introduction of the compact disc player in 1982. The compact disc (and all other optical storage systems) relies on the laser's unique attributes. The laser has become digital audio's link to improved recording quality. It is also used extensively as a light source for fiber-optic systems. As optical storage and fiber optics are increasingly incorporated into digital audio, the laser will become an increasingly important component. In short, an understanding of its operation is vital to the development of a comprehensive background of digital audio.

## A Brief History

The most important concepts leading to the invention of the laser appeared in the early twentieth century. In 1916, Albert Einstein proposed a new idea for describing the interaction of light and the atom. Prior to Einstein's proposal, the scientific community had recognized two types of light-atom interactions: absorption (raising the energy level of the atom) and spontaneous emission (resulting in the loss of energy from the atom). The concept proposed by Einstein took into consideration a third possibility—that an emission of energy from the atom can be forced by an external radiation field. This is recognized today as stimulated emission. In 1928, the first experimental evidence was presented showing that stimulated emission was not simply a theory but a physical reality.

It wasn't until 1951 that Einstein's theory of stimulated emission was put to use by an invention called the maser, designed by a Columbia University professor, Charles H. Townes. Maser is an acronym for *m*icrowave *a*mplification by *s*timulated *e*mission of *r*adiation. Simultaneously, in Moscow, Aleksander M. Prokhorov and Nikolai Basov developed detailed calculations of the conditions necessary for maser operation. In 1953, Townes and his associates completed a working maser and one year later, Prokhorov and Basov completed their calculations. In 1964, these three men shared the Nobel prize in physics for their contributions in the development of masers.

Townes continued pursuing masers throughout the 1950s, recognizing that it was possible to develop stimulated emission at shorter wavelengths near the optical spectrum, but that very different physical conditions were required. These conditions were outlined in a 1958 publication by Townes and Arthur L. Schawlow, a Bell Laboratories researcher. Prior to this publication, Townes and Schawlow had applied for a patent of their laser concept. Meanwhile, Gordan Gould, a former Columbia graduate student, had developed his own analysis of the required conditions for laser operation. Although he apparently wrote his proposals in 1957, he did not seek to publish his work until 1959, one year after Townes and Schawlow. He did eventually receive four patents in the 1970s and 1980s. In the 1960s, laser technology advanced at a tremendous pace, with the development of nearly a dozen different laser designs. The first person to construct an operating laser was Theodore H. Maiman, a physicist with Hughes Research Laboratories. In 1960, Maiman utilized a helical flashlamp surrounding a rod of synthetic ruby with reflective ends. The lamp would flash, causing the ruby to emit a red laser light. Soon afterward many other types of lasers were developed. The development of lasers into useful products has been slow, but some important contributions to society have resulted from the laser, as listed in Table 1.

**Table 1.** Applications of Lasers

| Laser Type | Application |
| --- | --- |
| Ruby | Holography, fusion research, optical radar, trimming resistors |
| Helium-Neon | Video disk playback, measurements (interferometers), laser printers, reading and scanning bar codes, holography, medicine (laser acupuncture and biostimulation), alignment and positioning in construction applications |
| Neodymium | Medicine, fiber-optic endoscopes, range finding, research applications |
| Semiconductor (laser diode) | Compact disc players, fiber optics, optical storage systems, reading bar codes, range finding, automated construction-alignment, laser pumping (neodymium lasers) |

| Laser Type | Application |
|---|---|
| Argon | Cutting master disks for video and audio, printing and publishing (exposing printing plates), laser pumping (dye lasers), medical surgery (opthamology), optical data storage requiring high speed |
| $CO_2$ | Cutting and welding, spectroscopy and photochemistry, surgery, heat-treating metals, optical radar, range finding |
| Dye | Laser light show, spectroscopy, medicine (cancer, eye, and skin disorders), time studies |
| Chemical | Research |
| Excimer | Spectroscopy, mask fabrication, photochemistry, isotope separation, patterning semiconductor integrated circuits, laser pumping (tunable dye lasers) |

**Table 1.** (cont.)

# Laser Physics

When analyzing the motion of an object such as a billiards ball, classical mechanics is applied along with its associated rules and postulates. Unfortunately, classical mechanics will not directly apply to atomic and subatomic particles such as molecules, atoms, electrons, protons, and neutrons. These particles are considered to follow the rules of quantum mechanics. Quantum mechanics is a very successful theory that explains the behavior of atomic and subatomic particles. An understanding of quantum mechanics is important in understanding laser operation with regard to energy levels, transitions, photons, and other topics.

The concept of energy levels of a system is a vital part of quantum mechanics and laser operation. Energy levels are used to describe the energy conditions of a given physical system such as an atom or molecule, as shown in Figure 1. The bottom level represents the ground state, which is the lowest possible energy level. A state or level (the terms are interchangeable) represents discrete amounts of energy contained by a system. The higher levels on the diagram correspond to greater amounts of energy within the system. When a system moves from one energy level to another a transition takes place. The system will interact with its surroundings by absorbing or emitting energy when a transition occurs. Energy is emitted when a downward transition

**Figure 1.** Basic energy level diagram.

occurs, and energy is absorbed when an upward transition occurs. The energy that interacts with the system is in the form of a photon. A photon is a localized bundle of concentrated light energy.

## Quantum Mechanics

There are two major postulates in quantum mechanics that apply to the behavior of photon-atom interactions. The first postulate, in its basic form, states that atoms and molecules can exist only in certain levels, characterized by definite amounts of energy. When an atom or molecule changes its energy level, it must absorb or emit an amount of energy just sufficient to bring it to another energy level.

According to the second postulate, when atoms or molecules absorb or emit light in the process of changing their energies, the wavelength of the light is related to the magnitude of the energy change by the equation

$$E = hc/\lambda,$$

where $h$ is Planck's constant ($6.626 \times 10^{-34}$ J·s) and $c$ is the speed of light ($3.0 \times 10^8$ m/s). It can be seen that a photon contains energy inversely proportional to its wavelength. The energy is therefore proportional to the frequency ($v$) of the light particle. This relationship is described in the following progression of equations:

$$E = hc/\lambda,$$
$$\lambda = c/v,$$
$$E = hc/(c/v) = hv.$$

# Laser Elements

There are three basic elements that are common to all lasers. First, the laser medium is a material from which the light is generated; it is often called the active medium. It can consist of one or several materials in either a liquid, solid, or gaseous state. The active medium interacts with an electromagnetic field by absorption and emission of energy. The medium absorbs energy causing upward transitions and emits energy as a result of downward transitions. Second, a power supply or pumping source provides energy to the medium for continued operation of the laser process. There are a variety of sources utilized, such as optical pumping, gas discharge pumping, or direct pumping by an electric current. Third, an optical cavity resonates, concentrates, and amplifies the light emitted from the medium. An optical resonator typically consists of two parallel mirrors at each end of the medium. One mirror is totally reflective, and the other is only partially reflective. Light that is generated within the cavity bounces between the mirrors, during which time it is amplified. When the gain becomes greater than the internal losses, light is released through the partially reflective mirror in the form of a laser beam.

# Laser Theory

## Spontaneous Emission

As its name implies, spontaneous emission occurs without influence from outside the atom. Figure 2a shows spontaneous emission occurring from energy level 2 to energy level 1 of a given system, such as an atom or molecule. When the system decays from energy level 2 to energy level 1, a photon is emitted, which results in the loss of energy equal to

$$E_2 - E_1 = h\nu,$$

where $E_1$ and $E_2$ represent the energy levels of states 1 and 2, respectively, $h$ is Planck's constant, and $\nu$ is the frequency of the emitted photon. Spontaneous emission is the most dominant source of emission in conventional light sources, such as light bulbs and sunlight.

Spontaneous emission may also be described in terms of transition rate. Given a medium with a certain number $N_2$ (often referred to as the population density) of atoms in level 2 and a certain number $N_1$ of atoms in level 1 at time $t = 0$, the spontaneous transition rate per unit time can be determined as

$$\frac{dN_2}{dt} = -A_{21}N_2 = -\frac{dN_1}{dt},$$

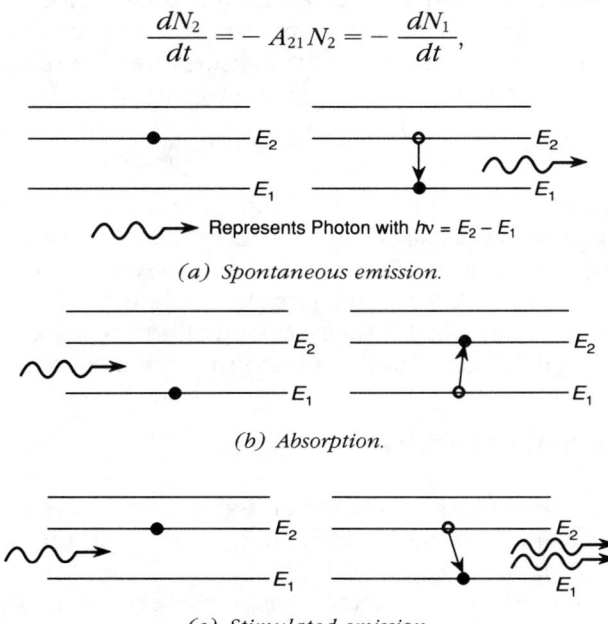

(a) Spontaneous emission.

(b) Absorption.

(c) Stimulated emission.

**Figure 2.** Photon-atom interactions.

where $A_{21}$ represents the probability of transition from state 2 to state 1 due to spontaneous emission. The negative sign shows that energy is being lost from the atom and that the transition is to a lower energy state. Therefore, $N_2$ is decreased by spontaneous emission, and $N_1$ is increased. If there are originally $N_2(0)$ atoms in energy level 2, then the number of atoms in energy level 2 decreases exponentially with time due to spontaneous emission:

$$N_2 = N_2(0) e^{-t/\tau_{sp}},$$

where $\tau_{sp}$ is a time constant (often called the spontaneous lifetime) equal to the reciprocal of the spontaneous transition probability:

$$\tau_{sp} = 1/A_{21}.$$

It can be observed that with no other processes influencing the energy levels of the medium, each atom in energy level 2 would ultimately fall to the lowest energy level after an initial average time constant, $\tau_{sp}$.

It is the process of excitation due to the absorption of energy which produces upward transitions from lower energy levels to higher energy levels.

## Induced Transitions

### Absorption

In the presence of a radiation field of frequency $v = E_2 - E_1/h$ and an energy density $\rho(v)$, an atom can undergo excitation from energy level 1 to energy level 2 by absorption of a photon with energy $hv$ from the field, as shown in Figure 2b. The probability of absorption occurring is represented by $B_{12}$. The transition rate of a medium due to absorption is

$$\frac{dN_2}{dt} = + B_{12} N_1 \rho(v) = - \frac{dN_1}{dt}.$$

This equation shows that the population density of level 2 increases and the population density of level 1 decreases due to absorption. The atoms gain energy, and transitions to higher energy levels occur as shown by the positive rate of absorption. When the population density of level 2 exceeds the population density of level 1 a state of population inversion exists. This condition is required for stimulated emission to occur as described below.

### Stimulated Emission

The process of stimulated emission also occurs in the presence of a radiation field in which the atom in energy level 2 produces a photon when stimulated to drop to energy level 1. The energy of the stimulating photon from the radiation field must equal the transition energy of the atom for stimulated emission to occur as shown in Figure 2c. The energy $hv$, given up in the form of a photon, has the same frequency, phase, polarization, and direction as the photon which induced the atom to undergo the transition. The rate of transition of a medium due to stimulated emission is dependent upon the population density of level 2 and the radiation field density $\rho(v)$:

$$\frac{dN_2}{dt} = - B_{21} N_2 \rho(v) = - \frac{dN_1}{dt},$$

where $B_{21}$ represents the probability of a transition from state 2 to state 1 due to stimulated emission.

It should be noted that both absorption and stimulated emission are proportional to the radiation field density $\rho(v)$, but spontaneous emission is independent of the radiation field. It would seem unlikely that an atom would encounter a photon of energy equal to the required transitional energy of the atom. On the contrary, it is an advantageous characteristic that common atoms produce and require photons of identical energy values due to their common energy level structure. In other words, when one or several atoms of a medium emit photons by spontaneous emission the emitted photons contain the exact amount of energy required to induce absorption and stimulated emission in other, like atoms. The spontaneously emitted photons trigger stimulated emission, increasing the number of photons available in the radiation field. In short, laser light is created.

# The Lineshape

The idea that atoms exist only in discrete energy levels, as stipulated by the first postulate of quantum mechanics, is a simple approximation of actuality. The spectral content of the radiation emitted due to spontaneous emission compared to our simple description of energy levels is shown in Figure 3. It can be seen that the energy levels are not discrete energy levels but actually narrow spectra with a peak at approximately the middle of the spectrum. The spectrum is called the lineshape and is defined as a function $g(v)$ such that $g(v) \, dv$ represents the probability that an emission from level 2 to level 1 results in a photon with a frequency between $v$ and $v + dv$.

The lineshape is an expression of how each atom, on average, will respond to radiation as a function of frequency. Specifically, the lineshape represents the probability that spontaneous emission will result in a photon with frequency between $v$ and $v + dv$, or that radiation at frequencies between $v$ and $v + dv$ will cause absorption by atoms in level 1 or stimulated emission by atoms in level 2. The lineshape is usually normalized according to the expression

$$\int_0^\infty g(v) \, dv = 1.$$

This accounts for all emitted photons in the medium. The lineshape applies to stimulated emission, absorption, and spontaneous emission and varies depending upon the atomic composition of the medium.

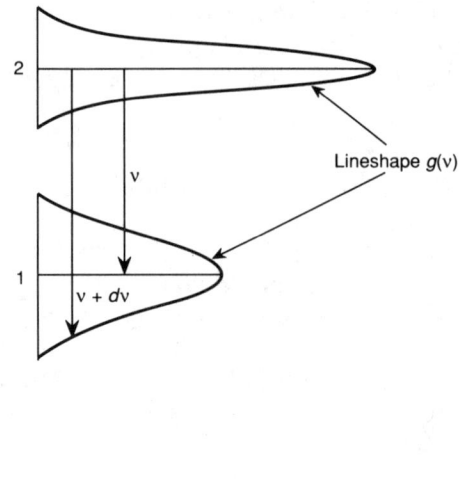

**Figure 3.** Lineshape—spectral content of transitions between two states.

## Amplification

The total rate of change of the population density of levels 1 and 2, including the effect of the lineshape $g(v)$, as determined by the above equations, is given below:

$$\frac{dN_2}{dt} = -\frac{dN_1}{dt} = B_{12}N_1\rho(v)g(v) - B_{21}N_2\rho(v)g(v) - A_{21}N_2g(v).$$

The rate of change of the energy density $\rho(v)$ of the radiation field is equal to the total rate of change of the population of levels 1 or 2 multiplied by the transition energy ($hv$):

$$\frac{d\rho(v)}{dt} = -\frac{dN_2}{dt}hv = \frac{dN_1}{dt}hv.$$

The energy density of the radiation field is the source of light for the laser. When the energy density increases, so does the laser's optical output power. The energy density of the radiation field increases when an atom gives up energy due to a transition from level 2 to level 1 and decreases when an atom absorbs energy due to a transition from level 1 to level 2. The rate of decrease in the energy density of the radiation field due to absorption is

$$\frac{d\rho(v)}{dt} = -B_{12}N_1\rho(v)g(v)hv.$$

The rate of increase in energy density of the radiation field due to stimulated emission is

$$\frac{d\rho(v)}{dt} = B_{21}N_2\rho(v)g(v)hv.$$

Einstein provided the following equality relating the probabilities of the two induced transitions:

$$B_{21} = B_{12} = B,$$

thus showing that the probability of absorption is equal to the probability of stimulated emission.

The effect of spontaneous emission on the energy density of the radiation field is considered negligible. The total change in energy density due to induced transitions, as determined by the previous three equations, is given by

$$\frac{d\rho(v)}{dt} = (N_2 - N_1)B\rho(v)g(v)\ hv.$$

Three different situations can exist involving the change in the energy density of the radiation field given the relative populations of levels 1 and 2. These scenarios are depicted in Figure 4.

1. $N_1 > N_2$. Here $d\rho(v)/dt$ is negative and attenuation of the energy density of the radiation field will result if the population of level 1 is greater than level 2. Energy is absorbed by the atoms from the radiation field, decreasing the energy of the radiation field.

2. $N_1 = N_2$. Here $d\rho(v)/dt$ is zero. The atoms of the medium have no effect on the radiation field. Neither absorption nor emission dominates in this case.

3. $N_2 > N_1$. Here $d\rho(v)/dt$ is positive and population inversion exists (a nonequilibrium condition when more atoms are in the excited state than in the ground state). Energy is emitted from the atoms in the medium, increasing the energy density of the radiation field. This condition corresponds to laser amplification. To produce population inversion, a pumping source is required to provide the energy to excite atoms into the second energy level, and a radiation field is required to interact with the atoms, extracting energy for laser amplification.

## Gain Saturation

Figure 5 shows an energy level diagram and transition rates due to pumping, decay, and induced transitions. The following equations describe the rate of change of the population density of levels 2 and 1, respectively, in the presence of a radiation field:

$$\frac{dN_2}{dt} = R_2 - \frac{N_2}{\tau_{20}} - \frac{N_2}{\tau_{sp}} - (N_2 - N_1)\,W_i(v),$$

$$\frac{dN_1}{dt} = R_1 - \frac{N_1}{\tau_{10}} + \frac{N_2}{\tau_{sp}} + (N_2 - N_1)\,W_i(v).$$

where $W_i(v)$ represents the induced transition rate:

$$W_i(v) = B\rho(v)g(v).$$

In the preceding equations $R_1$ and $R_2$ are the pumping rates into levels 1 and 2, respectively. Obviously, it is desirable for $R_2$ to be much larger than $R_1$ to create and maintain an inverted population. The second term of the equations describing $dN_2/dt$ and $dN_1/dt$ represents the change in population of the respective levels due to decay. The third term considers level 2 to level 1 transitions by spontaneous emission and is relatively small compared with the influence of the last term. The last term in each equation is the change in population density of the respective levels due to induced transitions. Once population inversion has been established by a pumping source, such as an injection current for a laser diode, a flashlamp for a pulsed laser, or energetic electrons in plasma-discharge gas lasers, the radiation field will cause induced absorption and stimulated emission. The induced level 2 to level 1 transitions and level 1 to level 2 transitions occur at an equal rate $[W_i(v)]$. Since $N_2$ is greater than $N_1$,

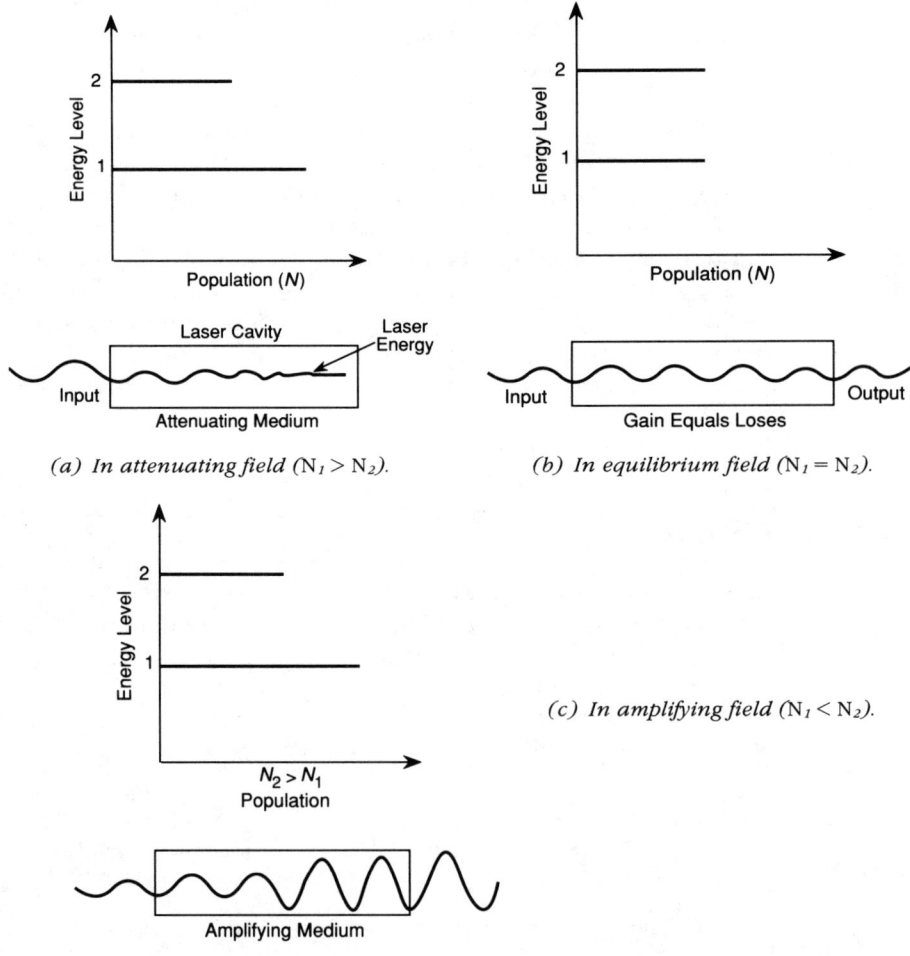

**Figure 4.** Traveling electromagnetic wave field conditions.

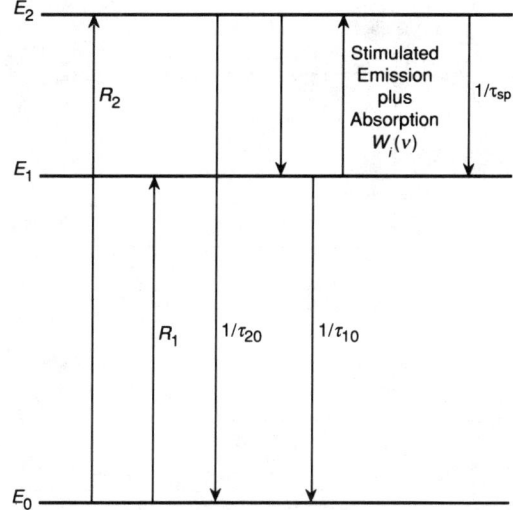

**Figure 5.** Transition rates of a three-level laser excited by a pumping mechanism.

the effect of the induced transitions, the last term of the equations describing $dN_2/dt$ and $dN_1/dt$ increases the population density of level 1 and decreases the population density of level 2. More level 2 to level 1 transitions occur than 1 to 2 transitions resulting in a reduced state of population inversion. The reduction in population inversion due to the presence of the radiation field is called gain saturation. Gain saturation limits the gain of the medium to equal the losses of the system. This condition allows for steady-state oscillation to exist within a laser. At steady-state conditions the rate of change of the population densities is zero, and the populations of levels 1 and 2 become constant.

# Oscillation

Population inversion is the only requirement for producing light via stimulated emission, but oscillation is necessary to produce a laser. Oscillation results from placing the laser medium within a feedback system such as a resonator. The resonator typically consists of two parallel mirrors, one of which is totally reflective and the other is only partially reflective, as shown in Figure 6. The shape of the mirrors may vary according to the requirements of the resonator. The mirrors are usually aligned perpendicularly to the optical axis of the active medium, allowing for amplification of the optical energy traversing the axis. This results in a large build-up of energy along the optical axis. The optical energy along the axis is in the form of a narrow light beam with low divergence. Energy moving in a direction other than along the axis is lost from the oscillator. The wavelength of the light output by the system is determined

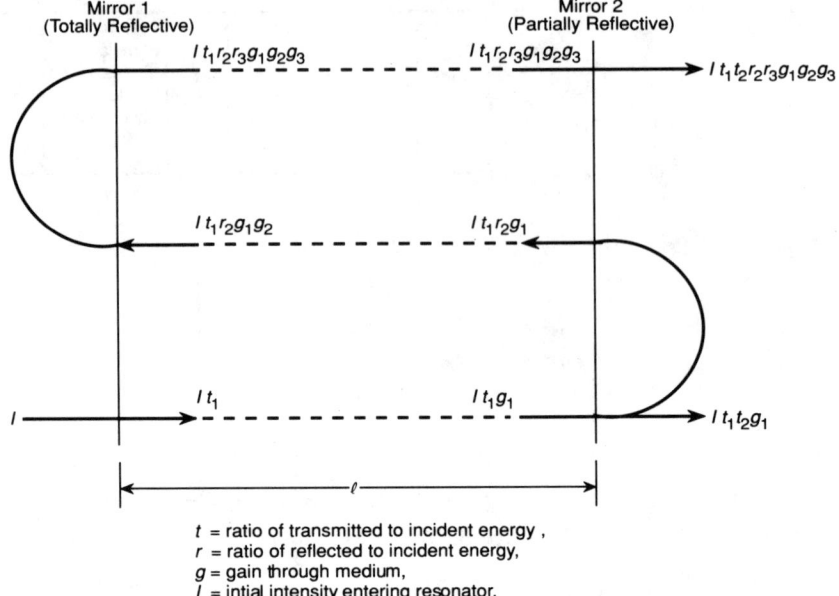

**Figure 6.** Optical resonator model.

by the length of the resonating cavity. The condition for oscillation at a particular wavelength exists when twice the cavity length ($L$) is equal to an integer multiple of the wavelength:

$$2L = N\lambda.$$

Each wavelength which meets this criterion defines a longitudinal mode of the laser cavity. Longitudinal mode is a term used to define the various wavelengths which may resonate within a cavity.

As the electromagnetic energy (optical energy in this case) of the appropriate wavelength oscillates within the resonator, it is amplified by stimulated emission while some of the energy is lost due to transmission out of the resonator. The gain of the electromagnetic energy per pass through the resonator ($g$) will increase until it equals the losses per pass ($t$) as shown in Figure 6. If the gain per pass exceeds the losses per pass, the total radiation field energy will increase. This condition cannot be maintained indefinitely due to gain saturation. The gain in terms of energy density of the radiation field can be expressed by

$$\frac{d\rho(v, x)}{dx} = \rho(v, x) e^{\delta(v)x},$$

where $x$ is the distance traveled into the medium and $\delta(v)$ is the gain coefficient of the laser.

The gain coefficient is proportional to the probability of emission from level 2. As the number of level 2 to level 1 transitions increases, the number of excited atoms available for level 2 emission decreases. The energy density of the medium increases and the gain coefficient becomes smaller. The gain coefficient thus decreases due to gain saturation, resulting in steady-state oscillation. The gain of the laser is a function of both the frequency of the radiation field and the length of the resonating cavity. Figure 7 shows the relationship of the gain coefficient of a medium, as well as the losses of the medium, as the energy density of the radiation field increases. The frequency dependence of the gain coefficient $\delta(v)$ is contained in the lineshape. If the gain is much larger than the losses, the range of frequencies above the loss line is great, providing a large spectral width for amplification. As the gain decreases, the spectral width becomes very small and the range of frequencies above the loss line decreases, ultimately creating a monochromatic light source.

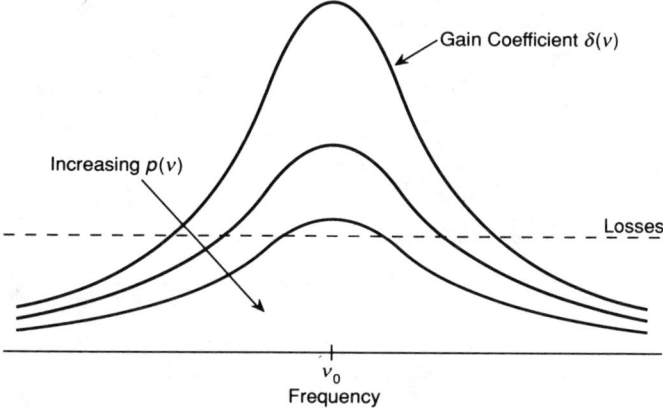

**Figure 7.** Variation of the gain coefficient with energy density.

Oscillation in the resonator begins with spontaneous emission. Spontaneous emission occurs in a large number of longitudinal modes. Most of the modes represent waves that are in the wrong direction (not toward the reflectors) and at the wrong frequency (not a frequency interval that coincides with the cavity resonance) and thus do not promote amplification. Only a few modes with the proper direction and frequency are available to cause stimulated emission and hence amplify the radiation field. Initially, as shown in Figure 8a, the modes are formed by spontaneous emission at frequencies corresponding to the cavity resonance. The gain coefficient is large over a broad spectrum. The electromagnetic energy produced by spontaneous emission is responsible for causing stimulated emission as the light energy bounces within the resonator cavity. The stimulated emission introduces light energy in phase, at the proper frequency, in the appropriate direction and with common polarization as the field that caused the stimulation. As a result, the modes closest to the center frequency ($v_0$) of the gain coefficient are greatly amplified. Other resonant frequencies are amplified to a lesser extent, according to their relationship to the center frequency of the gain coefficient. This is shown in Figure 8b.

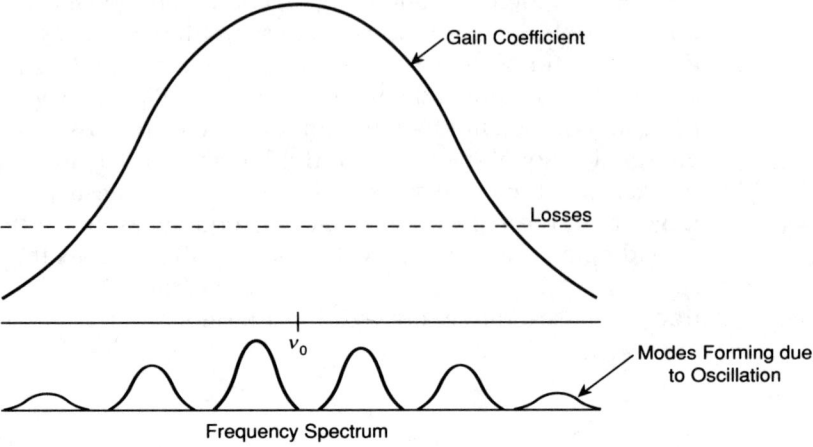

*(a) Many modes develop as a result of spontaneous emission.*

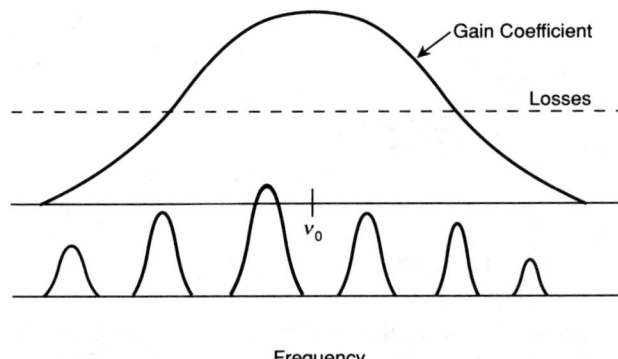

*(b) Amplification is greatest with modes near the middle of the gain spectrum.*

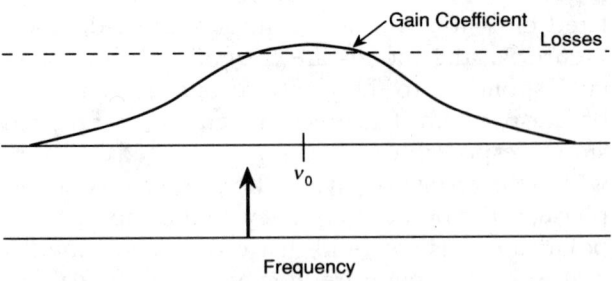

*(c) Gain saturation reduces the gain coefficient to equal the losses, limiting the number of amplified modes to one.*

**Figure 8.** Evolution of laser amplification.

After several trips through the cavity, the resonant frequencies are amplified with those closest to the center frequency being amplified at an accelerated rate. The intensity of the radiation field increases until an equilibrium exists where the number of level 2 to level 1 transitions equals the number of level 1 to level 2 transitions (gain saturation). From Figure 8c it can be seen that the gain coefficient decreases to approximately equal the losses as a result of gain saturation. The spectral shape is altered as the mode closest to the center frequency increases much more than the other resonant frequencies. The gain of the other resonant frequencies eventually falls below the losses. Thus oscillation occurs only at the mode closest to the center frequency, resulting in a monochromatic light source. In some lasers only one longitudinal mode falls under the gain curve. Some lasers emit in only one longitudinal mode at a time but are capable of hopping to another mode if operating conditions change.

# Laser Characteristics

There are several unique characteristics of laser light that distinguish it from other light sources. Some of the important characteristics of laser light are described below.

## Coherence

Laser beams are characterized by light waves that are phase coherent. The light generated by a laser is in phase as it exits the laser cavity as shown in Figure 9a. This is due to the creation of photons of common frequency and phase by stimulated emission which is reinforced by the resonating properties of the cavity. The time the laser remains coherent is the coherence time ($t_c$) and the distance over which laser light travels before losing its coherent nature is the coherence length ($l_c$). The coherence time and length are related by the speed of light ($c$), according to $l_c = ct_c$. The coherence length is generally only a fraction of a meter. The coherence length can also be related to the spectral bandwidth $\Delta v$ according to $l_c = c/\Delta v$. Lasers with narrower bandwidths thus have larger coherence lengths. Conventional light sources are incoherent. Conventional light is dominated by spontaneous emissions which tend to be random in frequency and phase and therefore no coherence exists.

## Monochromatic Light

Lasers produce light which is monochromatic, that is, has a single wavelength. Conventional light sources, on the other hand, emit light throughout the visible and infrared spectrum with no constraint on a specific range of wavelengths. This is because laser light is characterized by stimulated emission and conventional light by spontaneous emission.

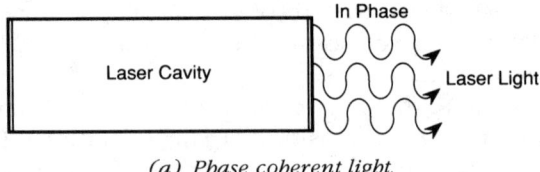

*(a) Phase coherent light.*

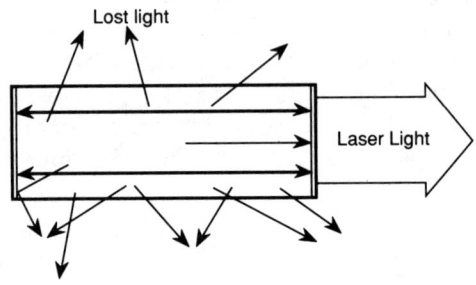

*(b) Directional light.*

**Figure 9.** Laser light is both phase coherent and directional.

## Beam Directionality

A laser's resonator provides only a single exit for the generated light. Only photons within the resonator that are emitted along the axis cause amplification and exit the laser at this output. These restrictions provide directionality to the laser beam as shown in Figure 9b. Conventional light is not restricted in its direction of propagation. For example, in a normal lightbulb light is allowed to exit the bulb in any direction.

## Beam Polarization

Polarization refers to the plane direction of the electric and magnetic fields of electromagnetic radiation such as light. This is shown in Figure 10. The electric vector is referred to as the vibration direction and the magnetic vector is referred to as the polarization direction. Although many lasers are made to produce polarized beams, not all lasers are made with this capability. In polarized light the plane of polarization is common throughout the beam. Conventional light does not maintain a common polarization characteristic. Polarization is an important laser characteristic because information can be carried by the laser light as a function of its polarization characteristics. Some fiber-optic systems utilize the polarization of the laser light source to carry information. Polarization is also used in the pickups of optical disc systems. These applications are described in Chapters 5 and 6, respectively.

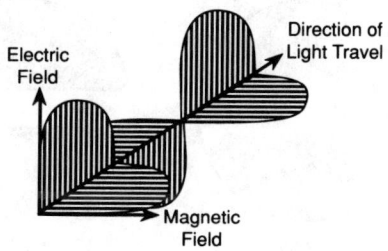

**Figure 10.** Laser beams consist of light with common polarization characteristics.

## Mode Structure

Laser resonators have two distinct types of modes, known as transversal modes and longitudinal modes. Transversal modes describe the cross-sectional profile of the laser beam's intensity. Lasers typically produce a very narrow, highly focused beam that diverges less than 1 mrad (one milliradian, which is 1 m divergence in 1 km). The laser beam's distribution is described by its transverse electrostatic mode (TEM). The TEM describes the cross-sectional distribution of the output beam by indicating the number of horizontal and vertical nulls of the beam, as shown in Figure 11. The first digit of the TEM number represents the number of vertical nulls in the beam, and the second digit represents the number of horizontal nulls. Longitudinal modes correspond to the different wavelengths that resonate within the gain bandwidth of the laser.

## Power

Lasers are capable of delivering high concentrations of energy to a small area. This is a positive attribute for welding and cutting and for such intricate operations as microscopic surgery or compact disc mastering. The power output varies with the particular type of laser and its design. Typical power outputs range from less than a milliwatt to tens of kilowatts with the greatest power outputs used in military laser weapons research at over a megawatt. The power density is also an important specification in determining a laser's ability to penetrate a material. The depth of cut into a material is determined by evaluating both the power density of the laser beam and the material composition. Power density is determined by dividing the output power by the area of the output beam. The power density must also be evaluated to determine the laser's ability to perform cutting on a particular material. A given laser may not be a relatively powerful light source, but if it is focused to a very narrow spot on a material, such as on a master disc, it may have sufficient power density to damage the disc substrate.

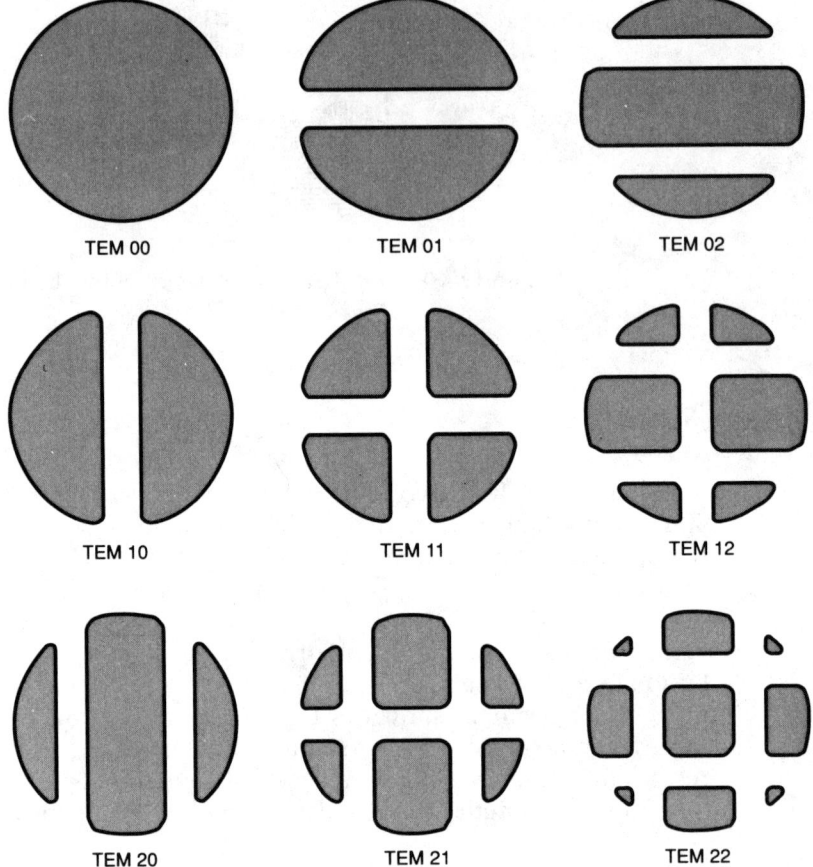

**Figure 11.** The distribution pattern of the laser beam is represented by a transversal electrostatic mode (TEM) number.

## Efficiency

Lasers tend to be very inefficient, usually converting between 0.1% and 30% of the input energy into useful optical power. Table 2 provides efficiency information for some common lasers. Much of the energy is dissipated in the form of heat. The excess heat may cause minor problems by slightly deteriorating the quality of the laser beam and shortening the laser's lifespan.

## Lifetime

The lifetime of a laser can be measured in several different ways depending on the type of laser evaluated. Most continuous-wave laser lifetimes are measured in numbers of hours of operation. A pulsed laser's lifetime is measured in the number of shots available. Table 2 provides typical operational lifetimes for several common types of lasers.

**Table 2.** Efficiencies of Some Common Lasers

| Laser Type | Efficiency | Operational Lifetime |
|---|---|---|
| $CO_2$ | 5%–25% | One to several thousand hours (continuous) |
| Chemical | 1%–2% | 1000–2000 hours (continuous) |
| Semiconductor (Laser Diode) | 0.5%–20% | 4500–100,000 hours (continuous) |
| Dye* | 2%–20% | $10^5$–$10^6$ shots (pulsed) |
| Excimer | 1%–4% | |
| Ion | 0.01%–0.001% | 1000–10,000 hours (continuous) |
| Neodymium* | 0.1%–1% | $10^6$–$10^7$ shots (pulsed) |
| Nitrogen* | 0.01%–0.1% | $10^7$–$10^8$ shots (pulsed) |
| Ruby* | 0.1%–1% | $10^6$ shots (pulsed) |
| Helium-Neon | 0.01–0.1% | 10,000–20,000 hours (continuous) |

*These lifetimes are associated with the flashlamps which provide the pumping source. The laser's lifetime can be extended on flashlamp replacement.

## Continuous-Wave Lasers

Many laser applications require a steady stream of light. Continuous-wave (CW) lasers emit photons at a steady, continuous rate. A laser that emits continually for more than a second is considered a continuous-wave laser. Heat buildup often limits some lasers to operation of less than a second. Most commercial lasers can emit continuously for much longer than a second; often days of continuous operation are possible without damaging the laser. Some common applications for continuous-wave lasers are cutting, welding, hardening metals, brazing and baring wires, soldering, marking, and recrystallization of various materials.

## Pulsed Lasers

Pulsed lasers emit photons in the form of a pulse or a train of pulses. The energy in a pulsed laser is concentrated and therefore potentially dangerous. Some types of lasers are only capable of operation as pulsed lasers because of the excessive heat buildup when operating. The time between pulses allows heat to dissipate preventing damage to the laser medium and cavity. Pulsed lasers are often used for spot welding, drilling, cutting wafers, and trimming electronic networks such as thick and thin film networks, as well as for military applications.

## Cooling Requirements

To minimize deterioration of the laser, special cooling requirements may apply to some high-power lasers as well as duty cycle limitations on pulsed lasers. Some lasers are cooled by forced air or flowing water. Often liquid nitrogen is utilized to cool the laser medium, particularly when cryogenic temperatures are necessary.

# The Laser Diode

In terms of economics and applications the laser diode (also known as the semiconductor laser) is the most successful laser in use today. The laser diode was invented in 1962 and was one of the first lasers to go from the research laboratory to the consumer market. This laser's technological growth has been stimulated by the development of fiber-optic systems and optical disk storage media. Several features make it attractive for use in digital audio: its ability to be driven by conventional electronics at low power, small size, relatively high efficiency, monolithic structure, compatibility with fiber-optic systems, and ability to be mass-produced efficiently and cheaply. Laser diodes are part of a family of lasers distinguished by the characteristic that each is composed of elements from columns III and V of the periodic table. Table 3 provides a listing of common laser diodes and their corresponding wavelengths.

**Table 3.** Common Laser Diodes and their Wavelengths

| Material | | | Wavelength (nm) |
| --- | --- | --- | --- |
| Active Layer | Passive Layer | Substrate | |
| $Ga_{1-x}Al_xAs$ | $Ga_{1-y}Al_yAs$ | GaAs | 700–900 |
| (GaIn)P | — | GaAs | 670 |
| Ga(AsP) | (GaIn)P | GaAs | 700 |
| (GaIn)(AsP) | InP | InP | 1150–1650 |
| Ga(AsSb) | (GaAl)(AsSb) | GaAs | 1000 |
| (GaAl)Sb | (GaAl)(AsSb) | GaSb | 1350 |
| (GaIn)(AsSb) | (GaAl)(AsSb) | GaSb | 1800 |

## Operation

The laser diode is designed similarly to a conventional diode, as shown in Figure 12. It relies on an injected current to stimulate recombination of electrons and holes in the pn junction, often referred to as the active layer. By applying an electric field in a forward-bias direction, as shown in the figure, a

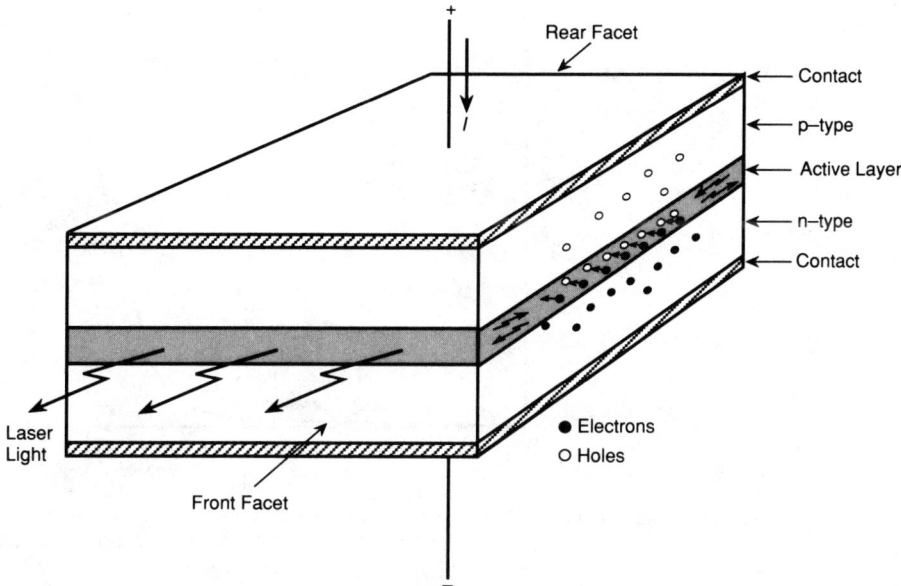

**Figure 12.** Basic laser diode structure.

current is created. An excess of carriers is developed in the junction. Electrons and holes begin to recombine, restoring equilibrium to the laser diode. Upon recombining, the carriers momentarily remain in an excited energy level before falling to a lower energy level due to spontaneous emission. If, while the carriers are still in the excited level, they can be stimulated to drop to a lower energy level, stimulated emission can dominate the emission process, allowing laser operation to persist. For stimulated emission to be prevalent in the pn junction, population inversion must be established within the junction.

High drive currents are required to produce population inversion in the laser diode. As the current through the pn junction is increased, the number of recombinations is also increased. A threshold current is established according to the compositional makeup of the semiconductor lattice. When the threshold is exceeded, the laser diode will produce and sustain laser action. Figure 13 shows the output power of a laser diode as a function of the input drive current. Only when the drive current exceeds the threshold current will the device operate as a laser. At lower currents the laser diode may operate as a light-emitting diode (LED). The laser diode differs from the LED because the laser diode has reflective coatings on opposing facets of the structure, providing feedback of the emitted photons, which in turn stimulates emission of photons from recombining carriers. Laser diodes operate at higher drive currents than do LEDs and are capable of producing greater optical power outputs.

Like all other lasers, the laser diode utilizes a resonator to oscillate the light. Opposite ends of the laser diode are cleaved (cut in parallel planes) to form partially reflective surfaces allowing the photons to bounce within the crystal lattice. A reflective coating may be applied to improve the reflective

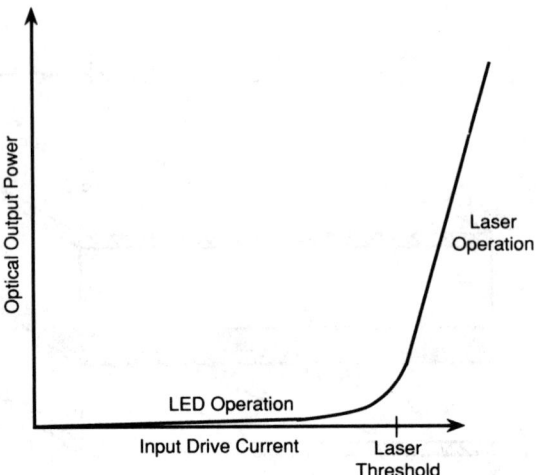

**Figure 13.** Output power versus drive current of a laser diode. Upon exceeding the threshold current, laser operation occurs.

qualities of the cleaved ends. As the photons bounce through the cavity they produce additional stimulated emission. The modes produced by the laser diode's stimulated emission are developed according to the intensity of the radiation at each particular mode. The dominating mode(s) is determined by the cavity length. Many different modes develop due to stimulated emission, but most modes eventually become attenuated. When the gain of one or more modes overcomes the losses of the cavity, laser action occurs. This condition can also be recognized by the drive current exceeding the threshold current.

The laser diode requires relatively little input power to operate. For example, a double-heterojunction laser diode requires a maximum input voltage of 2 V. Laser thresholds are typically between 10 and 100 mA with the average power requirement being well under 1 W for continuous lasers. Less input power is required for shorter wavelengths. These values are compatible with conventional electronics. The input power requirements of pulsed lasers vary greatly, up to several volts and tens of amperes for pulses lasting microseconds. Peak input powers may be as high as 100 W, with maximum output powers up to 20 W.

Most laser diodes are packaged in a housing that is designed to permit easy mounting with provisions for external connection to a power source. The transistor-style case is most common and provides a window for emission of laser light. A monolithic structure is available with a fiber-optic pigtail connector for coupling to optical fibers. The fiber-optic transmitter unit has provisions for internal cooling and either a pigtail connector or other fiber-optic interface. A light pen case comes with internal optics for collimating the laser beam. Often the laser diode is packaged with a built-in photodetector for the feedback used to control the drive current. Typically, external optics are required for focusing the laser beam.

Despite the laser diode's low operating power, the laser beam has the capability of causing damage to the eye. The eye's response to light decreases sharply above 700 nm, but when the light is tightly focused its energy density can be great enough to cause damage even beyond 1300 nm. The light emitted by laser diodes used in compact disc players is not visible. The diode is connected to an interlock; when the disc compartment is opened the laser will not function. Unless one is properly trained one should not override the interlock. When one is testing equipment such as compact disc players and fiber-optic systems, the optical beam should be monitored with a light meter and the eyes should remain at least 12 inches away from the focusing lens. Never look directly into the objective lens of an optical system.

## Structure

The laser diode is a pn junction developed by doping a semiconductor crystal with a p-type material on one side and an n-type material on the other side. Two opposing sides of the semiconductor crystal are made to reflect the light back into the medium to promote oscillation and amplification. The most common laser diodes used today, particularly in audio applications, are the GaAs and the GaAsAl lasers. These lasers can consist of many different structures, but the following are most common: homojunction, single heterojunction, double heterojunction, and stripe geometry. They emit light from only one facet but can be designed to allow laser emission from both ends of the laser.

The homojunction structure is a single semiconductor compound, such as GaAs doped with a p-type material on one side and an n-type material on the other side, as shown in Figure 14a. This structure has low tolerance to heat and is best operated in the pulsed mode. Its peak pulsed output power can be as high as 10 W. The homojunction laser diode is less efficient than other structures and therefore has been replaced by laser structures that provide improved laser characteristics.

Single-heterojunction or single-heterostructure (SH) laser diodes have junctions with doped compounds on either side of the active layer, as shown in Figure 14b. A single-heterojunction laser diode is so named because the bandgap of the n-type material is equal to that of the active layer while the bandgap of the p-type material is much greater. Therefore, a single jump in bandgaps exists. The thickness of the active layer is usually made small to improve efficiency and lower the threshold current density. This type of structure is used for high peak power applications where efficiency is not important and pulsed operation is required. Single-heterojunction laser diodes require high threshold currents and are not applicable where continuous laser action is required, such as in fiber-optic applications. The single-heterojunction structure is not used for digital audio applications at present.

The double-heterojunction (DH) laser diode has proven to be a successful light source for fiber-optic applications. The double-heterojunction structure is shown in Figure 14c. This structure consists of an active region bounded by

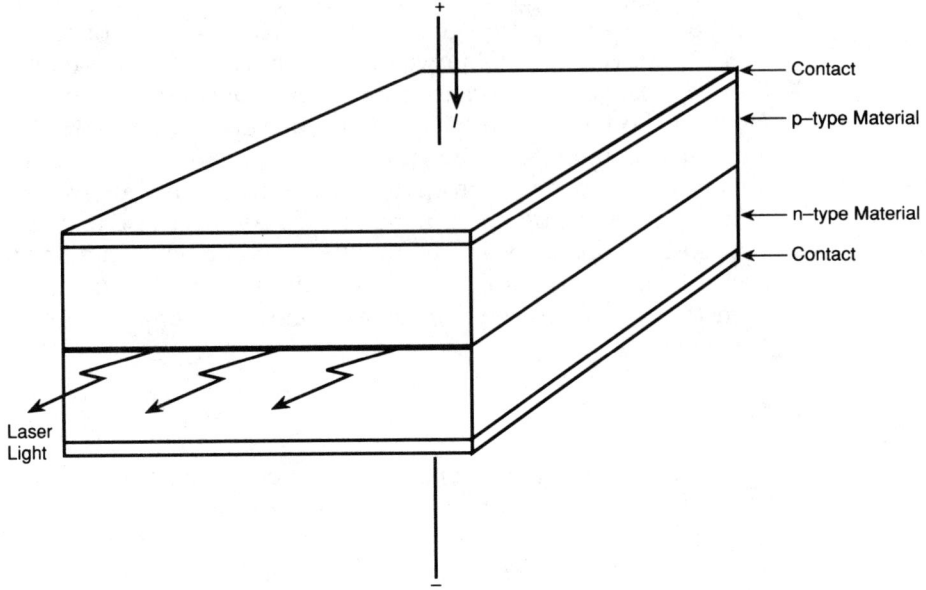

*(a) Homojunction structure.*

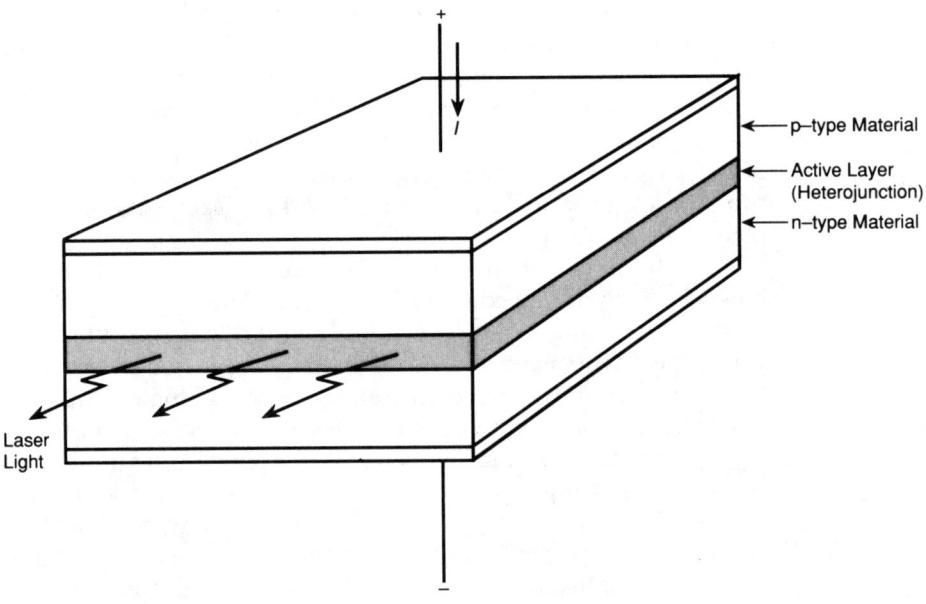

*(b) Single heterostructure.*

**Figure 14.** Four common laser diode structures.

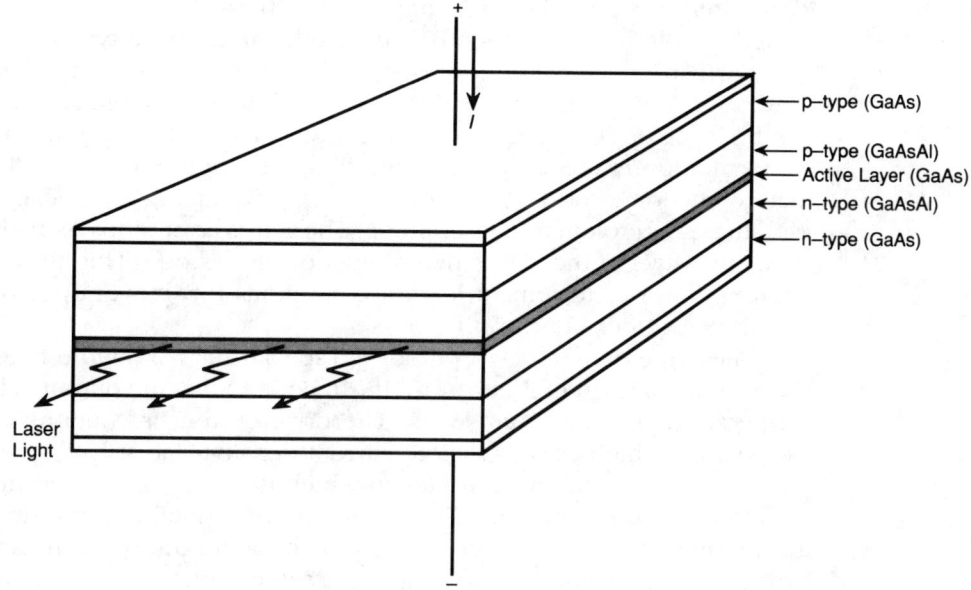

*(c) Double heterostructure.*

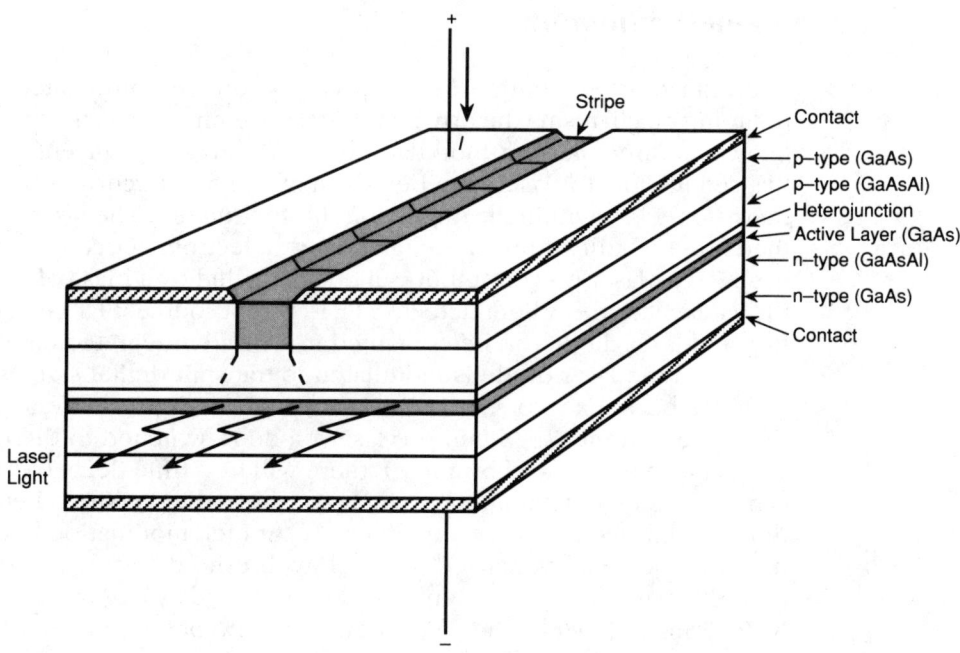

*(d) Stripe-geometry structure.*

**Figure 14.** (cont.)

two different materials, for example, GaAs between two GaAsAl layers. It is called double heterojunction because both bandgaps (n-type and p-type) are greater than the active layer's bandgap. Therefore, two bandgap jumps exist. This laser diode structure is efficient and has a low threshold current density. It confines the optical field and the injection current, resulting in a high photon concentration in the active layer and therefore high power gain. The double-heterojunction laser diode is not capable of producing high peak pulses as is the single-heterojunction laser diode. The active layer tends to be thinner than the active layer of the single-heterojunction laser diode. This allows threshold current densities to remain low for continuous low-power operation such as required for fiber-optic applications.

The stripe-geometry laser diode is a special class of double-heterojunction lasers and is shown in Figure 14d. It can be seen that the output is limited to a stripe approximately 5 $\mu$m wide. The drive current is contained within the stripe area, which creates a large current density. The stripe-geometry structure limits the number of modes in which it can oscillate, permitting a very defined frequency content. Since the output is contained in a small area the beam can be focused to a very small spot for better information handling. The stripe area aids in coupling to single mode fiber-optic cables. The stripe-geometry laser diode can be operated either as a pulsed laser or a continuous laser.

## Laser Diode Modulation

The laser diode is unique in that it can be directly modulated by the laser pumping mechanism (the driver current) up to microwave frequencies. This is known as direct current modulation. An advantage of current modulation is that semiconductor lasers can be driven by the output current of a field-effect transistor (FET) within the same monolithic structure. The laser and driver can therefore be built within an integrated optoelectronic circuit as shown in Figure 15. The laser power output can be controlled by a bias voltage applied to the gate electrode. A light detector, FET current source, FET current preamplifier, and laser diode can be contained in a single integrated circuit.

When the laser diode is modulated to transmit digital signals, the output switches between two states; $P_{off}$ and $P_{on}$. If the input current is below the threshold current ($I_{th}$) when the laser diode is switched to the on state (i.e. when a zero is followed by a one), there will be a time delay ($t_d$) allowing the carrier density in the junction to increase above the threshold before the laser diode will turn on, as shown in Figure 16. At high modulation frequencies the turn-on time delay associated with below-threshold biasing is too long for accurate transmission of the signal. As demonstrated by G. Arnold, P. Russer, and K. Petermann [1988], when $P_{off}$ is biased at 5% below threshold the turn-on time delay eliminates the first pulse of a 10111 input signal due to the excessive time delay required for the carrier density in the junction to increase above the threshold. When $P_{off}$ is biased at the threshold all of the pulses are present at the output, but some are distorted. At a bias of 5% above threshold,

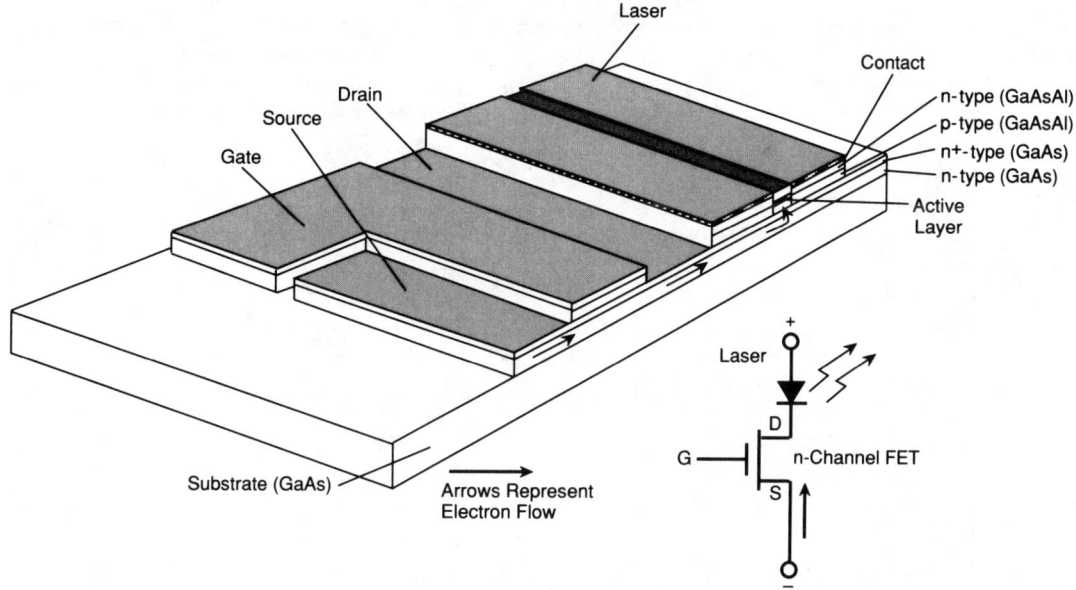

**Figure 15.** Integrated optoelectronic circuit containing laser-driver configuration. An n-channel FET is used to drive a double-heterojunction laser diode.

all pulses are present and of the same quality. Even when $P_{off}$ is biased above the threshold, a small time delay exists (typically under 100 ps) which is adequate for most digital signal transmission requirements. The higher the bit rate requirements for transmission, the higher the bias should be to ensure that an adequate signal is output from the laser.

## Manufacturing

Laser diodes are manufactured using the same methods as other integrated-circuit devices. The laser diode is built on a substrate wafer, grown in a conventional manner. It is important that the substrate be extremely pure and free from defects and doping nonuniformities because of the effect on the quality of the epitaxial growth. Table 4 lists the processes involved in the development of homostructure, single-heterostructure, double-heterostructure, and stripe laser diodes.

## Maintenance and Reliability

Maintenance is generally not required for laser diodes. As long as the laser is operated under normal conditions involving temperature, moisture, and drive current, the laser diode should operate without failure. Occasionally, it is

**Table 4.** Laser Diode Fabrication

| Order | Process | Homojunction | Single Heterostructure | Double Heterostructure | Stripe-Geometry |
|---|---|---|---|---|---|
| 1 | Selection and preparation of substrate wafer | ✓ | ✓ | ✓ | ✓ |
| 2 | Epitaxial growth of heterostructure layers | | | ✓ | ✓ |
| 3 | Epitaxial growth and junction diffusion | ✓ | ✓ | | |
| 4 | Reduction of wafer thickness by chemical polishing of facets | ✓ | ✓ | ✓ | ✓ |
| 5 | Delineation of the stripe by deposition of SiO$_2$ followed by photolithography for windows | | | | ✓ |
| 6 | Metallization of wafer on both sides | ✓ | ✓ | ✓ | ✓ |
| 7 | Cleave strip into bars in widths equal to the length of the laser diode | ✓ | ✓ | ✓ | ✓ |
| 8 | Apply protective facet coatings by evaporating Al$_2$O$_3$ | ✓ | ✓ | ✓ | ✓ |
| 9 | Separate bars into individual laser diode chips | ✓ | ✓ | ✓ | ✓ |
| 10 | Mount laser diode chip into header by low-temperature soldering | ✓ | ✓ | ✓ | ✓ |

advisable to incorporate some type of automatic feedback to control output stability or operating temperature. The temperature can be controlled with thermoelectric coolers. The output power can be stabilized by providing feedback to control the drive current. Some common failures in laser diodes include optical damage due to excessively high power densities, oxidation of optical facets from photochemical reactions with laser light, failure of soldering joints due to mechanical stress, and grown-in defects called dark-line defects, which are regions of the active layer that do not emit light.

Internal optical damage to the laser diode can be prevented by limiting the drive current to safe levels. Dark-line defects, solder failures, and degradation caused by oxidation depends mainly on manufacturing processes. These failures are rare, due to the implementation of quality controls by laser manufacturers. Laser diodes are usually replaced after malfunctioning because of the difficulty of repairing lasers and the relatively low cost of replacement.

Some laser diodes have a rated lifetime over 100,000 hours and can actually last up to 1,000,000 hours when operating under normal conditions. In general, the higher the power output, the shorter is the life expectancy. The lifetime also decreases with increasing wavelength, operating temperature, or

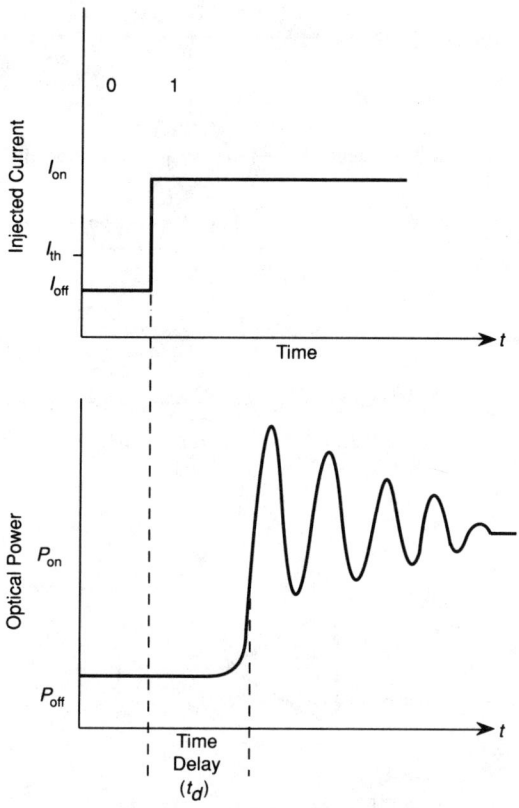

**Figure 16.** A time delay occurs as a laser responds to a step current as a result of $I_{off}$ being below the threshold current.

threshold current. Unfortunately, few laser specification sheets provide information on life expectancy.

## Driver Circuitry for the Semiconductor Laser

Most laser drivers are separate integrated circuits connected to the laser diode by a bond wire. Inductance problems associated with the connecting wire can be eliminated by combining the driver circuit and the laser diode on a common integrated circuit. Figure 17 shows two common driver circuits. The first circuit utilizes an emitter-follower configuration with two current sources. The second utilizes a monitor diode and feedback to limit the drive current. Figure 18 shows a circuit diagram for an optoelectronic driver/laser module that utilizes an emitter-coupled transistor pair for controlling the driver current. This circuit provides bias control which may be advantageous for modulating high-frequency digital signals. Figure 19 shows a common laser monitor control circuit for a compact disc player.

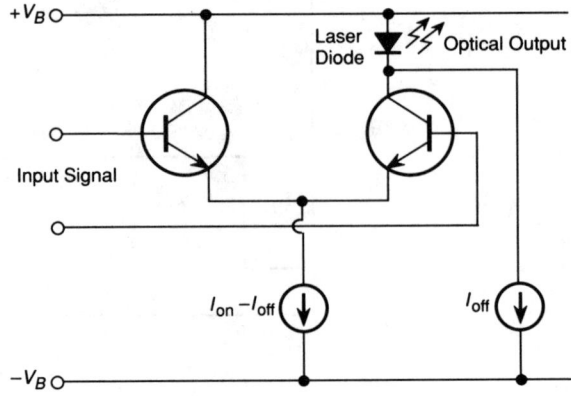

(a) Emitter follower utilizing two current sources.

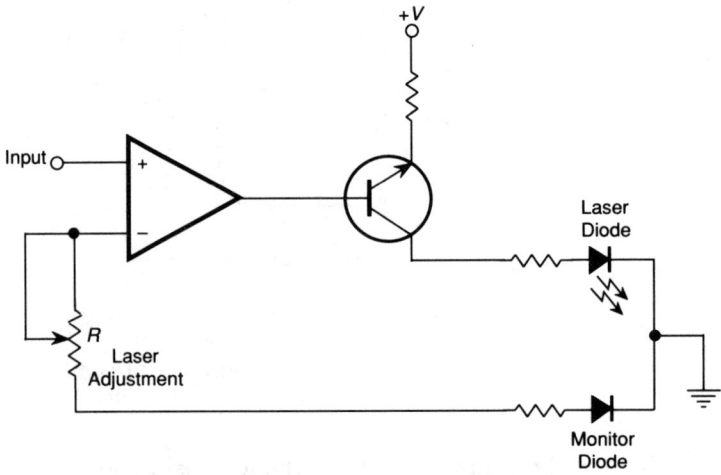

(b) Monitored laser driver circuit providing feedback for power control.

**Figure 17.** Laser driver circuits.

The output from the monitor laser is received at the operational amplifier and compared with an adjustable reference input. The reference should be set to allow for maximum drive current without causing excessive heat buildup and damage to the laser. If the laser diode output decreases, the laser monitor output also decreases. In this circuit the output of the operational amplifier therefore decreases, turning on transistor $Q_1$, providing higher drive current to increase the laser diode's output power.

## Audio Applications

There are many laser diode applications in digital audio. For example, a laser diode is utilized in compact disc players to transfer digital information from

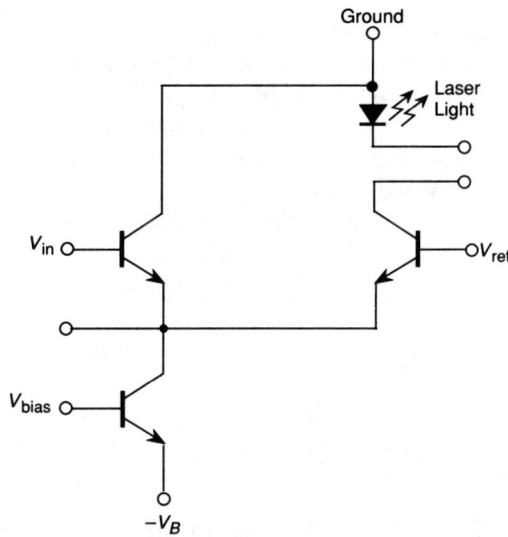

**Figure 18.** Optoelectronic laser/driver circuit.

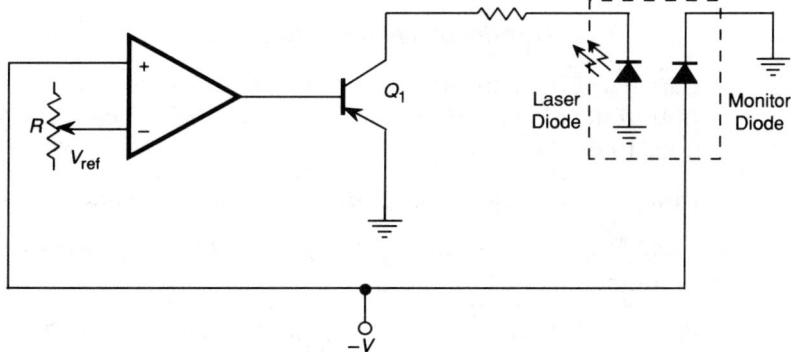

**Figure 19.** Monitored laser driver circuit commonly used in compact disc players, providing feedback for power control.

the compact disc medium to the laser pickup. It is also utilized for tracking and focusing the information on the disc. The compact disc player has been significant to the commercial success of the laser and introduced the laser into the homes of millions of people. As a result of the success of the compact disc, the production of laser diodes has increased significantly, justifying a sharp reduction in cost. The most common type of laser diode utilized in compact disc players is the GaAlAs double-heterojunction laser diode operating at a wavelength of 790 nm. Optical media are discussed in more detail in Chapter 6.

Prior to the introduction of the compact disc player, fiber optics provided the main impetus for laser diode development. The laser diode is used in fiber-optic systems as a light source for transferring information over optical fibers. A variety of laser diode types is utilized for fiber-optic applications. The laser diode is a light source that is capable of providing high-power signals at a variety of wavelengths. The first optical fibers operated at wavelengths of about 800 to 900 nm with GaAlAs lasers. Today 1300 and 1550 nm wavelength lasers, such as InGaAsP/InP type, are more widely used because of improved bandwidth characteristics and lower attenuation. Much of today's research is directed toward developing longer-wavelength laser diodes, although considerable effort has also gone into the development of visible-light diode lasers.

# References

Allard, F. C., *Fiber Optics Handbook for Engineers and Scientists*, McGraw-Hill Book Co., 1990.

Bauerle, D., *Laser Processing and Diagnostics*, Springer-Verlag New York, 1984.

Bell, T. E., "Communications," *IEEE Spectrum*, January, 1988.

Cheo, P. K., *Handbook of Solid-State Lasers*, Marcel Dekker, 1989.

Christian, N. L., and L. K. Passauer, *Fiber Optic Component Design, Fabrication, Testing, Operation, Reliability and Maintainability*, Noyes Data Corporation, 1989.

Doley, W. W., *Laser Processing and Analysis of Materials*, Plenum Press, 1983.

Ehrlich, D. J., *Laser Microfabrication, Thin Film Processing and Lithography*, Academic Press, 1989.

Eloy, J. F., *Power Lasers*, Ellis Horwood Limited, 1987.

Francon, M., *Laser Speckle and Applications in Optics*, Academic Press, 1979.

Halley, P., *Fibre Optic Systems*, John Wiley & Sons, 1987.

Hecht, J., and D. Teresi, *Laser, Supertool of the 80s*, Ticknor and Fields, 1982.

Hecht, J., *The Laser Guidebook*, McGraw-Hill Book Co., 1986.

Jones, W. B., Jr., *Introduction to Optical Fiber Communication Systems*, Holt, Reinhart and Winston, 1988

Katzman, M., *Laser Satellite Communications*, Prentice-Hall, 1987.

Koechner, W., *Solid-State Laser Engineering*, Springer Series in Optical Sciences, Springer-Verlag New York, 1988.

Marshall, F. G., *Laser Beam Scanners, Opto-Mechanical Devices, Systems, and Data Storage Optics*, Marcel Dekker, 1985.

Masterton, W. L., and E. J. Slowinski, *Chemical Principles*, 4th edition, W. B. Saunders Co., 1978.

Osgood, R. M., Brueck, S. R. J., and H. R. Schussberg, *Laser Diagnostics and Photochemical Processing for Semiconductor Devices*, Elsevier Science Publishers B.V., 1983.

Petermann, K., *Laser Diode Modulation and Noise*, Kluwer Academic Publisher, 1988.

Sakura, S., Onobuchi, S., Ito, M., and S. Katagiri, "Fiber Optics Links for Digital Audio Interface," *IEEE Transactions on Consumer Electronics*, vol. 34, no. 3, August, 1988.

Seippel, R. G., *Fiber Optics*, Reston Publishing Co., 1984.

Tang, C. L., *Methods of Experimental Physics: Quantum Electronics*, vol. 15, Academic Press, 1979.

Thompson, G., *Physics of Semiconductor Laser Devices*, John Wiley & Sons, 1980.

Tipler, P. A., *Modern Physics*, 2nd edition, Worth Publishing Inc., 1978.

Verdeyen, J. T., *Laser Electronics*, Prentice-Hall, 1981.

Winburn, D. C., *Practical Laser Safety*, Marcel Dekker, 1985.

Yariv, A., *Optical Electronics*, 3rd edition, CBS College Publishing, 1985.

*Chapter 5*

# Fiber Optics
### Brent A. Karley

## Introduction

It was in 1870 that British physicist John Tyndall demonstrated the bending property of light when influenced by a rush of pouring water. His experiment demonstrated the principle of total internal reflection, which is a characteristic of primary importance to fiber optics. Tyndall's experiment was the first to show the feasibility of an optical waveguide.

From this beginning came the invention of the optical fiber. An optical fiber is a waveguide designed for electromagnetic waves at optical wavelengths. The fiber provides a path for single or multiple beams of light. The use of fiber optics has grown because of its many advantages over its electrical counterpart, the copper cable. Although fiber optics are very capable of carrying analog signals they have become particularly important in digital audio applications because of its ability to transmit information at extremely large bandwidths over long distances without introducing distortion, interference or noise.

## Advantages

There are several reasons why fiber optics has become a leading technology in signal transmission. Optical fibers offer many advantages over conventional copper cable. Problems associated with copper cable include electromagnetic interference (EMI), radio-frequency interference (RFI), ground loops, and intercable crosstalk. Each of these problems are overcome by the transmission properties of fiber optics. Fiber-optic technology provides tremendous performance in transmission partly because it allows for a large bandwidth. Fiber-optic cable is very light in weight compared with copper cable weighing approximately 0.009 lb/ft, and it also requires less space. Because it is very difficult to tap into a fiber-optic cable, security is an added benefit. Finally, fiber-optic cable is safe due to its inherent light transmitting properties. Unlike conventional cable, optical fibers can neither shock nor cause dangerous sparks. It is expected that the use of fiber optics in professional digital audio applications will continue to expand as exemplified by the introduction of the fiber-optic snake and various local area network (LAN) systems designed to accommodate MIDI, SMPTE, and digital audio formats. The audio industry is utilizing fiber optics as demonstrated by the presence of fiber-optic connectors on many compact disc players and stereo preamplifiers.

## Disadvantages

There are several disadvantages to fiber optics which should be considered when analyzing its use for digital audio applications. At present a need exists for standardization of connectors, interfaces, splices, digital codes, and optical wavelengths. Another consideration is cost. Until mass production of fiber-optic systems begins, the initial investment will remain relatively high. The cost of maintenance should also be considered. The time and effort to splice damaged or broken optical cable depends upon the equipment available and the experience of the maintenance personnel. Unique tools are required for each of the various types of splices, and personnel need to be trained to use the tools to efficiently maintain the fiber-optic system. With regard to safety, individuals performing maintenance or installation should wear eye protection while working close to lasers or near the ends of optical fibers. Although light levels tend to be modest and the light spreads out rapidly from the fiber ends, it is good practice to wear eye protection to prevent damage to the eye.

## The Principles of Fiber Optics

A fiber-optic system consists of several basic components required to transmit, transfer, and receive information, as shown in Figure 1. A transmitter is used to convert an electrical signal into an optical signal. The optical signal is transferred over an optical fiber to a receiver which converts the optical signal back into an electrical signal. In short, fiber optics utilizes light to carry information over optical fibers.

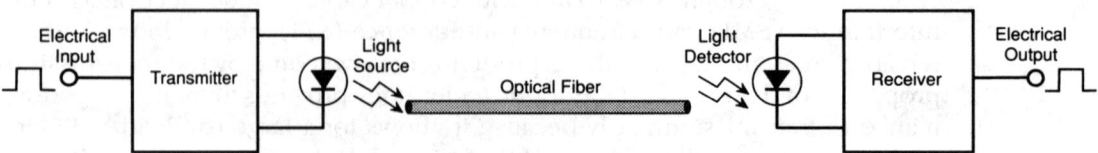

**Figure 1.** The basic components of a fiber-optic system.

Light has several unique properties that make it an efficient carrier of information. To provide a better understanding of how information is carried over optical fibers, a review of some important optical properties is presented.

## The Optical Spectrum

Light is composed of electromagnetic energy. Electromagnetic waves have long been utilized to transmit information at frequencies below the optical spectrum. Since information carried at lower frequencies limits the bandwidth, the amount of information that can be transmitted is also limited. The electromagnetic spectrum is shown in Figure 2.

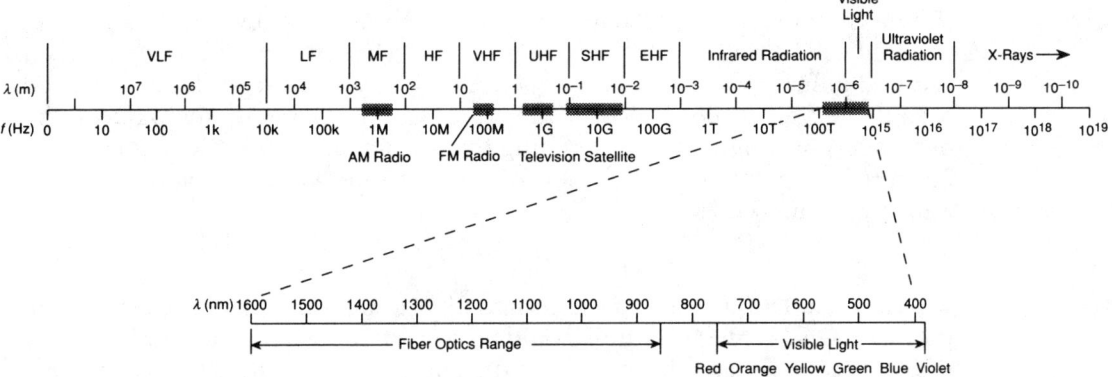

**Figure 2.** The electromagnetic spectrum.

The frequency content of the optical spectrum, although not standardized, is approximately between $7.5 \times 10^{10}$ Hz and $6.0 \times 10^{16}$ Hz and makes up only a small portion of the electromagnetic spectrum. The optical spectrum is divided into three bands called the ultraviolet band, visible band, and infrared band. The boundaries for these bands are not abrupt but are actually rather hazy. The ultraviolet band is not visible to the human eye and consists of frequencies from approximately $7.69 \times 10^{14}$ Hz to $6.0 \times 10^{16}$ Hz, which corresponds to wavelengths of 0.390 to 0.005 $\mu$m, respectively. The visible band is that part which can be seen by the human eye and is made up of frequencies from about $4.0 \times 10^{14}$ Hz to $7.69 \times 10^{14}$ Hz, which corresponds to wavelengths of 0.7500 to 0.3900 $\mu$m, respectively. The infrared band is not visible to the human eye and consists of frequencies from approximately $7.5 \times 10^{10}$ Hz to $4.0 \times 10^{14}$ Hz, which corresponds to 4000 to 0.7500 $\mu$m, respectively.

An overlapping color spectrum lies within the visible spectrum, with violet at the lower end (smaller wavelengths) and red at the upper end (larger wavelengths). Between these extremes are blue, green, yellow, and orange. Two beams of light of the same frequency are seen as the same color, but not all colors consist of a single beam of light of the corresponding frequency. It is possible for a mixture of two colors to create a third color; for example, red and green make blue.

Wavelength, as opposed to frequency, is a common way of describing electromagnetic waves. The wavelength is dependent on the frequency ($f$) and the velocity ($v$) of the wave:

$$\lambda = v/f,$$

where $\lambda$ is the wavelength of the wave. The velocity of an electromagnetic wave, such as light, in free space is approximately $3 \times 10^8$ m/s, although it will vary slightly in different materials.

## Diffraction

If light illuminates an obstacle in which there is a small opening or aperture, the light will be deflected at the edges as shown in Figure 3. This phenomenon is called diffraction and is a consequence of the wave nature of light. As can be seen in the figure, the shadow of the obstacle is not perfectly sharp due to the diffracted portion of light penetrating into the shadow area. The angle of diffraction is dependent on the ratio of the aperture width ($w$) and the wavelength of light ($\lambda$). Hence,

$$\theta = 2\lambda/w,$$

where $w > \lambda$. Diffraction is a concern to fiber optics designers because it constitutes optical power loss and should be considered when determining losses in fiber-optic cable, splices, connectors and couplers. Diffraction occurs not only with light but also with acoustical signals and other sources of wave propagation.

## Reflection and Refraction

When light strikes the surface of a medium different from which it originated, it is split into two components where it is both reflected from the surface and refracted as it enters the new medium. This is shown in Figure 4. The angle at which the light is reflected (angle of reflection: $\theta_r$) is equal to the incident angle, $\theta_i$. The angle of refraction is dependent on both the incident angle and the indexes of refraction ($n_1$ and $n_2$) of the two media. The index of refraction ($n$) is a dimensionless number representing the ratio of the velocity of light in free space to the velocity of light in a particular medium:

$$n = c/v,$$

where $c$ is the velocity of light in free space and $v$ is the velocity of light through a given medium.

It is also possible that no refraction (total reflection) may occur if the angle of incidence is less than the critical angle. The critical angle is defined to be

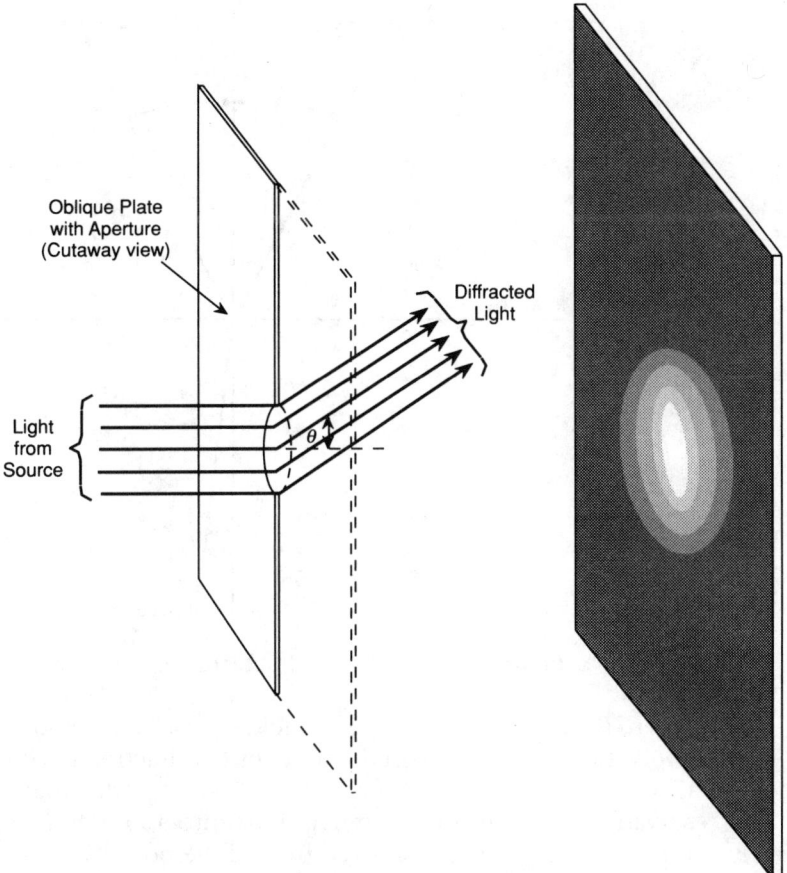

**Figure 3.** Diffraction of an incident light wave passing through a slit in an opaque plate.

the angle at which light is no longer refracted into the second medium. At the critical angle the refracted portion of the light is 90° from the boundary normal. The critical angle ($\theta_c$) is determined by the following equation:

$$\theta_c = \sin^{-1}(n_2/n_1),$$

where $n_1$ and $n_2$ are the indexes of refraction of media 1 and 2, respectively. If the angle of incidence is greater than the critical angle some of the light is refracted into the second medium at an angle of refraction dependent on the angle of incidence. The relationship between the angle of refraction and the angle of incidence was determined by a Dutch astronomer and professor of mathematics, Willebrod Snell. Snell deduced the law of refraction (Snell's law) as

$$n_1/n_2 = \sin\theta_2/\sin\theta_1,$$

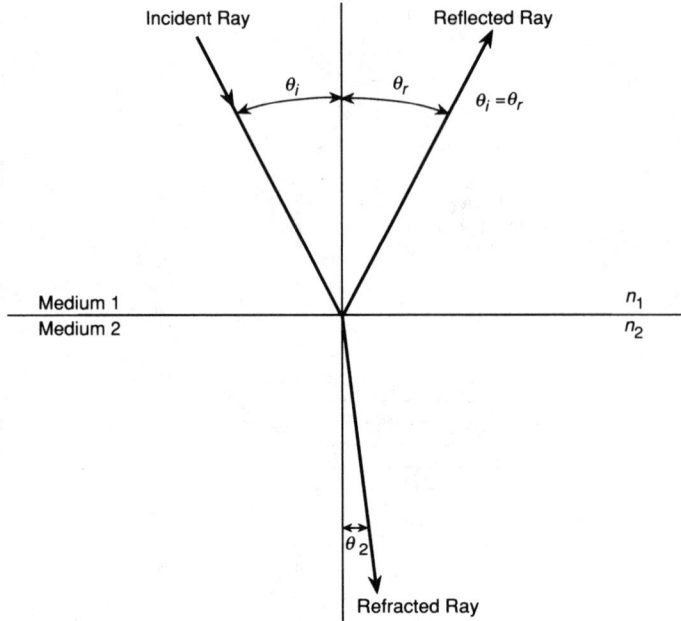

**Figure 4.** Reflection and refraction of an incident light wave.

where $\theta_1$ and $\theta_2$ are the angle of incidence and the angle of refraction, respectively. Because the ratio of the indexes of refraction is a constant, it can be seen that as the angle of incidence changes, so does the angle of refraction. An observation can be made that when light enters a medium with a higher index of refraction, the light is refracted toward the boundary normal, and if the index of refraction of the medium entered is less, the refraction is away from the boundary normal.

## Total Internal Reflection

Total internal reflection can be understood by analyzing light which strikes a medium of lesser density (lower index of refraction) as shown in Figure 5. As the angle of incidence increases a condition is reached (when $\theta_i = \theta_c$) where no refraction occurs into the second medium. Thus total internal reflection results. This forms the basis for light propagation in optical fibers. Figure 6 shows how light is guided through an optical fiber without any loss by refraction. This is accomplished by maintaining an incident angle greater than the critical angle, causing light to be totally reflected back toward the center of the fiber. Occasionally a physical deformation in an optical fiber called microbending occurs which affects a cable's ability to maintain total internal reflection. Microbending may cause the incident angle to be less than the critical angle, resulting in loss by refraction. Microbending occurs during manufacturing or by forces exerted on the fiber. Microbends caused by cable pressures exerted on the fiber can be mended by placing a loose fitting sleeve around the fiber to isolate it from the rest of the cable.

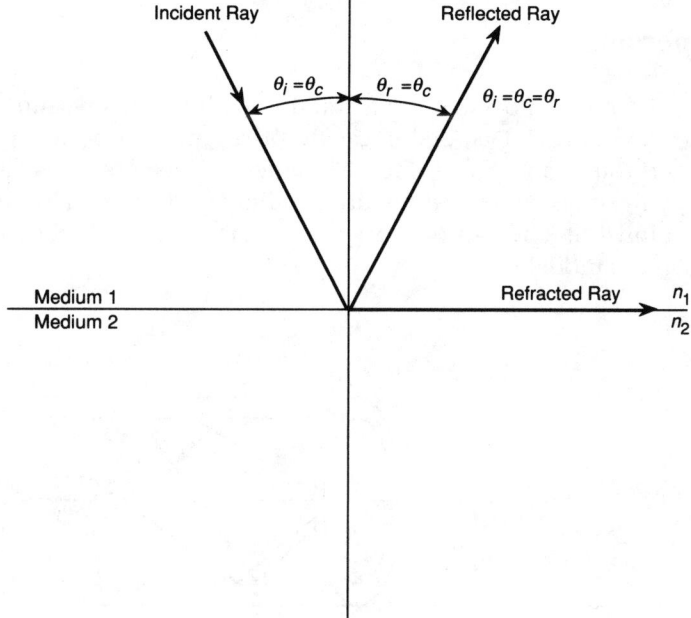

**Figure 5.** Total internal reflection occurs when the incident angle is equal to the critical angle.

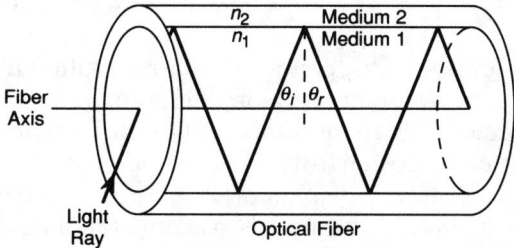

**Figure 6.** Light guided through a fiber. The incident angle is greater than the critical angle.

## The Numerical Aperture

The numerical aperture (NA) is a measure of the angle at which light may be accepted by an optical fiber. The greatest angle at which light may enter the fiber is known as the angle of acceptance. The numerical aperture is a dimensionless number determined as the sine of the angle of acceptance:

$$NA = \sin \theta.$$

At the output of a fiber, essentially 100% of the optical power is contained within this angle.

## Dispersion

Dispersion occurs due to varying path lengths within the fiber. This is shown in Figure 7. Dispersion causes broadening of optical pulses as they propagate through an optical fiber. The effect of dispersion is similar to low-pass filtering; it decreases bandwidth and limits data rates. Thus dispersion must be minimized. There are two types of dispersion in fiber optics: intermodal and intramodal.

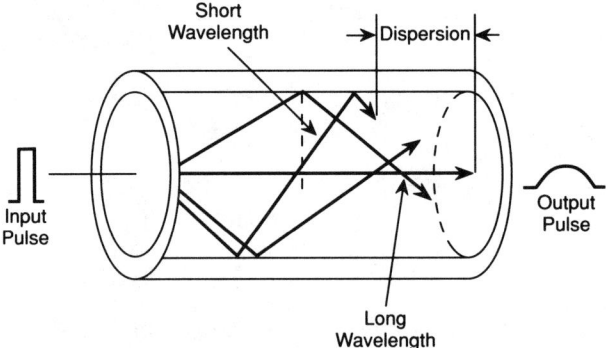

**Figure 7.** Dispersion causes pulse broadening, resulting in decreased bandwidth.

Intermodal dispersion (sometimes called modal dispersion) is due to the propagation of light rays along different paths in the optical fiber. In particular, this occurs only in multimode fiber cables where more than one path is available. A cable design called graded-index optical fiber limits intermodal dispersion. Intermodal dispersion does not occur in single-mode optical fibers because only a single path is available for optical transmission. These different fiber types are discussed below, in the section on optical fibers.

Intramodal dispersion occurs as a result of variations in the refractive index of the core and cladding as a function of wavelength. It can be traced to two sources: material dispersion and waveguide dispersion. Because the refractive indexes of the optical fiber's core and cladding are wavelength dependent, light will traverse the fiber at a rate dependent on the wavelength of the light. Dispersion results if the light source varies in wavelength. This is known as material dispersion. Material dispersion can be controlled by doping the core and cladding material for optimum performance at a desired wavelength. Waveguide dispersion is a consequence of the waveguide structure (core shape). The fiber's internal structure can be designed to cancel the effects of material and waveguide dispersions in some types of optical fibers.

## Absorption

Absorption is the process by which impurities in the fiber absorb light energy and dissipate it as small amounts of heat. Impurities such as water, copper, iron, cobalt, nickel, chromium, manganese, and vanadium cause absorption in fiber-optic cables. The impurity content in fiber-optic cables has decreased significantly since fiber optics' inception. Typical impurity concentrations are on the order of one part per billion. This is the result of improvements in clean room manufacturing techniques. At present, losses due to absorption are less than 0.2 dB/km.

## Scattering

Light energy is scattered in all directions when it strikes density irregularities in the optical fiber. A type of scattering known as Rayleigh scattering is the result of microirregularities in the fiber which occur during manufacturing. Unlike absorption which is wavelength dependent, scattering occurs at all wavelengths. The Rayleigh scattering law states that the attenuation ($a$) due to scattering as a function of wavelength varies inversely with the fourth power of the wavelength:

$$a \propto \frac{1}{\lambda^4}.$$

Thus scattering decreases at longer wavelengths. This can be seen by the Rayleigh attenuation curve shown in Figure 8. For example, at a wavelength of 1300 nm the scattering loss reaches only 18% of the scattering loss at 850 nm. It is therefore generally advantageous to utilize larger wavelengths (lower frequencies) in a fiber-optic cable.

## Polarization

Light can be either polarized or unpolarized depending upon the nature of its electric and magnetic fields. Light emanating from a conventional light bulb is unpolarized because its electric and magnetic fields oscillate randomly. The electric and magnetic fields of polarized light oscillate in a common pattern perpendicular to the direction of light travel. As shown in Figure 9, a phase difference of 0, $\lambda/2$, $3\lambda/2$, . . . , produces linearly polarized light. A phase shift $\lambda/4$,

$3\lambda/4, \ldots$, produces circularly polarized light. Elliptical polarization results from other phase displacements.

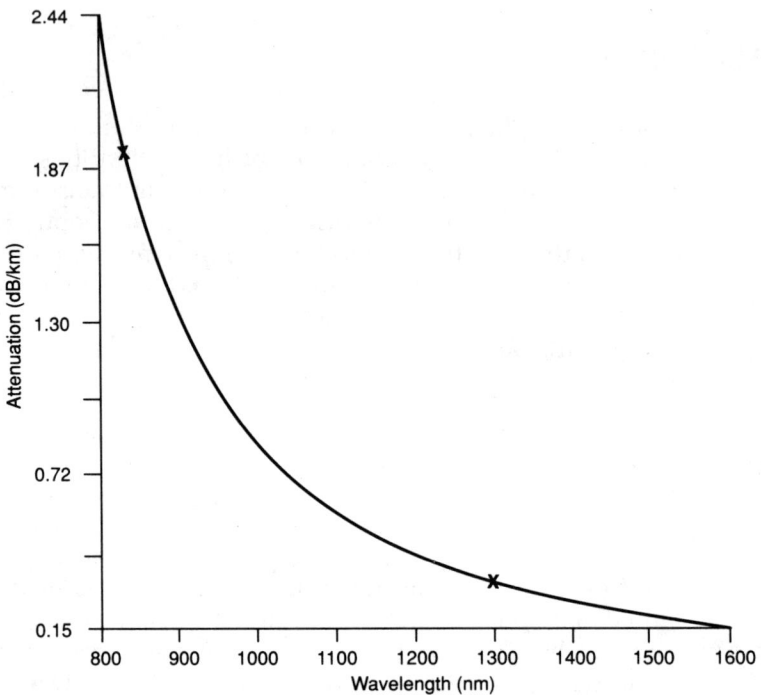

**Figure 8.** The Rayleigh attenuation curve shows attenuation as a function of wavelength.

# Optical Fibers and Cables

An optical fiber consists of three basic components called the core, cladding, and the buffer. A typical optical fiber is shown in Figure 10. The optical signal is transmitted through the core of the fiber, which has a higher index of refraction than the cladding. The greater index of refraction in the core provides continual reflection of the optical signal into the core (total internal reflection), thus minimizing losses due to refraction into the cladding. The buffer provides protection for the cladding and core. The buffer also adds additional mechanical strength to the fiber to prevent cracking and breaking.

## Optical Fibers

There are two types of optical fibers: multimode and single mode. Multimode fibers have a relatively large core diameter and provide several different paths

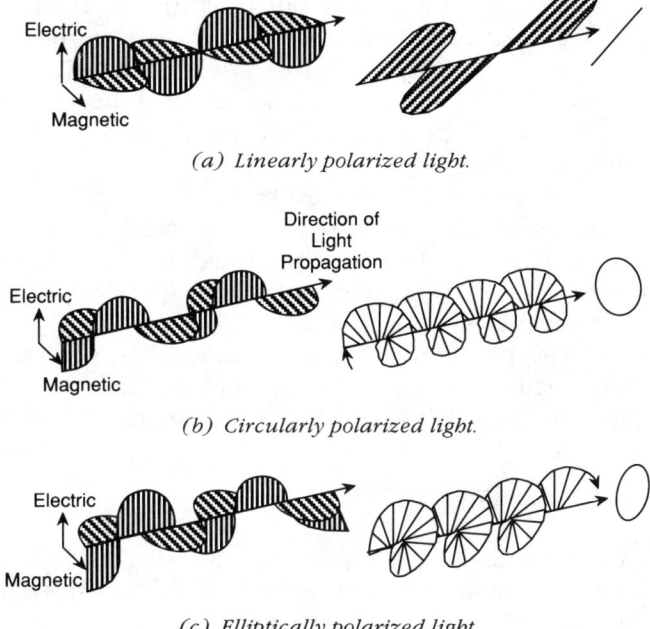

**Figure 9.** Types of polarized light.

for the optical signal to follow. The single-mode fiber is small in diameter and allows only a single path for the optical signal.

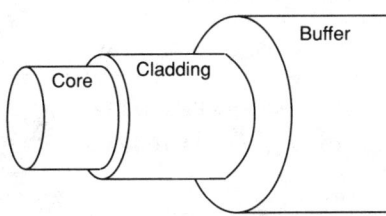

**Figure 10.** Optical fibers consist of a core, cladding, and a buffer.

## Fiber Types

Multimode fibers are available with two types of core structures (index profiles): step index and graded index. In a step-index multimode fiber the core is made of a material with a uniform index of refraction, and the cladding material has a different index of refraction. This creates an abrupt interface between the core and cladding. The advantage of this type of fiber is its ease of connection and splicing due to a large core size where alignment is not critical. A disadvantage of this fiber is that it suffers modal dispersion.

The core of a graded-index multimode fiber has a refractive index that gradually decreases toward the outer areas of the core. As light travels to the outer areas of the core it moves faster due to the lower index of refraction. This causes all modes to arrive at the output at the same time, reducing modal dispersion. This type of fiber is also simple to splice and connect due to its larger core diameter.

Single-mode fibers are normally available with a step-index profile, although some dispersion-shifted fibers are available but are not used very often. Since there is only one path through the core there is no modal dispersion. Single-mode fibers are used for high-speed applications and where long distance transmission is required. A single-mode fiber system is a highly reliable one because it utilizes fewer components, such as repeaters, than the multimode fibers. The disadvantage of single-mode fibers is the difficulty in connection and splicing due to the extremely small core diameter.

## Characteristics

A comparison of characteristics for single-mode and multimode fibers is shown in Table 1. The characteristics of the fiber determines the type of fiber best suited for a particular application. For example, the table shows that single-mode fibers are better suited for long distance transmission applications. This is because of better attenuation properties. Other important characteristics are the modal dispersion associated with multimode fiber, the greater size of multimode fiber and the option of step index or graded index for multimode fibers.

**Table 1.** Optical-Fiber Characteristics

|  | Single Mode | Multimode |
|---|---|---|
| Wavelength | 1300/1500 | 850/1300 |
| Operating Length | > 10 km | < 10 km |
| Index Profile | Step | Step, graded |
| Modal Dispersion | No | Yes |
| Attenuation | 1.0 dB/km | 3.0 dB/km |
| Typical Core/Cladding Diameter | 8/125 $\mu$m | 50/125, 65/125, 85/125, 100/140 $\mu$m |

## Fiber-Optic Cable Design

A fiber-optic cable can contain one or more optical fibers. The many possible applications of fiber optics require a variety of cable constructions. A cable consists of optical fibers, a strength rod made of steel or fiberglass, plastic tape to hold the fibers together, filler, an inner jacket, Kevlar™, and an outer jacket.

One reason why optical fibers are packaged in cable is to provide mechanical strength. High attenuation may occur due to fiber stress, tensile bending, and torsion strain. Cable is designed to protect the optical fiber from these effects as well as the effects of the environment, such as excessive heat and humidity. Cable structures are typically tailored to specific types of applications.

Tight tube or tight buffer cables have a buffer coating (sheathing) connected directly to the optical fiber as shown in Figure 11. The fibers, along with a filler material, are normally wrapped around a central strength member, which is typically made of steel or a dielectric material. Some indoor fiber-optic cables are designed without the central strength member. A layer of plastic strands covers the optical fibers, and an outer jacket is placed over the cabled structure. The tight tube design is a very compact cable, lightweight, very flexible, and crush resistant. It is most commonly used for military applications and should neither be buried nor used for aerial applications. The tight tube design is prone to microbending and typically experiences higher attenuation than other cable structures. The sheathing is difficult to remove, which makes it hard to install connections, splices, or couplers.

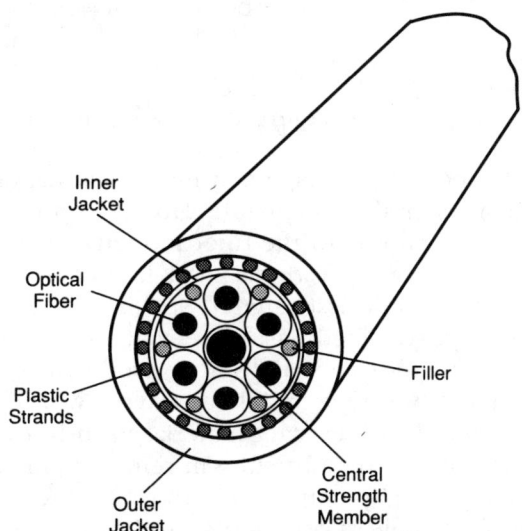

**Figure 11.** Tight tube cable structure. Commonly used for military applications.

In a loose tube cable the optical fiber is allowed to move within the cable. This freedom of movement reduces the chance of damage due to microbends or fiber expansion caused by high temperatures. Figure 12 shows two loose tube cable constructions. The first construction (Figure 12a) is a multimode loose tube design. The second construction (Figure 12b) is a single-mode loose tube design. The loose tubes are wrapped around a central strength member. An outer jacket is placed around the assembly for protection. This type of cable is commonly used for aerial installations or with additional armoring in direct buried or self-supporting aerial installations. It is possible to accommodate up to 144 fibers in a loose tube cable with provisions for up to 600 in experimental designs.

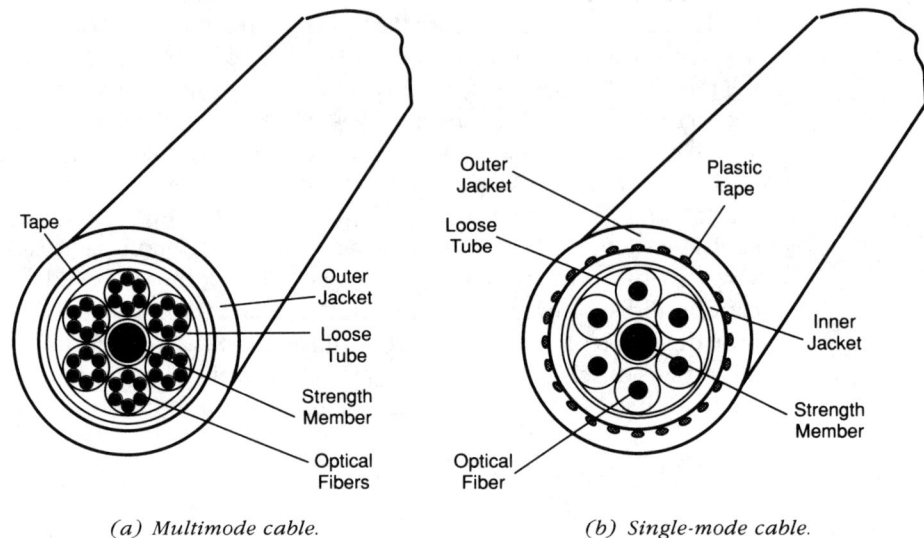

(a) Multimode cable.  (b) Single-mode cable.

**Figure 12.** Loose tube cable structures.

The loose tube design may make termination more difficult, particularly if the tube material is too pliable. However, if it is too rigid, bending or kinking may occur and render the fiber inoperative. The major disadvantages of this design are long repair times and a lack of compatible connector styles.

Plenum cable is designed specifically for use in air-handling ducts or above suspended ceilings; it meets the requirements of the National Electric Code for fire resistance and smoke producing characteristics. The cable can be either loose tube type or tight tube type. What distinguishes a cable as plenum is the special fire-resistant material used in the jacket. One disadvantage of plenum cable is the difficulty in removing the jacket for maintenance; heavy duty cutting equipment is required.

Ribbon cable is commonly used in telephone trunk lines. The fibers are packaged inside a cable sheath and held in place by a plastic tape, which is extruded over the fibers. Ribbon cable is shown in Figure 13. The ribbon cable is a weaker structure than the loose tube or the tight tube cable designs. It is not designed for applications where it may get crushed. The fibers of the ribbon cable can be connected or spliced individually or simultaneously by an array splice.

A submarine cable is designed for use under a body of water, specifically a lake, river, or ocean. It is a loose tube cable with a steel-wire armor wrapped around the cable. Connection usually takes place at a point near the shore and to a nonsubmersible loose tube cable. Special tools are required for removal of the armor sheathing. Often, submarine cable is laid in pairs because it is faster

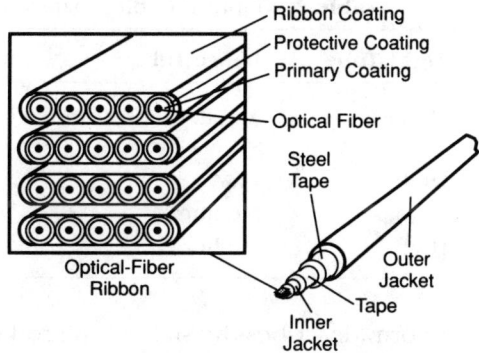

*(a) Parallel structure cable.*

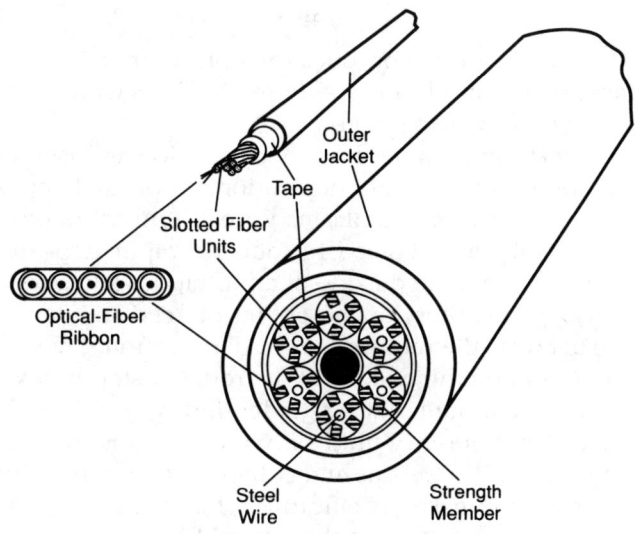

*(b) Concentric structure cable.*

**Figure 13.** Ribbon cable structures.

and easier to disconnect a malfunctioning cable and use a redundant cable than to repair a cable underwater.

Table 2 lists various characteristics for the different types of optical cable. Common applications of each cable type are provided along with mechanical properties.

## The Manufacture of Fiber-Optic Cable

In 1970, a breakthrough occurred in the production of optical fibers with the introduction of vapor phase deposition methods. This is the process of depositing layers of chemical vapor onto a preform rod in the production of optical

**Table 2.** Optical-Cable Characteristics

|  | **Tight Tube** | **Loose Tube** | **Plenum** | **Ribbon** |
|---|---|---|---|---|
| Applications | Military | Aerial, duct, buried, submarine | Duct | Aerial, duct, buried, submarine |
| Crush Resistance | High | Low | High | Medium |
| Impact Resistance | High | Medium | High | Low |
| Flexibility | High | Medium | High | Medium |

fibers. Preform describes the state of the rod before it is drawn into an optical fiber. Ultrapure silica is produced by deposition of $SiO_2$ from silicon tetrachloride by combining it with $O_2$ to produce chlorine gas:

$$SiCl_4 + O_2 \rightarrow SiO_2 + 2Cl_2.$$

The $SiO_2$ is used to produce the optical fiber. This technique is used to manufacture multimode and single-mode fibers with large bandwidths, low attenuation, and low dispersion.

Four common methods of deposition are used in the manufacture of optical fibers: outside vapor deposition, vapor axial deposition, modified chemical vapor deposition and plasma vapor chemical deposition.

As shown in Figure 14, outside vapor deposition (OVD) utilizes a gas burner to heat the outside of an ultrapure silica rod that is rotated on a lathe. Oxygen and other dopants are introduced and deposited on the outside of the rod to create the required index of refraction. These elements combine to form a fine soot on the outside of the rod. For step-index fibers the doping concentration is constant, but for graded-index fibers the doping concentration is reduced continuously until only pure $SiO_2$ is deposited. The rod is heated to 1400° to 1600° Celsius and collapsed to a thin, solid glass rod while flooding the rod with gaseous chlorine. This dries any moisture within the rod thus reducing attenuation in the optical fiber.

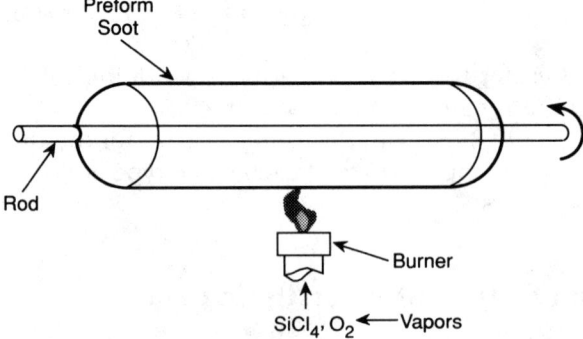

**Figure 14.** Outside vapor deposition. An ultrapure rod is coated with preform soot at specified doping levels to produce an optical fiber with the appropriate index of refraction.

The vapor axial deposition (VAD) method is very similar to the OVD method except that the burner is introduced at the end face of the rotating silica glass rod. This creates a porous preform that is drawn up the rod as shown in Figure 15. The advantage of the VAD method is the ability to produce refractive index profiles that are impossible to manufacture with the other vapor deposition methods. By utilizing two burners rather than one and varying their distances from the rod, unique index profiles are possible.

The modified chemical vapor deposition (MCVD) method also utilizes a burner to heat the glass tube, but the oxygen and gaseous dopants flow through the tube rather than onto the outside of the tube. The dopants do not react with the flame as with the OVD and VAD techniques, but rather collect inside the tube to form the inner part of the cladding and the core. Again, the type of fiber (step-index or graded-index) is determined by the doping characteristics during the deposition process. The rod is then heated to 2000° Celsius, causing it to shrink. As long as the rod is kept free from hydrogen there is no need to perform special drying procedures.

The plasma chemical vapor deposition (PCVD) method is similar to the MCVD method except for the reaction technique utilized. Figure 16 shows the PCVD method in which a plasma is created from the excitation of ionized gases. The dopant material reacts with the oxygen to form $SiO_2$. The soot particles are deposited directly on the inside of the tube, creating the cladding and

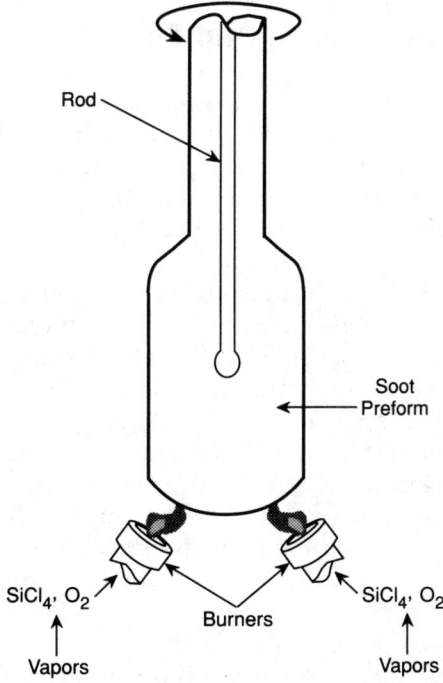

**Figure 15.** Vapor axial deposition. This method creates a porous preform unique compared with other deposition methods.

core as in the MCVD method. The resulting rod is heated, which causes it to shrink. The rod is then attached to a mount in a draw tower and heated so it can be drawn from the melting preform. To maintain a constant diameter, both the drawing speed and feed mechanism must be properly adjusted. Following the drawing of the fiber a protective coating is applied to improve the fiber's strength and to protect it from microbending.

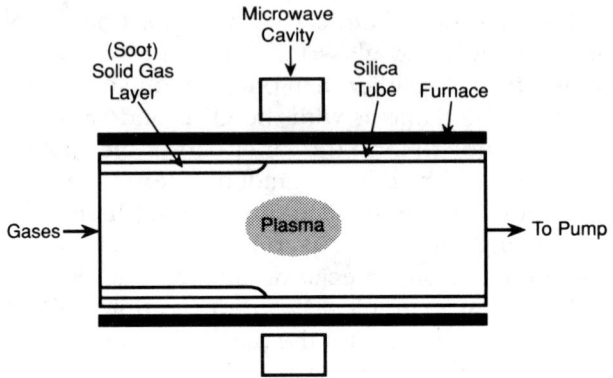

**Figure 16.** Plasma vapor deposition. The soot particles form on the inside of the tube.

## Common Failures of Optical Fibers

Several types of failures may occur in optical fibers. The most prevalent failure is fracturing. Fractures may develop during the fabrication or installation of the cable, but they occur most commonly due to construction crews and external agents such as rodents that may chew up the cable. Stress and fatigue may also cause fractures. During the cabling process the fibers are stressed; this may cause microcracks which in turn develop into fractures during installation or operation if the cable is placed under any tension. If the cable sheathing becomes damaged during installation and moisture penetrates to the fiber, the fiber lifetime can be significantly reduced. Moisture aids the propagation of microcracks which cause catastrophic failure due to mechanical breaks in the fibers. Excessive attenuation of the optical signal can be caused by stress and hydrogen diffusion which occurs when water penetrates the cable.

During installation the cable may experience excessive friction when pulled through conduit or other casings. Mechanical pullers or heavy equipment may administer excessive tension or force the cable to exceed the recommended bend radius—all of which contribute to the possible failure of the optical fiber.

A long-term concern is the effect of the environment. Excessively low or high temperatures may swell or shrink the cable jacket, which may cause mechanical failure of the fibers. The jacket may crack or break under conditions of cycling temperatures that may allow moisture to be introduced into the cable.

Jackets may be designed to prevent shrinking, swelling, or cracking and should be used for applications involving excessively harsh environments. If the application exposes the cable to the outdoors the cable jacket becomes an important component to consider when choosing a cable design. Weather, temperature, and wildlife may all contribute to the failure of the optical fibers. The failure rate of optical cable is comparable to that of coaxial transmission cable at about one failure per 146 km of cable.

## Maintenance

When failures occur in fiber-optic cables, swift detection of the point of failure is critical. An optical time domain reflectometer (OTDR) is a common diagnostics tool that identifies malfunctioning fibers and locates the exact point of failure. It is possible to repair a damaged fiber without interrupting the operation of the other fibers in the cable. Typically, if all fibers are damaged, a temporary repair may be administered; a permanent repair can be performed on each fiber while the others remain in operation. Preventative maintenance is generally not required with fiber optics after the design and installation have been completed.

# Transmitters

The fiber-optic transmitter functions as a transducer. It provides a source of light representative of the input electrical signal. The transmitter consists of four major components: emitter, coupling system, monitoring photodetector, and cooling system. An emitter generates light when given an electric current as its input. The emitter must be capable of transmitting light at an angle that provides maximum power transfer to the optical fiber. The performance of the emitter affects the entire fiber-optic system. A problem with the emitter or its coupling system will result in inaccurate output information. The most extensively used emitters for fiber-optics applications are the light-emitting diode (LED) and the laser diode. Both of these emitters are small, require low drive currents, and are capable of emitting light at the appropriate rate, wavelength, and brightness.

## The Light-Emitting Diode

The light emitted by a light-emitting diode (LED) is generated by spontaneous emission. This results in nondirectional random emission of photons at a wavelength determined by the energy gap of the semiconductor material. The LED has a longer life expectancy, provides for greater stability, operates under a wider temperature range and is less expensive than the laser diode.

## Structure

An LED is formed by a pn junction in a semiconductor material as shown in Figure 17a. Also shown in this figure are three basic LED structures: the planer-surface LED, the etched-well LED, and the edge-emitter LED.

Light is emitted from the hole shown on the semiconductor on the planar-surface type LED as shown in Figure 17b. The isolation layer is grown on top of the substrate, leaving a hole with a diameter which defines the emitting area of the LED. A resistive layer is added to provide a higher concentration of injection electrons available for recombination. The active layer is added to define the wavelength according to the compound makeup. The junction is formed at the window layer to create a larger energy gap at the n-type material than at the p-type material. The n-type material is thus transparent to light allowing light to be directed out of the device.

Because of the small emitting area, heat dissipation is an important factor in limiting internal degradation of the LED. By increasing the emitting area, heat generation is decreased along with the optical coupling efficiency. The disadvantage of this structure is that it requires a large emitting area (typically 50 to 100 $\mu$m wide) to provide adequate output power. Thus the large dimensions make it inappropriate for use with small fibers and lengthy systems. This is a low-cost device and is typically used in digital applications with data rates less than 10 Mb/s.

The etched-well LED, as shown in Figure 17c, also known as the Burrus LED after its developer, provides improved heat transfer and dissipation characteristics and therefore can operate at higher current densities. Much like the planar-surface LED, the etched-well structure contains a window, an active

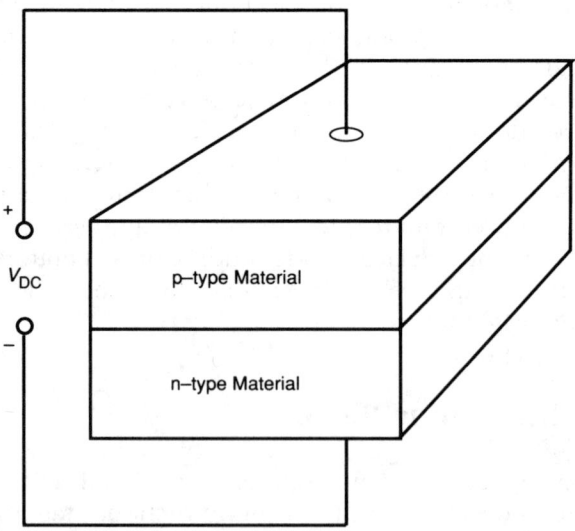

*(a) Basic LED structure.*

**Figure 17.** The light-emitting diode.

layer, and a confinement layer. The metal contact is located at the bottom, with an isolation layer maintaining a small area for electron-hole combinations. A contact layer provides good connection between the outer contact and the active layer. A well in the substrate over the isolation opening reaches the window layer, where an optical fiber is connected with an organic cement.

The etched-well LED has a small emitting area allowing for operation at higher data rates of 10 to 20 MHz. Bandwidth may be increased by controlling the impurity concentration and by making the active layer thinner but at the expense of output power. Since the active layer is close to the heat sink, dissipation is increased; this permits higher current densities to be utilized for

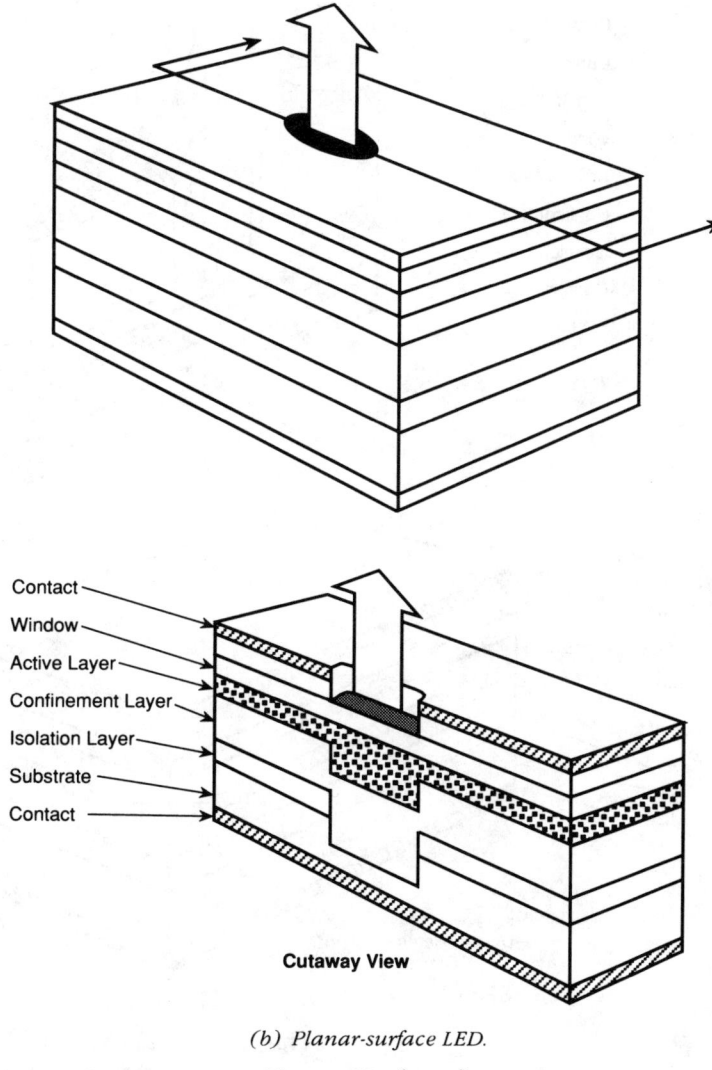

(b) *Planar-surface LED.*

**Figure 17.** (cont.)

improved output signal performance. The etched-well LED requires a manufacturing process called dual-sided alignment which increases the cost slightly over that of the planar-surface LED.

The edge-emitter double-heterojunction LED, as shown in Figure 17d, is a highly directional LED designed to provide more power than the planar-surface or the etched-well LEDs. Light is emitted from the edge rather than the surface of the LED as in the planar-surface and etched-well designs. The benefit of this structure is a higher concentration of carriers and recombinations, which are required for outputs of higher optical power.

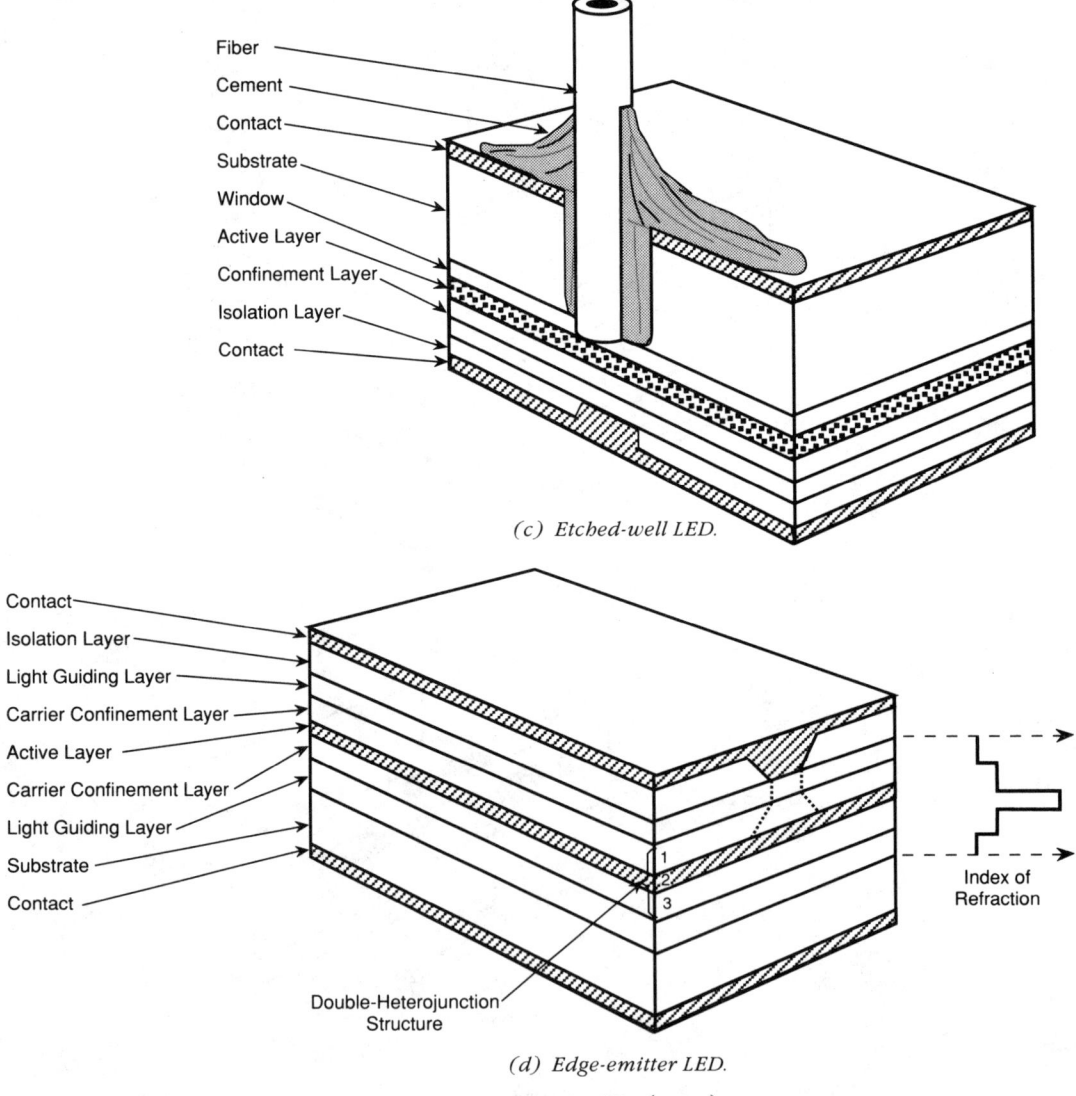

(c) Etched-well LED.

(d) Edge-emitter LED.

**Figure 17.** (cont.)

Electron-hole recombinations are limited to an area just below the isolation layer stripe which runs the length of the LED. The optical output is thus also limited to this small area producing a directional characteristic comparable to the laser diode but at 10 to 15 dB less power. This provides better coupling capability and allows higher data rates of up to 100 MHz.

The characteristics most generally considered when choosing an LED are summarized in Table 3. Light-emitting diodes used for fiber optics commonly operate at frequencies of 850, 1200, 1300, or 1550 nm.

**Table 3.** Light-Emitting Diode Characteristics

|  | AlGaAs/GaAs | GaAs | InGaAs/InP |
|---|---|---|---|
| Wavelength | 800–885 nm | 880–950 nm | 1300/1550 nm |
| Structure | Double heterostructure | Planar | Double heterostructure |
| Spectral Width | 45 nm | 50 nm | 100 nm |
| Rise Time | 10–20 ns | 20–100 ns | 5–20 ns |
| Transmission Distance at Bit Rate | 5 km at 50–100 Mb/s | 0.1 km at 10 Mb/s | 5 km at 50–200 Mb/s |

## The Laser Diode

A laser diode uses stimulated emission to generate a light source that is highly directional, coherent, monochromatic and has a relatively high power capability. The laser diode is also known as a semiconductor diode laser and injection diode laser (IDL). Discussions of the operation and various structures are provided in Chapter 4.

The characteristics most generally considered when choosing a laser transmitter for fiber-optics applications are provided in Table 4. The laser diode is commonly operated in the 850 to 880 nm wavelength range. Lasers that operate at wavelengths above 1100 nm are in development.

**Table 4.** Laser Diode Characteristics

|  | GaAlAs/GaAs | GaInAs/InP |
|---|---|---|
| Wavelength | 850–885 nm | 1300/1550 nm |
| Structure | Double heterostructure | Double heterostructure |
| Spectral Width | 3–5 nm | 3–5 nm |
| Rise Time | < 1 ns | < 1 ns |
| Transmission Distance at Bit Rate | 5–20 km at 565 Mb/s | 35 km at 1200 Mb/s |

## Transmitter Couplers

A coupler is a necessary component of the transmitter and is designed to provide maximum coupling efficiency between the emitter and the optical fiber, as well as maintain a reliable connection. Three types of transmitter couplers are commonly used.

The first type of coupler utilizes discrete optical lenses to focus the light output on the optical fiber. A collimating lens is placed in front of the emitter and a graded-index lens is positioned to focus the optical signal onto the optical fiber. This type of coupler suffers from high reflectivity back into the emitter, decreasing stability and coupling efficiency. In addition, this type of coupler tends to be more difficult to manufacture because of the many different elements and the importance of maintaining alignment. A second type of coupler utilizes only the optical properties of the fiber. In this design the fiber tip is rounded, which significantly reduces reflections. Tapering increases the coupling efficiency by increasing the area in which the light strikes the fiber. This type of coupling requires significant physical constraints in order to maintain its higher efficiency. A third type of coupler is made by butting the optical fiber against the emitter. This provides the lowest efficiency of the three couplers at approximately 10% to 12%.

## The Photodetector

The photodetector provides feedback from a laser source to the laser's drive circuitry, allowing for optimal output power without damaging the transmitter. A monitor photodiode is placed at the rear face of the laser where it monitors the light that leaves this face, which is proportional to the light at the front face. Drive current is thus altered by a feedback circuit to adjust for increasing threshold currents with aging and temperature fluctuations.

## Cooling Systems

The cooling system extracts thermal energy from the active layer junction of the emitter to prevent internal damage. Under constant power conditions the temperature of the device will rise due to the high power outputs. As the temperature of the device increases, its efficiency drops and output power decreases. The input current will rise to accommodate the drop in output power as the system strives to maintain a constant power output. This rise in input current continues the cycle by increasing the temperature, which decreases the lifetime of the emitter. An external cooler is provided to extract the excess heat from the emitter; and this results in a longer life expectancy for the emitter.

# Receivers

The function of the receiver is to convert optical signals into electrical signals for further amplification or processing. The main component of the receiver is the photodetector. Two types of photodetectors are most often utilized with fiber-optic receivers: the positive-intrinsic-negative (PIN) photodetector and the avalanche photodiode (APD).

## The PIN Photodetector

The positive-intrinsic-negative (PIN) photodetector is most commonly used for fiber optics applications. The PIN diode operates by generating electron-hole pairs due to the absorption of photons of light from the optical fiber. A common PIN structure is shown in Figure 18. Both the n- and p-type layers are heavily doped and separated by a lightly doped intrinsic layer. Since a depletion region can easily extend into a lightly doped region the PIN photodetector has a large depletion area, which increases the absorption coefficient for the device. The photodiode can then operate at a faster speed, with lower noise, and greater efficiency than conventional photodetectors. In addition the carriers may be separated by an applied electric field (5 to 12 V DC), increasing the device sensitivity. Thus an output current results from the absorption of incident photons.

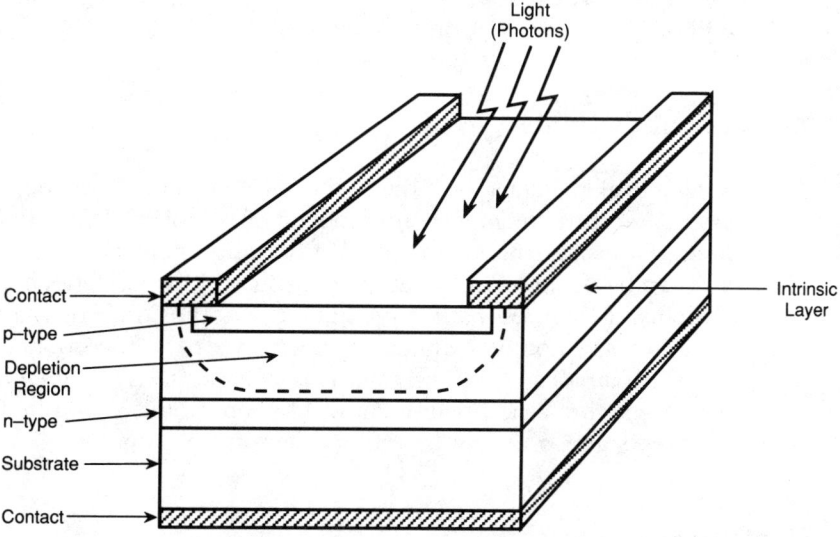

**Figure 18.** The PIN photodetector.

### The Avalanche Photodiode

The avalanche photodetector (APD) provides amplification through an applied reverse-bias electric field which provides energy to the carriers created by impinging photons. The carriers create new carriers by impact ionization. The new carriers are energized by the applied electric field and create more new carriers. This is called the avalanche effect. The amount of new carriers is controlled by the applied electric voltage. The APD structure is shown in Figure 19 along with the associated electric field distribution. Photons enter the APD at the p+ layer where electron-hole pairs are created. The carriers are separated by the weak electric field at the p+ layer and drift to the avalanche multiplication region due to an increasing electric field, resulting in current gain. For gain to occur, the applied electric field must be very high (100 to 400 V, depending upon the material).

## Fiber-Optic Switches

There are several types of optical switches available with unique advantages and disadvantages, and many new switch designs are still in developmental stages that have yet to be introduced. Most optical switches fall into one of two categories: mechanical and electro-optic. Electrical switches are also available which require the optical signal to be transformed into an electrical signal prior to switching and then returned to the optical domain after switching. Electrical switches are covered extensively in other literature and therefore will not be considered in this text.

### Mechanical Switches

Mechanical switches were the first optical switches to be developed for optical systems and are the most common type of switch available. Mechanical switches operate on either the principle of fiber alignment or prism/mirror movement and can be designed for a variety of switching combinations. In alignment type switches the optical path is changed to another fiber by a magnetic solenoid, sometimes remotely operated. Prisms may also be incorporated into a switch to alter the path of an optical signal. This switch acts as a single-pole, double-throw switch in which the prism is moved by a remotely controlled magnet to intercept the signal and reroute it to a different fiber output.

### Electro-Optic Switches

The electro-optic switch (often referred to as a photonic switch) is a recent development, and therefore few models are available. Figure 20 shows an

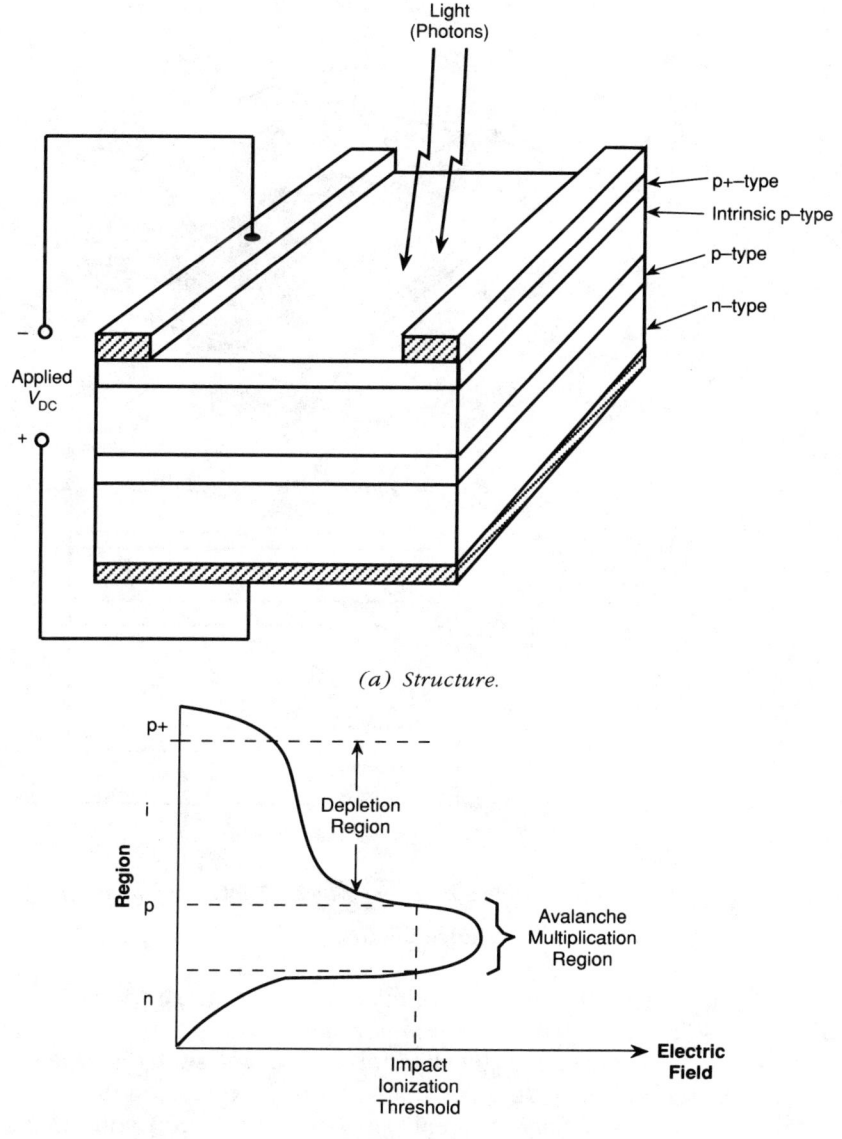

*(a) Structure.*

*(b) Electric field strength.*

**Figure 19.** Avalanche photodetector and electric field distribution.

electro-optic switch with a substrate made of $LiNbO_3$, which changes its index of refraction when placed in an electric field. The channel is doped with a substance such as titanium; this makes the index of refraction of the channel greater than the index of refraction of the substrate. In absence of an applied voltage at the electrodes the channel acts as a normal optical waveguide passing the input signal directly through to the output. With a voltage applied to the electrodes the channel index of refraction becomes less than that of the

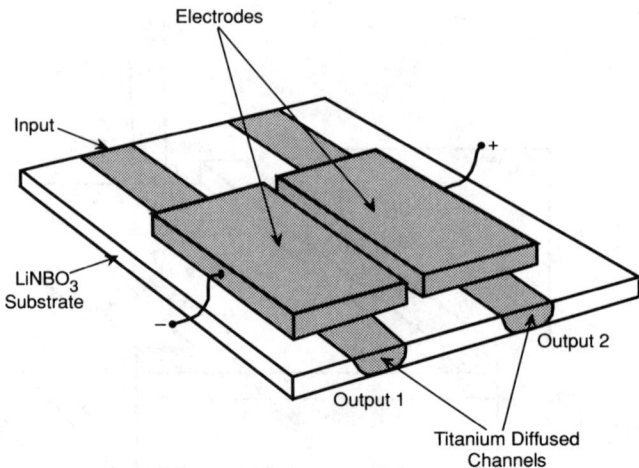

*(a) Titanium diffused channels within LiNbO₃ substrate.*

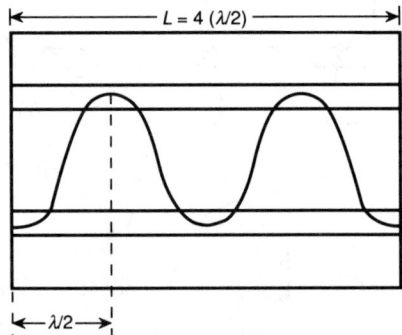

*(b) Channel length is an even multiple (4) of the signal wavelength divided by two.*

**Figure 20.** Integrated electro-optic switch.

substrate, allowing for optical coupling to occur between the channels of the switch. The channels no longer maintain waveguide characteristics. The propagation constant of the channels are altered such that one will increase and the other will decrease. The mathematical expression describing the propagation constant consists of a real part (the attenuation constant) and the imaginary part (the phase constant), hence two effects may be noted when the voltage is applied: The amount of power coupled between the two channels decreases, and the light entering the switch experiences a phase change which resists propagation to the extent that switching is allowed to occur. The length of the substrate is designed to be an even multiple of the optical wavelength divided by two. In this way when no voltage is applied the input signal exits the switch in the same channel. When the length of the substrate is an odd multiple of the wavelength divided by two, such as when an electric field is applied, the optical signal is switched to the adjacent channel. Figure 21 shows the switching effect when the voltage is applied to the electrodes.

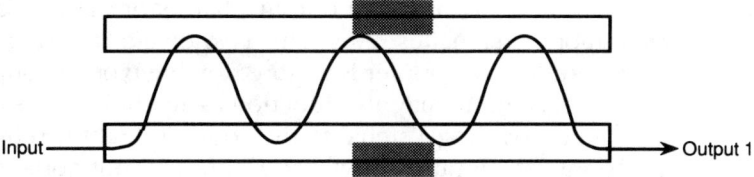

*(a) Without an applied electric field.*

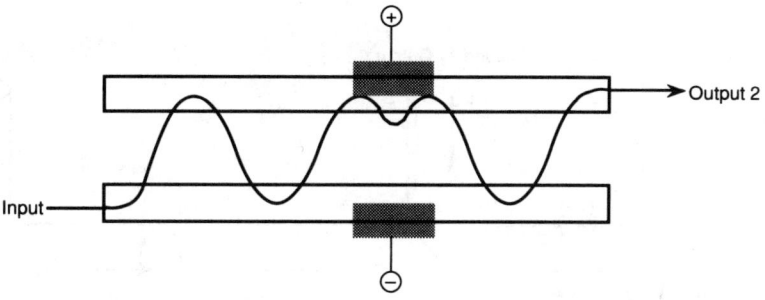

*(b) With an applied electric field.*

**Figure 21.** Spatial coupling between adjacent channels in an electro-optic switch.

# Optical Couplers

There are two general types of optical couplers: directional couplers and star couplers. Directional couplers connect three or four ports to separate or combine signals. In a three-port coupler a single input is split into two outputs by a partial mirror as shown in Figure 22a. It allows part of the signal to pass through one output while part of the signal is reflected to the other output. The fibers are joined to the coupler with optical connectors. This type of coupler can be utilized for either step-index or graded-index fibers. It can also be used to combine two signals as shown in Figure 22b. This coupler is often used to monitor a signal in an optical fiber. It may be used to tap into a fiber, providing a secondary output with only 5% to 10% of the input signal being routed to the secondary output. This secondary signal may be used to monitor the input signal or provide feedback to a control source.

Star couplers are designed to evenly distribute the input signal(s) among many outputs. They are most commonly used in data networks as central distribution points. There are two types of star couplers: transmissive and reflective. A transmissive coupler confines the input signal and distributes it over the surface of the output plate, where it is received by the output ports as shown in Figure 23a. This allows all output ports to receive all inputs to the coupler.

Since all of the light at the output plate is not received by an output port a fraction of the signal escapes. This is known as packing fraction loss.

The reflective coupler is a more flexible type of coupler. As shown in Figure 23b, each input may also function as an output. This coupler operates with one input providing a signal to the coupler. Light is reflected off the rear surface toward the input face, where all of the input ports receive equal portions of the reflected signal. The input ports also act as output ports. A disadvantage of this concept is an increase in noise due to optical reflections sent back to optical sources. Packing fraction losses also occur in reflective couplers.

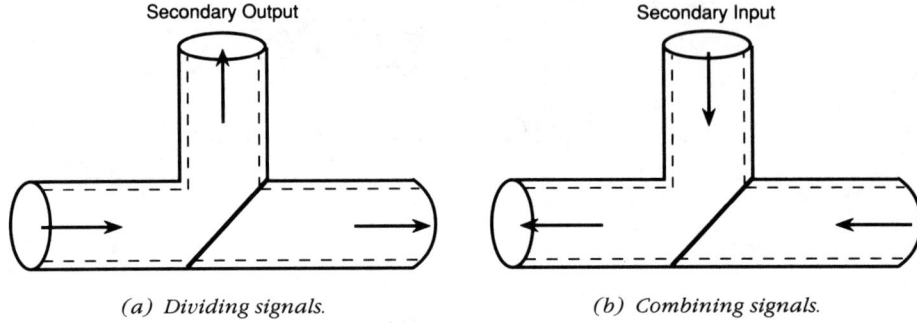

(a) Dividing signals.    (b) Combining signals.

**Figure 22.** Three-port coupler.

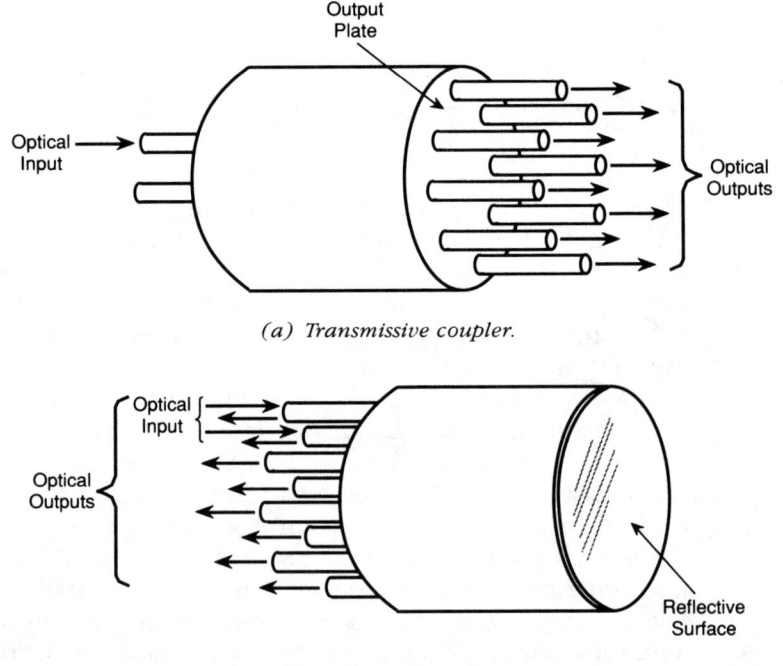

(a) Transmissive coupler.

(b) Reflective coupler.

**Figure 23.** Optical couplers.

# Multiplexers and Demultiplexers

Optical multiplexers and demultiplexers are similar to couplers and are referred to as couplers in some fiber-optics literature. Both coupling and multiplexing involve splitting and combining light beams, but couplers do so without regard to wavelength whereas multiplexers do so according to wavelength. Optical multiplexers have several optical inputs and only one optical output. All input signals are combined and transferred over a single output fiber. The demultiplexer reverses the process performed by the multiplexer. Signals transferred over a single fiber are separated according to wavelength at the demultiplexer and distributed to several output fibers. The advantages of optical multiplexing are savings in cable cost, cable weight, and cable redundancy. The multiplexing system can operate in one direction as described above or in two directions where the devices at each end of the fiber are capable of acting as both multiplexer and demultiplexer.

Optical demultiplexers perform their functions by wavelength division multiplexing (WDM). With this technique several signals of varying wavelengths may simultaneously be transmitted by an optical fiber and separated according to wavelength at the system output as shown in Figure 24. There are several ways to perform wavelength division multiplexing as described below.

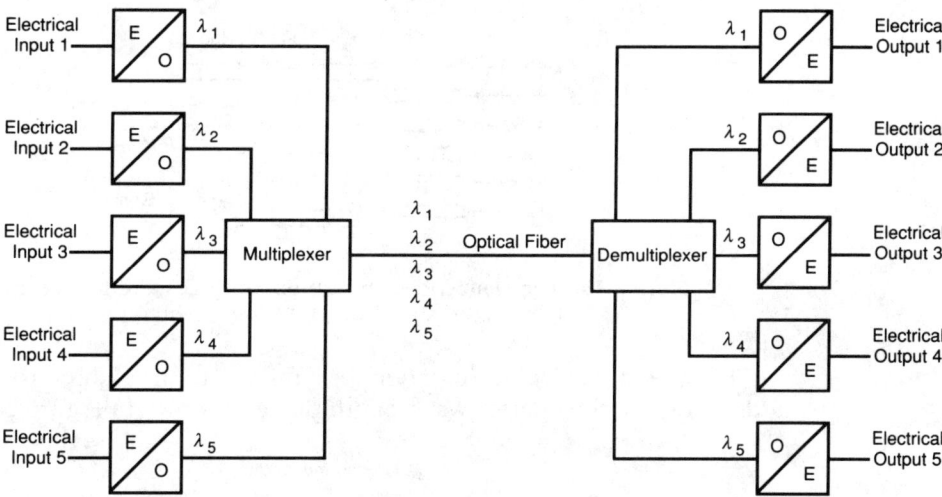

**Figure 24.** Wavelength division multiplexing allows simultaneous transmission of signals with differing wavelengths.

## Wavelength Dispersive Devices

Wavelength division multiplexing may be performed by a wavelength dispersive device. It should be noted that wavelength sensitive components are not required to combine beams of differing wavelengths (multiplexers), but they are required to separate beams of differing wavelengths (demultiplexers). By utilizing prisms and gratings, light can be separated into its spectral components. Wavelength dispersive demultiplexers are relatively uncomplicated and compact devices. They require only a few simple components to control up to three separate signals. A disadvantage of wavelength dispersive demultiplexers is their inability to achieve low losses at critical wavelengths, such as 800 nm and 1300 nm. Two types of wavelength dispersive devices are the slant rod and the prism grating.

The slant rod device shown in Figure 25 as a demultiplexer utilizes a mirrored grating plane positioned at an angle on the end of a graded-index glass rod. The length of the glass rod is a quarter wavelength; This allows the graded-index rod to collimate incident light and focus it to individual output fibers according to wavelength. As light enters the graded-index rod it is angularly dispersed as a function of wavelength due to the nature of a graded index. The light is reflected from the mirror and diffracted to various output fibers according to wavelength. The slanted rod demultiplexer operates most efficiently at a wavelength four times the length of the graded-index rod. The output at this wavelength is focused, whereas the outputs of other wavelengths experience chromatic aberrations because the focal point of these wavelengths is in front or behind the output fibers. This reduces coupling efficiency at these wavelengths.

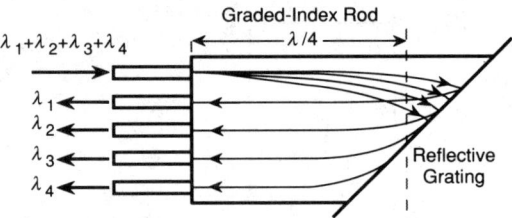

**Figure 25.** The slant rod demultiplexer utilizes a reflective grating to separate signals of differing wavelengths.

The prism grating device operates similarly to the slanted rod device, but adds a prism at the quarter-wavelength plane as shown in Figure 26. The prism

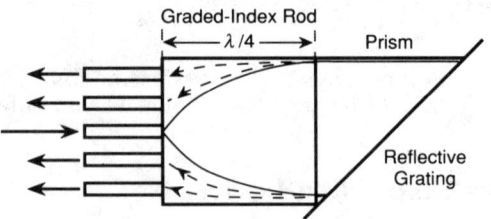

**Figure 26.** The prism grating demultiplexer incorporates a prism to improve coupling efficiency.

decreases the amount of aberration that occurs at wavelengths that do not correspond to the length of the graded-index rod. The angular dispersion is much less than the dispersion in a graded-index rod of the same length. The prism grating device therefore provides improved coupling efficiency over the slanted rod device.

## Wavelength Selective Devices

Optical filters can also be used to function as a multiplexer or demultiplexer. As shown in Figure 27, a filter can be used to separate a short-wavelength signal from a long-wavelength signal. The filter is made up of alternating layers of dielectric materials of high and low refractive indexes. The cutoff wavelength is dependent upon the angle at which the light strikes the filter. The split signals are directed to output fibers.

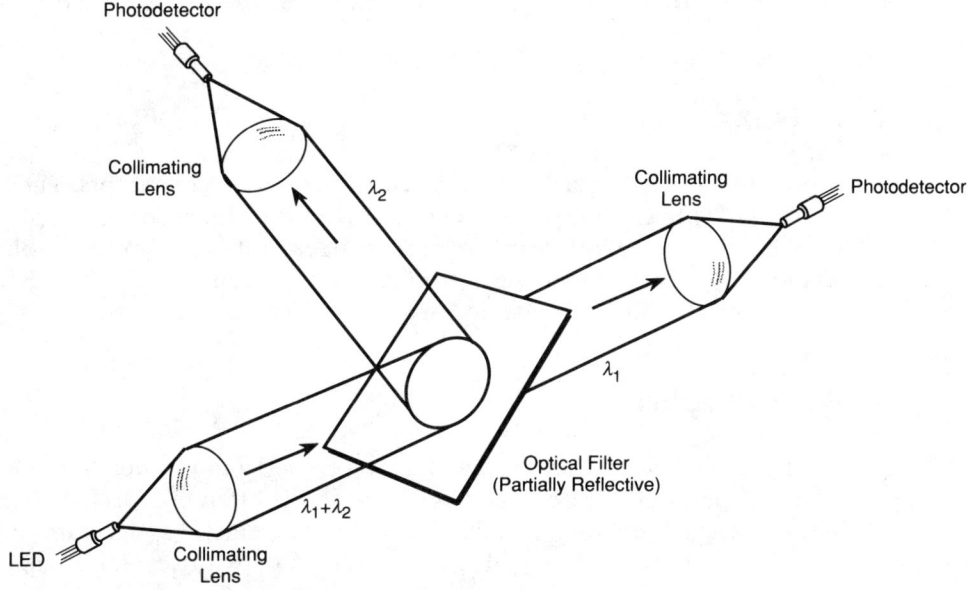

**Figure 27.** Optical filters are capable of separating signals of differing wavelengths.

A band-pass interference filter can function as a demultiplexer as shown in Figure 28. The band-pass filter is capable of providing a narrow bandwidth of less than 1 nm, allowing up to six-channel demultiplexing. The filter-type multiplexers and demultiplexers are capable of handling more channels than the wavelength dispersive devices, but they are more complex and require more precise alignment.

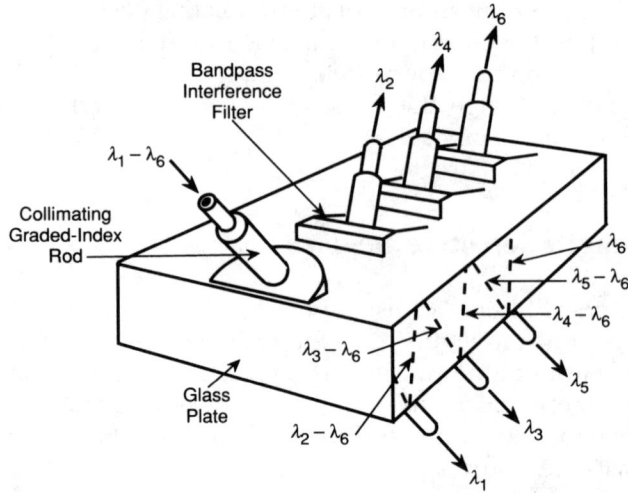

**Figure 28.** A band-pass interference filter operates as a demultiplexer.

# Optical Splicing

An optical splice is a physical connection between two fibers. There are many types of splices and splicing techniques, but the most common splices fall into one of two categories: mechanical splices and fused splices. Some requirements for good splices are low power loss (typically 0.01 to 0.5 dB), quick and easy installation, light weight, smallness, and strong construction.

## Mechanical Splices

The mechanical splice was the first to be used and is much faster and easier to make than other splices; however, today it is often used as a temporary splice to be replaced with a fused splice. The mechanical splice can be performed either as an individual splice or a mass splice with several fibers. It operates just as well with single-mode fibers as with multimode fibers. There are several types of mechanical splices for a variety of applications. For multimode single fibers three types of mechanical splices are common: V-groove splice, elastic splice, and loose tube splice.

The V-groove splice utilizes a sheet of metal that wraps around the optical fibers as shown in Figure 29. The metal is formed into a V shape and positioned to meet the fibers to be spliced together. The fibers are placed at an angle to create a force pushing the fibers together; an index-matching adhesive is applied to secure the connection. The V groove is held in place by a spring mechanism and crimped onto the outer jacket.

The elastic splicing technique is advantageous for splicing fibers of different diameters because of its inherent ability to center fibers. As shown in Figure 30, the fiber is slipped between two elastic members, one of which has a V groove centering the fiber. The fibers are secured by a cement compound within the members. The elastic members are then contained by a glass sleeve.

The loose tube splice utilizes a rectangular tube to contain the cut and polished fibers. The fibers are bent at an angle forcing the ends to butt together and settle in the interior corner of the tube. The fibers are then connected with an index-matching epoxy.

For mechanical multifiber multimode splices, two commonly used types are the ribbon splice and the silicon chip array. Figure 31 shows a ribbon splice applied to a multifiber cable. The fibers are laid in grooves on a plastic tray and held in place by a vacuum while a cover plate with identical grooves is placed over the fibers and holder. An index-matching gel is applied through a slot in the cover plate and a cover is placed over the slot.

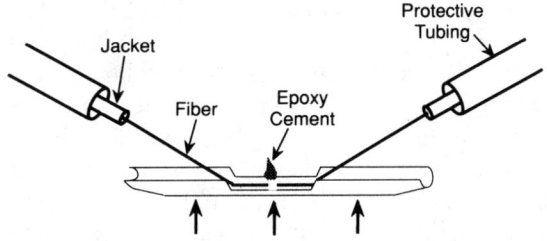

*(a) Applying adhesive to fibers.*

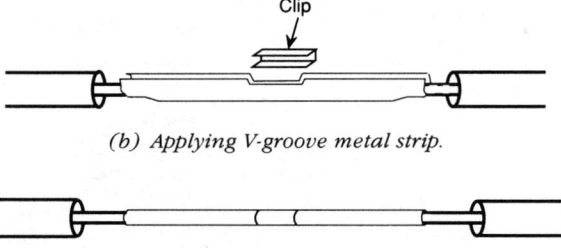

*(b) Applying V-groove metal strip.*

*(c) Finished splice.*

**Figure 29.** V-groove splice.

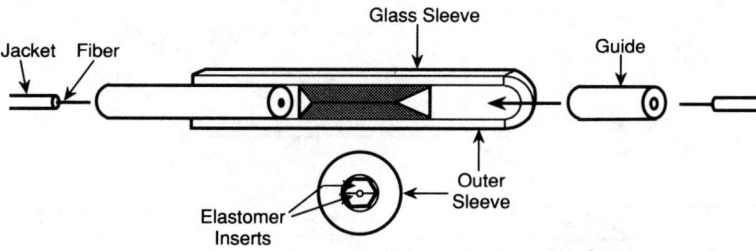

**Figure 30.** The elastic splice.

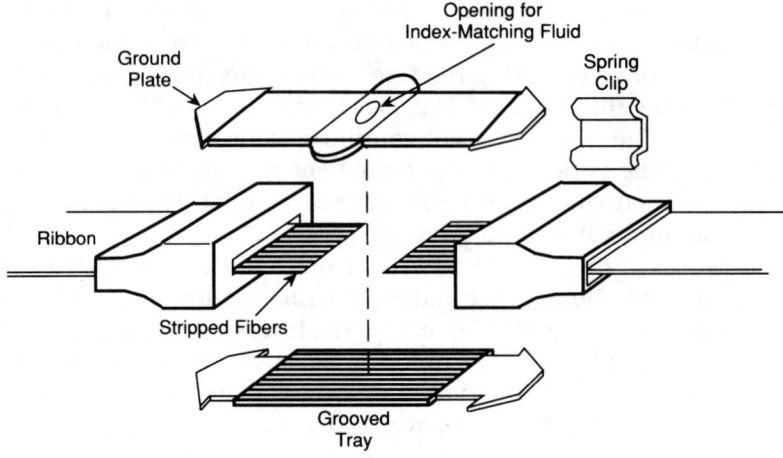

**Figure 31.** Ribbon splices are used for multifiber cables.

The silicon chip array is also used for ribbon cables; it is shown in Figure 32. The ribbon is stripped and the fibers laid in V grooves in a silicon chip with the ends of the fibers protruding from the chip. A second, identical chip is laid on top of the fibers and clamped with a metal clip. The same operation is performed on a second ribbon. The fiber ends are ground and polished and butted together within a silicon chip casing. This type of splice is used extensively in the telephone industry because the arrays can be assembled and stored until a

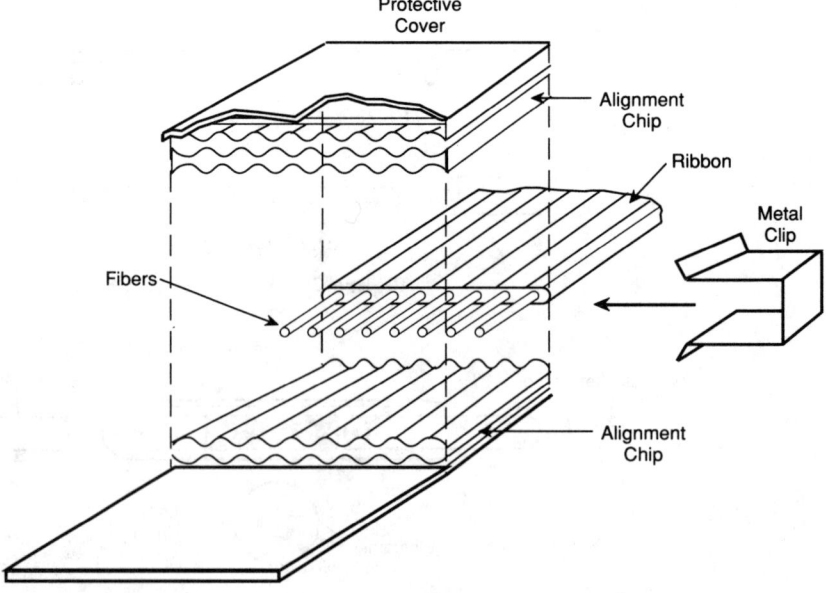

**Figure 32.** An exploded view of the silicon chip array.

repair is necessary. When preassembled the chip array splice can save time on fiber repairs.

The principal type of mechanical splice used for single-mode fibers is the rotary splice. A rotary splice is used for single-mode fibers because it is capable of accurately aligning narrow fibers. The fibers are placed in ferrules encased in an alignment sleeve which allow the ferrules to be independently rotated. By rotating the ferrules the cores of the fibers can be aligned as shown in Figure 33. An optical time domain reflectometer (OTDR) monitors the fiber while the ferrules are rotated until losses are minimized. An adhesive is used to cement the fibers in the ferrules and an index-matching epoxy is used to connect the fibers to each other. The modified three glass-rod alignment sleeve helps to provide alignment accuracy on the order of hundredths of a micrometer. The three-rod sleeve has a built-in offset so that as the ferrules rotate, the centers of their cores cross each other and alignment is detected by the OTDR.

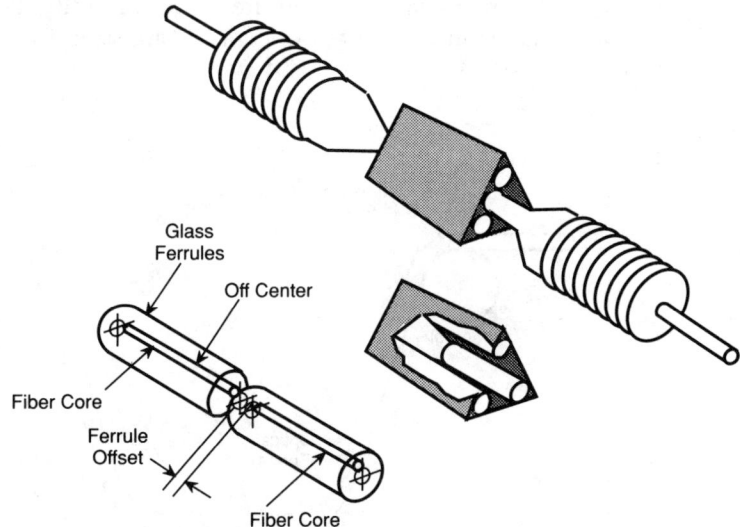

**Figure 33.** A rotary splice is used for single-mode fibers.

## Fused Splices

As noted, the fused splice is a permanent splice in which two fibers are fused into a continuous fiber. Typically, the fiber coating is removed, the fiber ends prepared and butted together and quickly heated to the fiber's melting point with a gas flame or electric arc. At the melting point fusion occurs between the two fibers. Slight core deformation will develop due to the fusion process, and unavoidable small losses will result. Fusion splicing is performed by an automated machine which typically utilizes small electrodes to create an arc across the interface of the fibers. A flame-type fusion splicer is under development and has the advantage of being usable in ocean cable systems.

Fusion splicing falls into one of two categories: profile alignment systems (PAS) and local injection and detection (LID). The PAS method was first designed to be used for single-mode splicing. It is a precise splicing technique; it aligns the fiber cores with a very low loss (0.05 dB). Two cameras are used along with a computer, a mirror, and a light source. Collimated light passes radially through the cores of both fibers in an $x$ and $y$ direction and is detected by the two cameras. The information is analyzed by computer to locate the core centers. The computer automatically aligns the fiber cores, and the fibers are accurately fused.

The LID method is shown in Figure 34. This system utilizes the principle of microbending to inject light through the cladding and buffer material. This same principle is used to detect light after it passes through the splice. A photodetector is used to measure light at the output to maximize light transmission through the splice. When the fibers are aligned to allow maximum transmission, the fibers are fused together. Fusion splices are the smallest type of splice and remain stable in both heat and humidity. Fusion splices are designed to be faster and easier to perform and are superior to mechanical splices in terms of losses.

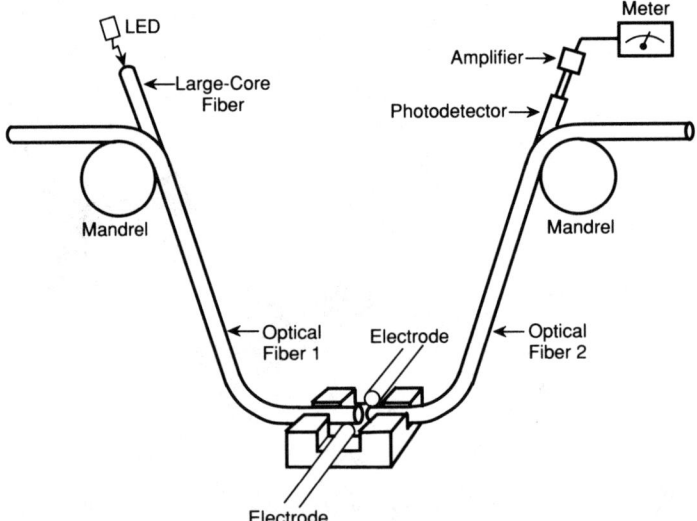

**Figure 34.** Local injection and detection (LID) arrangement used in fusion splicing.

## Connectors

Connectors, unlike splices, are designed for simple disconnection. Connector losses tend to be greater than splice losses, typically falling between 0.5 and 2.0 dB. To minimize losses it is important for connectors to be properly aligned. As shown in Figure 35, losses are related to axial offset, angular

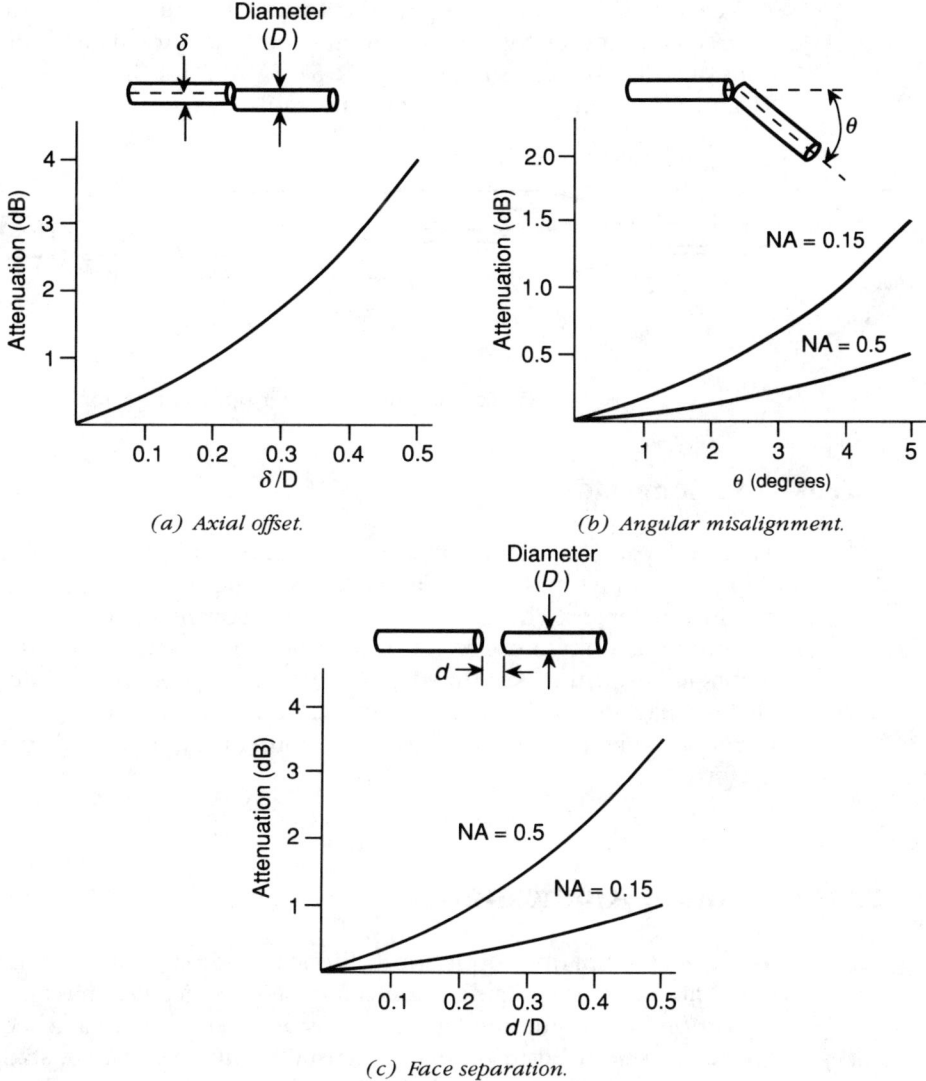

**Figure 35.** Connector losses.

misalignment, and face separation; these must be reduced to provide maximum transmission performance. Two types of connectors are commonly used: the butt-type connector and the lens connector.

## Butt-type Connectors

The butt-type connector is the most widely used connector; Figure 36 shows a typical butt connector. The ends of the fibers are ground and polished, and each is placed into a separate cylindrical ferrule. The ferrules are easily connected together with an alignment sleeve adapter. There are several variations of butt connectors including the ST, SMA, D4, FC, and PC styles. Only the ST and SMA connectors are interchangeable.

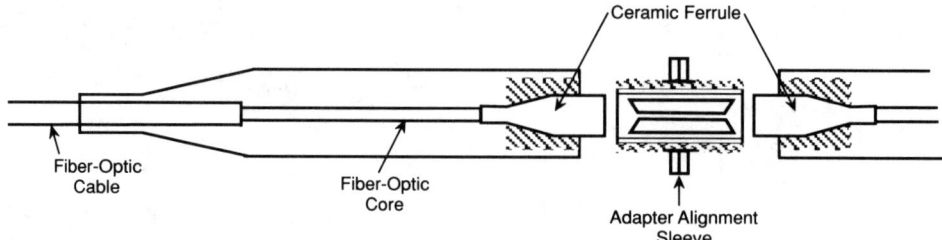

**Figure 36.** Butt-type fiber-optic connector.

## Lens-type Connectors

The lens-type connector transforms the emitted light from the sending fiber into parallel rays and then concentrates the light into a receiving fiber, using two in-line lenses. This type of connector is advantageous because larger fiber separation is tolerable. Fresnel reflections, however, cause loss, and if an index-matching fluid is utilized between lenses, debris can become trapped, also increasing losses. For most applications there are no practical advantages to using the lens-type connector; therefore, butt-type connectors are typically preferred.

# Fiber-Optic Audio Applications

There are several fiber-optic applications in audio that are currently fulfilled by conventional copper cable often referred to as snakes. Snakes carry audio information from equipment at one location to equipment at a second location. Snakes can be used in permanent installations in theaters, stadiums, arenas, amusement parks, concert halls, broadcast stations, recording studios, and postproduction studios. Concert tours often require snakes as temporary installations. Each of these venues can benefit from fiber-optic technology, particularly with the advent of digital audio.

One system designed to accommodate multiple applications provides up to 96 audio channels with 24 returns. Thus two-way transmission is available on a single fiber-optic link. The AES/EBU standard is applied to this system, providing 16-bit (or more) resolution at a standard sampling frequency. Specifications for this fiber-optic system state a signal-to-noise (S/N) ratio of 94 dB, frequency response of ±1 dB at 20 Hz to 20 kHz, and 0.01% distortion. This system can be configured as a direct, bus, star, or ring local area network (LAN) to accommodate all digital audio applications and formats.

Chapter 5: Fiber Optics

# A Design Example

A community concert hall is outfitted with a small recording studio, as shown in Figure 37, to provide recording services for recitals and other musical performances. Several problems exist in the audio system including occasional interference from a local radio station and electromagnetic interference induced by nearby power cables and overhead lighting. It is determined that the cable connecting the control room console to the microphone panel on the stage should be replaced. A fiber-optic system is selected to replace the original electrical cable. It is also determined that provisions should be made in the design to include connections in the concert hall for sound reinforcement.

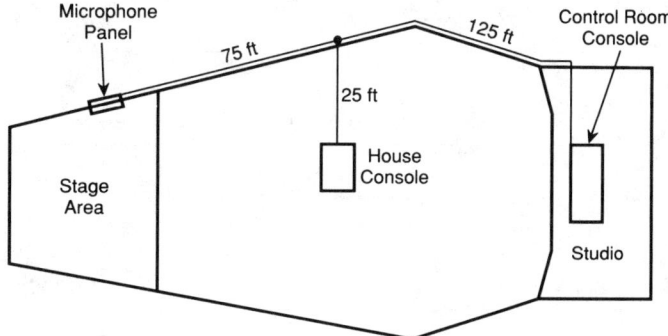

**Figure 37.** Community concert hall layout.

The system specifications should be defined prior to evaluating the individual components that will make up the system. Table 5 lists the system specifications for the community concert hall. Most of the specifications are met with the fiber-optic system's input and output electronic interfaces. The input impedance and the output impedance as well as the input and output levels are specified to allow for connections to microphones at the input and the console at the output. Errors occurring in the digital system are of primary concern. The bit error rate (BER) describes the transmission efficiency of the digital signal. If one error occurs for every $10^9$ bits transmitted, the BER is $10^{-9}$. When transmitting digital information, bit errors are dependent on the signal-to-noise (S/N) ratio. Table 6 provides values for BER in relation to the S/N ratio of the transmitted signal. The S/N ratio refers to the noise in the digital signal being optically transmitted and should not be confused with the S/N ratio in the audio signal. The optical-fiber length is an important parameter in determining the loss of the system. A fiber contributes a loss per unit length. Therefore, longer fibers contribute larger losses. Each of the optical components (switches, couplers, splices, and multiplexers) also contribute to the loss of the system.

Figure 38 shows a diagram for the system needed to transmit the audio signal to the control room console and the concert hall console. The audio

**Table 5.** System Specifications

| | |
|---|---|
| Input Impedance | 2–3 k$\Omega$ |
| Output Impedance | 50–200 k$\Omega$ |
| Bit Error Rate | $10^{-9}$ |
| Input Level | −60 dB to +4 dB |
| Output Level | +4 dB |
| Distance from Microphone Panel: | |
|     To the control room console | 200 ft |
|     To the house console | 100 ft |
| Transmission | Digital |
| A/D Converter | 18-bit sigma-delta |
| D/A Converter | 18-bit PCM |

**Table 6.** Bit Error Rate versus S/N Ratio

| Bit Error Rate | S/N Ratio |
|---|---|
| $10^{-2}$ | 13.5 |
| $10^{-3}$ | 16.0 |
| $10^{-4}$ | 17.5 |
| $10^{-5}$ | 18.7 |
| $10^{-6}$ | 19.6 |
| $10^{-7}$ | 20.3 |
| $10^{-8}$ | 21.0 |
| $10^{-9}$ | 21.6 |
| $10^{-10}$ | 22.0 |
| $10^{-11}$ | 22.2 |

signal will be processed by analog-to-digital (A/D) converters and transmitted digitally from the microphone panel to the console where it will be converted back to an analog signal by digital-to-analog (D/A) converters. The system involves more than just optical components, and there are several ways to realize the system. The designer must choose the optimal system based on performance and cost. For example, multiplexing can be performed in the optical domain but would require a transmitter and receiver for every channel, thus increasing the cost of the system and decreasing the reliability with additional hardware.

The optical system consisting of the transmitter, optical cable, coupler, connectors, and receivers is provided an input electrical signal from a multiplexer. The transmitter receives the signal and converts it into an optical signal to be transferred by optical fiber to the coupler which splits the signal into two paths. One path leads to a receiver at the control room console, and the other leads to the concert hall console. Chart 1 provides a summary of the optical components selected for the community concert hall.

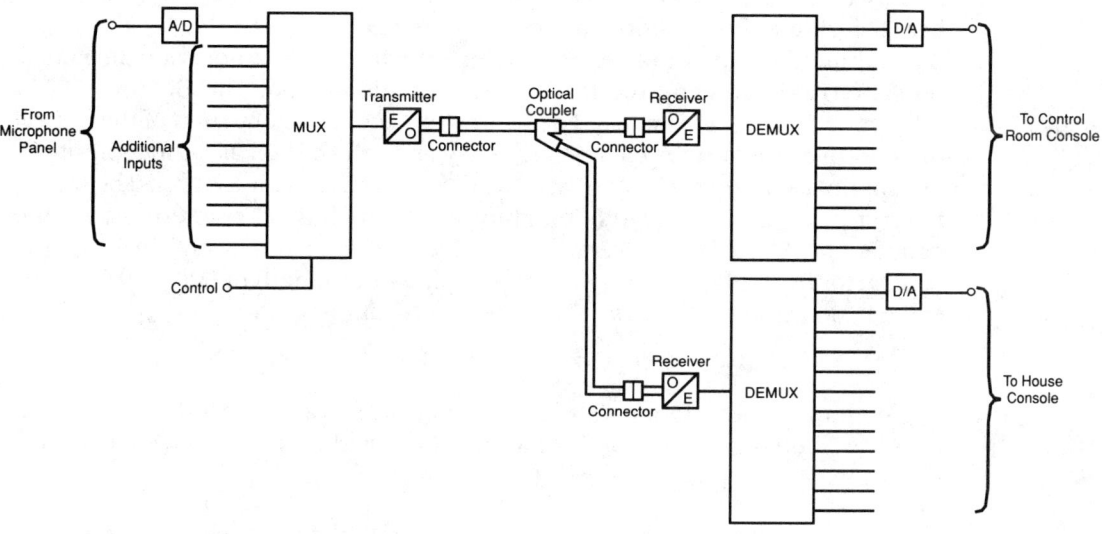

**Figure 38.** Block diagram of community concert hall fiber-optic system.

**Chart 1.** Community Concert Hall Fiber-Optic Components

| | | | | |
|---|---|---|---|---|
| Fiber-Optic Cable | Manufacturer and Part Number: Belden 227101 | Core/Cladding Diameter: 50/125 μm | Attenuation: 5.5 dB/km | Operating Wavelength: 850 nm | Bandfactor: 400 MHz · km |
| Transmitter | Manufacturer and Part Number: TXES 488 | Source: LED | Wavelength: 850 nm | Bandwidth: 44 Mb/s | Output Power: −8 dB |
| Receiver | Manufacturer and Part Number: Augat 698 127 DG1 | Detector Type: Pin | Optical Sensitivity: −26 dBm | Data Format: 40 Mb/s | NRZ |
| Coupler | Type: Directional splitter | Loss 4.0 dB | | |
| Connector | Type: Cylindrical sleeve | Loss: 1.0 dB | | |

There are several specifications to consider when choosing an optical cable. The index type, mode, core diameter, bandwidth, attenuation and wavelength specifications should be evaluated to determine the proper cable for the application. For this system a digital signal is transmitted in one direction in a serial format. A single fiber can easily handle this load. If required, returns could be incorporated in the same fiber. The multimode step-index fiber is subject to dispersion and therefore not appropriate for this application. The multimode graded-index fiber is designed for short and intermediate length systems and limits dispersion to an adequate level for this application. The step-index fiber is designed for long distance systems and would function adequately in the community concert hall application, but is more expensive than the multimode fiber. The single-mode fiber is difficult to connect and splice. For these reasons the multimode graded-index fiber is selected. The fiber must provide the largest possible bandwidth and the least amount of attenuation. The Belden 227101 cable meets the requirements. The bandwidth and the attenuation of the optical fiber are important fiber parameters. They are determined as a function of the optical-fiber length from information provided by the manufacturer. The manufacturer provides a bandwidth-length product which is often referred to as the bandfactor. The fiber's overall bandwidth is determined by dividing the bandfactor by the length of optical fiber. For a fiber length of 300 ft (0.091 km) and a bandfactor of 400 MHz · km, the bandwidth is

$$BW = 400 \text{ MHz} \cdot \text{km}/0.091 \text{ km} = 4.396 \text{ GHz}.$$

The electrical equivalent bandwidth is equal to the optical bandwidth divided by 1.41. Therefore, the cable's electrical bandwidth is 3.118 GHz. The total system bandwidth is determined by

$$\frac{1}{BW_{tot}^2} = \frac{1}{BW_T^2} + \frac{1}{BW_R^2} + \frac{1}{BW_C^2},$$

where $BW_T$, $BW_R$, and $BW_C$ represent the bandwidths of the transmitter, receiver, and cable, respectively. The term $BW_{tot}$ represents the total bandwidth of the system. Here, $BW_T$ is 40 MHz, $BW_R$ is 44 MHz, and $BW_C$ is determined to be 3.118 GHz. The total bandwidth, $BW_{tot}$ can be calculated to be 29.6 MHz. Obviously, this would easily accommodate the needs of the community center concert hall. The power of the system must be budgeted to ensure adequate signal at the receiver. The power input is determined by the transmitter manufacturer, and the required power at the receiver is specified by the receiver's manufacturer. The attenuation of the signal is determined as follows.

The attenuation in the fiber as a function of fiber length is provided by the manufacturer as a specification. The attenuation for the selected cable is 5.5 dB/km. The total attenuation is therefore

$$5.5 \text{ dB/km} \times 0.091 \text{ km} = 0.5 \text{ dB}.$$

In evaluating a transmitter it is vital that it meets or exceeds the provided specifications. In many cases an LED can source enough power for the system and provide stability and reliability. For example, the Texas Instruments TXES 488 provides adequate bandwidth at a wavelength of 850 nm. This is an LED-type transmitter which has a TTL input; therefore, its input impedance matches the output impedance of the multiplexer.

Similarly, the digital receiver must be evaluated according to its characteristics. The receiver to be used for the concert hall system is the Augat 698 127 DG1. This receiver provides the benefits of high sensitivity and a bandwidth large enough to match the rest of the system, and it meets the specification of operating with the nonreturn to zero (NRZ) data format. Further evaluation is required to determine the overall efficiency of the system.

The losses due to the connectors are typically supplied by the manufacturer. Typical connector losses are 0.5 to 2.0 dB. The system requires three connectors. The connector losses are determined as follows: The total connector loss is equal to the number of connectors in system times the loss per connector. In this case the total connector loss is

$$3 \times 1.0 \text{ dB} = 3.0 \text{ dB}.$$

This system has no splices, so the splice loss for this system is zero. Manufacturers of splicing equipment provide loss-per-splice information, and the total splice loss for a system is determined by multiplying the loss per splice by the number of splices. In this case the coupler which splits the signal between the house console and studio adds a 3 dB loss. An additional 1 dB loss is due to coupler connections per the manufacturer's specifications.

There are additional losses to be accounted for, such as coupling losses of the detectors and system degradation due to aging and temperature variations. Typical losses are as follows:

$$\text{detector coupling loss} = 1.0 \text{ dB/detector},$$
$$\text{detector coupling loss} = 2 \text{ detectors} \times 1.0 \text{ dB/detector} = 2.0 \text{ dB},$$
$$\text{temperature degradation loss} = 3.0 \text{ dB},$$
$$\text{aging degradation loss} = 3.0 \text{ dB}.$$

The total system loss includes the fiber loss, connector loss, splice loss, detector coupling loss, temperature degradation loss, and aging degradation loss. In this example, the total system loss is

$$0.5 \text{ dB} + 3.0 \text{ dB} + 0.0 \text{ dB} + 4.0 \text{ dB} + 2.0 \text{ dB} + 3.0 \text{ dB} + 3.0 \text{ dB} = 15.5 \text{ dB}.$$

The efficiency of the system is determined by subtracting the total system loss from the power margin. The power margin is a term used to describe the difference between the adjusted output power and the total loss of the system. The adjusted output power is determined by subtracting the detector sensitivity from the transmitter power output. Thus the adjusted output power is

$$-8.0 \text{ dB} - (-26.0 \text{ dB}) = 18.0 \text{ dB}.$$

The power margin is thus

$$18.0 \text{ dB} - 15.5 \text{ dB} = 2.5 \text{ dB}.$$

A negative power margin would mean the design is inadequate and must be changed to either utilize a better transmitter that can source more light or a receiver with greater sensitivity. Other alternatives would be a reduction of the system losses by choosing a fiber that provides better attenuation characteristics or reduction of the number of connectors and splices. The evaluation has shown that the system is adequate from an attenuation standpoint with a 2.5 dB power margin. In addition, the bandwidth of the system is sufficient. A cost evaluation may also be required but will not be considered in this example.

By incorporating fiber optics into this studio application, several common problems have been alleviated: electromagnetic interference, crosstalk, and ground loops. Bandwidth has been expanded as well as data rates. Propagation delays have been reduced along with bit error rates.

# Conclusion

As demonstrated by the preceding example, fiber optics may provide simple solutions to common audio problems that have historically been difficult to solve. In audio, where noise, transmission speed, and bandwidth are important considerations, the fiber-optic system has the upper hand on conventional copper wire. Many off-the-shelf and packaged systems are available specifically for digital audio purposes. Fiber-optic systems designed to transfer and control MIDI information, time code, computer information, and digital audio signals are becoming more prevalent, particularly in large, multistudio facilities.

One system, installed in a multistudio setting, can utilize fiber optics to link all mastering, mixing, digital editing, copy, and machine rooms. Several digital interfaces are available in packaged fiber-optic systems. The fiber-optic systems eliminate phase noise that is normally associated with digital transmission over coaxial cable. Another application would link two studios together, allowing them to expand their capabilities. Fiber optics make it possible to transmit digital information at high rates eliminating delay problems between synchronized machines in the two facilities. This application would not be possible with conventional copper wire.

Fiber-optic snakes provide all the qualities of fiber-optic transmission and digital sound quality plus are lightweight and small, which makes them easy to install and remove. Fiber optics are thus advantageous for sound reinforcement applications.

Most studios will find that fiber-optic systems can allow greater communication capabilities between pieces of studio equipment, such as keyboards, drum machines, signal processors, and computers. A single cable can carry the information necessary to simultaneously control each of these devices. Fiber optics extend the possible number of MIDI channels beyond 16 and make it possible to extend MIDI cable runs far beyond the present 50 ft limit.

The cost of fiber-optic equipment and cable is currently greater than copper wire systems, particularly for short cable runs. On a large scale, however, the two costs converge. The many benefits provided by fiber optics in digital audio applications enhance the future of both technologies and will ultimately find more applications and hence reduce the cost as the audio industry strives to meet the need for improved audio quality and transmission flexibility.

# References

Ajemian, R. G., and A. R. Grundy, "Fiber Optics, The New Medium for Audio: A Tutorial," *Journal of the Audio Engineering Society*, vol. 38, no. 3, March 1990.

Allard, F. C., *Fiber Optics Handbook for Engineers and Scientists*, McGraw-Hill Book Co., 1990.

Baack, C., *Optical Wideband Transmission Systems*, CRC Press, 1986.

Baker, D. G., *Fiber Optic Design and Applications*, Reston Publishing Co., 1985.

——, *Monomode Fiber Optic Design with Local-Area Network Applications*, Van Nostrand Reinhold, 1987.

Belden Wire and Cable, *Guide to Fiber Optics System Design*, 1987.

Cancellieri, G., and U. Ravaiola, *Measurements of Optical Fibers and Devices: Theory and Experiments*, Artech House, 1984.

Chaffee, C. D., *The Rewiring of America: The Fiber Optics Revolution*, Academic Press, 1988.

Cheo, P. K., *Fiber Optics, Devices and Systems*, Prentice-Hall, 1985.

Cherin, A. H., *An Introduction to Optical Fibers*, McGraw-Hill Book Co., 1983.

Christian, N. L., and L. K. Passauer, *Fiber Optic Component Design, Fabrication, Testing, Operation, Reliability and Maintainability*, Noyes Data Corporation, 1989.

Daley, J. C., *Fiber Optics*, CRC Press, 1984.

Geckeler, S., *Optical Transmission Systems*, Artech House, 1987.

Halley, P., *Fibre Optic Systems*, John Wiley & Sons, 1987.

Haupt, H., *Optical Communications, ECOC 1984*, Elsevier Science Publishers B.V., 1984.

Howes, M. J., and D. V. Morgan, *Optical Fibre Communications*, John Wiley & Sons, 1980.

Izawa, T., and S. Sudo, *Optical Fibers: Materials and Fabrication*, KTK Scientific Publishers, 1987.

Jeunhomme, L. B., *Single-Mode Fiber Optics, Principles and Applications*, Marcel Dekker, 1983.

Jones, W. B., Jr., *Introduction to Optical Fiber Communication Systems*, Holt, Reinhart and Winston, 1988.

Karp, S., Gagliardi, R. M., Moran, S. E., and L. B. Stotts, *Optical Channels—Fibers, Clouds, Water, and the Atmosphere*, Plenum Press, 1988.

Keiser, G., *Optical Fiber Communications*, McGraw-Hill Book Co., 1983.

Lester Audio Laboratories, *DAS 2000 Specification Sheet*.

Lin, C., *Optoelectronic Technology and Lightwave Communications Systems*, Van Nostrand Reinhold, 1989.

Mahlke, G., and P. Gossing, *Fiber Optic Cables: Fundamentals, Cable Technology, Installation Practice*, John Wiley & Sons, 1987.

Midwinter, J. E., *Optical Fibers for Transmission*, John Wiley & Sons, 1979.

Miller, C. M., *Optical Fiber Splices and Connectors, Theory and Method*, Marcel Dekker, 1986.

Mims, F.M., III, *Light-Beam Communications*, Howard W. Sams & Co., 1975.

Monster Cable Products, Inc., *OptoDigital Lightspeed 12 System*, Specification Sheet.

Morris, D. J., *Pulse Code Formats for Fiber Optical Data Communications*, Marcel Dekker, 1983.

Murata, H., *Handbook of Optical Fibers and Cables*, Marcel Dekker, 1988.

National Photonics, *Sidewinder E.N.G. Video System*, Data Sheet.

Palais, J. C., *Fiber Optic Communications*, Prentice-Hall, 1984.

Personick, S. D., *Fiber Optics Technology and Applications*, Plenum Press, 1985.

Personick, S. D., *Optical Fiber Transmission Systems*, Plenum Press, 1981.

Pohlmann, K. C., *Principles of Digital Audio*, 2nd edition, Howard W. Sams & Co., 1989.

Sakura, *et al.*, "Fiber Optics Links for Digital Audio Interface," *IEEE Transactions on Consumer Electronics*, vol. CE-34, August, 1988.

Seippel, R. G., *Fiber Optics*, Reston Publishing Co., 1984.

Senior, J. M., *Optical Fiber Communications, Principles and Practice*, Prentice-Hall, 1985.

Seumatsu, Y., "Optical Devices and Fibers," *Japan Annual Review in Electronics, Computers and Telecommunications*, vol. 5, Ohmsha Ltd and North-Holland Publishing Co., 1983.

Wadia Digital Corp., *Wadia DigiLink 20 Fiber Optic Transmission System*, Specification Sheet.

*Chapter 6*

# Optical Disc Technology
### Brent A. Karley

## Introduction

The means for recording and reproducing information have evolved through several stages including mechanical, magnetic, and, more recently, optical technology. Advancements in technology have led to improvements in the quality of music reproduction and in the efficiency of recording systems. In pursuit of the ideal recording system, the following characteristics have been targeted for improvement: reproduction quality, ease of recording and reproduction, recording capacity, cost, durability of recording medium, access time, physical size of system, and transportability of medium.

## Recording Systems

The Edison cylinder was the first mechanical recording system. Mechanical recording systems have improved with the application of electronics. For example, Edison's original wax cylinder evolved into a more efficient medium—the vinyl disc. The development of a stereo disc format and increased information density advanced the mechanical medium to a technically sophisticated level. However, this medium has probably reached the limit of its information capacity. The mechanical disc recorder is no longer utilized as a means for conventional recording, and the vinyl disc is rapidly declining as a major distribution medium.

Contemporary recordings have been dominated by magnetic recording systems, such as the multitrack recorder, the cassette recorder, and other two-track systems. Magnetic recorders provide improved recording quality, repeated recording capability, ease of recording, multitrack recording capabilities, and editing capabilities. Magnetic media have proved to be versatile, and they have been utilized in a variety of applications in the audio industry. Digital audio was introduced on a magnetic medium in 1967. The first PCM information was encoded on a 1-inch magnetic tape by a two-head helical scan video tape recorder. The introduction of digital audio tape (DAT) has brought magnetic media to the forefront of digital audio technology. Magnetic media (such as the floppy disk and hard drive) dominate the computer industry and have also been successfully incorporated into contemporary audio recording settings.

The most recently introduced means of recording is optical recording. This technology utilizes unique properties of light to read, write, and/or erase information stored on optical discs, tapes, or cards. Discs are the preferred optical medium for digital audio applications primarily because of their random access capabilities and portability. Optical discs are capable of storing a variety of digital information, such as computer software or data, digital audio information, and MIDI or time code information.

Magnetic systems are years ahead of optical systems in technical development. Optical systems, however, provide several advantages over their magnetic counterpart. The storage density of optical media is about thirty times greater than magnetic media, and the cost of storage is less. The optical medium is more durable as it is less susceptible to heat, humidity, dust, fingerprints, magnetic fields, and vibrational shocks. The lifetime of the optical medium is much greater than magnetic discs and tapes. Optical systems also allow for portability of large amounts of information, providing convenience to the user. Above all, high data densities are provided by optical media. A single optical disc may be capable of storing as much data as some multidisk hard drives.

The compact disc is the audio standard for optical playback systems. In addition, some optical recorders are designed to conform to its standards and format so that the recorded information may be reproduced on a conventional compact disc player.

## The Compact Disc

The compact disc has introduced more people to digital audio than any other medium. Technological advancements in optoelectronics, digital signal processing, error correction, optical scanning, modulation, and laser technology have been incorporated into the compact disc system, providing remarkable performance specifications and accuracy in sound reproduction.

### The Compact Disc Message

A compact disc contains digitally encoded audio information in the form of pits impressed into its surface. The information on the disc is read by the player's optical pickup, decoded, processed, and ultimately converted into acoustical energy. The disc is designed to allow easy access to the information by the optical system as well as provide protection for the encoded information.

## The Compact Disc Medium

Figure 1a shows the dimensions of the compact disc. The diameter of the disc is 12 cm (4.72 in) with a thickness of approximately 1.2 mm (0.047 in). The center hole is 15 mm (0.59 in) in diameter and allows for the disc to be placed on a CD player's spindle motor shaft. The compact disc player's laser beam is guided across the disc from the inside to the outside beginning at the lead-in area moving outward through the program area and ending at the outer edge with the lead-out area. The lead-in and lead-out areas are designated to provide information to control the player. The lead-in area contains a table of contents which provides information to the player, such as the number of musical selections as well as the starting points and duration of each selection. The lead-out area informs the player that the end of the disc has been reached. Data is recorded on a radius 35.5 mm (1.39 in) across (not including the lead-in and lead-out areas) and provides a maximum available recording time of 74 minutes and 33 seconds when CD specifications are followed. Although each particular CD rotates at a fixed, constant linear velocity (CLV), the CLVs used for different discs may vary from 1.2 m/s (42.7 in/s) to 1.4 m/s (55.1 in/s). The length of the recorded program is typically a determinant for establishing the linear velocity of a CD. A CLV of 1.4 m/s is commonly used for for programs less than 60 minutes and 1.2 m/s is commonly used for longer programs. The disc's angular velocity decreases as the optical pickup moves toward the outer tracks of the disc. At a linear velocity of 1.2 m/s the angular velocity varies between 486 and 196 rpm. At 1.4 m/s the angular velocity varies between 568 and 228 rpm. Data is stored in pit formations, as shown in Figure 1b. These pits vary in length from 0.833 to 3.054 $\mu$m depending on the encoded data and the linear velocity of the disc. The information contained in the pit structure on the disc surface is coded so that the edge of each pit represents a 1 and all spaces between the edges represent 0 as shown in Figure 1c. The width and depth of the pits are approximately 0.5 and 0.11 $\mu$m, respectively. The pits are placed in a spiral track with a pitch of 1.6 $\mu$m. The track runs circumferentially from the inside of the disc to the outside. As shown in Figure 1d, approximately sixty CD tracks can be contained within one LP groove. The total number of spiral revolutions contained on a disc is 20,625.

The specifications for the compact disc were jointly developed by Philips Corporation and Sony Corporation and are documented in what is known as the Red Book. The standards for the compact disc are also documented in the IEC (International Electrotechnical Commission) standard BNN15-83-095, *Compact Disc Digital Audio System*.

## Data Format

The compact disc stores stereo audio information with provisions for four channels. The audio information is sampled at 44.1 kHz with 16-bit linear

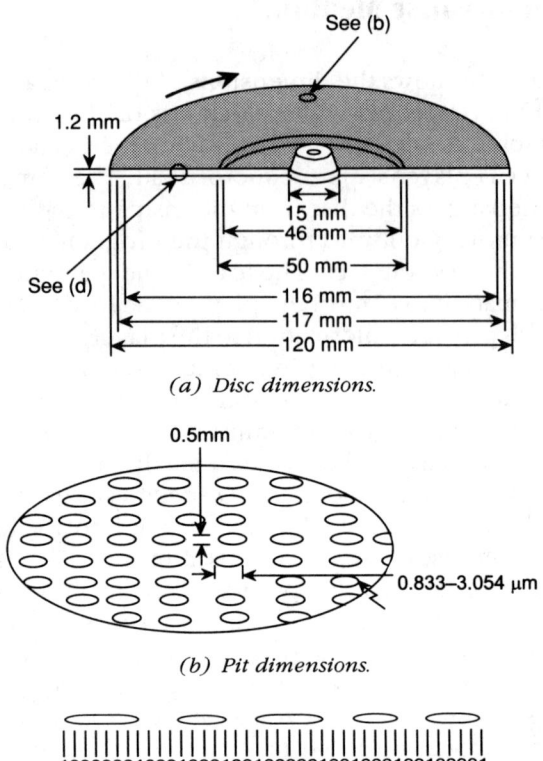

(a) Disc dimensions.

(b) Pit dimensions.

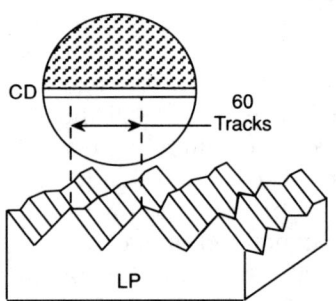

(c) Pits on the disc represent digital data via EFM modulation.

(d) Sixty CD tracks can be contained in an LP groove.

**Figure 1.** The compact disc is a high-density data storage medium.

quantization. The audio output rate of the player is thus 1.41 Mb/s (44.1 kHz × 16 bits × 2 channels). The information is encoded with a cross-interleaved Reed-Solomon code (CIRC) for error correction and modulated by eight-to-fourteen modulation (EFM) for increased information density. After synchronization and subcode information are added, only about one third of the bits represent audio information. Data encoding is covered later in this chapter.

## Structure

The compact disc medium consists of a transparent polycarbonate substrate covered by a reflective material, which in turn is covered by a protective layer. The label is placed over the protective layer. This is shown in Figure 2a. The substrate allows the laser beam to penetrate to the reflective layer. The spot size of the laser beam is focused from an initial diameter of 0.8 mm at the disc surface to 1.7 µm at the reflective surface. Accurate control of the focusing system causes dust, scratches, or fingerprints on the surface of the disc to appear out of focus to the reading laser.

The pits appear as bumps from underneath where the laser enters the medium. The wavelength of the laser in air is 780 nm. Upon entering the substrate with a refractive index of 1.55, the wavelength is reduced to approximately 500 nm. The depth of the pits are between 110 and 130 nm and are designed to be approximately one fourth of the laser's wavelength. The pit depth of $\lambda/4$ creates a diffraction stucture such that reflected light undergoes destructive interference between the zero- and first-order reflected rays. This interference thus decreases the intensity of light returned to the pickup lens. The presence of pits and land areas are detected in terms of changing light intensity by

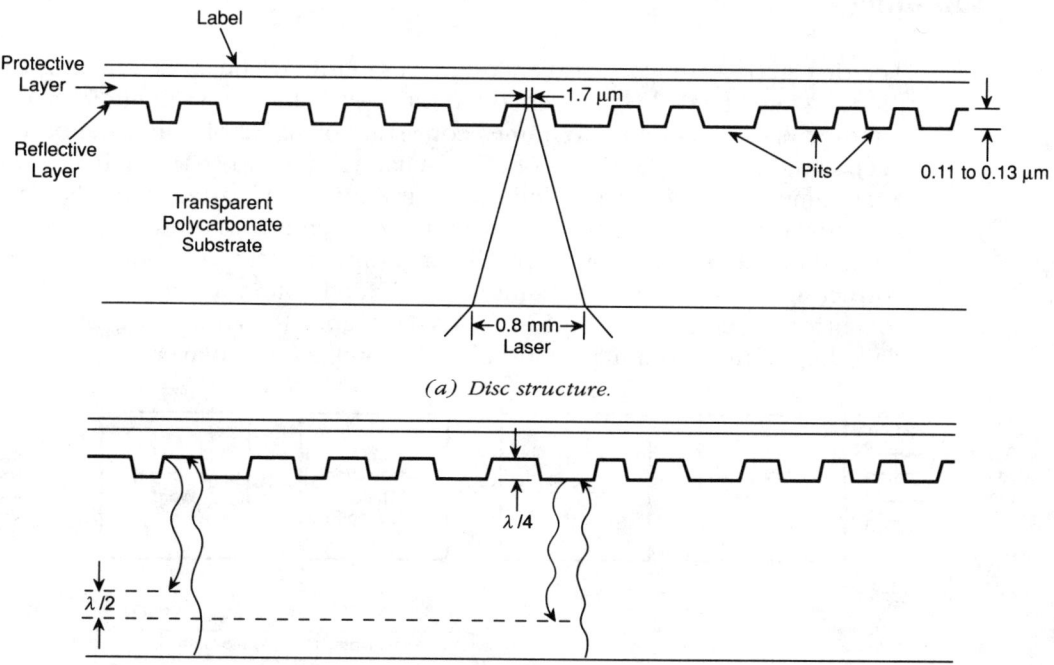

*(a) Disc structure.*

*(b) The CD data surface acts as a reflective grating. Destructive interference, as a result of diffraction, conveys pit length via light reflected back into the pickup.*

**Figure 2.** Compact disc cross section.

photodetectors. The light signal is converted to a corresponding electrical signal by the photodetectors.

Information is stored on only one side of the compact disc. It is possible to store information on both sides of the disc, but, due to the excessive cost of manufacturing and the added expense of extra optical hardware required for the compact disc player, this option is not economically feasible.

## Compact Disc Encoding

A substantial amount of additional information is added prior to disc mastering. As shown in Figure 3, the encoding process is elaborate and includes CIRC encoding, EFM modulation, as well as the addition of control and synchronization words. As shown in the figure, the encoded data is formatted by frames which provides a means of distinguishing the type of encoded data by its position in the frame. Figure 4 provides a detailed algorithm describing the encoding process of the audio signal for the compact disc system. It will be helpful to refer to this figure while reviewing the encoding process.

### Sampling

The first step in compact disc encoding is sampling. The audio signal is sampled at a rate of 44.1 kHz and encoded into a PCM format. (See Figure 4.) The sampled data is divided into frames consisting of six 32-bit sampling periods. Each sampling period consists of 16 left-channel bits and 16 right-channel bits. Each sampling period is divided into four 8-bit words called symbols. To ensure proper decoding of the data from the disc an error correction scheme is applied in the encoding process. The encoding process incorporates an error correcting scheme called cross-interleaved Reed-Solomon coding (CIRC). Error correcting is required to preserve data integrity, thus providing for accurate decoding of the audio information by the compact disc player.

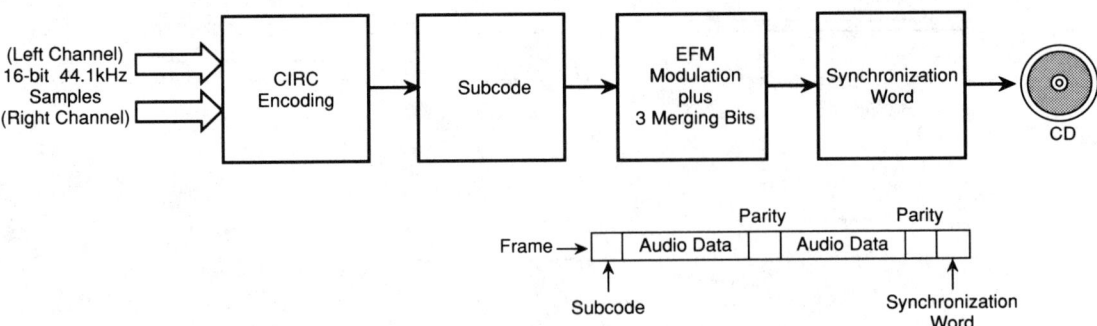

**Figure 3.** Compact disc encoding process.

# Chapter 6: Optical Disc Technology

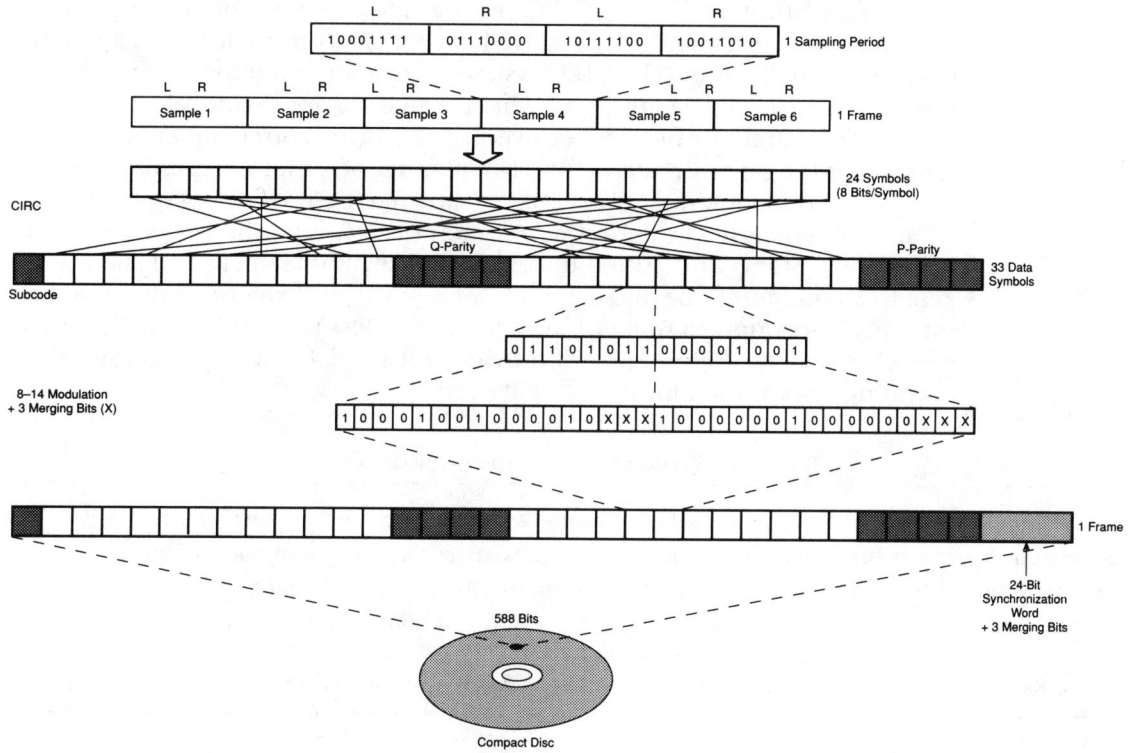

**Figure 4.** Encoding algorithm for compact disc.

## CIRC Encoding

Errors may occur in the decoding process due to a variety of anomalies. Errors may result from manufacturing defects in photoresist, plating, or the transparent layer of the disc or as a result of surface obstructions such as scratches or dust. Errors are categorized as burst errors or random errors. Burst errors affect a large number of bits whereas random bit errors are generally single-bit errors. It should be noted that since bits are placed on the disc in a spiraling track an annular scratch has more severe consequences than a radial scratch. The bits affected by an annular scratch are consecutive and thus related to one another whereas bits affected by radial scratches are normally unrelated, and therefore the effect can be more easily tolerated.

The CIRC is a combination of several error correction techniques and is designed to detect and correct both burst errors and random errors. Any errors which are uncorrectable by CIRC are marked, allowing them to be concealed by interpolation (the process of approximating missing information using the surrounding information). Errors that are not corrected or concealed may produce an audible click in the reproduced audio signal.

The specifications for the CIRC encoder include a maximum correctable burst error length of 4000 consecutive bits, which corresponds to an approximate length of 2.5 mm. The CIRC system is capable of interpolating burst errors up to 12,300 consecutive bits, which is an error approximately 7.7 mm in length. The number of errors received by the error correcting system is referred to as the bit error rate (BER). The number of samples per unit time that require interpolation for a given BER is shown to be one every ten hours at a BER of $10^{-4}$. The lower the required interpolation rate, the better the error correction system. Undetected errors can cause audible clicks and should be kept to a minimum. The undetected error rate is less than one undetected error every 750 hours at a BER of $10^{-3}$. If the BER is less than $10^{-4}$, the undetected errors become negligible. Table 1 provides a list of specifications for the CIRC encoding system used for the compact disc.

**Table 1.** Specifications for the Compact Disc

| | |
|---|---|
| Maximum Correctable Burst Error Length | 4000 data bits (i.e. 2.5 mm track length) |
| Maximum Interpolatable Burst Error Length | 12,300 data bits (i.e. 7.7 mm track length) |
| Sample Interpolation Rate | 1 sample/10 hours at a BER of $10^{-4}$ |
| Undetected Error Samples | < 1 error/750 hours at a BER of $10^{-3}$ (negligible at a BER of $10^{-4}$) |
| Code Rate | On average, 4 bits are recorded for every 3 data bits |

Figure 5 shows the CIRC encoder algorithm. Six samples from each of the left and right channels are divided into 8-bit words (symbols) and stored in the RAM of the CIRC encoder. Twenty-four 8-bit symbols are accepted for encoding and interleaved. Even samples are delayed by two symbols to increase the scrambling of the data. Twenty-four symbols are input to the $C_2$ encoder (Reed-Solomon (28, 24) encoder). Twenty-eight symbols are output from the $C_2$ encoder stage including four $Q$-parity symbols. The $Q$-parity symbols are inserted in the incoming 24 symbols. Between the $C_2$ and $C_1$ encoders unequal delays are applied to the 28 symbols output from the $C_2$ encoder. By delaying designated words the data is dispersed and interleaved to reduce the impact of burst errors. The 28 symbols are input into the $C_1$ encoder (Reed-Solomon (32, 28) encoder) where four symbols of $P$ parity are produced. Even samples are subject to a single-symbol delay stage after the $C_1$ encoder. The $P$- and $Q$-parity symbols are inverted to ensure that nonzero parity symbols exist with zero input data. Thirty-two symbols sequentially leave the CIRC encoder to form a frame. The 32 symbols consist of 24 scrambled 8-bit data symbols plus eight parity symbols.

## Subcode Word

One 8-bit symbol called a subcode word is added to every 32 symbols departing the CIRC encoder. The subcode word contains information on where

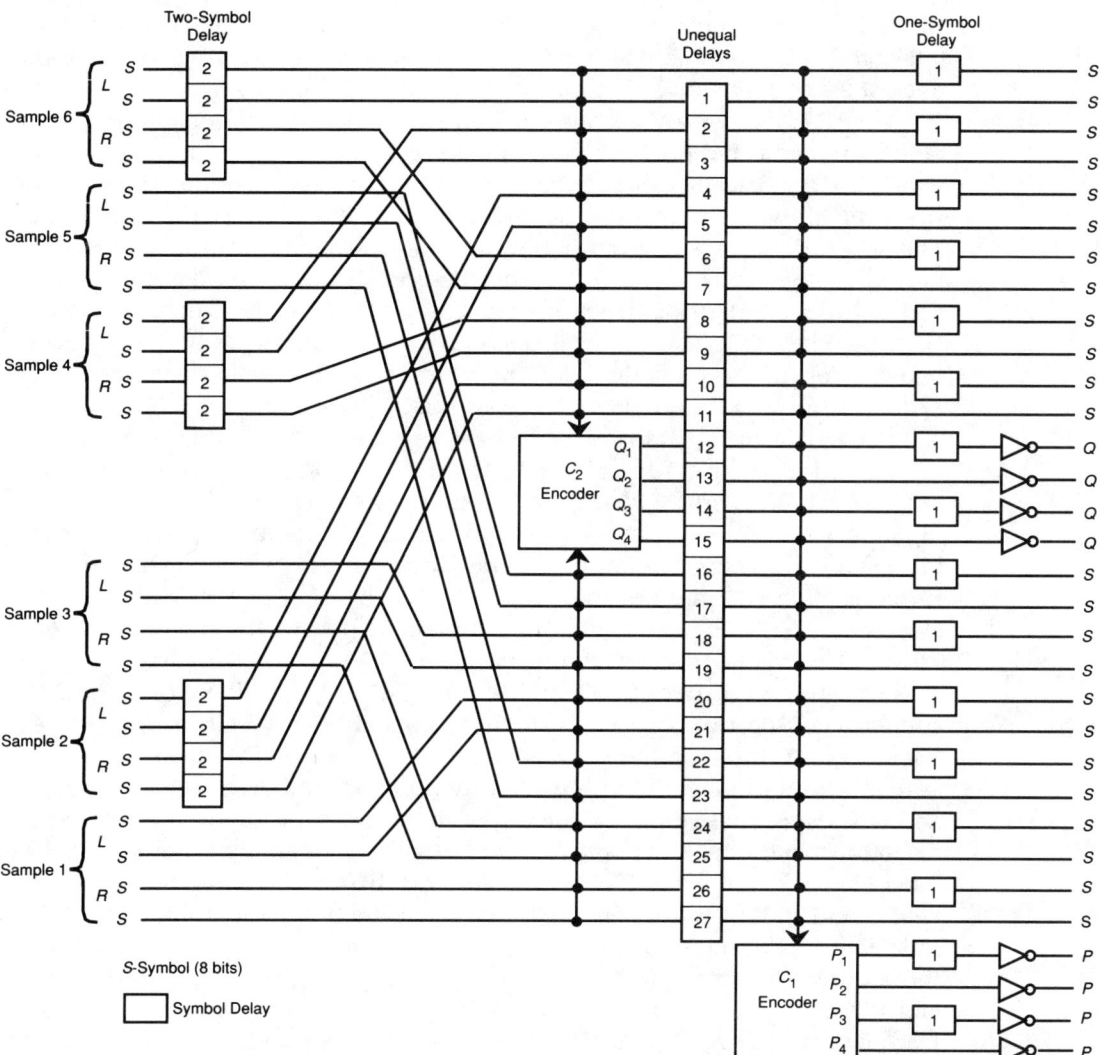

**Figure 5.** Cross-interleaved Reed-Solomon code (CIRC) encoder.

tracks begin and end, track numbers, disc timing, and index points. Each bit is referred to as a subcode bit and is designated *P, Q, R, S, T, U, V,* or *W*. The subcode bits are removed from the audio data upon playback and combined to form a subcode word 98 bits long. Therefore 98 frames must be read in order to form a subcode word for each of *P, Q, R, S, T, U, V,* and *W* as shown in Figure 6. Only the *P* and *Q* (not to be confused with the *P-* and *Q*-parity symbols of the CIRC encoder system) channels are specified in the basic CD audio format. The other subcode user bits may carry information such as graphics and text but are normally recorded with zeros on audio discs.

The *P* subcode is used to designate the start and stop points between music tracks. During music tracks the *P* subcode is normally 0, and between music tracks it is a 1. The *P* subcode switches between 0 and 1 during the lead-out track at a frequency of 2 Hz to indicate the end of the disc.

The *Q* subcode provides information such as track number, address, error detection, and time. Figure 6 also shows the content of the 98-bit *Q*-subcode word. The first two bits, $S_0$ and $S_1$, are used for synchronization of the subcode word and allow the player to distinguish the subcode word in the data stream. The subcode bits are used to denote timings, the number of channels, copy prohibit, pre-emphasis, and other information. The last 16 bits of the *Q*-subcode word are an error correction code using a cyclic redundancy check code (CRCC). The CRCC is a cyclic block code that generates a parity check word to detect errors in the subcode field. The address bits designate the mode for the *Q* data. Three modes are used in the compact disc system.

Mode 1 is designated by an address of 0001 and is utilized for lead-in, music, and lead-out tracks. Mode 1 provides two formats: lead-in format and music/lead-out format. During the lead-in track the *Q* subcode is as shown in Figure 6a. The 72-bit *Q*-data section consists of nine 8-bit parts. The first eight bits contain the track number (TNO) ranging from 0 to 99. The track number is stored in binary coded decimal (BCD) form. The TNO in the lead-in track is always set to 00, indicating that the number is part of the table of contents (TOC). During the first music selection (track) it changes to 01, in the second music section it changes to 02, and so on. POINT, PMIN, PSEC, and PFRAME contain TOC information relating to the track number and starting time of each track. The POINT parameter indicates the number of the music selections available, and the corresponding starting times are indicated by PMIN, PSEC, and PFRAME. The beginning time for each track is repeated in three consecutive frames in the lead-in track. After each track has been identified the POINT parameter indicates A0. PMIN indicates the first track number on the disc (PSEC and PFRAME are zero). The POINT parameter sets to A1, and PMIN indicates the last track number of the disc. A2 in the POINT parameter refers to the starting time of the lead-out track, which is then given in the PMIN, PSEC and PFRAME parameter. The TOC is continuously repeated throughout the lead-in track. The zero parameter of the lead-in track provides eight bits, which are all 0. The MIN, SEC, and FRAME parameters maintain a running time for the lead-in track.

For music and lead-out tracks the 72-bit *Q*-data field is formatted differently from the lead-in track. TNO indicates the current track number in BCD form (01–99). During the lead-out track the TNO is set to AA. POINT designates index numbers within each track in BCD form (01–99). POINT is set to 00 during pauses between tracks and is set to 01 during lead-out. MIN, SEC, and FRAME indicate running time of each track beginning at zero at the start of a track and continually increase to the end of the track. The running time decreases during pauses, reaching zero at the end of the pause. AMIN, ASEC, and AFRAME indicate absolute time from the beginning of the first music track to the end of the lead-out track.

Mode 2 is designated by an address of 0010 and provides a catalog number.

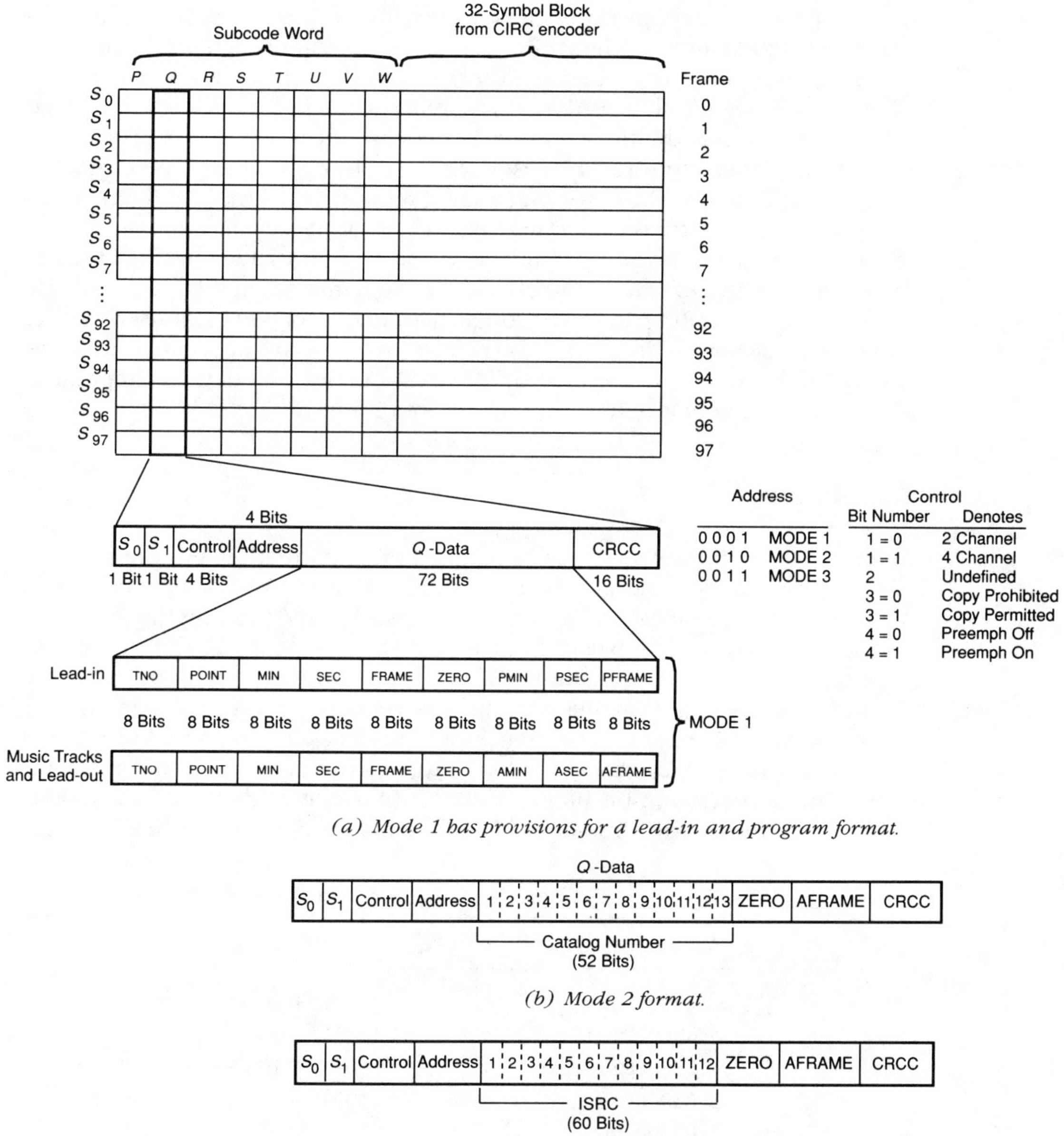

# Chapter 6: Optical Disc Technology

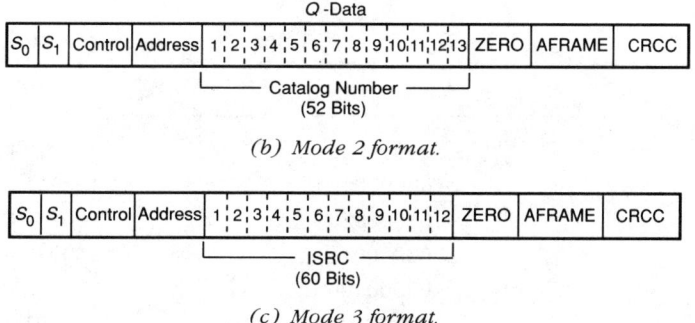

(a) Mode 1 has provisions for a lead-in and program format.

(b) Mode 2 format.

(c) Mode 3 format.

**Figure 6.** The $Q$ subcode consists of 98 bits from 98 consecutive frames.

The format for the 98-bit $Q$-subcode word is shown in Figure 6b. The synchronization bits, control, address, and the CRCC error correction information are the same as mode 1. The catalog number is represented as 13 digits in BCD form according to the Universal Product Code (UPC) standard for bar coding.

Mode 2 provides continuous absolute time with the adjacent mode 1 frames by the AFRAME parameter. The ZERO parameter contains 12 consecutive 0s.

Mode 3 is designated by an address of 0011 and provides a 12-character International Standard Recording Code (ISRC) for each recording on the disc. Characters 1 and 2 identify a country code; characters 3, 4, and 5 provide the owner code; characters 6 and 7 designate the year of recording, and characters 8 through 12 provide the recording's serial number. As shown in Figure 6c, 60 bits are designated for the 12-character ISRC information with two bits always being 0. Characters 1 through 5 utilize a 6-bit format. Characters 6 through 12 follow a 4-bit BCD format. The absolute time is maintained by the AFRAME parameter. The ZERO parameter contains four consecutive 0s. If used, modes 2 and 3 must appear in at least one out of every 100 consecutive subcode words. Modes 2 and 3 do not appear in the lead-in or lead-out tracks and can be deleted completely from the subcode if not required.

## EFM Modulation

After the audio information has undergone error correction and the subcode data has been added, eight-to-fourteen modulation (EFM) converts each 8-bit block of data into blocks of 14 bits. The purpose of modulating the code is to shape the spectrum and dynamic range of the process to match the characteristics of the medium. For example, EFM requires that more bits be used, but because the size of the smallest pit (highest frequency) required to encode the information is increased, fewer transitions are necessary to convey the data. A lower track velocity can therefore be used, allowing for longer playing time. The conversion from 8-bit blocks to 14-bit blocks is performed by a lookup table stored in read-only memory (ROM) as shown in Table 2. The advantages

**Table 2.** Eight-to-Fourteen Modulation Partial Lookup Table

| 8-Bit Symbol | 14-Bit Word |
|---|---|
| 00000000 | 01001000100000 |
| 00000001 | 10000100000000 |
| 00000010 | 10010000100000 |
| 00000011 | 10001000100000 |
| 00000100 | 01000100000000 |
| 00000101 | 00000100010000 |
| 00000110 | 00010000100000 |
| 00000111 | 00100100000000 |
| 00001000 | 01001001000000 |
| 00001001 | 10000001000000 |

of EFM are reduction in information bandwidth and DC content (in current disc systems, DC content may hinder servo clocking capabilities) greater information density, additional synchronization, and error correction information. It is important to utilize the most advantageous 14-bit words to represent the 8-bit symbols. The 8-bit symbols require $2^8$ (256) different patterns. There are $2^{14}$ (16,384) possible words available in a 14-bit code, but only 256 are utilized in the lookup table. Fourteen-bit words are selected that contain at least two but no more than ten consecutive zeros. In this way the bandwidth is reduced (reduction in the highest frequency of transition) and the DC content is minimized. The many unique word patterns available in a 14-bit code increases the error correction capability of the system.

Three additional bits, called merging bits, are added to each 14-bit word to further reduce DC content. Two of the three bits are always 0. The third bit is designated 1 or 0 according to the preceding and succeeding words to maintain clock synchronization and to further suppress DC content.

After EFM processing the signal is converted from a nonreturn-to-zero (NRZ) signal to a nonreturn-to-zero inverted (NRZI) signal as shown in Figure 7. NRZI signals provide fewer transitions than NRZ signals, simplifying the pit structure on the disc. By utilizing a NRZI signal the minimum pit length (or land length) is three clock periods ($3T$) and the longest pit length (or land length) is eleven clock periods ($11T$).

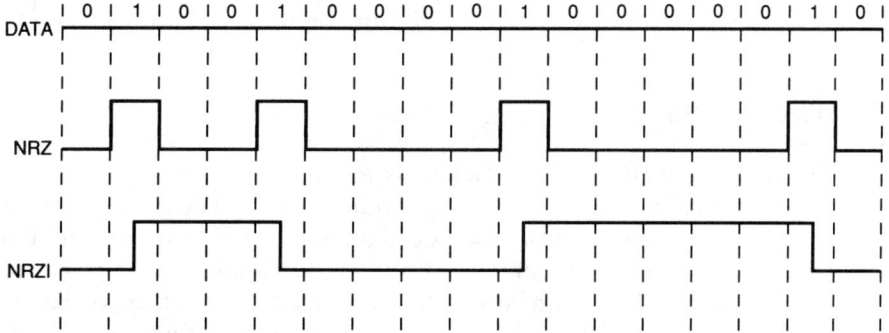

**Figure 7.** The digital data is converted from NRZ to NRZI.

## Synchronization Word

Following EFM processing, the partial data frame consists of 561 bits and becomes a complete frame after the addition of a 24-bit synchronization word plus three merging bits. The synchronization word designates the beginning of a frame and provides a means for self-clocking. The synchronization word is uniquely defined as 100000000010000000000010. The total number of bits in a frame is therefore 588.

## The Compact Disc Player

The audio compact disc (known as CD-A or CD-DA) was the first optical system to provide digital audio recorded material for use by consumers. Other optical systems such as CD-ROM, CD-I, CD-V, CD-WO, Erasable CD, CD+G, CD+MIDI, and CD-3 have been developed from the compact disc utilizing much of the same technology. For example, each of these systems takes advantage of optoelectronics technology. Optoelectronics provides compact disc systems with a means for transferring digital information via modulated light from the optical medium. The compact disc also utilizes highly sophisticated electronic systems to process the information to yield an accurate reproduction of the original signal.

## The Optical System

Thanks to the development of optoelectronics and the application of sophisticated optical components, the compact disc has become a successful means for storing and reproducing audio information. The compact disc utilizes an optical pickup to read data from the optical medium. The optical pickup is a complex system consisting of the optoelectronic components required to focus a laser beam on the medium and track the pit data structure.

### The Pickup

The function of the pickup is to transfer the encoded information from the optical disc to the player's decoding circuit. The pickup is required to track the information on the disc, focus a laser beam, and read the information as the disc rotates. The entire lens assembly is allowed to move across the disc as directed by the tracking information taken from the disc and programming information provided by the user. The pickup must respond accurately under adverse conditions such as playing damaged and dirty discs or while experiencing vibrations or shock.

Compact disc players use either a three-beam pickup or a single-beam pickup. Figure 8 shows an example of a typical three-beam pickup. A laser diode (often referred to as a semiconductor laser) functions as the optical source for the pickup. A detailed description of the laser diode and its operation is provided in Chapter 4. The AlGaAs laser is commonly used in compact disc players. It provides a beam with a 780 nm wavelength.

The laser beam is split by a diffraction grating into multiple beams. Diffraction gratings are plates with slits placed only a few wavelengths apart as shown in Figure 9a. As the beam passes through the grating it diffracts in different directions resulting in an intense main beam (primary beam) with successively less intense beams on either side. Only the primary and secondary

Chapter 6: Optical Disc Technology

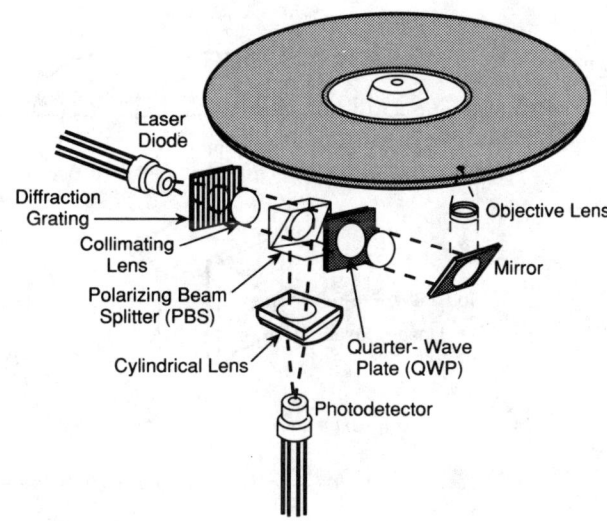

**Figure 8.** The compact disc three-beam optical system.

beams are used in the optical system of a compact disc player. The primary beam is used for reading data and focusing the beam. The outer two beams (secondary beams) are used for tracking. The light is subjected to a collimator lens that converges the previously divergent light into a parallel path. The light is conditionally passed through a polarization beam splitter (PBS). The PBS acts as a one-way mirror allowing only vertically polarized light to pass to the disc; it reflects all other light. (See Chapter 5, for a review of polarization.) The light is directed toward a quarter-wave plate (QWP). As shown in Figure 9b, the quarter-wave plate is an anisotropic crystal material designed to rotate the plane of polarization of linearly polarized light by 45°. The laser light is then focused onto the disc by an objective lens. The objective lens converges the impinging light to the focal point at a distance ($d$) from the lens called the focal length. This is shown in Figure 9c. The objective lens of a compact disc

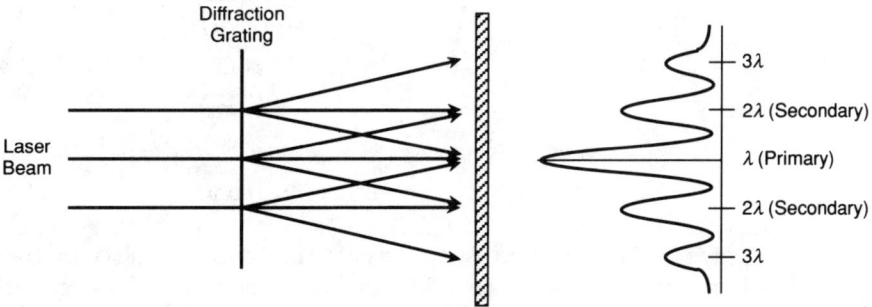

*(a) A diffraction grating splits the laser beam into multiple beams.*

**Figure 9.** The optical pickup is made up of several components.

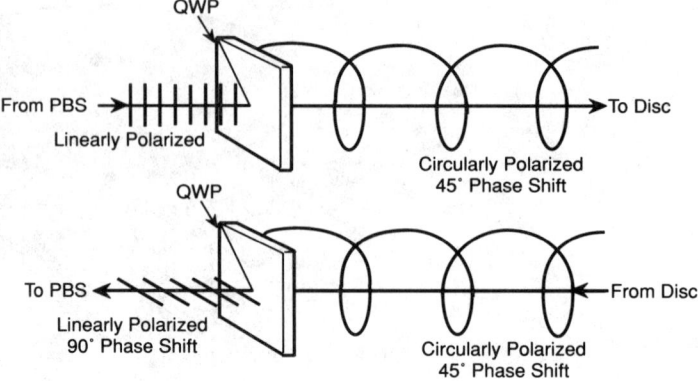

*(b) The quarter-wave plate rotates the plane of polarization 45°.*

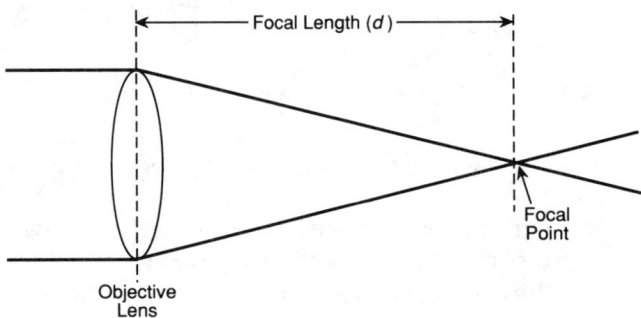

*(c) The objective lens converges the light.*

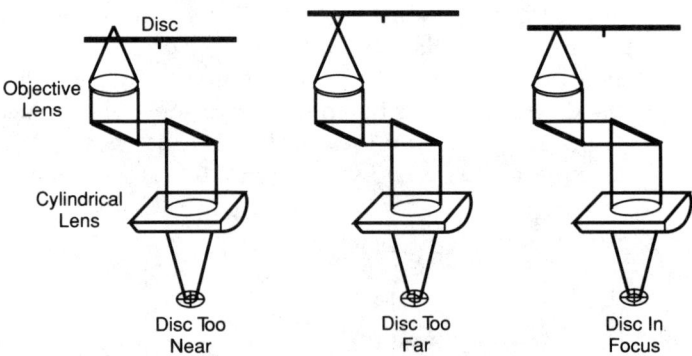

*(d) Astigmatism is measured by photodetectors to determine the focusing error.*

**Figure 9.** (cont.)

player is mounted on a two-axis actuator that is controlled by focus and tracking servos. The spot size of the primary beam on the disc surface is 0.8 mm and is further reduced in size to 1.7 $\mu$m at the reflective surface of the disc. This is due to the converging effect of the objective lens.

The light is reflected from the disc and again passes through the objective lens. The light converges as it passes through the objective lens and is again phase shifted 45° by the QWP. The plane of polarization of the reflected light is now at a right angle to its original state; it is horizontally polarized. Since the light is horizontally polarized it is reflected by the PBS toward a cylindrical lens. The cylindrical lens utilizes an astigmatic property to reveal focusing errors in the optical system as shown in Figure 9d. Light passes through the cylindrical lens and is received by an array of photodetectors (typically a four-quadrant photodetector).

## Focusing

A perfectly focused beam places the focal point of the light on the photodetectors where the shape of the image on the photodetectors is correspondingly circular. When the focal point is in front of the photodetectors an elliptical image is projected on the photodetectors at an angle. If the focal point moves behind the photodetectors, the elliptical image is rotated 90°. The photodetectors act as transducers converting the impinging light signals into corresponding electrical signals. The electrical signals thus contain information for both focusing and tracking the laser beam, as well as audio information.

Figure 10 shows the four-quadrant photodetector (labeled A, B, C, D) used to control the focus servo and to transfer the audio signal to the decoding circuit. The focus circuit provides control for the vertical positioning of the two-axis objective lens actuator. The focus correction signal $(A + C) - (B + D)$ is equal to zero when the laser is focused correctly on the disc, and the shape of the image on the photodetector is correspondingly circular. When the optical system is out of focus the focus correction signal is a nonzero value. The focus correction signal provides feedback to the focus servo circuit, which moves the objective lens up or down accordingly until the laser is focused and the shape of the image on the photodetectors becomes circular. The audio information is

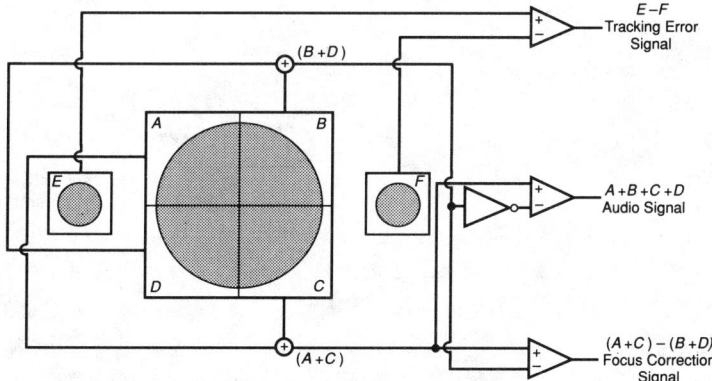

**Figure 10.** The signal from the four-quadrant photodetector controls focus and carries the audio signal. The signal at the outer detectors controls tracking.

collected by summing the signals from the four photodiodes $(A + B + C + D)$. The audio signal must exceed a threshold level prior to activation of the focus servo. When the disc is first placed in the player the distance between the objective lens and the disc is large, and therefore the audio signal is below the threshold and the focus servo is inactive. A focus search circuit initially moves the lens closer to the disc, causing the audio signal to increase. When the audio signal exceeds the threshold the focus servo is activated.

## Tracking

The tracking servo is controlled by the signals received at two outer photodetectors ($E$ and $F$ in Figure 10). The secondary beams are directed to these photodetectors. Figure 11 shows the positioning of the three-beam system on the disc. The lead beam's signal is electronically delayed and compared with the corresponding lagging beam. The two outer photodetectors generate a tracking error signal $(E - F)$. The tracking system detects mistracking to the left or right and returns a tracking error signal to the tracking servo. The servo moves the pickup accordingly to correct the tracking error. The tracking error signal controls the horizontal movement of the two-axis objective lens actuator. The tracking servo continually moves the objective lens in the appropriate direction to reduce the tracking error.

## Single-Beam Pickups

The single-beam pickup uses a wedge lens to split the laser beam into two beams as shown in Figure 12. The split laser beam uses the same optical compo-

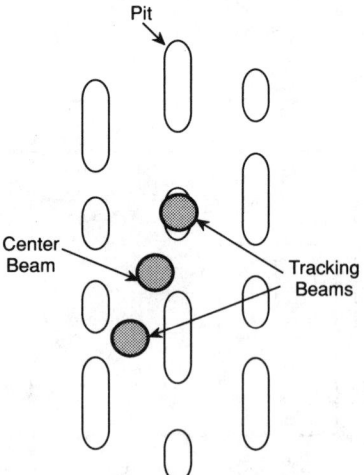

*(a) Mistracking to the left.*

**Figure 11.** The tracking system recognizes three possible conditions.

nents as the three-beam pickup, except there are four photodetectors ($G, H, I, J$) rather than six. The focusing error equation is $(G+J) - (H+I)$; three focusing conditions exist for the single-beam pickup. The first condition exists when the focal point lies behind the disc. The second condition exists when the focal point lies in front of the disc. The third condition results when the focal point is focused on the disc. The tracking error equation is $(G+H) - (I+J)$. Relative light intensity on photodiode pairs $G/H$ and $I/J$ is used to determine tracking status. If more light impinges on $G$ and $H$, the track shifts left. If the image covers more of $I$ and $J$, the track shifts right. Perfect tracking is designated by equal images on $G, H, I$, and $J$. The four photodetectors also provide audio information by summing the signal from all four detectors $(G+H+I+J)$.

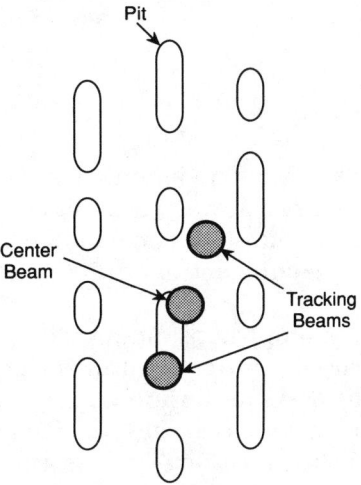

*(b) Mistracking to the right.*

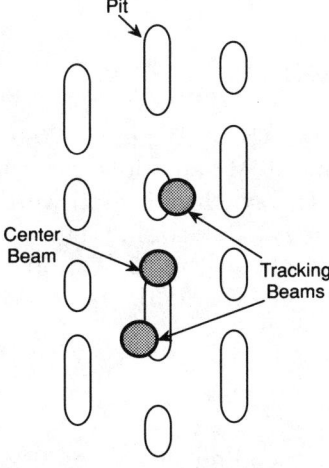

*(c) Correct tracking.*

**Figure 11.** (cont.)

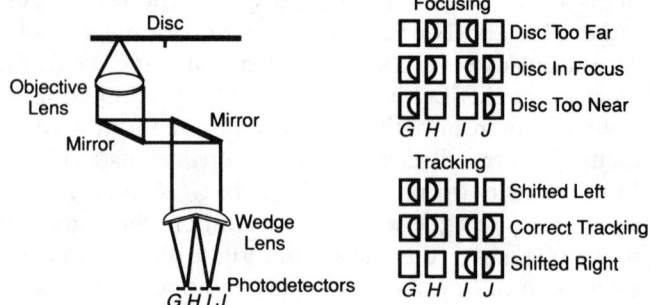

**Figure 12.** Single-beam pickups utilize a wedge lens to split the laser beam, which is detected by a four-photodetector array.

## Pickup Control

Three-beam pickups are mounted on a sled that moves radially across the disc, providing coarse tracking capabilities. Tracking signals are derived from the signal used to control the two-axis objective lens actuator. The sled servo operation is also contingent upon the level of the primary beam exceeding the threshold.

Single-beam pickups mount on a pivoting arm which guides the pickup across the disc in an arc. A coil and magnet are used to manipulate the pivoting arm, which allows for positioning anywhere across the disc.

Precise tracking is always provided by the tracking servo and corresponding control circuit. During fast forward or reverse a microprocessor takes control of the tracking servo to increase locating speed.

# The Electrical System

Figure 13 shows a block diagram of the compact disc's decoding and electrical output systems. EFM demodulation returns the data from 14-bit words back to 8-bit words. The synchronization word at the beginning of each frame is removed prior to CIRC decoding. The electrical output system consists of a digital filter, digital-to-analog (D/A) converters, and low-pass filters. Analog and digital outputs are provided.

## Decoding

Compact disc decoding consists of EFM demodulation, frame synchronization, and error correction. The decoding system simply reverses the encoding

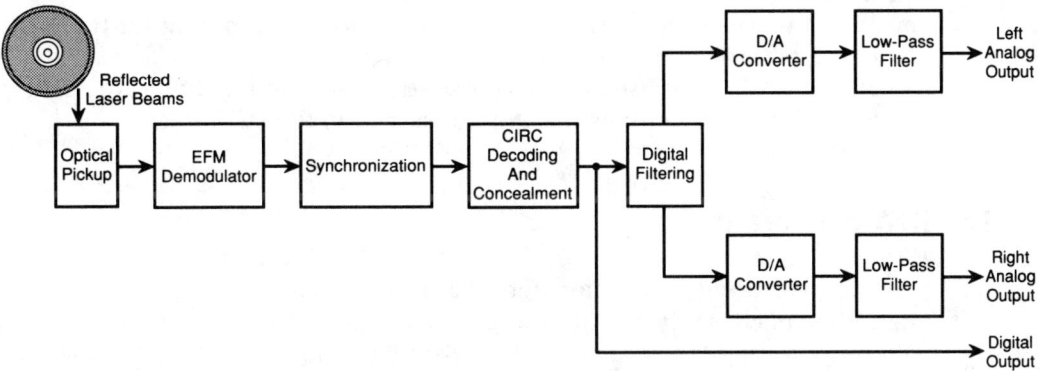

**Figure 13.** Compact disc decoding block diagram.

process to extract the original audio data. The NRZI digital signal is demodulated back to a NRZ signal. The synchronization word is removed from the EFM signal, and the resulting 17-bit EFM words are demodulated back to eight bits. The synchronizing word is used to control the disc rotating speed as well as indicate the beginning of frames for timing purposes within the compact disc player. A ROM lookup table is often used to demodulate the EFM signal. Following EFM demodulation each frame contains subcode, parity, and audio data. The data is submitted to the CIRC decoder for error detection and correction. Not all players apply the same CIRC decoding method; therefore, varying degrees of error correction can be expected. If the decoder cannot correct an error, a flag marks the uncorrected data. Interpolation and muting circuits are used to conceal uncorrectable data. Concealment circuits reconstruct the erroneous data through evaluation of the surrounding data and interpolation. If the data is uncorrectable by the CIRC decoder or unconcealable by interpolation, muting is used to prevent audible clicks. Valid data that is not flagged by the CIRC decoder passes through the concealment circuits unaltered. The subcode is removed and organized into 98-bit blocks for processing to recover table of contents (TOC), timing, and pointing information along with CRCC error correction, addressing and control information. The data is then applied to the output electronics which consist of digital oversampling filters, D/A converters, and anti-imaging low-pass filters.

## The Digital Filter

Digital oversampling filters interpolate additional samples within the existing data, raising the sampling rate. As a result of digital filtering, the output image spectra are raised well beyond the audio band, allowing gentle low-pass filters to remove undesired out-of-band high-frequency images without adding phase distortion. The operation and design of digital oversampling filters is discussed in Chapter 2, and the detailed theory is discussed in Chapter 10. In many CD players, digital data is output prior to oversampling for transfer to other digital processing equipment. A standard serial interface format such as SPDIF or AES/EBU is used for this purpose. Oversampling filters are also beneficial for decreasing requantization noise within the audio band.

### The D/A Converter

Digital-to-analog (D/A) converters change the digital signal to an analog signal. Some players rely on only a single D/A converter to convert both channels although most provide dual D/A converters (one for each audio channel). Players utilizing a single D/A converter must compensate for the resulting delay between channels. If the time delay is not compensated, stereo imaging can become altered. Multibit digital-to-analog converters are covered extensively in Chapter 3, and low-bit converters are discussed in Chapter 12.

### The Low-Pass Filter

The output low-pass filter is often referred to as a smoothing filter because it removes the high-frequency content of the pulse amplitude modulation signal output by the D/A converters, leaving only the original continuous waveform. Even though the frequencies which are filtered are above the audible spectrum, filtering is essential to prevent adverse effects as a result of high-frequency modulation in other equipment. The audio signal is output from the compact disc player in its original analog form; then it is amplified and reproduced as an acoustic audio signal.

## Optical Recording Systems

Since the inception of the compact disc, recording to optical media has been an expensive process typically performed by large disc-mastering and duplicating facilities. This situation is rapidly changing. Optical recording can be performed by the end user utilizing a write-once or erasable system. Write-once technology provides for a one-time recording onto a disc without provision for erasing although the information can be read many times. This is also referred to as WORM (write-once, read-many) technology. Write-once drives require both recording and playback capabilities. Write-once systems which conform to the CD format are called CD-WO (compact-disc, write-once). Erasable optical systems provide for recording, erasing, and playback. Both of these technologies have developed with the assistance of the computer industry and are therefore available in formats other than compact disc.

# Write-Once Systems

Write-once systems are available in a variety of formats with differing disc dimensions, encoding schemes, data formats, and error correction processes. Obviously, not all write-once systems are compatible. For audio applications the most important format is CD-WO. The CD-WO disc is compatible with conventional compact disc CD players and is therefore identical with an audio CD in disc dimensions, encoding, data format, and error correction.

## Optical Recording Methods

The means for recording with optical write-once systems include pit formation, bubble formation, phase change, dye polymer, and texture change. Figure 14 shows examples of each form of recording.

Pit formation by ablation of the medium is the most common form of recording. Permanent holes are burned into the reflective layer of a disc either by melting or by vaporizing the material with a laser beam. Melting is preferred to vaporization because the latter leaves metal deposits on the surface which may cause problems in reading the information. The power of the laser is usually about 6 to 9 mW. The pits create the necessary change in reflectivity, allowing the disc to be read by a conventional CD player.

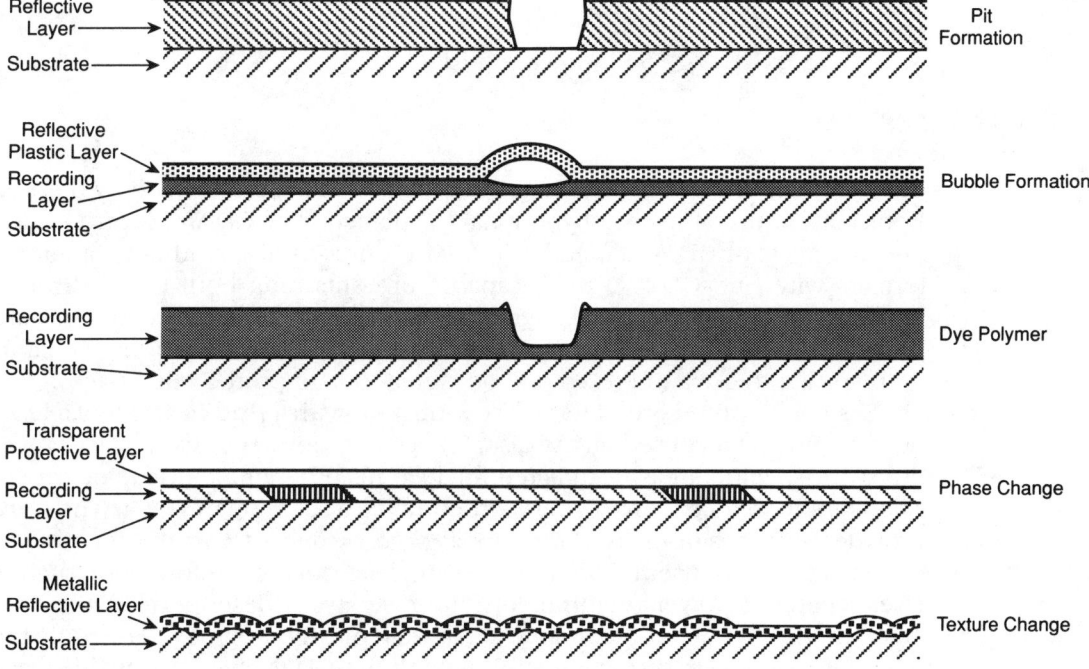

**Figure 14.** Optical recording technologies.

A variation of this method is bubble formation, which vaporizes an active layer (also referred to as the recording layer), causing a bubble to form in an enveloping plastic layer. The bubble disperses the impinging read laser beam, varying its reflective intensity. The bubble may be designed to burst; leaving a pit similar to the ablation method.

The dye polymer method also physically changes the active layer by leaving a depression when the disc is heated by a laser. The active layer is a plastic material containing an organic dye. The laser softens the active layer, and when this layer cools a depression forms. The dye is chosen according to its absorption characteristics at the particular wavelengths produced by the laser. Dye polymer discs are less expensive than metal film discs, thus making them a more desirable medium for recording.

The phase change method of recording does not cause the metal film to undergo a change in shape but a change in atomic structure. By changing the physical structure of a material from a crystalline to an amorphous state the reflectivity of the material can be altered. A metal film is highly reflective when in its crystalline state and less reflective when in its amorphous state.

The texture change method uses a substrate impressed with tiny aberrations that are smaller than the size of the spot formed by a laser beam. The aberrations in the substrate located at repeated intervals cause the laser beam to be dispersed. Because its texture is similar to that of a moth's eye, the material is often referred to by this term. The surface is melted by a laser beam which creates smooth faces in the metal layer, causing the surface reflectivity to increase.

## CD-WO Discs

CD-WO discs are designed to meet the specifications established by the Red Book and are playable on conventional CD players. The CD-WO format implements aspects of CD-Audio and CD-ROM (compact- disc read-only memory) formats with one CD-WO track capable of containing both CD-Audio and CD-ROM information. User programmable areas and prerecorded areas are provided on a CD-WO disc in one of two formats. The first format contains prerecorded tracks and the second does not contain prerecorded tracks.

Discs with prerecorded tracks are formatted with a lead-in area and a lead-out area already recorded on the disc. Figure 15a shows the layout of a CD-WO disc with specific areas designated for lead-in, program, and lead-out areas. The lead-in area contains a prerecorded table of contents (TOC). The TOC provides start positions for the optional prerecorded user area, user table of contents (UTOC), user recordable area and lead-out area. Information such as laser recording power requirements and disc layout description is also provided in the TOC. User recordable areas are typically pregrooved to guide the write laser and to ensure compatibility with the CD format. Two widths have been standardized for the pregroove and are simply referred to as the narrow and wide pregroove. A CD-WO disc may also be provided without prerecorded information. The layout for this disc is shown in Figure 15b. The optional prerecorded user area is not included in the program area of this disc.

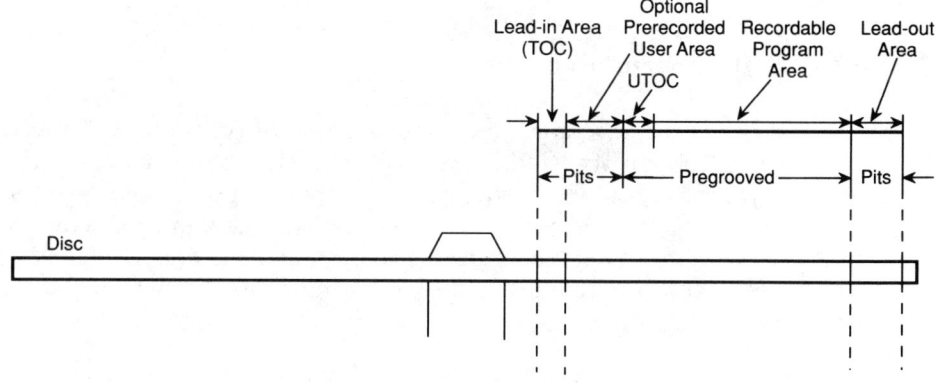

(a) Disc with preformatting.

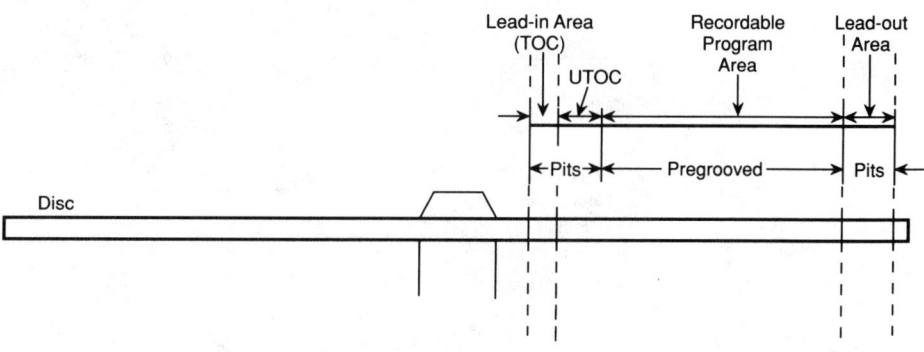

(b) Disc without preformatting.

**Figure 15.** CD-WO formats.

Constant linear velocity (CLV) clocking information is provided by a low-frequency radial groove wobble which permits the servo controlling rotation of the disc to properly set the appropriate CLV at any given radial position on the disc. The groove wobble motion is modulated to create a timecode signal throughout the grooved track. CD-WO discs do not require continuous recording across the disc. New data may be recorded at the end of a previously recorded track.

All prerecorded information is recorded with pits. All user data is stored as changes in reflectivity. A change in reflectivity occurs in the recording layer during the recording process. A typical in-groove reflectivity is 72%. The narrow-pregroove system utilizes a high-to-low reflective transitional material, and the wide-pregroove system uses a low-to-high transitional material to simplify servo operation. All tracks on the disc, including prerecorded and user tracks, are numbered, with a maximum of 99 tracks. The tracks may be of any length. Up to 650 MB of data can be held on a disc corresponding to 74 minutes and 33 seconds of audio information. A carrier-to-noise ratio of 47 to 50 dB is common. Shelf-life expectancy is over ten years under proper conditions.

## CD-WO Disc Structure

CD-WO disc structures are similar to that of conventional compact discs, with the exception of an additional recording layer as shown in Figure 16. A reflective layer is sandwiched between the substrate and protective layer. An organic dye recording layer is placed between the substrate and reflective layer. The substrate is made of a transparent polycarbonate material designed to allow the laser beam to penetrate to the recording layer. CD-WO discs are single-sided media.

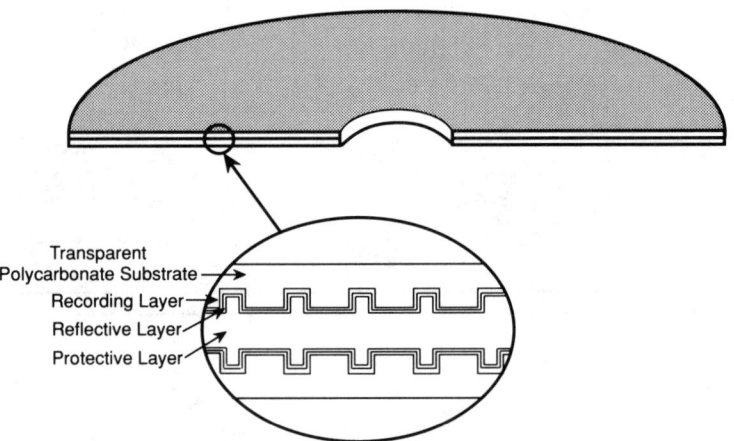

**Figure 16.** CD-WO disc structure.

## CD-WO Applications

Artists and producers are choosing to master their productions to CD because of its high audio reproduction quality and its world wide acceptance. Prior to the introduction of CD-WO, artists and producers waited several weeks to hear their production on CD. The introduction of CD-WO products makes it possible to produce a reference CD at the studio after completing the editing session. Reference copies may be provided for artists and producers to review and approve. CD-WO discs are able to provide reference copies identical with the end product; this was previously not possible.

Often CD mastering and replication is too expensive or time consuming.

Projects that are not expected to require mass production will find CD-WO a viable option. CD-WO also provides a good medium for archiving projects.

CD-WO may also find applications in recording movie sound effects and music for postproduction. Voice tracks for computer aided instruction (CAI) and endless-loop tapes for public address systems, as well as tapes for radio jingles, commercials, and background music, may be replaced by CD-WO.

Two CD-WO systems have recently been developed providing custom recording capabilities. The first of these systems converts pulse code modulation (PCM) signals from a PCM-1630 processor or other source to EFM signals. A *PQ*-subcode generator formats user entered information, such as song titles, catalog numbers, SMPTE time code information, the artist's name, and other comments. The information is processed and edited prior to being optically recorded to disc. The optical recorder achieves a scanning speed of 1.3 m/s (4.26 ft/s) and provides 60 minutes of recording time with provisions for up to 68 minutes. This system uses a gallium-arsenide (GaAs) diode laser operable over a range of wavelengths from 765 to 795 nm and an output power of 0.5 W. Over ten years of archival use can be expected from the disc. This system is provided by Yamaha and Gotham Audio.

A second system designed for the CD format also uses an EFM encoder and a CD format processor controlled by computer. Several optical recorders may be included in the package for multiple recording capabilities. All projects are normally transferred to a master tape or hard disk with indexing and coding. A *PQ* subcode generator adds *PQ* encoding information and the optical drives record the information onto disc. The discs used for this system utilize gold for the reflective layer because it lasts longer than aluminum and does not undergo oxidation. A green dye is placed under the reflective layer to absorb the laser energy. The heat generated by the laser decomposes the dye, which causes the polycarbonate substrate layer to expand and mix with the dye. The polycarbonate and dye mixture form a pit, causing a decrease in reflectivity of the disc. The change in reflectivity can be read by conventional compact disc players. This CD-WO system is produced by Sony Taiyo Advanced Research Technology (START Labs).

Both systems are capable of duplicating anything that can be recorded to the master tape, including CD-ROM, CD-I, and CD-Graphics information.

## Other Write-Once Systems

Other write-once formats are available with a large range of disc diameters. Write-once discs exist in 14-, 12-, 8-, 5.25-, 3.25-, and 3-inch sizes. The thickness of these discs varies depending on the disc's structure. These discs do not conform to the CD format and are used primarily in computer applications. Scanning velocity corresponds to the formatting of the tracks on the disc. Scanning at a constant linear velocity (CLV) is required if the data is contained in a spiraling track as with the CD standard. The rotational speed of the disc is varied to maintain a constant data rate. A disc formatted for constant angular velocity (CAV) rotates at a constant speed. The data on the outer tracks is recorded at a lower density than on the inside tracks. The overall data density of the CAV format is lower than that of the CLV format. The CAV discs are often divided into concentric data sectors. Figure 17 shows both CLV and CAV disc formatting.

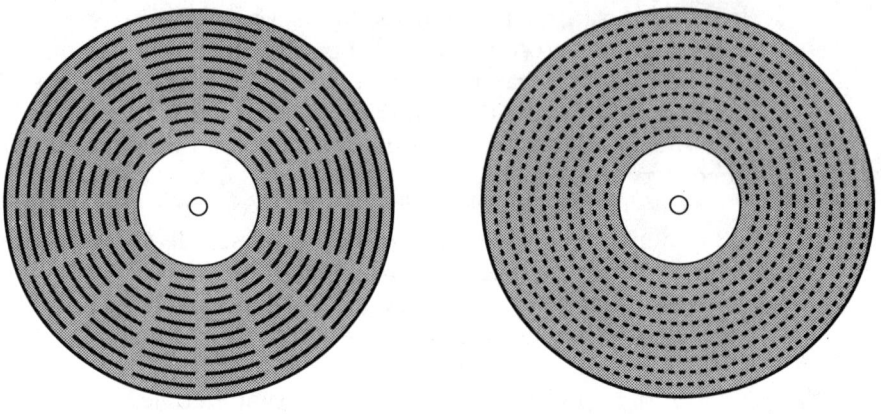

*(a) Constant angular velocity (CAV) format.*  *(b) Constant linear velocity (CLV) format.*

**Figure 17.** Optical disc formatting.

Modulating schemes vary in write-once systems. Two-to-seven modulation is adapted from magnetic hard drive for many write-once systems, and four-to-fifteen modulation is also common. The Reed-Solomon long-distance and product codes are typically utilized for error correction. Corrected error rates of $10^{-14}$ are claimed given a raw bit error rate of $10^{-4}$ from the disc.

The CD standard utilizes pit length modulation to represent data on the disc. Some write-once discs apply mark interval modulation as shown in Figure 18 due to difficulties in controlling pit shapes in some write-once media. Mark length modulation provides greater recording density but also demands greater accuracy in pit edge positioning.

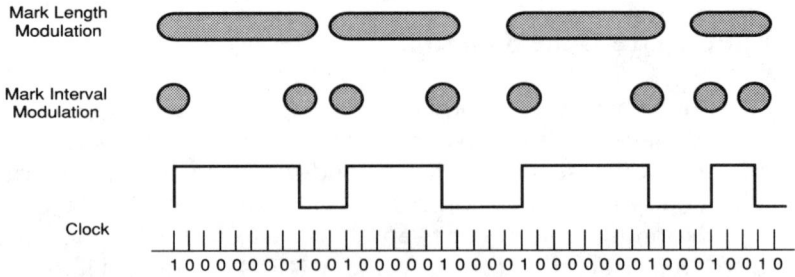

**Figure 18.** Pit modulation techniques. Compact disc uses mark length modulation. Some write-once systems use mark interval modulation.

Many write-once drives provide interfacing via a small computer serial interface (SCSI). This allows for data transfer between the host system and the write-once drive.

## Structures

Write-once disc structures vary depending on the technology used for recording. Typically, a reflective material, such as aluminum, is sandwiched between a substrate (composed of toughened glass, polycarbonate plastic, or aluminum) and a protective layer. Write-once discs can be designed with either an air-incident or substrate-incident structure. The former requires that the laser beam access the reflective layer through the transparent protective layer. In this case the substrate may either be opaque or transparent. The latter requires the reflective layer be accessed through the substrate. The substrate must therefore be transparent; the protective layer need not be transparent. Tellurium and its alloys are the most common materials utilized for the reflective material because of their low melting point and high sensitivity. For large discs, glass substrates are commonly used because of their strength and stability although plastic is attractive due to its lower cost. Smaller discs almost exclusively utilize plastic substrates.

Some write-once discs provide a symmetrical construction allowing double-sided recording even though most drives do not have two-sided recording capabilities. The extra layers are beneficial in providing additional strength and for reducing disc deformation. An air gap often exists between the two halves of a dual-sided disc which may be sealed or left open to prevent distortion due to external pressure changes.

## Audio Applications

Audio applications for write-once media include archiving of musical performances. Artists choose to archive master recordings on optical media because of the large storage capacity, long life expectancy, and durability. A 12-inch double sided disc is capable of storing over 2 GB of data or three hours of stereo digital audio. Figure 19 shows a typical arrangement for archiving recorded musical performances. A magnetic hard drive is utilized for editing digital information prior to its transfer to optical disc. A computer provides control over editing, and DAT copies are used to audition audio material and as reference tapes for the artist. Write-once optical media should be carefully stored under manufacturer's recommendations. Table 3 shows commonly recommended storage requirements as well as conditions for transporting write-once discs. There are also many applications for write-once systems in the computer industry. Storage of business transactions, computer images, and employee data bases are a few possible applications.

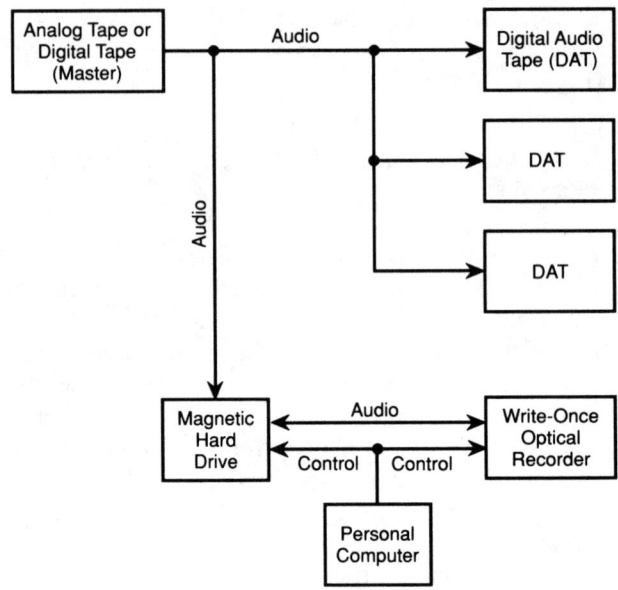

**Figure 19.** Arrangement for archiving audio to write-once optical discs.

**Table 3.** Write-Once Disc Permanent Storage and Transporting Requirements

**Storage Conditions**

| | |
|---|---|
| Temperature Range | 14°F to 120°F |
| Relative Humidity | 10% to 90% |
| Atmospheric Pressure | 75 kPa to 105 kPa |

**Transporting Conditions**

| | |
|---|---|
| Temperature Range | 14°F to 140°F |
| Relative Humidity | 5% to 90% |
| Atmospheric Pressure | 75 kPa to 105 kPa |

# Erasable Optical Systems

Erasable optical media have been the aim of much research. Manufacturers seek to incorporate the familiar benefits of magnetic tape into optical media while maintaining the advantageous characteristics of the latter. The ability to record, play back, erase, and record again has kept magnetic media at the forefront of recording technology. Optical systems provide many advantages over magnetic media, but recordability has previously been their weakness.

Several erasable optical technologies have been introduced, with three types in principle use: magneto-optics, dye polymer, and phase change.

The performance of erasable media has slowly improved with advances in medium technology. The first erasable discs were found to be limited to only a few hundred recording cycles due to fatigue. In some technologies disc fatigue limits the ability to record on a disc but does not effect data already recorded on a disc. Over one million recording cycles are reported in recently introduced discs. Specifically, dye polymer and phase change methods are currently limited to a few thousand recording cycles, whereas magneto-optical discs are able to provide an almost unlimited number of recording cycles. The lifetime of the recorded data is estimated to be greater than ten years.

## Magneto-Optical Recording

Magneto-optical recording (MOR) utilizes magnetic properties and optical technology to perform erasing and recording on a magneto-optical disc (MOD). It is often referred to as optically assisted magnetic recording. It uses a vertically magnetized medium as shown in Figure 20a. The advantage of a vertically magnetized medium is increased recording density and shorter recorded wavelengths.

A magneto-optical disc consists of a recording layer sandwiched between a substrate and a protective layer (Figure 20a). As with write-once discs the laser may penetrate either the substrate or the protective layer. The only stipulation is that the laser reaches the recording layer. Typically, a glass substrate is used with a recording material made of an amorphous, thin-film, magnetic material such as terbium (Tb) and iron (Fe) alloyed with gadolinium (Gd). A coercivity between 2500 and 5000 Oe is common. An adhesive layer made of silicon dioxide gives mechanical strength to the disc.

Magneto-optical recording utilizes a magnetic field and a laser beam to influence the magnetic orientation of particles within the recording layer of the optical medium. Data is encoded within the magnetic structure of the particles. The magnetic field is not powerful enough to effect the magnetic orientation of the particles unless a laser beam is simultaneously applied to the medium. As the laser beam heats the medium to its Curie temperature, the coercivity (magnetic force required to erase a medium) decreases, thus allowing the magnetic field to affect the magnetic orientation of the particles. The recording area is limited to the laser spot on the medium. After removing the laser, the spot cools to a temperature below the Curie temperature where it can no longer be affected by an external magnetic force.

The Kerr effect is used to play back the information stored on the disc. The Kerr effect describes the rotational properties of polarized light as it reflects from a magnetized medium. The laser beam of the magneto-optical recorder is initially linearly polarized. The plane of polarization rotates after the light is reflected from the medium. The angle of rotation of the plane of polarization depends on the alignment of the magnetized particles. The angle of rotation

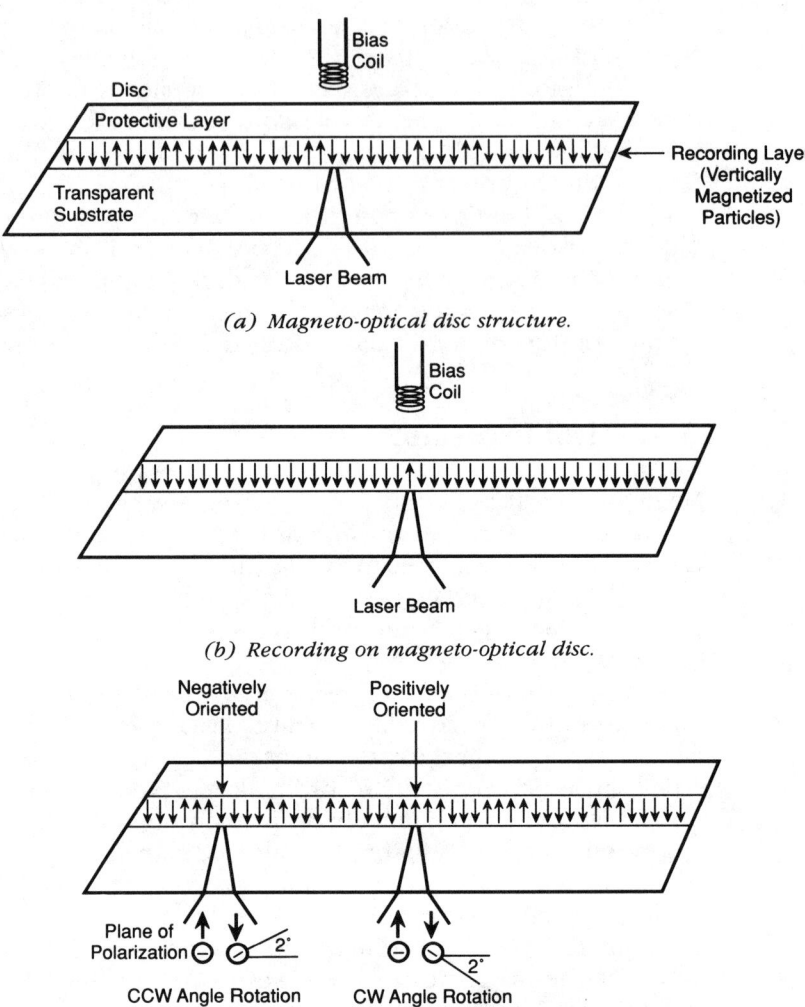

*(a) Magneto-optical disc structure.*

*(b) Recording on magneto-optical disc.*

*(c) Playback on magneto-optical disc.*

**Figure 20.** Magneto-optical recording.

differs for particles positively oriented from that for particles negatively oriented. The rotation angle is only about 2°. A total difference of 4° separates a clockwise rotation from a counterclockwise rotation. This small angle is enough to distinguish the two rotational directions. The reflected light is monitored to determine the angle of rotation and therefore the direction of magnetization. Figures 20b and 20c show the process for recording and playback of a magneto-optical disc, respectively. Additional optical components are required to convert the change in polarization into a change in intensity. The data can then be read with conventional optical detectors, decoded, and processed.

Magneto-optical recorders may be designed for playback of conventional CDs, but unfortunately magneto-optical discs are not readable by conventional CD players without modification. The reflectivity of the magneto-optical discs is about 20%, whereas the CD standard requires a minimum reflectivity of 70% (most CDs are actually closer to 90%), hindering complete interchangeability between magneto-optical systems and the conventional compact disc system.

Blank magneto-optical discs contain prerecorded addressing information to achieve compatibility with optical recorders. The addressing information is nonerasable and allows any magneto-optical player of like format to read the data. Both constant linear velocity (CLV) and constant angular velocity (CAV) formats are available. A pregrooved photopolymer layer may define the spiraling or concentric tracks to provide compatibility between magneto-optical recorders.

One magneto-optical recorder utilizes a 3.5-inch disc housed in a cartridge with a protective shutter similar to a 3.5-inch floppy disk. The data format is the same as CD-audio, and approximately 20 minutes of continuous audio recording is possible. This system utilizes a constant linear velocity format. A laser spot approximately 1 $\mu$m in diameter and a magnetic recording head provides a bias field of 200 to 300 gauss. During erasing the laser is constantly powered to provide uniform erasure and the recording head is reverse biased, returning all magnetic particles to a zero state. During recording the head is forward biased. Only the objective lens, tracking and focusing actuator, and mirror are mounted on the pickup carriage. All other optical components are mounted on a stationary baseplate. This provides fast access times. The laser diode used for this system has a wavelength of 780 nm and an output power of 35 mW, although typical laser settings are 4 to 6 mW for erasing and recording and 1 to 1.5 mW for reading. This system can easily be converted to 120 mm (4.72 in) compact discs by decreasing the linear velocity. Over one hour of recording is then possible. This recorder was developed by the Nakamichi Corporation.

A magneto-optical recorder designed by Thomson Consumer Electronics uses a disc with a recording layer made from a thin layer of terbium-iron-cobalt (TeFeCo) alloy. During recording the laser remains powered and the magnetic field is controlled by the digital data stream. Erasing and recording can be performed in the same sweep of the laser beam. The laser outputs 25 mW during recording and 0.8 mW during playback of magneto-optical discs and 0.4 mW for conventional CDs. An optional data compression technique is provided. The data is compressed using 4-bit quantization, which extends the recording time to four hours for digital stereo information on a 120 mm disc. Sixteen-bit A/D conversion with 64-times oversampling is provided at the input and 8-times oversampling with 16-bit D/A conversion at the output. This system has performed very well under test. Frequency response in the playback mode is flat between 20 Hz and 20 kHz. Linearity, dynamic range, and total harmonic distortion plus noise are comparable to those of medium- to high-priced CD players. In record/playback mode the frequency response is equivalent to that of the playback mode. The total harmonic distortion plus noise for the record/playback mode, though not as low as for the playback mode, is equal to that of many high-end compact disc players. Linearity is excellent in the record/playback mode. The D/A converters are designed to compensate in the playback mode for any slight linearity error introduced by the A/D converters in the recording mode.

## Dye Polymer Recording

The dye polymer method for recording is very similar to dye polymer technology used in write-once optical systems. Erasable dye polymer media require a bilayer structure consisting of an expansion layer and a retention layer. Each layer reacts to a different wavelength of laser light. The temperature of the expansion layer increases when absorbing light of a particular wavelength, causing the layer to expand. The retention layer is transparent to a laser beam but is forced to deform physically by the adjacent expanding layer. When the laser is removed the temperature of the medium decreases, but a bump remains in the retention layer. The bump causes laser scattering and a corresponding decrease in reflected light intensity. The data is encoded within the bumps on the disc and modulated by the reflected light intensity. The data is erased by applying a laser beam at a wavelength that is absorbed by the retention layer. The temperature of the retention layer then increases, which causes the layer to soften and return to its original form.

## Phase Change Recording

Erasable phase change systems use technology similar to that of the write-once phase change systems. Materials such as tellurium (Te) alloyed with germanium (Ge) or indium (In) have been used for the recording layer of phase change erasable media. Gallium antimonide (GaSb) and indium antimonide (InSb) have also been used in the recording layer. These materials change phase from a crystalline state to an amorphous state when heated above their melting point. When reheated to a temperature just below the melting point, the material returns to the crystalline state. Crystallization can also be performed by reheating the material above the melting point and allowing it to cool slowly so that the crystalline structure can develop. This method of crystallization can be performed by heating a larger area, for example by defocusing the laser spot and increasing the laser intensity. This slows down the cooling process, and crystallization occurs because this is the more stable state. These characteristics form the basis for recording and erasing respectively. The amorphous state exhibits low reflectivity, whereas the crystalline state exhibits high reflectivity. The encoded data is read by a laser passing over the surface of the disc. The data is encoded within the phase variations on the disc and modulated by the reflected light intensity.

## Disc Drives

Many manufacturers offer a multifunctional drive capable of reading several types of optical discs including write-once and erasable. Clearly, the standards must be the same for each type of optical recording medium. The compact disc format provides the standard for audio, but the standards for 2.5-, 3.5-, 5.25-, and 8-inch drives are diverse.

## Erasable Optical System Applications

Recording studios are implementing optical recorders as secondary storage systems with magnetic hard drives as the primary storage and editing media. Erasable optical recorders are often used as sound libraries. Sounds are copied from the optical disc to hard drive for editing.

Reduced access times and improved editing capabilities are required before erasable optical systems replace magnetic hard drives as the primary storage and editing media in the studio. Improved data compression is required before optical recorders will be able to record more than four channels on a disc. In the future single-disc multichannel optical recorders (possibly up to 24 tracks) may be realized.

Erasable optical systems are expected to replace write-once technology involving several applications where repetitive disc use is advantageous. Erasable reference discs and demo discs can be cheaper than write-once media because they can be reused. It is not likely that erasable optical storage systems will replace write-once media in archiving digital audio data. Write-once discs are more stable and have longer expected lifetimes than erasable discs.

# Conclusion

Many products have resulted from the most recent technical advances in optical media. Write-once and erasable optical systems now provide convenient recording capabilities like those provided by magnetic tape and hard drives. Several models already been introduced into the professional market. Other developments for recordable optical disc systems include multifunctional optical players with multichannel recording and editing features, as well as recordable/erasable mini-disc formats for consumer applications.

# References

Baert, L., Theunrsson, L., and G. Verfult, *Digital Audio and Compact Disc Technology*, Heinemann Professional Publishing, 1988.

Bradley, A., *Optical Storage for Computers: Technology and Application*, Ellis Horwood, 1989.

Bouwhuis, G., Braat, J., Huijer, A., Pasman, J., van Rasmalen, G., and K. S. Immink, *Principles of Optical Disc Systems*, Adam Hilger, 1985.

Feldman, L., "Exclusive U.S. Test: Thomson CD Recorder," *Audio Magazine*, March, 1990.

Freese, R. P., "Optical Disks Become Erasable," *IEEE Spectrum*, February, 1988.

Gosch, J., "From Thomson, A CD Player that Erases and Records," *Electronics*, March 17, 1988.

Isailovic, J., *Videodisc and Optical Memory Systems*, Prentice-Hall, 1985.

Jamieson, R. S., "Optical Firmware," *NASA Technical Briefs*, vol. 13, no. 5, item 132, 1989.

Joch, A., "Sony, Alphatronix Enter Race to Debut Erasable-Optical Disk Drives," *Mini-Micro Systems*, January, 1989.

Jurgen, R. K., "Consumer Electronics," *IEEE Spectrum*, January, 1988.

Lambert, S., and S. Ropiequet, *CD-ROM, The New Papyrus*, Microsoft Press, 1986.

Lenk, J. D., *Troubleshooting and Repair of Audio Equipment*, Howard W. Sams & Co., 1987.

Marshall, G. F., *Laser Beam Scanning, Opto-Mechanical Devices, Systems, and Data Storage Optics*, Marcel Dekker, 1985.

Mascenik, S. A., "A Magneto-Optical Disk Digital Audio Recorder," *Audio in Digital Times: AES 7th Conference Proceedings*, May, 1989.

Murata, S., Tanifuji, T., Yamamoto, H., Sumi, S., and K. Torazawa, "Multimedia Type Digital Audio Disc System," *IEEE Transactions on Consumer Electronics*, vol. 35, no. 3, August, 1989.

Oppenheim, C., *CD-ROM Fundamentals to Applications*, Butterworth, 1988.

Pohlmann, K. C., *The Compact Disc: A Handbook of Theory and Use*, A-R Editions, 1989.

———, *Principles of Digital Audio*, 2nd edition, Howard W. Sams & Co., 1989.

Roth, J. P., *Essential Guide to CD-ROM*, Meckler Publishing, 1986.

Verkaik, W., "Compact Disc Mastering, An Industrial Process," *Digital Audio Collected Papers*, Audio Engineering Society, 1983.

Yamamoto, T., "Optical Recording Technology: Can Optical Discs Be Applicable to Digital Audio Workstations?" *Audio in Digital Times: AES 7th Conference Proceedings*, May, 1989.

*Chapter 7*

# Digital Audio for Video and Film

### Michael Shawn Ballman

## Introduction

Earlier chapters presented conversion and storage techniques that apply to digital data where audio is the only information being represented. This concept has recently been extended into multimedia systems as well. In particular, several standards have been developed for the use of digital audio with video and film. This chapter discusses the various formats employing digital audio in both video and film.

## Digital Audio for Video

Since the development of video recorders, the quality of accompanying audio sound tracks has been sacrificed because the primary consideration has been the video signal. With the advent of digital audio systems, however, a strong emphasis has been placed on achieving that same digital audio fidelity in video systems. From the professional component and composite video systems to the 8 mm video consumer format and the industry standard 1-inch Type C format, the implementation of digital audio in video media has become widespread.

### D1—Component Digital Video

The D1 or 4:2:2 format was the first worldwide standard for component digital television recording. Years of dedication, effort, and cooperation from groups such as SMPTE, EBU, CCIR, AES, and a number of private corporations have culminated in what has been called the most significant technological achievement in television history.

## Cassette Shell and Magnetic Tape Specifications

Since the development of the first videocassette format, cassette designs have offered many advantages over tape reel formats. Cassettes allow automatic threading through the transport, enable the tape to be removed from a machine without winding to an end, and protect the tape from impurities and careless handling. The D1 cassette shell uses double doors to completely protect the tape when it is outside of the machine. When the cassette is inserted into the machine, outer lid lock tabs are released, exposing the tape to the transport elements.

The standard cassette dimensions call for three sizes, small, medium, and large, labelled D1-S, D1-M, and D1-L, respectively. Figure 1 gives the dimensions of a D1 videocassette. Specific measurements for each size are given. Small, medium, and large cassette shells have varying lengths and widths. This implies that digital component video machines must be flexible enough to adjust for these differences. More significantly, the distance between the supply and takeup hubs is also different for each cassette size. These variations necessitate the mechanical complexity of movable supply and takeup motors. It should also be noted that parameters such as the window size and outer lid extension vary slightly among the D1-S, D1-M, and D1-L cassettes.

Each D1 videocassette housing has four coding holes and four user holes. The coding holes indicate tape parameters such as tape type and thickness and are set by the manufacturer at the time of production. The user holes, on the other hand, are accessible and allow information and status about the recording to be set by hand. Table 1 details the significance of both the coding and user holes. It should be noted, not apparent from the diagram, that individual user plugs are set differently. Plugs 1 and 2 require the use of a small screwdriver, while plugs 3 and 4 can be merely pushed from one side of the cassette or the other.

The D1 format supports tape thicknesses of 16 and 13 $\mu$m (see Table 1). The 16-$\mu$m tape thickness has playing times of 11, 34, and 76 minutes. The less common tape thickness of 13 $\mu$m yields running times of 13, 42, and 94 minutes. Many tape manufacturers also offer variable tape lengths running from 3 to over 200 minutes. This has been favorably accepted in the video industry because tape lengths can more precisely approximate program material from short commercial spots to full-length features without unnecessary waste of this expensive medium.

Tape formulation is metal oxide with a polyester base. The tape coercivity is 850 Oe, and the width of the tape is 19.010 mm, which is almost identical with the U-matic ¾-inch format. Standard D1 cassettes also feature transparent leader and trailer segments for detecting the beginning and end of a tape. These tape characteristics are currently implemented. Other types, thicknesses, and coercivity levels may be implemented in the future; variances will be detected with available coding hole configurations.

Chapter 7: Digital Audio for Video and Film

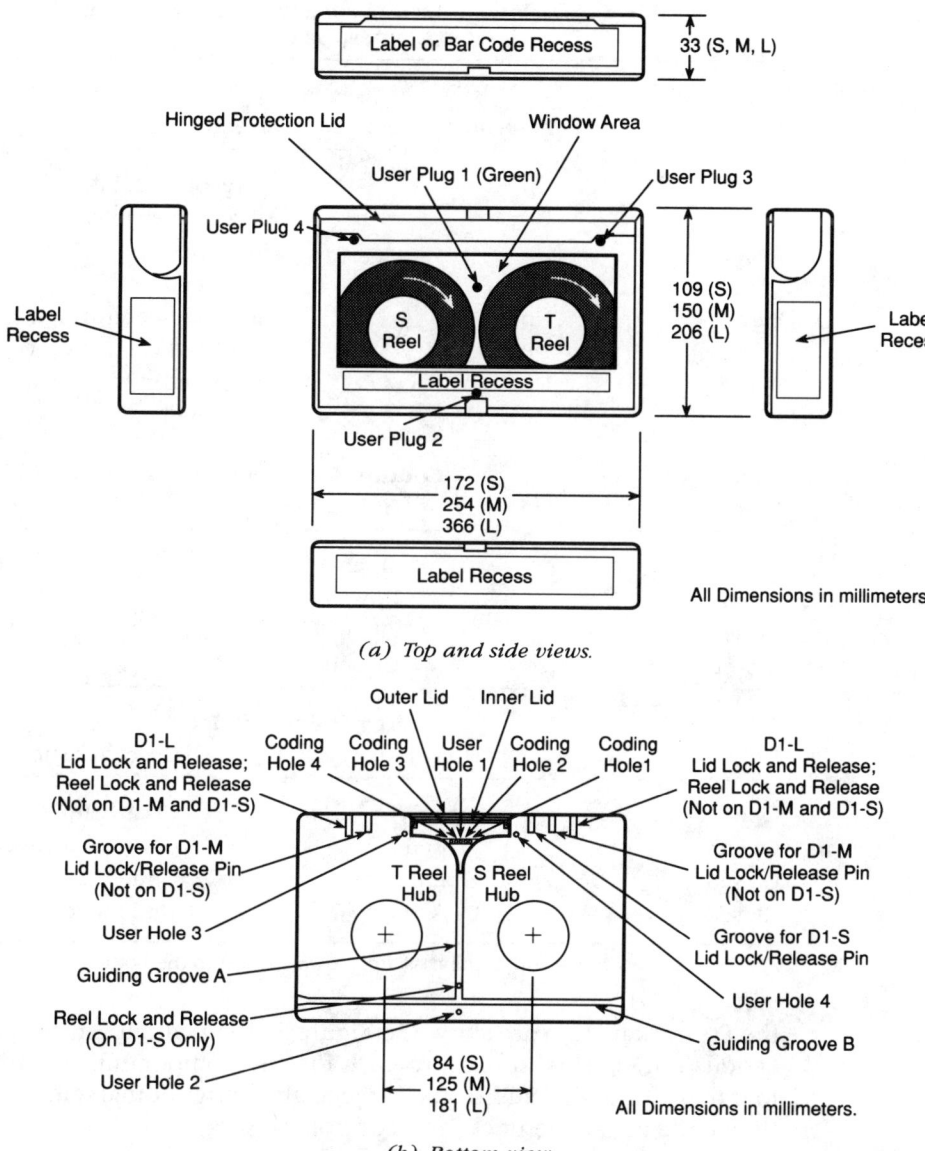

**Figure 1.** D1 videocassette physical dimensions. *(Courtesy Sony Corp.)*

## Track Pattern and Transport

The design of the track pattern for a new video format is of great importance because adherence to its parameters ensures that tapes will be compatible between machines from different manufacturers. The layout and dimensions of

**Table 1.** D1 Videocassette User and Coding Hole Configurations
(Courtesy Sony Corp.)

| Plug | Position* | User Plugs Effect |
|---|---|---|
| Plug 1 | Up | All record inhibit |
|  | Down | All record enable |
| Plug 2 | Up | Undefined |
|  | Down | Undefined |
| Plug 3 | Up | Video and control signal record inhibit |
|  | Down | Video and control signal record enable |
| Plug 4 | Up | Undefined |
|  | Down | Undefined |

| Hole 1 | Hole 2 | Characteristic |
|---|---|---|
| | **Coding Holes 1 and 2** | |
| Open | Open | 16 $\mu$m tape |
| Open | Closed | 13 $\mu$m tape |
| Closed | Open | Undefined |
| Closed | Closed | Undefined |

| Hole 3 | Hole 4 | Characteristic |
|---|---|---|
| | **Coding Holes 3 and 4** | |
| Open | Open | Class 850 Oe coercivity |
| Open | Closed | Class 1500 Oe coercivity (D2) |
| Closed | Open | Undefined |
| Closed | Closed | Undefined |

*Up: closest to the top of the cassette. Down: closest to the bottom of the cassette.

the D1 track pattern are shown in Figure 2 along with tolerance figures for the various tracks. This format provides for one component digital video signal, four tracks of digital audio, and three analog longitudinal tracks consisting of a time code track, cue track, and control track.

Servo control is provided by the control track. The control signal consists of a 150 Hz reference pulse for governing tape speed and optional interspersed pulses which mark the start of an audio, video, and color frame. Their frequencies depend on the frame rate of the system in use. Although monitoring digital audio tracks at varying shuttle speeds is possible, it has not yet been effectively implemented. Thus, the D1 component digital video format also includes an analog cue track for monitoring when not at normal speed. Any combination of the digital audio tracks can be mixed down to the cue track in real time during playback or record. A longitudinal time code track is located at the bottom of the tape. Unlike conventional time code tracks, this track

# Chapter 7: Digital Audio for Video and Film

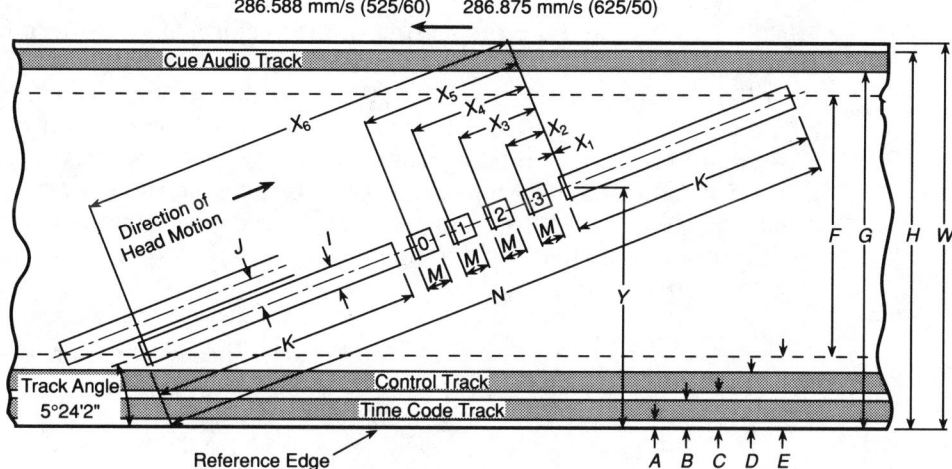

(a) Layout diagram.

| | Dimensions | Nominal (mm) | Tolerance (mm) |
|---|---|---|---|
| A | Time-code track lower edge | 0.2 | ± 0.1 |
| B | Time-code track upper edge | 0.7 | ± 0.1 |
| C | Control track lower edge | 1.0 | ± 0.1 |
| D | Control track upper edge | 1.5 | ± 0.05 |
| E | Lower edge of program area | 1.8 | (Derived) |
| F | Program area width | 16.00 | (Derived) |
| G | Cue audio track lower edge | 18.1 | ± 0.15 |
| H | Cue audio track upper edge | 18.8 | ± 0.2 |
| I | Program track width | 0.040 | ± 0/0.005 |
| J | Program track pitch | 0.045 | Basic |
| K | Video sector length | 77.78 | (Derived) |
| M | Audio sector length | 2.56 | (Derived) |
| N | Program track total length | 170.00 | (Derived) |
| W | Tape width | 19.010 | 0.015 |
| $X_1$ | Location of start of upper video sector | 0.0 | ± 0.1 |
| $X_2$ | Location of start of audio sector 3 | 3.4 | ± 0.1 |
| $X_3$ | Location of start of audio sector 2 | 6.8 | ± 0.1 |
| $X_4$ | Location of start of audio sector 1 | 10.2 | ± 0.1 |
| $X_5$ | Location of start of audio sector 0 | 13.6 | ± 0.1 |
| $X_6$ | Location of start of lower video sector | 92.2 | ± 0.1 |
| Y | Program track reference location | 10.490 | Basic |

(b) Dimensions.

**Figure 2.** D1 track pattern showing layout and dimensions.
(Courtesy Sony Corp.)

incorporates the newest revision in the time code standard to allow for dual timing references. In this way, an internal continuous reference can be made while time code numbers from the source material are retained. This eliminates the need to use an audio track to record time code corresponding to source video.

Although location of the digital audio tracks in the center of the tape improves robustness of the data, it causes the video tracks to be segmented. However, segmented digital video tracks do not present any of the problems inherent in analog segmented video tracks. Each section of recorded video is called a sector and the processing of the sector is unique. To reduce the rate and quantity of the data to be processed, component digital video machines utilize four individual processing blocks and video heads each of which handle a sector. To facilitate recording and editing, information contained in a field must be represented by a multiple of four video sectors. A field of data in the 525/60 (NTSC) system requires 20 sectors or 10 video tracks, while the 625/50 (PAL/SECAM) system requires the use of 24 sectors or 12 video tracks. The start and finish of a video field along with the ordering of sectors for both systems is shown in Figure 3.

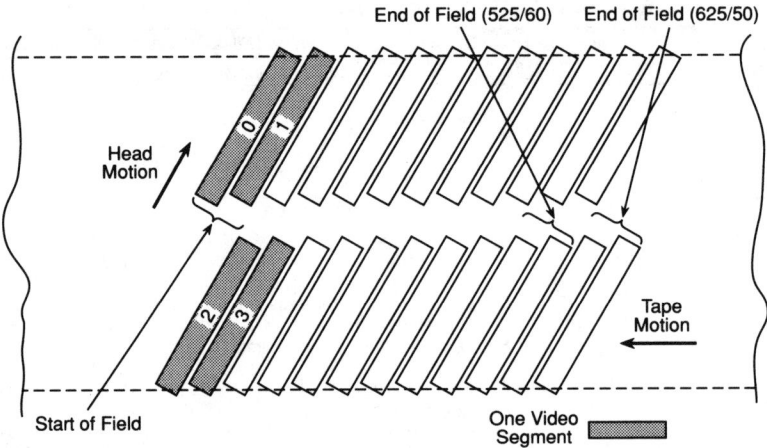

**Figure 3.** D1 video sector/field relationship for the 525/60 and 625/50 television standards. *(Courtesy Sony Corp.)*

Like video, audio information is processed in four sectors simultaneously. However, while each video sector contains unique data, there is 100% redundancy among audio sectors. The orientation and numbering of audio tracks within a video field are shown in Figure 4. The four shaded sectors represent information for audio channel number 1 at the start of a field. Two identical odd samples and two identical even samples along with error correction information comprise the four sectors. This track layout provides complete protection from tape edge problems. A complete duplication of audio information is contained in one processed segment; this multiplexing scheme among audio channels ensures that each sector in a processed segment is recorded and played back by a different processing block and head and placed at a different distance from any one tape edge. It is interesting to note that three audio segments exactly correspond to a video field in the 625/50 system, while two and a half audio segments correspond to a video field in the 525/60 system. Thus audio and video fields are only correlated at the end of a frame in the 525/60 system.

The standardization of a track pattern without specific constraints on transport mechanisms and head configurations allows the manufacturer to deter-

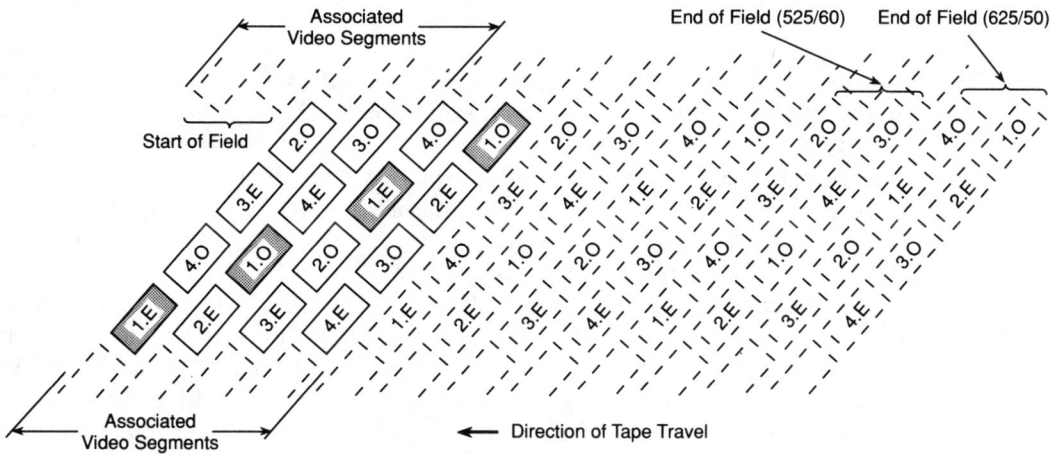

**Figure 4.** D1 audio sector/video field relationship in the 525/60 and 625/50 television standards. *(Courtesy Sony Corp.)*

mine what type of mechanical layout will be used. Any number of drum sizes and head configurations could be used to realize the D1 track pattern format with varying drum rotation rates and time compression/expansion techniques. Figure 5 shows the head/drum configuration for the world's first production component digital video tape recorder, the Sony DVR-1000.

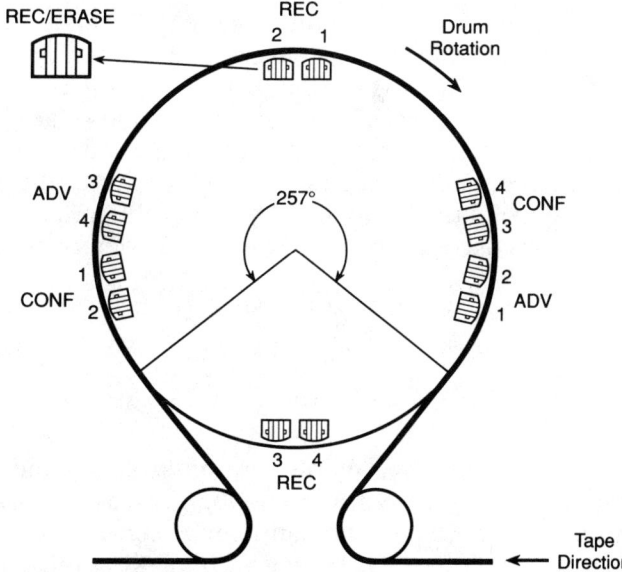

**Figure 5.** Sony DVR-1000 component DVTR head/drum configuration. *(Courtesy Sony Corp.)*

## Video Processing

The word component is derived from the method used for video recording in the D1 system. Color and luminance components are kept separated throughout digital processing of the video signal. The alternate name of 4:2:2 arose from the ratios of frequencies used for sampling these different video components. Eight-bit samples provide enough quantization levels for both the luminance and color components. The first component, the luminance ($Y$) component, is sampled at a frequency of 13.5 MHz. The next component, the color-difference component ($R - Y$) is sampled at 6.75 MHz. Likewise, the last component, the color-difference component ($B - Y$) is sampled at 6.75 MHz. The reason why this is referred to as 4:2:2 and not 2:1:1 is because CCIR Recommendation 601, which details the structure for digital video data, defines the fundamental reference frequency to be 3.375 MHz. Other applications in this document utilize ratios of 2:1:1, 3:1:1, 4:1:1, and even 4:4:4 for the high-quality *RGB* digital system.

CCIR Recommendations 601 and 656, EBU technical documents 3246 and 3247, and SMPTE Recommended Practice 125 serve as the standards for parallel and serial digital video interfacing for both 525/60 and 625/50 systems. D1 component digital video tape recorders use the parallel interface format as the basis for video processing. This interface utilizes nine balanced signal pairs of a standard 25-pin connector cable; the first eight pairs carry the digital video words, and the remaining pair conveys a synchronous clock signal. Digital data received serially is converted from an 8-bit to a 9-bit coding scheme suggested by the CCIR. At the output, parallel data is reserialized by the inverse scheme if needed. Where analog video signals are present at the input, some studio machines feature A/D converters for digitizing the signal. Likewise, D/A converters provide analog video signals at the output.

The amplitude of the luminance component is approximately 50% to 70% less than the color-difference signals on average. For this reason, the color-difference signals are weighted before coding occurs. Reduced color-difference signals are notated as *CR* and *CB*. Specifically, the following expressions dictate weighting factors for the color-difference samples and the system of equations used for recovering the *RGB* components:

$$Y = 0.299R + 0.587G + 0.114B,$$
$$CR = 0.500R - 0.419G - 0.081B = 0.713(R - Y),$$
$$CB = -0.169R - 0.331G + 0.500B = 0.564(B - Y).$$

After weighting, samples are multiplexed and clocked onto the parallel video bus at 27 Mword/s. Remembering that luminance is sampled at twice the color-difference rate, the multiplexed signal takes the form of | *CB Y CR Y* | *CB Y CR Y* | *CB Y CR Y* | with *CB* marking the start of an active video line. There are 720 *Y* samples, and 360 samples each of the *CB* and *CR* components, which results in 1440 multiplexed samples per active digital video line. A horizontal blanking period completes the timing relationship among lines. Both the 525/60 system and the 625/50 system provide space in the data stream for auxiliary data. Although not defined in the basic standard, this data may contain ID information or equipment checking data. While all of this data is generated, only the 1440 bytes representing the active part of the line are recorded. Thus, 250 lines per field are recorded in the 525/60 system, while 300 lines per field are recorded in the 625/50 system. The 720 pixels (picture elements) per television line are regenerated using the corresponding *Y* component and its neighboring *CB* and *CR* components.

As noted, analog and serial digital video signals are converted to parallel digital form at the input and reconverted (if needed) at the output. It is in the parallel form that the heart of the D1 video processing takes place. Decoding the signal requires separation of the 1440 samples of the active video line from timing and auxiliary data. Demultiplexing is more complex because the active video line data must be rearranged into four separate streams corresponding to the four individual processing blocks and video heads.

Source mapping is the next major processing step in the video data path. This process serves to boost the integrity of data on a pixel-by-pixel basis. Eight-bit values corresponding to *Y, CB*, and *CR* components are mapped into new 8-bit values. These new values ensure that translation back to the original 8-bit value will result in a minimum deviation from the original sample if errors are encountered during the encoding or decoding process. This form of error concealment works well as minor visual errors are less perceptible than their aural counterparts. Common components, all *CR* components for example, are grouped together by the process of intraline shuffling. Samples within a video line are separated into like columns of 30 samples. These columns are used to generate two Reed-Solomon check words per column via the outer error coder.

Sector array shuffling is the last measure of processing before the digital video data is multiplexed with the audio data and sent to the channel coding section of the DVTR. In sector array shuffling, active video data and outer code words are interleaved to form a new array of data. This array forms the basic data blocks that are clocked out and multiplexed with the audio blocks. In addition, inner code words are formed across the rows of this array. As video data blocks are sent to the channel coder, inner code words are appended.

## Audio Processing

The input and output of digital audio signals in the D1 format adheres to the serial AES/EBU format. For analog audio input signals, A/D converters sampling at 48 kHz are used and correspondingly there are D/A converters at the output. The D1 format also supports the various 16- to 24-bit modes outlined by the AES/EBU format. Although an AES/EBU interface allows a pair of digital signals to be carried over a single cable, the D1 format specifies the implementation of four individual AES/EBU interfaces. In this way, digital audio signals can originate at different sources. If, for some reason, there are more than four signals available at the input (with four interfaces there can be up to eight), appropriate signals will be selectable. Technical information about this standard is contained in the ANSI S4.40 1985 and AES3-1985 documents and EBU technical document 3250-E.

Digital audio signals enter the main processing blocks of the component digital video recorder in their AES/EBU form. Audio processing is very similar to video processing. The first step of decoding and demultiplexing serves the general purpose of separating raw sample data into four separate streams, one for each processing channel. Until they are separated, however, channels are dealt with two at a time, with channels 1 and 2, and 3 and 4 as a pair. Figure 6 shows a more detailed block diagram which outlines processing for a pair of signals at this stage. The bi-phase mark decoder in conjunction with the clock regenerator recovers the serial audio data and generates a sync block for the serial-to-parallel converter and demultiplexer. The serial-to-parallel converter provides a parallel stream of data which is demultiplexed into two separate parallel streams, one for each channel. During the demultiplexing process, 24-bit audio samples are truncated to 20 bits. Information for each channel is then fed into a word former, which adds any interface, processing, or user control data that may be needed.

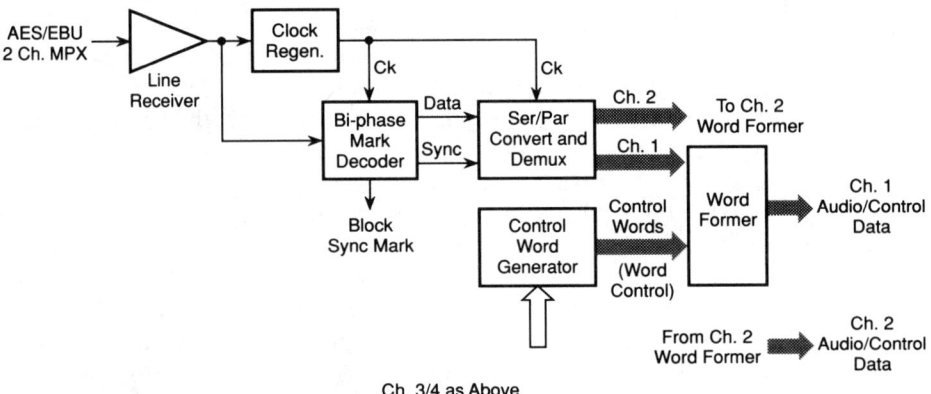

**Figure 6.** D1 audio decode and demultiplex. *(Courtesy Sony Corp.)*

Following separation into four data streams, data enters the intragroup shuffling portion of the circuit. Intragroup shuffling is the audio complement of the intraline shuffle for video processing. Odd and even samples, along with the added control data, are placed in an array which is used by the outer error coder. Reed-Solomon check words are formed along the columns of the array as in video processing.

Data shuffling is performed before the audio signal is multiplexed with the video signal. Like the sector array shuffling step in video processing, audio data is reorganized into a new array consisting of audio samples, added control data, and Reed-Solomon check words. Rows of the array are used to generate inner error codes which are appended to data blocks as they are sent to the channel coder.

## Channel Coding

Channel coding is the final processing step before sending the signals to the heads and tape. At the start of the channel coding section, video and audio data received on a total of eight parallel busses (four for each) are multiplexed into a single set of four busses to maintain consistency with the processing blocks and video heads. With video and audio data interspersed, an array is formed for constructing a final level of Reed-Solomon correction check bytes. A data randomizer is used to reduce DC content caused from a long string of ones or zeros at the output of the data serializer. The randomized video and audio data is multiplexed with synchronization and identification data from the sync and ID generator. Finally, these complete data blocks are converted to four serial data streams using an NRZ code.

Information sent to the heads is delineated in sync blocks. The sync block format implemented in D1 is shown in Figure 7. This 134-byte block consists of two blocks of sync and two blocks of identification information followed by a pair of 60-byte data and four-byte error code segments. Sixty-byte data blocks are filled with video information for video sectors and audio information for audio sectors. As shown in Figure 8, one rotary track consists of two 160-block video sectors surrounding four 5-block audio sectors. Each sector has a pre and postamble and is separated from neighboring sectors by an edit gap. A complete diagram of the audio and video record and playback processing paths used in the D1 format is shown in Figure 9.

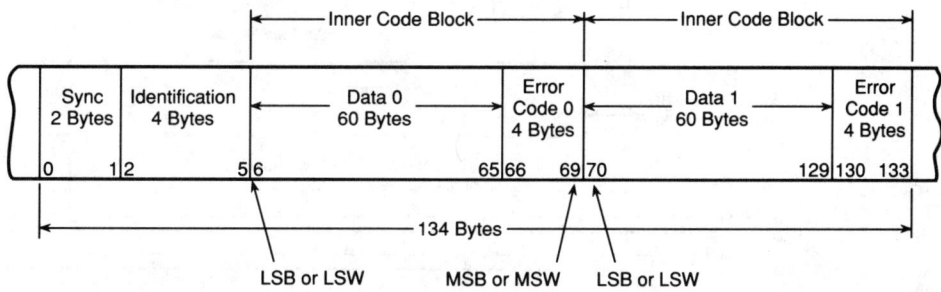

**Figure 7.** D1 sync block format. *(Courtesy Sony Corp.)*

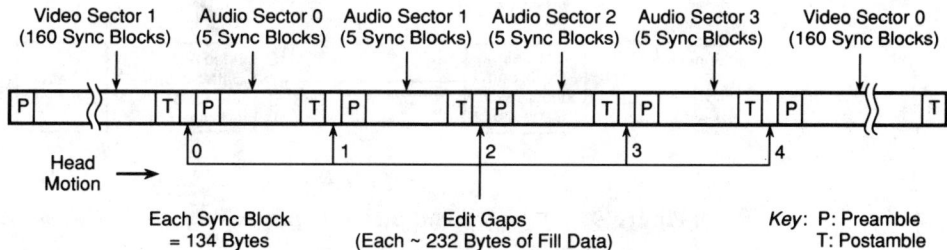

**Figure 8.** D1 sync block/rotary track relationship. *(Courtesy Sony Corp.)*

## D2—Composite Digital Video

The development and standardization of the D1 and D2 formats occurred simultaneously; thus they share many of the same specifications and technologies. The D2 format is discussed here with reference to the D1 format previously described. Significant differences are explained in detail, while commonalities are mentioned briefly.

### Cassette Shell and Magnetic Tape Specifications

The D2 format uses the same cassette shell as specified for component digital video. Coding and user hole configurations are identical. However, as the coding hole options suggest (see Table 1), only tape with a coercivity of 1500 Oe or the equivalent is used for composite digital video recording. Also, because the D2 linear tape speed is slower than that in D1, cassette playing times are about two and half times longer.

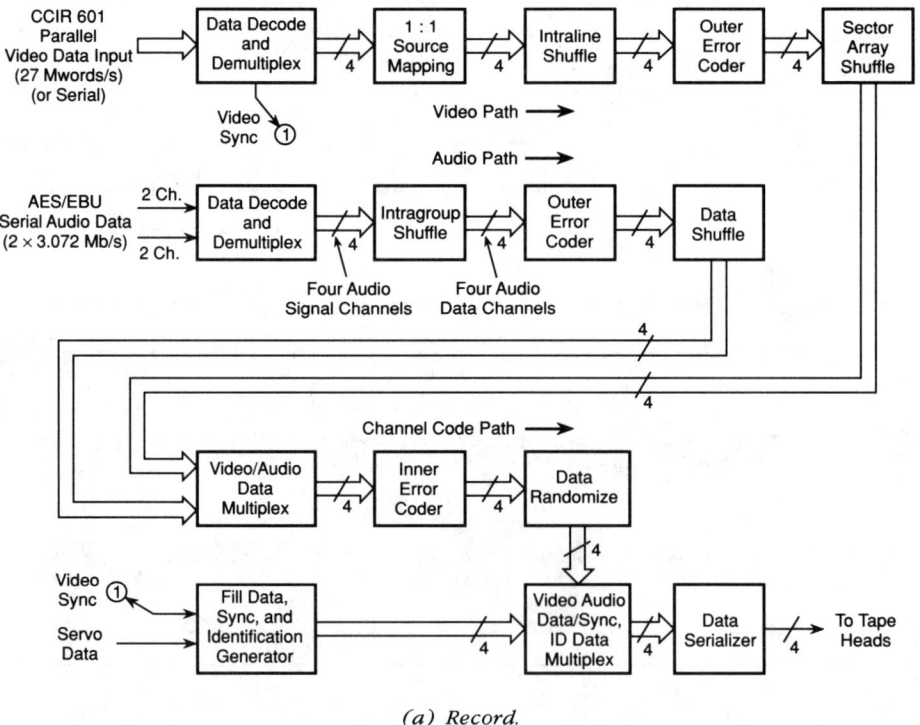

(a) Record.

**Figure 9.** D1 record and playback processing. *(Courtesy Sony Corp.)*

## Track Pattern and Transport

Most of the major differences between the D1 and D2 formats can be seen in the track layout. As shown in Figure 10, the basic encoded information is the same—one digital video signal, four PCM audio signals, cue, control, and time code analog tracks. However, the area reserved for digital video and audio signals is utilized differently.

Video information is in the center of the band, while audio data is recorded at both the top and bottom edges. Placement of the audio sectors at the ends of the rotary tracks causes audio data to be more susceptible to tape edge and tracking problems. Although audio data has complete 100% redundancy as in the D1 format, its integrity falls short of that found in the D1 format. Like D1, D2 is designed so that duplicated audio sectors are recorded at different distances from any one tape edge and by a different video head. Figure 11 shows the relationship between audio and video sectors recorded in one head pass.

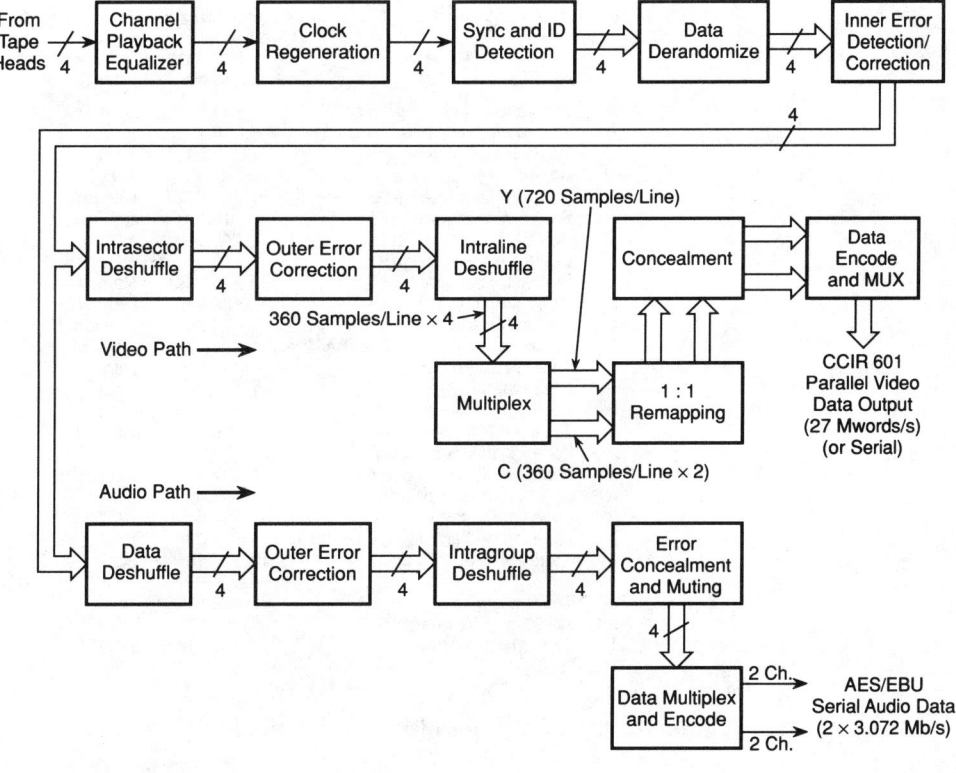

(b) *Playback.*

**Figure 9.** (cont.)

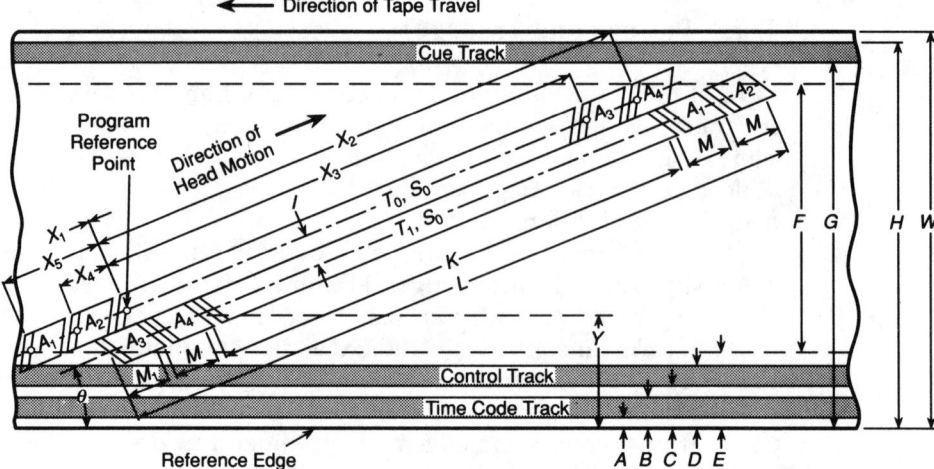

Notes: 1. $A_1, A_2, A_3, A_4$ are audio sectors.
2. $T_0, T_1$ are track numbers, $S_0$ is segment number (typical).

*(a) Track pattern layout, with tape viewed from magnetic coating side.*

| | Dimensions | Nominal (mm) | Tolerance (mm) |
|---|---|---|---|
| A | Time code track lower edge | 0.2 | ± 0.1 |
| B | Time code track upper edge | 0.7 | ± 0.1 |
| C | Control track lower edge | 1.0 | ± 0.1 |
| D | Control track upper edge | 1.5 | ± 0.05 |
| E | Program area lower edge | 1.807 | Derived |
| F | Program area width | 16.1 | Derived |
| G | Cue track lower edge | 18.2 | ± 0.1 |
| H | Cue track upper edge | 18.9 | ± 0.1 |
| I | Helical track pitch | 0.0391 | Nominal |
| K | Video sector length | 132.49 | Derived |
| L | Helical track total length | 150.78 | Derived |
| $M_1$ | Audio sector $A_1$ track 0 and $A_3$ track 1 | 4.13 | Derived |
| M | All other audio sectors | 4.01 | Derived |
| $P_1$ | Control track | 107.66 | ± 0.3 |
| $P_2$ | Cue/time code track | 108.41 | ± 0.3 |
| W | Tape width | 19.01 | ± 0.015 |
| $X_1$ | Location of video sector | 0.0 | ± 0.1 |
| $X_2$ | Location of start of audio sector $A_4$ | 137.57 | ± 0.1 |
| $X_3$ | Location of start of audio sector $A_3$ | 133.03 | ± 0.1 |
| $X_4$ | Location of start of audio sector $A_2$ | 4.54 | ± 0.1 |
| $X_5$ | Location of start of audio sector $A_1$ | 9.08 | ± 0.1 |
| Y | Program reference point | 2.80 | Basic |
| $\theta$ | Track angle | 6.1296° | |
| 0 | Azimuth angle (track 0) | +14.97° | ± 0.17° |
| 1 | Azimuth angle (track 1) | −15.03° | ± 0.17° |

*(b) Dimensions.*

**Figure 10.** D2 track pattern layout and dimensions. *(Courtesy Sony Corp.)*

As in D1, a pair of rotary heads are active each pass to record two adjacent tracks simultaneously.

Composite digital video information is recorded and processed on a sector basis as in the D1 format. Similarly, four individual processing channels and

Chapter 7: Digital Audio for Video and Film

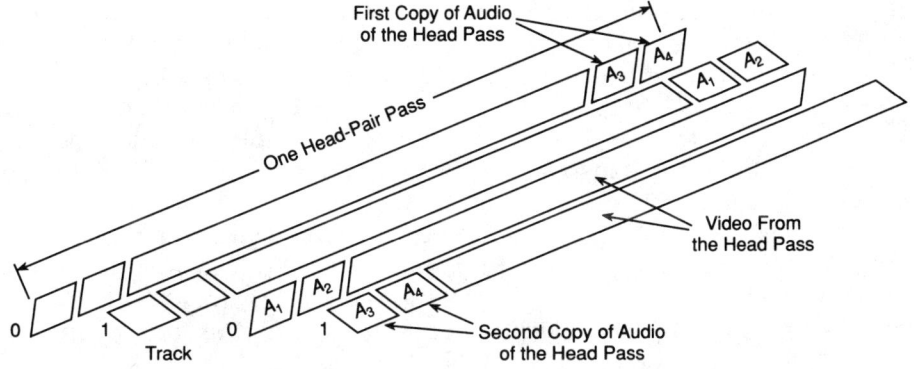

**Figure 11.** D2 audio/video sector relationship of one head pass.
*(Courtesy Sony Corp.)*

heads are used to reduce the rate and quantity of data to be processed. Unlike the component digital format, however, the composite format does not require that a field be made up of a multiple of four video sectors. Because one head-pair pass produces only two video tracks, field data need only be comprised of a multiple of two video sectors. The numbering of audio and video sectors and their relationship to the start and end of a video field is shown in Figure 12. The 525/60 system needs only six video sectors to represent a field, while the 625/50 system requires ten.

Another major difference between the composite and component digital video systems is the azimuth with which helical tracks are recorded. In D1, all rotary heads are parallel and digital tracks are recorded with the same azimuth (see Figure 2). In D2, rotary heads of a pair are aligned differently so that heli-

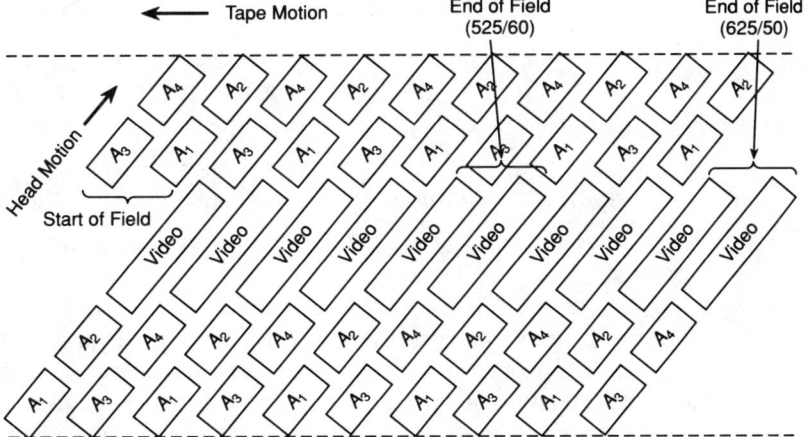

**Figure 12.** D2 audio and video sector/video field relationship in the 525/60 and 625/50 television standards.

cal tracks can be recorded with different azimuths (see Figure 10). This allows for the overlapping of tracks as signals picked up by the incorrect head will be attenuated because of azimuth differences. The 8 mm video and DAT formats use an 8/10 channel code to compensate for low-frequency problems due to azimuth recording, while D2 systems employ a Miller code. Although head, drum, and transport specifications are not standardized, the drum diameter is typically 96.444 mm with head pairs placed at opposite sides of the drum as shown in Figure 13. Such a design would yield rotation rates of 89.91 rpm in the 525/60 system and 125 rpm in the 625/50 system. While almost all characteristics of the longitudinal tracks are nearly identical in both the component and composite digital video formats, differing servo requirements rely on a 180 Hz control pulse in the D2 format.

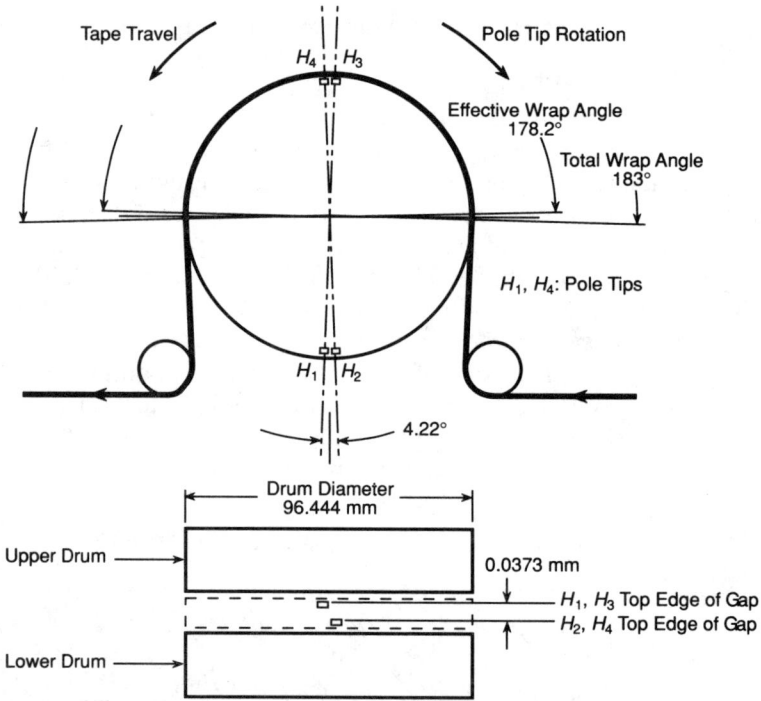

**Figure 13.** Typical D2 head/drum configuration. *(Courtesy Sony Corp.)*

## Processing

The overall processing scheme for D2 is very similar to that used in D1. A complete D2 processing block diagram is shown in Figure 14. Although explanations of individual processing stages can be found in the D1 section of this chapter, some specific differences must be addressed. A composite digital video signal is inherently different from a component digital video signal. Rather than sampling a combination of luminance and chrominance compo-

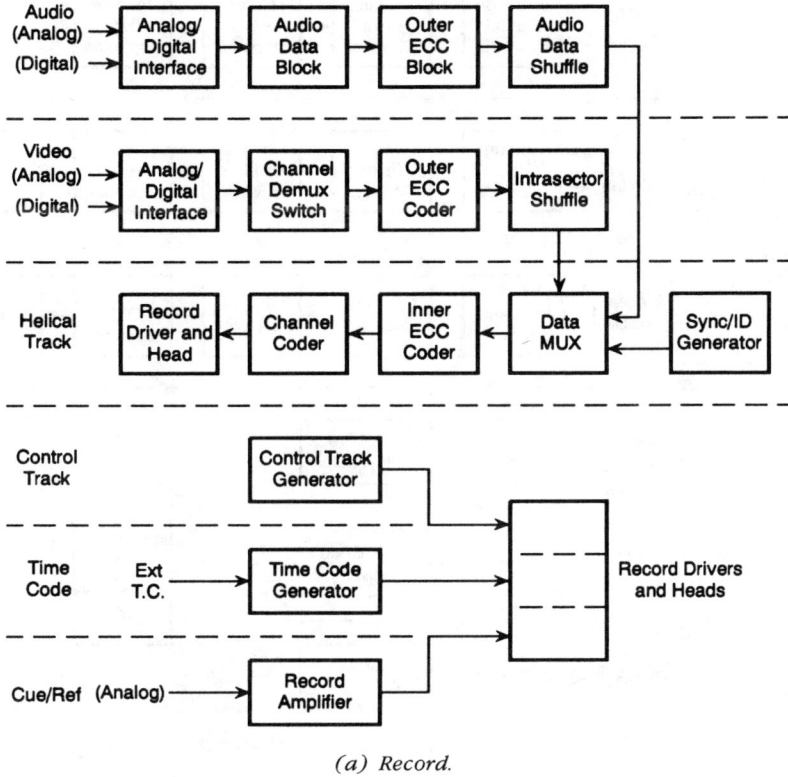

(a) Record.

**Figure 14.** D2 record and playback processing. *(Courtesy Sony Corp.)*
*(cont. p. 220)*

nents and then regenerating the picture, single 8-bit samples are used to represent each element of a conventional composite video signal. The sampling rate is 14.32 MHz with 910 samples per video line; 768 samples correspond to the active digital line, while there are 142 samples for the digital blanking period.

Analog and digital video inputs conform to standards outlined by CCIR Report 624-3, CCIR Recommendation 601, SMPTE Recommended Practice 125, and SMPTE Proposal V16.941. Analog signals are typically sampled as 10-bit words and then converted to 8-bit words.

Aside from different track distribution on tape, audio processing is virtually identical between the D1 and D2 formats. There are slight differences in the number of samples used to form arrays for code word generation and so forth, but features and specifications of the audio signal are the same as for D1. Analog and digital inputs are provided with internal processing conforming to the standard AES/EBU format.

Video and audio data is independently demultiplexed, shuffled, and processed through the outer error coder. Data is then multiplexed, and identification and sync information is added along with additional Reed-Solomon error check words generated by the channel coder. This process is equivalent to video, au-

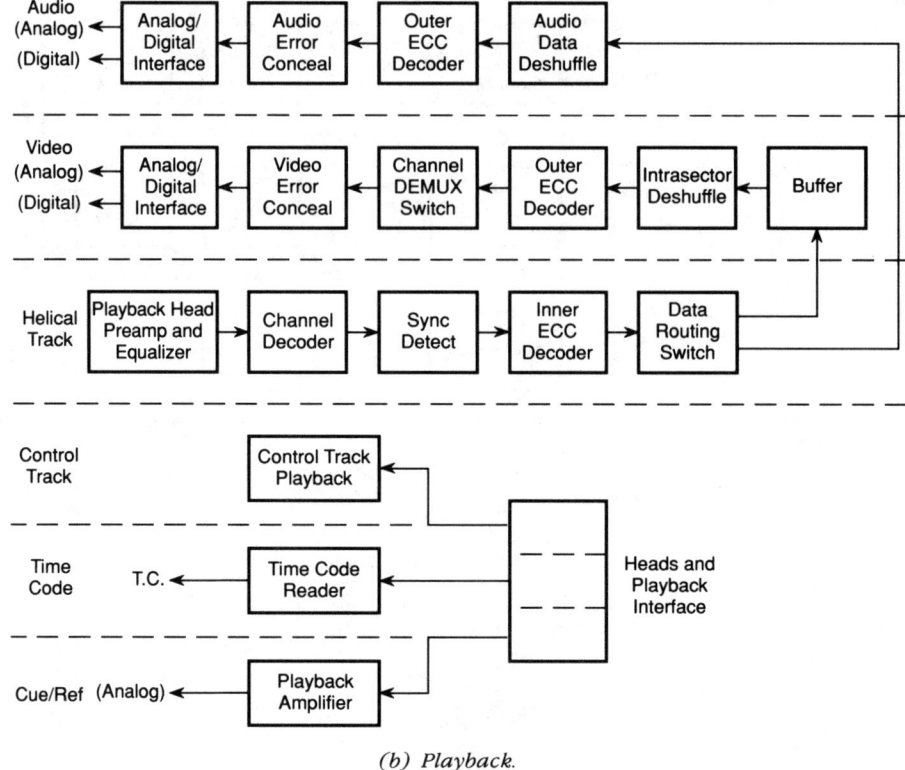

*(b) Playback.*

**Figure 14.** (cont.)

dio, and channel processing in the D1 format. However, at the output of the channel coder, sync blocks are produced which are unique to the D2 format.

The sync block format used in D2 is shown in Figure 15. This 190-byte block consists of two blocks of sync information and two blocks of identification information followed by two sets of 85-byte data and eight inner check bytes. The 85 bytes of data contain either audio or video data depending on the sector being recorded. With this in mind, helical tracks can be defined in terms of recorded sync blocks. As shown in Figure 16, one slant track is made up of two bottom edge audio sectors of six sync blocks each, a 204 sync block video sector, and two upper edge six-block audio sectors. Each sector has a pre- and postamble with edit gaps to separate the sectors.

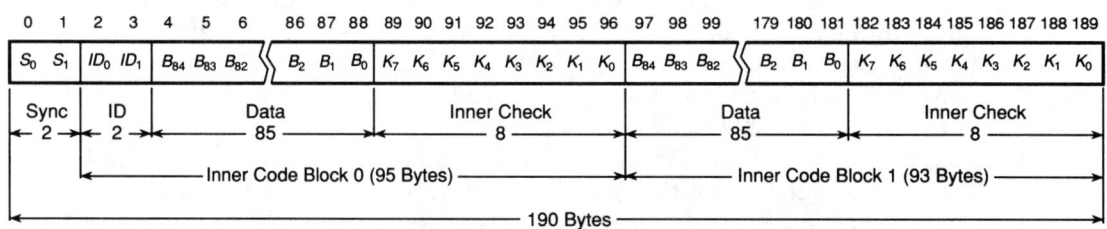

**Figure 15.** D2 sync block format. *(Courtesy Sony Corp.)*

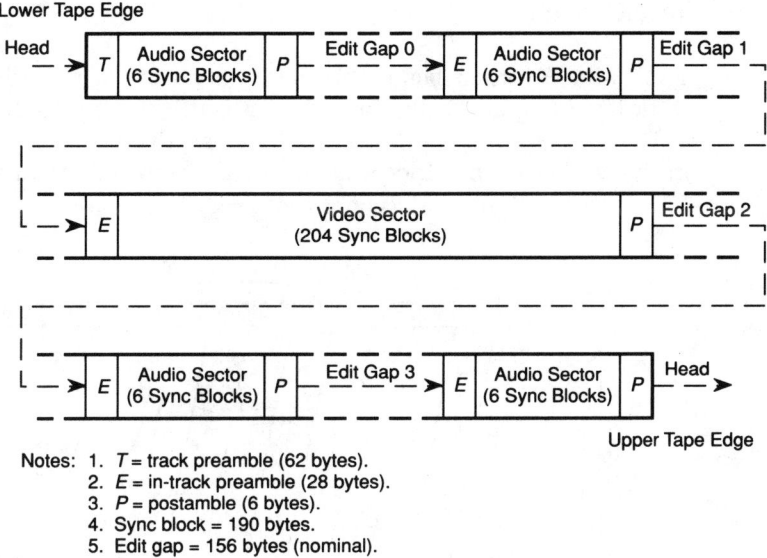

**Figure 16.** D2 sync block/helical track relationship. *(Courtesy Sony Corp.)*

## 8 mm Video

Video 8, or 8 mm video, is a videocassette format which utilizes tape that is 8 millimeters in width. It is the first consumer videocassette format to record stereo digital sound simultaneously with the picture. Furthermore, its exceptional video quality, small size, and economical tape consumption make it an accepted format in many professional situations as well.

### Physical Properties

Like many of its predecessors, the 8 mm video format employs rotary head recording. The head drum, which is 40 mm in diameter, has an angular velocity of 1500 rpm. This, along with a metal-particle, high-coercivity tape (1450 Oe), permits a relatively slow linear tape speed. Tape speeds of 20.051 mm/s (SP mode) and 10.058 mm/s (LP mode) are provided. Together, these measurements yield an impressive relative head-to-tape scanning speed. The SP and LP modes provide 3.12 m of tape scanned each second. However, the track is narrower in the LP mode.

A narrow video track width is another feature of the Video 8 format that makes its tape consumption very economical. A 17.2-$\mu$m track width in the LP mode (34.4-$\mu$m for SP) corresponds to four hours of playing time on one cassette. The 8 mm videocassette shell, which has double doors for protection and holes for identification of two tape types (metal powder or evaporated) and two thicknesses (13 $\mu$m or 10 $\mu$m), has dimensions of 95 $\times$ 62.5 $\times$ 15 mm, making it nearly identical in size with a standard analog audiocassette tape.

## Track Format

The Video 8 format has incorporated a variety of recording schemes into its track layout. Specifically, in addition to the video signal, there are provisions for two longitudinal audio tracks, one mono audio signal which is frequency modulated onto the video waveform, and stereo PCM audio recording. Figure 17 shows the 8 mm video tape format.

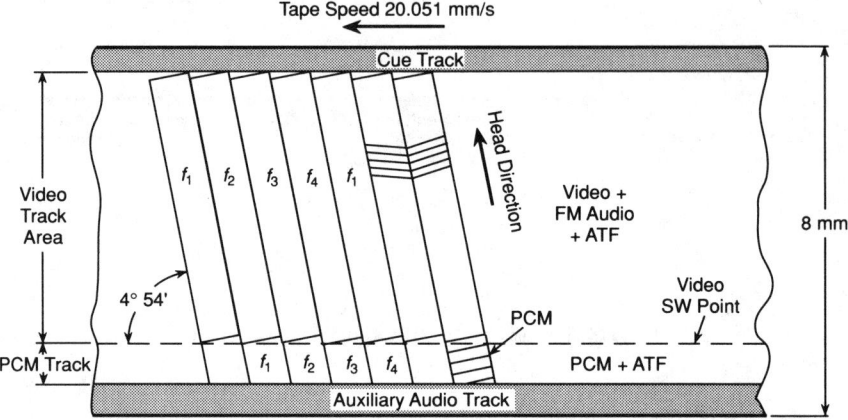

**Figure 17.** 8 mm video track pattern showing layout and dimensions. *(Courtesy Sony Corp.)*

Audio track widths are narrower than those in previous videocassette formats with the auxiliary audio and cue tracks having a width of 0.5 mm each. Video tracks reside in a 5.35-mm band which is only half of the area utilized in the VHS format. Tape consumption is further economized by a zero gap width between adjacent video fields.

Stereo PCM data, which is recorded by the rotary video heads, is placed into a tape band that is approximately 1.65 mm wide. This requires the use of an extended Omega wrap of the tape around the head drum. As shown in Figure 18, the total wrap around the drum is 221°. Conventionally, there are 180° during which the heads are switched in for record/playback of the video signal. The addition of the preceding 41° of rotation consists of 31° delegated to digital audio data and a 5° overlap margin at the head and tail of the PCM segment.

Continuous video recording is preserved by time sharing the video heads. This is done equally during the drum rotation with the changeover occurring during the vertical interval. However, recording of the PCM audio tracks requires that both heads be active simultaneously for 36° per revolution. While head $A$ finishes the last 36° of its rotation, the time compressed audio, which corresponds to the last 36° of head $B$'s previous video field plus the portion of video recorded thus far by head $A$, is recorded by head $B$. Figure 19 shows this relationship.

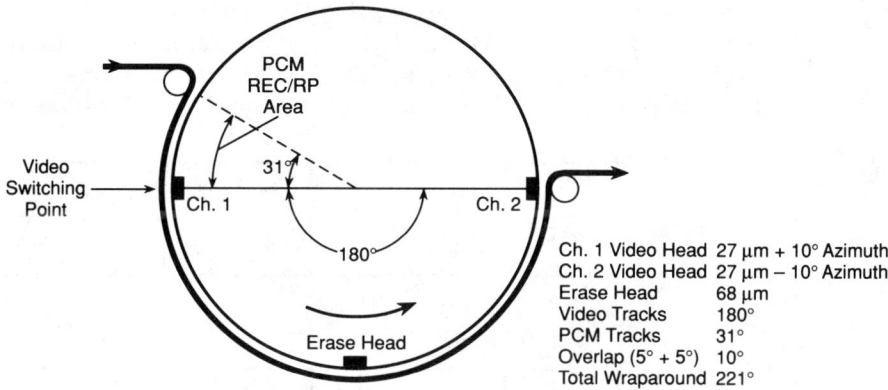

**Figure 18.** 8 mm head/drum configuration. *(Courtesy Sony Corp.)*

The head switching circuits used to select between audio and video signals flexibly provide exclusive access to either sound or picture segments. Therefore, independent recording of either audio or video tracks is feasible. Stereo digital audio can be dubbed synchronously to a prerecorded video segment. Likewise, video may be edited to an existing audio track. Many 8 mm VCR decks offer a built-in mixing function which combines line and microphone input audio signals from both the front and rear panels. Such features provide the home enthusiast with audio/video production and editing capabilities.

## Audio Conversion Process

A block diagram for the overall audio conversion process used in the 8 mm Video system is shown in Figure 20. Input analog audio signals are level compressed and fed through a low-pass filter to avoid aliasing. A 10-bit A/D con-

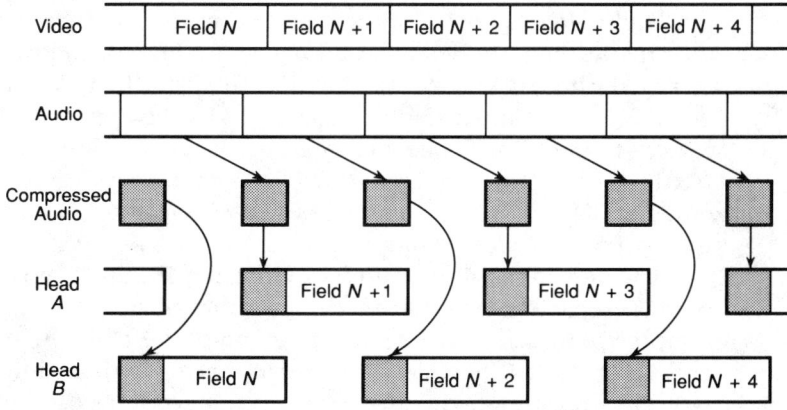

**Figure 19.** 8 mm audio/video track relationship. *(Courtesy Sony Corp.)*

verter samples this signal at twice the line rate of the television in use (31.5 kHz for NTSC and 31.25 kHz for PAL/SECAM). Samples are then digitally compressed into a nonlinear 8-bit form. Table 2 shows mathematical representations for both recording and playback with this nonlinear conversion method.

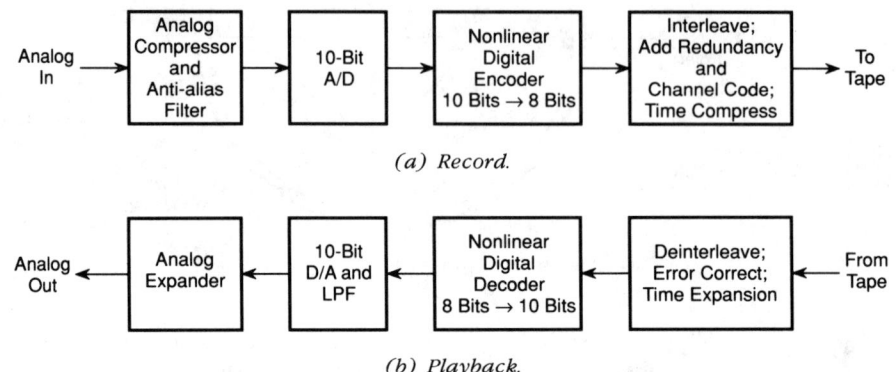

**Figure 20.** 8 mm audio record and playback processing.
*(Courtesy Sony Corp.)*

**Table 2.** 8 mm Digital Audio Compression/Expansion Scheme

| In $X$ | 10/8 Compression | Out $Y$ | ... | In $Y$ | 8/10 Expansion | Out $X$ |
|---|---|---|---|---|---|---|
| 0–15 | $Y = X$ | 0–15 | ... | 0–15 | $X = Y$ | 0–15 |
| 16–63 | $Y = X/2 + 8$ | 16–39 | ... | 16–39 | $X = 2(Y - 8)$ | 16–63 |
| 64–319 | $Y = X/4 + 24$ | 40–103 | ... | 40–103 | $X = 4(Y - 24)$ | 64–319 |
| 320–511 | $Y = X/8 + 44$ | 104–127 | ... | 104–127 | $X = 8(Y - 44)$ | 320–511 |
|  |  |  | ... |  |  |  |

Higher levels can accommodate a compression factor of one eighth, while the smallest amplitudes must go without compression to maintain low-level clarity during playback. Specifically, the first 16 levels (0–15) remain unchanged, while the next 48 levels (16–63) are compressed by one half into 24 levels (16–39). The 256 levels ranging from 64–319 are compressed by a fourth into a 40–103 range and the last 208 levels are compressed by an eighth from a range of 320–511 to 104–127. In both 10/8 compression and 8/10 expansion, the sign bit is added to the magnitude of the sample after coding.

The new 8-bit samples undergo interleaving and time compression before being recorded to tape. Because the system records the same number of stereo samples per track as lines per video frame, there are 1050 samples for NTSC and 1250 for PAL/SECAM. These samples are written into an array used for typical cross-interleaving. Columns in the array are used to form sync blocks and a CRC code is added at the end of a column. The format of a sync block used in 8 mm digital audio recording is shown in Figure 21. There are 132 blocks per field in the NTSC system and 157 blocks per field in the PAL/SECAM system. Additional P and Q code words are formed on diagonals of the array for error correction and an MFM channel code is used.

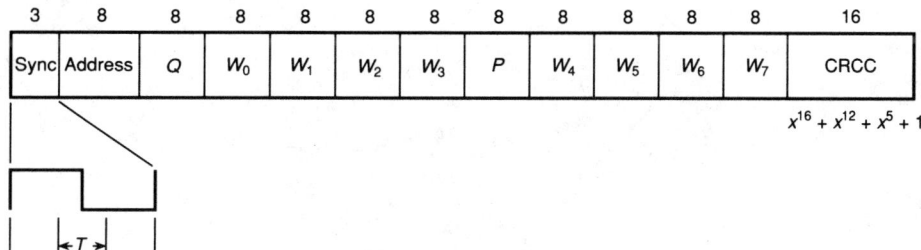

**Figure 21.** 8 mm sync block format. *(Courtesy Sony Corp.)*

Upon playback, the process is inverted. Data blocks are time expanded and error corrected. Audio samples are derived from interleave, code word, and redundancy information for 8/10 bit digital expansion. Ten-bit samples are returned to analog form and fed to a low-pass filter and a complementary level expander at the output.

Despite the tremendous amount of data reduction in this system, audio performance is very good. Signal-to-noise ratios of more than 60 dB and a dynamic range greater than 85 dB can be achieved.

## Automatic Track Following

The capstan in the 8 mm video format is controlled with automatic track following (ATF) signals. These signals (see Figure 17) are placed in both the video and PCM track portions. Four unique tones which lie outside both the audio and video recorded bandwidths are recorded in a repeated sequence, one with each track. The exact ATF frequencies for both the NTSC and PAL/SECAM systems are shown in Figure 22. During tape scan, the head travels over a track in a manner which is also shown in Figure 22. Because of head overlap, the ATF signals from adjacent tracks are detected, and difference tones are formed with the main tone. The tone series is cleverly devised so that any track will produce a 16 kHz and 45 kHz pair of difference tones with its adjacent tracks. When tracking properly, the levels of both the 16 kHz and 45 kHz tones should be identical.

This type of tracking servo has proved to be very accurate; it has facilitated editing in this system. In addition to tracking, the ATF signals can be used to determine which system (NTSC or PAL/SECAM) was used in making a recording, which mode (SP or LP) was used, and in the case of the multitrack mode, in which direction (forward or reverse) a track was recorded.

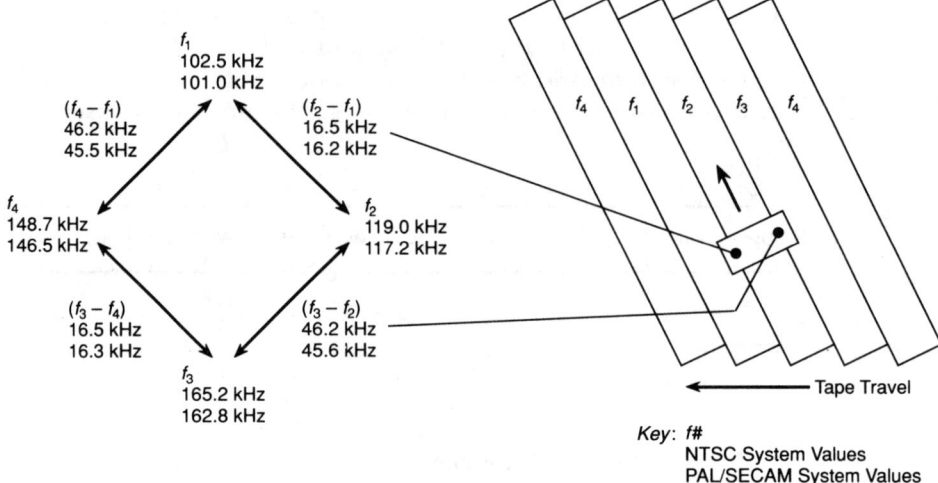

**Figure 22.** 8 mm rotary head scanning with ATF frequencies and difference tones. *(Courtesy Sony Corp.)*

## "Multitrack" Mode

One useful feature of the 8 mm video system is the "multitrack" mode. Developers of this feature associated the term multitrack with its operation although it does not function as a true multitrack machine. In this mode, the video band of the tape is replaced with five additional PCM stereo tracks. This modification to the track pattern and head delegation is shown in Figure 23. Odd-numbered tracks are recorded and played in the forward direction, while even-numbered tracks are scanned with the tape traveling in reverse. At the end of a track, the tape travel changes direction. This can be accomplished through the ATF signals or by decoding information contained in ID words that are recorded with each track. Identification (ID) word information, which is repeated within segments to facilitate indexing and locating while winding, is shown in Figure 24. Head timing and switching is realized with a phase-locked loop and decoder circuit. A complete circuit block diagram for the PCM audio-only mode is shown in Figure 25. To avoid confusion, some people refer to the multitrack feature as the audio-only mode. In this respect, an understanding of this mode becomes easier. Although six individual recordings are made, it actually performs as a stereo digital audio machine with extended playing time. The LP mode can provide 24 hours of PCM audio on one cassette.

Because tracks are recorded separately and ATF is performed with signals embedded in those tracks, tracks 1 through 6 may not be aligned with each other on tape. Furthermore, forward and reverse tape travel suggests orientation differences between odd and even tracks. Even with the addition of a control track and one direction recording, realization of a true multitrack recorder would have to overcome the limitations of having one PCM audio input/output circuit.

Chapter 7: Digital Audio for Video and Film

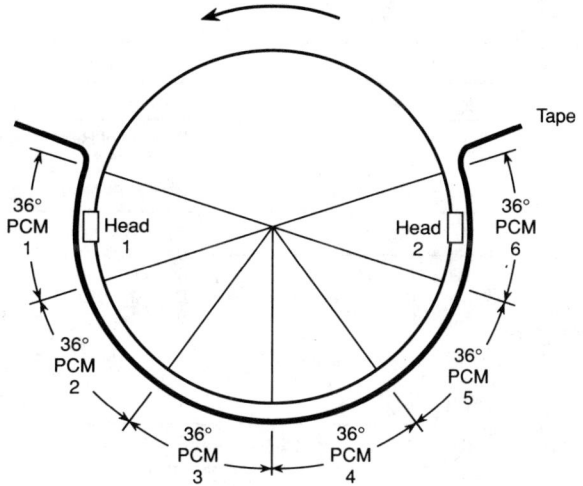

Note: PCM tracks 2–6 occupy normal video segment

*(a) Head delegation.*

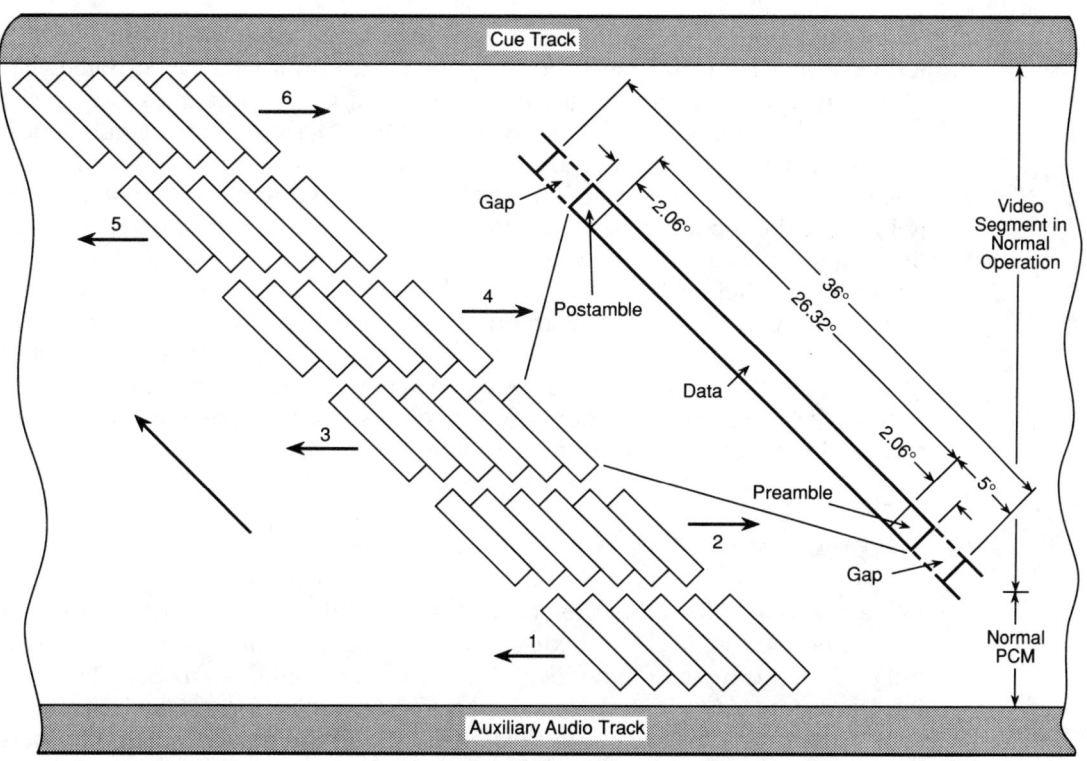

*(b) Track pattern.*

**Figure 23.** 8 mm "multitrack" mode head delegation and track pattern.
*(Courtesy Sony Corp.)*

| | | | |
|---|---|---|---|
| $B_7$ | 0 | | 1 |
| $B_6$ | 1 = Cue Flag | | 0 = Forward  1 = Reverse |
| $B_5$ | $B_4$ $B_5$ | | Position of Recording on Next Track |
|  | 1   0 | Valid Audio | 0 = Same as Current  1 = Beginning of Next Track |
| $B_4$ | 1   1 | Mute | Always 1 |
| $B_3$ | Always 1 | | $B_1$ $B_2$ $B_3$   Next Track No. |
|  | | | 0   0   0   — |
| $B_2$ | $B_1$ $B_2$ | | 0   0   1   3 |
|  | 0   0 | Standard Speed | 0   1   0   5 |
| $B_1$ | 0   1 | Reserved | 0   1   1   1 |
|  | 1   0 | Half Speed | 1   0   0   6 |
|  | 1   1 | Reserved | 1   0   1   2 |
|  | | | 1   1   0   4 |
|  | | | 1   1   1   Increment |
| $B_0$ | 0 = Forward  1 = Reverse | | 0 = Forward  1 = Reverse |

**Figure 24.** 8 mm ID word information. *(Courtesy Sony Corp.)*

## Video Performance

Video operations in the 8 mm format are very similar to those in established VCR formats. Because it is a newer format, however, 8 mm video incorporates advances in video head and magnetic tape technology. Smaller and more precise video heads along with high-coercivity metal tape permit narrower video tracks and slower tape speeds, optimizing tape consumption. Even with a 37% reduction in overall tape width from the VHS format, the 8 mm video format manages to increase the recorded video bandwidth and produce 20 more lines of resolution than standard VHS recorders. Increased luminance and chrominance recording frequencies along with crosstalk cancellation methods also enhance the video quality of this format. Furthermore, the quality of video editing for the consumer is improved through the use of a 68 $\mu$m flying erase head located on the outer drum perpendicular to the video heads. By erasing two tracks simultaneously, seamless assemble edit inserts and noise-free picture transitions can be achieved without the color flutter problems of previous VCRs.

## 1-Inch Type C Video

Just a few years after initial experiments with video tape recorders, a format was introduced which is still considered the standard for broadcast television today. Succeeding the quad-type video recorders, 1-inch open-reel VTRs were introduced in 1960. Due to improved transistor technology, 1-inch machines solved the problems of size and expense inherent in previous video formats. The 1-inch open-reel family of VTRs was the first to employ helical scanning. This now commonplace method pioneered the features of slow motion, stop motion, freeze frame, and frame by frame, which became vital in making video a workable and editable medium.

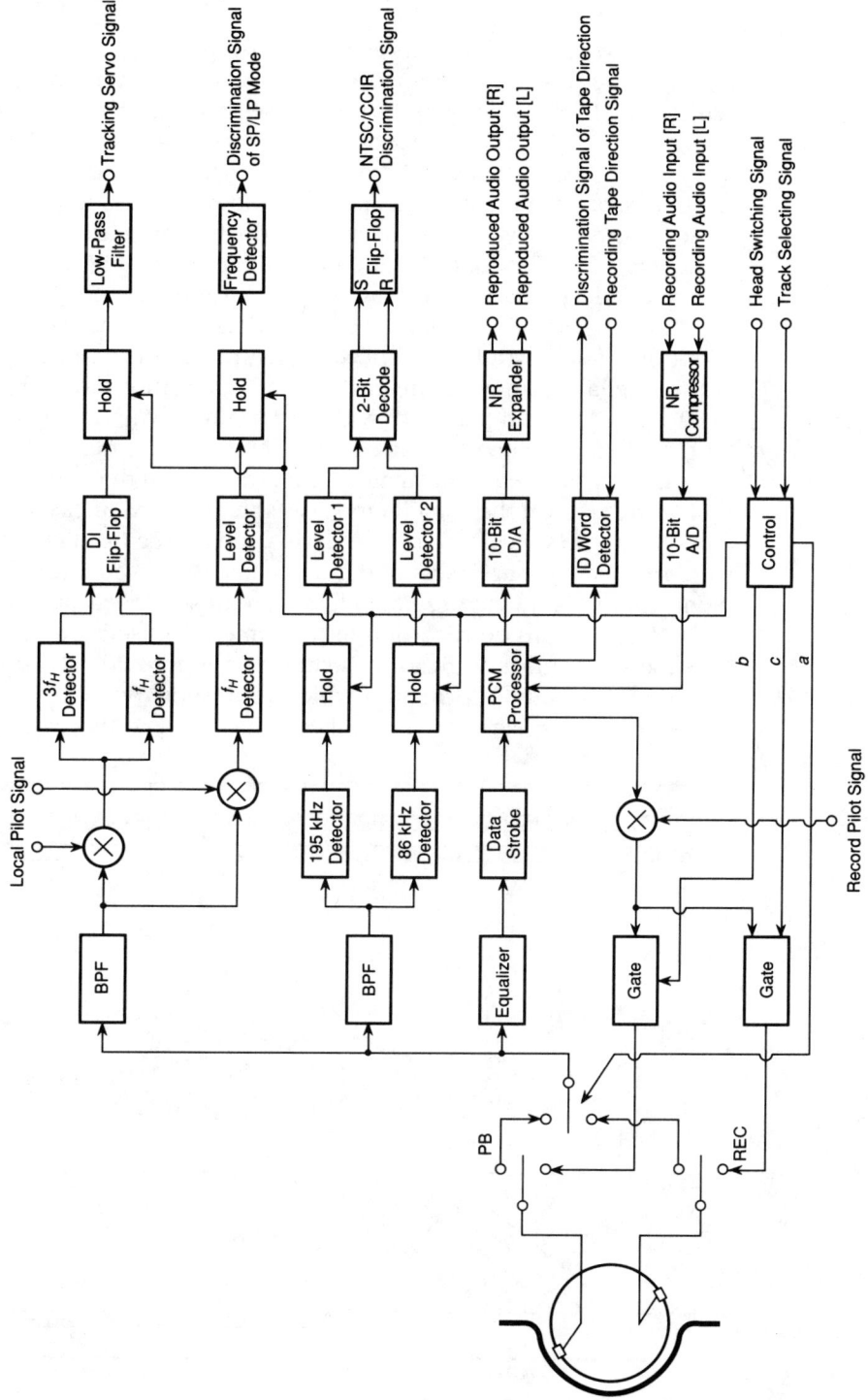

**Figure 25.** 8 mm "multitrack" mode circuit diagram. *(Courtesy Sony Corp.)*

Among the 1-inch formats, Type B and Type C were most readily accepted. However, because the Type B format resorted to segmented video methods, its use eventually died out with its transverse scanning predecessors. The Type C format emerged as the format of choice among industry professionals and remains an industry standard. Like any standard, the Type C format has encountered its share of new formats and technologies aimed to replace it. Although these recorders have included many formats which are markedly more efficient, cost-effective, flexible, and offer higher audio and video quality, the industry's widespread investment in this medium has not been overcome. In addition, the Type C format has combatted its shortcomings in audio quality by adopting a track pattern modification that permits the recording of two channels of digital audio while maintaining compatibility with the original format.

To preserve compatibility with the original Type C format, it was mandatory that digital audio tracks leave intact essential elements of the track pattern such as the video segment, analog audio tracks, and the control track. The only remaining area on the 1-inch tape available for PCM tracks was the sync track area. The three sync heads on the drum used for recording the vertical interval were replaced by three sets of digital heads. Because the track width requirements for digital tracks are much less than those for analog tracks, three tracks of PCM data could be recorded in the same area used for just one sync track. This track pattern modification is shown in Figure 26. Three PCM tracks per video field contain information which represents two channels of audio and redundancy. Displaced by 120°, the digital heads only read and/or write for

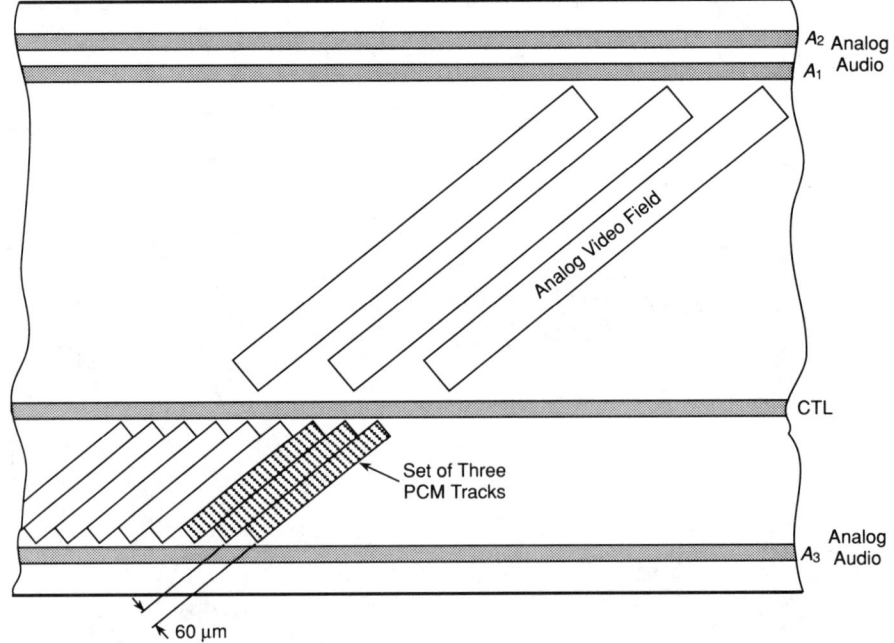

**Figure 26.** 1-inch Type C video modified track pattern. *(Courtesy Sony Corp.)*

22° per drum rotation. This requires the use of time compression and prohibits monitoring the digital audio data at any speed other than the standard; analog tracks can still be used for cueing at other speeds. However, three sets of pre-read, record, and confidence PCM heads offer simultaneous playback during recording or editing and ensure proper lip-sync.

A conventional 1-inch Type C VTR interprets the digital tracks as a non-recorded sync track. This is accomplished by adding 15° of azimuth tilt to the digital heads for PCM recording and playback. This slight off-angle displacement and the canceling tendency of parallel digital tracks when read by one common head, provide enough attenuation to appear transparent to the video sync heads.

The PCM mode uses an extensive cross-interleave process to form sync blocks. After 16-bit samples are error corrected with a cross-interleaved Reed-Solomon code, a multiplexed data stream of the two PCM channels is fed into a 192- × 32-byte array by rows. Three 64-byte columns correspond to the data blocks sent to each head, and each block also represents a sync block. This cross-interleave method allows code words to be formed from vertical and diagonal array information which aids in cross-referenced error correction. With 800 recorded samples per field per audio channel, interleaving is limited to one drum revolution, which facilitates editing. The same 8/10 channel code used in the DAT format is employed here. Sampling rates of 44.1 and 48 kHz yield a dynamic range of 90 dB. The five-track recorder (two PCM and three analog—usually one for time code) also comes equipped with an AES/EBU interface.

# Digital Audio for Film

Several methods for providing digital audio soundtracks with motion picture film have been suggested. These techniques include synchronizing a separate digital medium such as tape or disk, using a strip of magnetic tape along the side of the film for recording and playback of digital soundtracks, and utilizing space on the film itself for optical digital soundtracks. In special applications, the first two techniques have been implemented. However, little effort has been made to standardize either of these methods. Optical digital soundtracks and the concept of having both the audio and the picture in a single-medium system have received the most attention from the motion picture and film community because it is clearly the most efficient.

## Design Issues for Digital Sound on Film

Before discussing the different methods available for realizing digital optical tracks, it is important to discuss the major issues involved in designing such a system: number of channels, bandwidth, dynamic range, resistivity to film wear, and the ability to be copied at high speeds.

## Number of Channels

In considering the number of sound channels needed for motion picture film, the channels used for localizing on-screen sound are the most important. With speakers at the far right and far left behind the screen, lateral localization can be obtained through conventional stereo. In practice, however, localization of center-screen sound is not accurate enough with two channels. Employing a third channel in the center provides needed localization cues and achieves an overall enhancement of localization sensitivity across the horizontal spread of the screen. Adding more behind-screen speakers does not significantly improve on-screen localization of sound. Furthermore, it becomes more and more difficult to mix as channels are added, and the packing density of digital information on the film must be kept at a minimum.

Surround sound is the next consideration. Surround channels are primarily used for ambience, not localization. Channels used for surround sound generally should not be discernible as separate sound sources by any listener in the theater. Keeping in mind that minimizing the number of digital channels is essential, an efficient method would minimize the number of channels needed to achieve this surround ambience. Although one channel can be used to achieve this goal, the sense of ambience is noticeably better with two channels. Little improvement is gained by adding more channels.

Another consideration, and an accepted practice, is use of an additional channel for low frequencies. Because low frequencies are nearly impossible to localize, only one channel is necessary. This subwoofer channel usually handles frequencies below 100 Hz and contributes a fuller sound to motion picture audio.

These six speakers (three behind-screen, two surround, one subwoofer) constitute the minimum array of channels needed to fulfill the requirements of motion picture sound in the theater. Additional digital tracks which may be considered include a time code track, ASCII data subtitle tracks, foreign language dialogue tracks, and any tracks that may be needed for theater automation control signals.

## Bandwidth

Digital audio usually implies an audio bandwidth of at least 20 to 20,000 Hz. Ideally this is the kind of fidelity that theatrical presentations should provide. It was thought that because of high-frequency attenuation and the difficulty of dispersing higher frequencies in the open-air space of a theater, the upper bandwidth of digital audio for film could be limited to approximately 15 kHz. Listening tests reveal that limiting the bandwidth in such a way results in a loss of audio quality. For this reason, bandwidth considerations remain the same as in most digital audio applications: 20 to 20,000 Hz. This frequency range represents the overall response, and it should be noted that a reduced bandwidth may be a design consideration on specific channels, such as the ones used for the subwoofer, dialogue, time code, ASCII data, and automation controls.

## Dynamic Range

Although the goal is the same high-fidelity sound as in the home, the issue of dynamic range must be treated differently in motion picture applications. Rather than focus on an overall dynamic range, it is best to look at the dynamic range needed in each octave band. The factors which determine the required dynamic range are the noise floor and the maximum program level in a given band. With these criteria in mind, the system can be designed with a noise floor just below the lowest noise level in the theater and with a maximum output just above the highest level to be encountered on a soundtrack. Tomlinson Holman and the SMPTE Study Group on Digital Audio for Motion Pictures performed studies which indicated that a dynamic range of only 78 dB is needed for the lowest octave band while a dynamic range of 105 dB is required for the highest band. These results, which reflect the higher levels of theater noise at lower frequencies, suggest that encoding schemes similar to those used in compact disc technology are necessary to achieve such a dynamic range.

## Durability and High-Speed Duplication

For commercial success it is critical that optical digital tracks are durable and have the capability to be copied accurately at very high speeds. A typical film will be played hundreds and perhaps thousands of times. In that time, wear from handling and projecion cannot be avoided. It is important that the digital information on the film retain the integrity required for playback within the constraints of the error correction code. In addition, digital data which may entail physical bit sizes on the order of a few micrometers must be able to withstand the high speeds of present-day film copiers.

## More Specific Considerations

The various design issues govern more specific considerations and will lead to a standard format for encoding digital audio on film. Specific issues such as sampling rate, number of bits per sample, bit size, track widths, and layout must be agreed upon to establish compatibility between manufacturers of film equipment. Likewise, standards on the use of time code, sync tracks, and clocking signals are required. The use of companders and nonlinear encoding, noise reduction, and error correction schemes will be more flexible from system to system, but a standard for digital audio on film will determine the actual parameters involved.

## Fluorescent Layer System

The use of a clear fluorescent layer on the base side of film is one method for achieving an optical digital storage medium. The properties of this fluorescent layer are such that when its particles are exposed to visible light they are colorless. When exposed to ultraviolet light, however, the particles become fluorescent. Binary information is represented as a series of fluorescent and nonfluorescent images. Recording the digital data on this fluorescent layer is a complex electrochemical process. First, the data must be recorded onto a photoconductive master medium. From this recording, a charge must be transferred electrostatically to the back side of the film. This dielectric color print film is then treated with fluorescent toning particles and is heat fused. The use of a photoconductive master and a fluorescent layer which is colorless in visible light allows the digital tracks to be recorded across the entire width of the film, which in turn allows a greater amount of information to be stored. However, the processes and costs involved in producing such a soundtrack may outweigh these benefits when compared with other methods of achieving digital optical audio tracks.

The playback mechanism for the fluorescent system is very similar in principle to the mechanism used in the imaging dye technique described below. With the fluorescent system, ultraviolet light acts as the light source which illuminates the bit stream. Lenses are used to focus the light on the film and on an array of sensory devices designed to detect fluorescent images. Conversion and decoding of the digital soundtracks can then proceed. Peter Custer has proposed this method under the name Cinedigital (formerly Fluorescentsound) and claims eight-track capability with 100% redundancy using bit sizes of 25 $\mu$m.

## Imaging Dye Technique

Of all the methods proposed for placing optical digital audio tracks on film, the imaging dye technique shows the most promise. This method encodes binary data as a series of opaque and transparent images. Conventional print film imaging dyes and layers can be used so that special film is not necessary. This single-medium technique also allows sound tracks to be printed from the negative and audio tracks to be copied at high speeds using traditional methods of film transfer. There is no need for separate treatment of audio when applying soundtracks. These advantages result in more efficient and less expensive motion picture production. The mechanism needed to record digital data on film must utilize an accurate light-exposing device in order to encode the bits on the imaging layers. Possible devices include light-emitting diodes, gas lasers, laser diodes, and cathode-ray tubes. Figure 27 shows an imaging dye digital encoding system that employs an array of light-emitting diodes. Light from the array is focused onto the film using a lens. The array can be used to properly locate the data on the film and to separate track information.

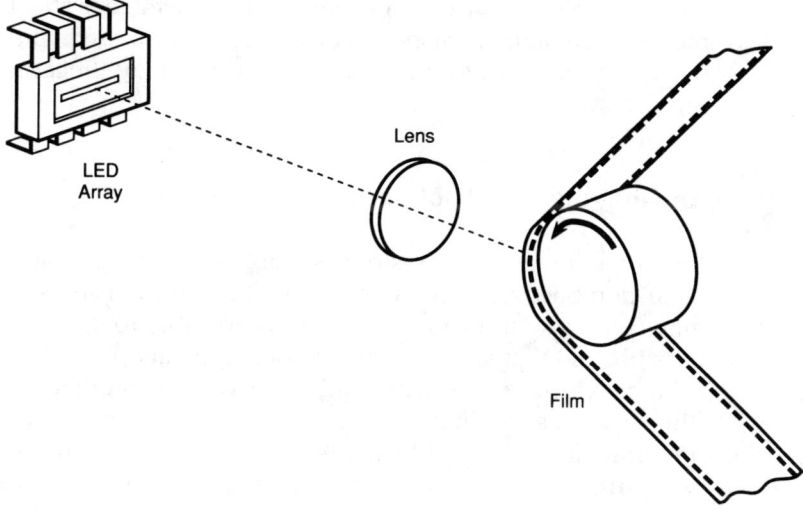

**Figure 27.** Record mechanism for imaging dye system.
*(Courtesy Optical Radiation Corp.)*

A digital imaging dye playback system is shown in Figure 28. A light source from one side of the film is focused onto the data surface via a lens. Another lens on the opposite side focuses the image on an array of sensory devices; this is the first step in converting and decoding the data back into sound.

Conversion from light pulses to digital electronic signals has been performed for many years, so the conversion of digital audio data on film does not

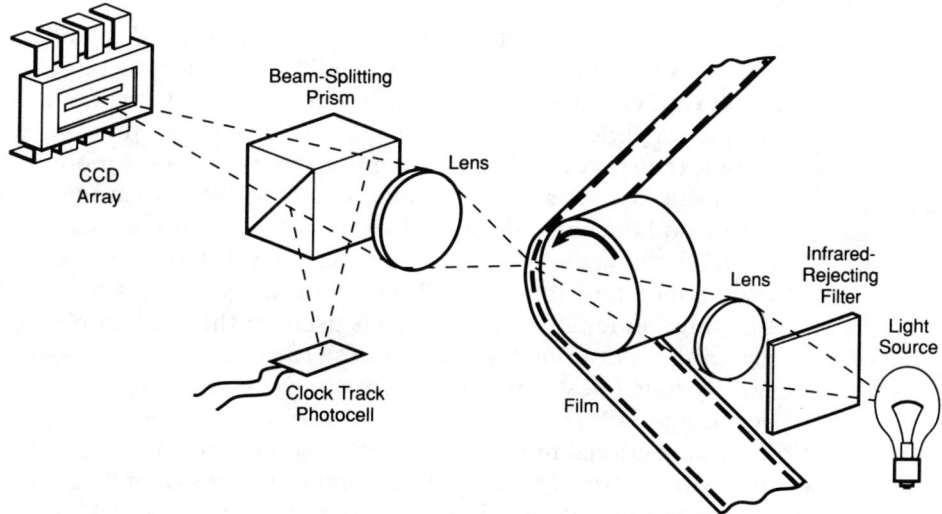

**Figure 28.** Playback mechanism for imaging dye system.
*(Courtesy Optical Radiation Corp.)*

pose a problem. However, because dirt and scratches affect transparent images more than opaque images, the error rate encountered while decoding opaque bits is less than that for transparent ones; this may affect the design of an imaging dye system.

## An Imaging Dye Application

Prototype imaging dye systems suitable for coding digital audio on film have been demonstrated. Tests show that packing densities necessary to achieve high-fidelity sound and error rates comparable to the compact disc format are possible and that wear on film is not significant. One such system utilizes the 0.1 inch wide area conventionally reserved for sound in the 35 mm and 70 mm film standards for digital audio tracks and data tracks. Six audio channels are implemented. Five full-bandwidth audio channels (three behind screen and two surround channels) are applied to the input of the system as 16-bit samples at 44.1 kHz. Samples are data compressed into 12-bit words. One in every 32 samples retains its original 16-bit form to provide an accurate reference every 726 $\mu$s. The subwoofer channel does not employ data compression. Instead the sample rate is decreased to 1378 Hz, yielding an upper audio bandwidth of 114 Hz with anti-aliasing and anti-imaging strategies applied in the remainder of the frequency range.

In addition to the six digital audio channels, three data/control channels are provided. One SMPTE time code channel and another channel for MIDI control signals offer flexibility in performing theater automation or external synchronization of equipment. The third data channel, an identification channel, may be used to record a variety of user-defined parameters specific to the film.

Because the data rate is 5.8 million bits per second, significant error detection and correction is required. A Reed-Solomon block code is used with additional CRC characters for error correction. Interleaving of odd and even audio samples is performed to protect against burst errors. Just as in audio tape machines, transport problems with guides, tension, and supply and takeup reels can result in vertical or horizontal weave. As bit sizes are only 14 $\mu$m, precise timing and tracking is essential. Horizontal tracking is provided by a 76 MHz digital servo, while vertical timing is accomplished with an algorithm which is written into the format itself. Rows of data are scanned horizontally, thus a self-clocking run length limited code is used for this error correction. A 6-to-8 bit mapping is performed upon encoding to ensure that each 8-bit word contains exactly four ones. This form of parity works well in correcting errors upon decoding.

Conventional methods presently used to mix sx sound for motion picture film will not have to change to accommodate the digital audio cinema format. However, the audio tracks must adhere to a 16-bit, 44.1 kHz digital audio format before they can be delivered to the system's encoders. Testing has demonstrated that the system meets the fidelity standards established by the compact disc format. In addition to a 20 kHz bandwidth, the system provides a dynamic range of 96 dB which is not affected by film wear. Total harmonic distortion is 0.01% with 100 dB separation between channels.

Optical Radiation Corporation, in cooperation with Eastman Kodak Company, has implemented Cinema Digital Sound (CDS) optical digital audio format in some of its commercial products. This patented system may find applications in the motion picture film industry.

# References

ANSI S4.40 1985: AES Recommended Practice for Digital Audio Engineering—Serial Transmission Format for Linearly Represented Digital Audio Data.

Baldwin, J. L. E., "The Evolution of the Digital Television Recording Format," *IEE Conference Publication*, 1986.

Beeching, S., *Videocassette Recorders: a Servicing Guide*, 3rd edition, Heinemann Professional Publishing, 1988.

CCIR Recommendation 601: Encoding Parameters of Digital Television for Studios.

CCIR Recommendation 656: Interfaces for Digital Component Video Signals in 525-line and 625-line Television Systems.

CCIR Recommendation 657: Digital Television Tape Recording.

CCIR Report 624-3: Encoding Parameters of Digital Television for Studios.

*Cinema Digital Sound*, (Information Package), Optical Radiation Corp. and Eastman Kodak Co., 1990.

EBU Tech. 3246-E: Parallel Interface for 625-line Digital Video Signals.

EBU Tech. 3247-E: Serial Interface for 625-line Digital Video Signals.

EBU Tech. 3250-E: Specification of the Digital Audio Interface.

EBU Tech. 3252-E: Standard for Recording Digital Television Signals on Magnetic Tape in Cassettes.

Eguchi, T., and J. H. Wilkinson, "The 4:2:2 Component Digital VTR," *International Broadcast Engineer*, September, 1986.

Gregory, S., *Introduction to the 4:2:2 Digital Tape Recorder*, Pentech Press, 1989.

Gross, L. S., *The New Television Technologies*, William C. Brown, 1983.

Huber, D. M., *Audio Production Techniques for Video*, Howard W. Sams & Co., 1989.

Itoh, S., et al., "Multi-Track PCM Audio Utilizing 8 mm Video System," *IEEE Transactions on Consumer Electronics*, vol. CE-31, August, 1985.

Lambert, M., "Digital Audio for Film," *Mix*, September, 1990.

Pensinger, G., editor, *4:2:2 Digital Video: Background and Implementation*, Society of Motion Picture and Television Engineers, 1989.

Pohlmann, K. C., *Principles of Digital Audio*, 2nd edition, Howard W. Sams & Co., 1989.

Rumsey, F., "Did Somebody Say Eight Bit?" *Broadcast Systems Engineering*, February, 1986.

Shibita, Y., and Y. Machida, "The 8 mm Format and How It Was Established," *Journal of Imaging Technology*, 1986.

SMPTE EG10: Proposed SMPTE Engineering Guide for Component Digital Audio and Video Recording 19 mm Type D1 Cassette Format—Tape Transport Geometry Parameters.

SMPTE EG11: Proposed SMPTE Engineering Guide for Component Digital Television Recording 19 mm Type D1 Cassette Format—Format Nomenclature.

SMPTE Recommended Practice RP 125: Bit-Parallel Digital Interface for Component Video Signals.

SMPTE T14.227: Parallel Conductor Transmission Format for Linearly Represented Digital Audio Data.

SMPTE V16.83: Proposed National American Standard for Digital Television Tape Recorder for Composite Digital Video Recording 19 mm Type D2 Format—Tape Record.

SMPTE V16.84: Proposed National American Standard for Digital Television Tape Recorder for Composite Digital Video Recording 19 mm Type D2 Format—Video Magnetic Tape.

SMPTE V16.85: Proposed National American Standard for Digital Television Tape Recorder for Composite Digital Video Recording 19 mm Type D2 Format—Tape Transport and Geometry Parameters.

SMPTE V16.87: Proposed National American Standard for Digital Television Tape Recorder for Composite Digital Video Recording 19 mm Type D2 Format—Helical Data and Control Records.

SMPTE V16.88: Proposed National American Standard for Digital Television Tape Recorder for Composite Digital Video Recording 19 mm Type D2 Format—Cue Record and Time and Control Record.

SMPTE V16.89: Proposed National American Standard for Digital Television Tape Recorder for Component or Composite Digital Video Recording 19 mm Type D1 and Type D2 Format—Nomenclature.

SMPTE V16.940: Proposed National American Standard for Digital Television Tape Recorder for Composite Digital Video Recording 19 mm Type D2 Format—Description and Index.

SMPTE V16.941: Proposed National American Standard for Digital Television Tape Recorder for Composite Digital Video Recording 19 mm Type D2 Format—Encoding of Video Signals (System M/NTSC).

SMPTE V16.942: Proposed National American Standard for Digital Television Tape Recorder for Composite Digital Video Recording 19 mm Type D2 Format—Bit-Parallel Digital Interface for Composite Television Signals.

SMPTE 224M: Proposed National American Standard for Component Digital Television Recording 19 mm Type D1 Cassette Format—Tape Record.

SMPTE 225M: Proposed National American Standard for Component Digital Television Recording 19 mm Type D1 Cassette Format—Principal Properties of Magnetic Tape.

SMPTE 226M: Proposed National American Standard for Component or Composite Digital Television Recording 19 mm Type D1 and Type D2 Cassette Formats—Dimensions of Tape Cassettes.

SMPTE 227M: Proposed National American Standard for Component Digital Television Recording 19mm Type D1 Cassette Format—Signal Content of Helical Data Records and of Associated Control Record.

SMPTE 228M: Proposed National American Standard for Component Digital Television Recording 19mm Type D1 Cassette Format—Signal Content of Cue and Time Code Longitudinal Records.

Takahashi, N., "High-Quality Picture Transmission in a Digital Audio System," *Digital Audio Collected Papers*, Audio Engineering Society, 1983.

Takayama, J., and S. P. Burgess, "Enhancement to One-Inch VTRs," *IEE Conference Publication*, 1986.

*Time Code Handbook*, Cipher Digital, Inc., 1987.

Uhlig, R. E., "Feasibility of Digital Sound on Motion Picture Film," *Audio in Digital Times: AES 7th Conference Proceedings*, May, 1989.

Varela, A., "Film Sound Goes Digital," *Post*, September, 1990.

Watkinson, J., *The Art of Digital Audio*, Focal Press, 1988.

*Chapter 8*

# Data Compression
**Michael Shawn Ballman**

## Introduction

The advent of digital audio and its rapidly advancing capabilities have placed greater demands on storage and conversion systems. In terms of storage, higher data rates and longer wordlengths require increased storage capacity. In addition, processors involved in conversion are expected to perform faster and with more precision. One solution is to develop more powerful processors and more efficient and densely packed storage media. Although this seems like the obvious answer, this option usually solves the problem only temporarily, requires a significant effort to develop, and is not very cost-effective.

An alternative and more sophisticated solution is data compression. Data compression reduces the number of bits needed to represent a data set ideally without affecting the information it conveys. The importance of this idea is seen not only in digital audio but also in other areas, such as computer systems, broadcasting, and telecommunications. There are a variety of approaches and parameters that can be utilized for achieving effective data compression. This chapter discusses basic theory and looks at specific data compression techniques used in digital audio.

## A Short History of Data Compression

Although the history of data compression dates back to 1898, when W. F. Sheppards presented his theory of rounding, it is still considered to be a young science. This is because implementation of data compression systems has just recently become widespread. Storage, bandwidth, and channel limitations that were temporarily overcome by enhanced physical designs have proven to be too demanding to be solved solely by hardware advancements. This has caused many developers to turn to data compression. By reducing the number of bits needed to represent data, data compression relieves some of the demands placed on memory, processing, and transmission systems.

Early applications of data compression centered on text compression. Teletypewriter operators who were charged by the letter for telegraphed messages gave birth to the idea. The classic example of CN U RD THS SNTCE? is a 3:2 compression of "Can you read this sentence?" and serves as a basis for discussing elementary principles of data compression. Twenty-seven characters were reduced to eighteen by simply removing some of the vowels. Although the message is noticeably different, the information it conveys is retained.

## The Measurement of Data Compression

It is interesting to note that data compression may be measured either empirically or subjectively. Empirically, it is common to use a mathematical ratio or percentage to express the amount of data compression. Subjectively, however, data compression is more difficult to measure due to differences in the way people respond. On a general scale, a subjective measure of data compression may be whether or not the original information can be recovered from the compressed data. For example, some people may be unable to decipher the text compression example above. More specifically, subjective measurements may be used to describe the fidelity or tolerability of the compressed data. In digital audio, the effects of data compression on an audio signal may go unnoticed by one listener while another listener detects degradation.

Therefore, the goal of data compression in digital audio is to obtain the most simplified form of data possible that will yield no intolerable loss of sonic fidelity. Analogous measures of the effects of data compression on fidelity can be made in digital applications such as computer systems and networks, radio and television broadcasting and transmission, telecommunications, speech and image processing, pattern recognition, information retrieval, storage, cryptography, space digital telemetry, facsimile transmission, biomedicine, and remote sensing.

## Data Compression in Theory

Like most processes used in digital systems, data compression can be modeled by mathematical theories from a variety of disciplines. Information theory or communications theory, coding and transmission theory, estimation theory, and other mathematical and logic-based theories comprise the majority of ingredients which shape the foundation of data compression. Among these areas, information theory and in particular the work of Claude E. Shannon are considered the principles of digital data compression.

## Information Theory and Data Compression

As shown in Figure 1, a communication system can be modeled as a source of information, an information channel, and a receiver of information. For the purpose of discussing data compression, the source is considered to be capable of producing a stream of data, whether it be a teletyped message or a digitized signal transmission. The receiver may be thought of as any element desirous of the original signal. While the information channel may represent an actual transmission channel, it may also be a storage device or a processor. In general, any model of a transmission channel must also include a noise source; this recognizes the likelihood that noise will be added to the signal during transmission. If this system were to incorporate techniques of data compression and expansion, the block diagram would be modified as in Figure 2. The source encoder and decoder represent the data compression and decompression devices, respectively. In a typical digital system, the information channel, despite its function, would consist of a channel encoder and modulator at the input of the channel and a channel demodulator and decoder at the output of the information channel. These additions to the information channel represent an area of study referred to as channel coding and error control. Although essential to many systems, they lie outside the subject of data compression and are therefore discussed in little detail in this treatment.

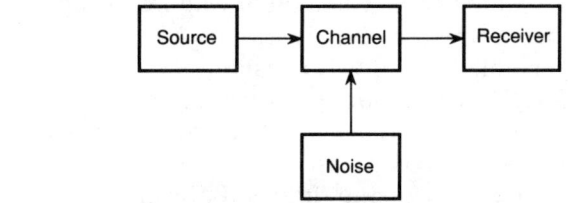

**Figure 1.** An elementary model of a communication system.

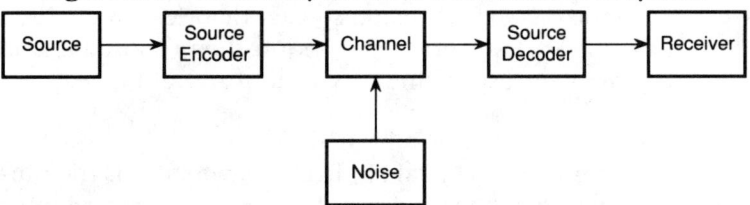

**Figure 2.** A communication system with data compression.

The source is considered to produce a random sequence of symbols from a finite alphabet,

$$A = \{a_1, a_2, \ldots, a_i, \ldots, a_r\}.$$

The number $r$ of elements in $A$ is called the radix. These $r$ elements could be combined amongst themselves to form groups of elements called symbols, denoted by the set $S$:

$$S = \{s_1, s_2, \ldots, s_i, \ldots, s_q\}.$$

The total number of symbols $q$ depends on the radix $r$ and the number of elements allowed in a symbol. For example, if $A$ is a binary set $\{0, 1\}$ and $S$ is the set of source symbols with a length 3 then we could have the following symbols:

$S_1 = 000 \quad S_5 = 100$
$S_2 = 001 \quad S_6 = 101$
$S_3 = 010 \quad S_7 = 110$
$S_4 = 011 \quad S_8 = 111$

In this case $q = 8$ and $r = 2$. In the case of digital systems, the alphabet contains the symbols 0 and 1 and $r = 2$. However, the set of symbols $S = \{S_1, S_2, \ldots, S_8\}$ need not necessarily represent numerical values shown in binary; these symbols could be representative symbols for any kind of information.

Associated with each element of the set $S$ is a probability $P_i$ which describes the likelihood of an element's occurrence. In digital communications it is often safe to assume that the occurrence probabilities of 0 and 1 are equal. However, if $S$ contained the letters of the English alphabet and the source output text messages, the letter "a" would be much more likely to occur than the letter "z". Predictability of this type can be considered an important design issue of a data compression system. As audio signals are generally periodic and therefore somewhat predictable, design of a digital audio data compression system may center around algorithms based on probability models.

In general, the source is considered to produce a discrete stream of data at a fixed rate. If a continuous or variable-rate signal is encountered, it is assumed to be discretely sampled and converted at a constant rate. In addition to these assumptions, data compression theory relies on freedom from any taxing limitations such as processing time and power.

Given these initial considerations mathematician and engineer Claude E. Shannon [1948] formulated several theories which describe data compression principles. First among his theories he proposed that the amount of information $I(s_i)$ corresponding to a symbol $s_i$ could be expressed as

$$I(s_i) = \log_r(1/P_i).$$

As noted, in digital applications where $A$ is the binary set, typically $r = 2$. Shannon also theorized that the average amount of information or the entropy of a source as described above could be expressed as

$$H_r(S) = \sum_i P_i \log_r(1/P_i),$$

where $i$ ranges from 1 to $q$.

Rather than examining one symbol at a time, entropy can also be measured on $m$-tuple words constructed from the symbols of $S$. In this case, the entropy of such words, $S^m$, is given by

$$H(S^m) = mH(S).$$

These equations add a mathematical dimension to Shannon's initial conceptions of source coding. For example, Shannon theorized that if noiseless source coding was implemented, not only would the process be invertible but the entropy rates of both the original and coded sourcewords would be equal. On the other hand, if noisy data compression, or entropy encoding was used, the process would be irreversible and coded sourcewords would have a lower level of entropy than their original counterparts. Ultimately, optimal source codes would have an entropy rate approximately equal to but not less than the source entropy. Noiseless and noisy coding are discussed in detail below.

As Shannon was a pioneer and an expert of information theory, many of the principles in this discipline originated with him. Although other information theorists who followed Shannon considered more complex source models, Shannon's work represents the basis for data compression theory.

## Distortion Rate Theory

As mentioned earlier, entropy encoding or noisy encoding is an irreversible process that yields an approximation of the source data upon decoding. In any application of this type, there will be a measurable error in the decoded output. Errors, however, can have varying degrees of importance. For example, an 8-bit binary word with an erroneous least significant bit would result in a difference of 1 in the reproduced value, while the same word with an error in the most significant bit would result in a difference of 128. Therefore, a more useful measure of the reproduced difference which takes into account the relative importance of the errors is needed. This is called the distortion measure.

If $S = \{s_1, s_2, \ldots, s_i, \ldots, s_q\}$ is the source symbol and $C = \{c_1, c_2, \ldots, c_i, \ldots, c_m\}$ is the reproduction symbol of the decoder then a distortion measure of $s_i$ reproduced by some $c_i$ can be represented by the function notation $d(s_i, c_j)$. If $k$-tuple words of $S$ are decoded into $k$-tuple words of $C$ then the distortion measure is notated as $d_k(s_i, c_j)$.

Because there are different distortion measures for each possible reproduction of a $k$-tuple source word, it is often helpful to evaluate the average distortion. The average distortion $D_k$ of a $k$-tuple $S$ word reproduced by a $k$-tuple $C$-ary word is

$$D_k = \sum_i \sum_j P(s_i, c_j) d_k(s_i, c_j),$$

where $i = 1, \ldots, q$, $j = 1, \ldots, m$, and where $P(s_i, c_j)$ is the joint probability that a $k$-tuple $S$ word is reproduced by a certain $k$-tuple $C$ word.

Theoretical distortion measures of this type are useful when designing a data compression system. In practice, there is a design trade-off between the distortion and the rate of the reproduced output. If a certain rate is to be output then design goals are set to obtain the minimum amount of distortion subject to that rate constraint. If a certain maximum level of distortion is acceptable then the output rate is minimized subject to that fidelity constraint. The first situation is referred to as distortion-rate function (DRF) minimization, while the second is called rate-distortion function (RDF) minimization.

Examination of DRFs and RDFs can reveal insight into what is known as the optimal performance theoretically attainable (OPTA) for a data compression system subject to particular design constraints. In this way, the engineer is aware of limitations before implementation begins. Furthermore, this standard can be used to compare and test the performance of a data compression scheme after a system is developed.

Along these lines, OPTAs have been evaluated for many quantization and modulation data compression schemes. The difficulty arises in the subjective measurement of tolerable distortion levels. As data compression techniques are considered in the design of digital audio systems, the most illusive definition remaining is the fine line between what is and is not an acceptable distortion measure.

## Data Compression in Practice

There are many different parameters that characterize practical data compression systems. Specific implementation of these parameters distinguishes one data compression technique from another. Therefore, before discussing specific data compression techniques, it is important to understand the terminology pertaining to the design approaches and parameters of data compression practices.

### Noiseless versus Noisy Coding

At the highest level, a data compression system can be characterized as either noiseless or noisy. Noiseless coding is essentially a mapping of source data into data of another form which, when reconverted, restores original source data exactly. This type of encoding relies heavily on redundancy removal, is invertible, and ensures an average distortion of zero. Noiseless encoding is also called exact coding, information preserving coding, or distortion-free redundancy removal.

Noisy coding is an irreversible operation that produces an approximation of the original source data. There will always be some distortion in a data compression system of this type. However, because noiseless coding rarely produces any significant reduction in data, data compression systems predominantly rely on noisy coding schemes.

## Block versus Nonblock Compression

Block compression relies on the mapping of successive fixed-length segments of data into other fixed-length segments. In this way, data is examined a block at a time without respect to previous or following data. Compression schemes of this type are always synchronous and do not require elaborate memory buffer systems for examining data.

Nonblock compression takes into account past data redundancy by examining overlapping segments of data. This is very similar in operation to nonlinear discrete-time filters and convolutional codes. Data segments are of varying length and therefore require elaborate memory buffer and synchronization systems.

The division between these two differing approaches is not always clear as there are instances in which certain techniques rely on properties of both block and nonblock compression.

## Quantization

Analogous to rounding or truncation, quantization is the simplest form of data compression. In quantization, source data values are mapped into a discrete number of quantization levels. As this is an approximation technique, it falls into the noisy coding category. Quantization distortion is easy to measure, and its relationship to the number of quantization levels is evident. In digital systems, quantization is used as a matter of fact in converting analog signals to a digital form before processing or a more complex method of data compression is used.

In most cases, quantization is performed on a scalar basis. This means that signal values are quantized one at a time. Vector quantization, on the other hand, examines a group or a block of values simultaneously and then reduces that data to a single value. The resulting sample acts as a vector which describes the original data block.

Vector quantization relies extensively on the linear, nonlinear, and probabilistic dependencies among data. This type of relational processing falls into the category of pattern-matching or pattern-recognition. Consequently, areas such as speech coding and image processing have found vector quantization an efficient method of redundancy removal. As an example, vocoders (voice coders) used in speech coding have achieved compression ratios of 12:1 and more when employing vector quantization. For the most part, however, digital audio and visual processing systems have refrained from using vector quantizers due to the presently unjustifiable expense of elaborate memory buffers and multibit real-time processors.

## Prediction versus Interpolation

Prediction is the concept of using past information to predict the value of the next data segment. Interpolation, on the other hand, uses both past and future information to predict the value of the intermediate sample. Figure 3 shows a block diagram of a basic predictive/interpolative data compressor. In both instances, a comparison is made between the actual and predicted sample to determine if the difference falls within a tolerable error window. If the error is small enough, the sample is taken to be redundant and is discarded as it has been shown to be accurately predictable at the output. If the error is too large, either the sample or a measure of the error is transmitted to ensure that this value is used at the output when the incorrect prediction is made again by the decoder. Obviously, the efficiency of the predictor/interpolator corresponds directly to the amount of compression that will be obtained. Moreover, as the majority of samples are not coded, the compression effect is obvious in terms of both bits and bandwidth. The only major problems arise in keeping transmitted samples synchronous with decoding, and in overcoming errors that occur between the encoding and decoding process or that are inherent in the prediction/interpolation method itself.

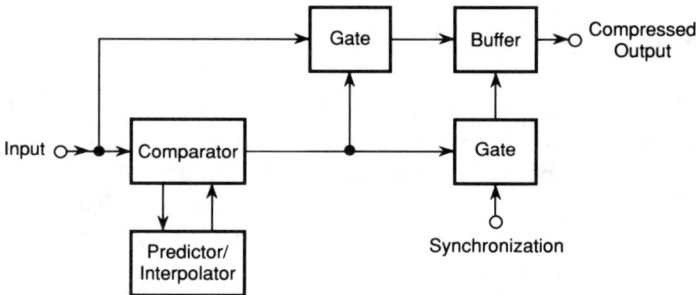

**Figure 3.** A predictive/interpolative data compression system.

In practice, prediction and interpolation algorithms commonly employ $n$th-order polynomials for making predictions. An $n$th-order polynomial will utilize $n + 1$ data points in predicting the current value. Of course, as higher order polynomials are more accurate they are also much more difficult to implement. Fortunately, first-order and even zero-order polynomials are often sufficient in predicting sample values and thus avoiding use of higher-order polynomials.

## Adaptation

Most data compression systems are designed to process a certain kind of data. For example, a data compression system used in music coding may be optimized for compressing typical periodic musical waveforms. There are instances in any system, however, when atypical data is presented to the data compressor. When this happens, a data compression scheme may become overburdened and introduce errors. Some systems have been known to employ compression algorithms that in certain situations actually increase the amount of transmitted data rather than reduce the data rate. To protect against such encounters, many data compression systems employ the idea of adaptation.

Almost all data compression techniques have the ability to be adaptive. Being adaptive implies being able to change or modify the compression algorithm if inefficient and potentially invalid data compression occurs. A predictive data compression system which modifies the coefficients of its polynomial during operation can be viewed as an example of adaptive data compression. As with any process involved in encoding, identical integrity needs to be maintained at the decoding stage.

# Data Compression Techniques

A wide variety of coding methods has been devised, each with particular strengths and weaknesses. The classical noiseless Huffman coding scheme may be considered optimal for some cases of text block coding. Arithmetic coding, on the other hand, allows a greater level of data compression, one that is closer to the theoretical bound. Ziv-Lempel coding is well suited for long, repeated sequences of data strings. Modulation methods such as differential pulse code modulation, delta modulation, adaptive delta modulation, and adaptive delta pulse code modulation all can provide significant compression over PCM methods. Some transformations implemented in DSP may provide very efficient compression performance.

## Huffman Coding

In 1951, David A. Huffman [1952] developed a simple noiseless coding scheme which is optimal for data companding using fixed-length or variable-length block encoders. This statistical encoding technique is optimal in the sense that it yields the shortest average code length for representing symbols of a given alphabet. Although Huffman coding can be used with any number of symbols in the coding alphabet, in digital applications the codes are commonly constructed from the binary set. For the purpose of this discussion, attention will focus on coding examples using the 0 and 1 symbols only.

Huffman coding relies primarily on the examination of the relative frequency of occurrence of source data blocks. The most frequently occurring source words or symbols are coded with the shortest code words while infrequently occurring source data values are encoded with longer code words. To accomplish this, a nonduplicating prefix system must be employed. This means that a shorter code word cannot be the beginning of a longer code word. For example, 101 and 1010001 cannot both be code words.

The construction of a Huffman code is a simple and iterative process if the relative frequency of occurrence is known for each possible source word. An example of the Huffman code construction for a source with the possible output words of

$$S = \{s_1, s_2, s_3, s_4, s_5, s_6, s_7\}$$

and their associated probabilities of occurrence

$$P = \{0.10, 0.05, 0.20, 0.15, 0.15, 0.25, 0.10\}$$

is shown in Figure 4. First, sourcewords are arranged in descending order of occurring frequency. Then, starting at the bottom of the table, the formation of a binary tree begins by grouping together the two lowest frequencies of occurrence and adding their probabilities of occurrence to obtain a new composite frequency. This process continues until the tree is completed. At this point, one branch of each node is given a value 0 while the other is designated 1. The initial decision is arbitrary, but consistency must be kept throughout the tree. Code words can be read by tracing the path of 1s and 0s from the root node to the source words at the leaves.

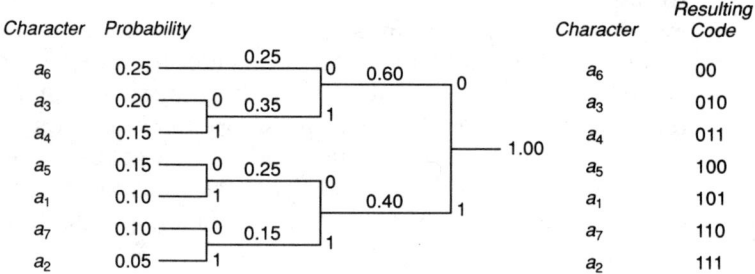

**Figure 4.** An example of Huffman code construction.

The average code length is easily obtained as it is the summation of each code wordlength times its corresponding frequency of occurrence. For the example in Figure 4, the average code length is $2(0.25) + 3(0.75) = 2.75$ bits. The number of bits needed to encode a particular source word using the Huffman technique can be computed prior to actual code construction using the formula

$$b_i = f(-\log_2 P_i),$$

where $P_i$ is the probability of occurrence of the sourceword in question and $f(x)$ is a function which yields the closest integer greater than or equal to $x$.

In the example of Figure 4, prior to code construction, it could have been computed that two bits would be needed to encode the source symbol with an occurrence frequency of 0.25: $(-\log_2 0.25 = 2)$.

There are three major disadvantages of Huffman coding. The first is the fact that the relative frequency of occurrence for each sourceword is usually not known prior to encoding. Second, if Huffman encoding is attempted without prior knowledge of probability frequencies then memory buffers used to analyze sourcewords and generate code words may fail when encountering atypical data. In addition, if all possible source words occur with approximately the same frequency then Huffman coding will not be efficient. From these arguments it can be seen that Huffman coding is better suited for text compression rather than music coding.

As noted, the Huffman coding technique translates each received symbol into a minimum integral number of bits. For example, if the probability of occurrence of a particular symbol dictates that 3.4 bits are theoretically needed for encoding, the Huffman coding algorithm would use a 4-bit code to represent it. As a worst case scenario, if the probability of a single symbol among an alphabet approaches one, Huffman coding may yield an encoded message much longer than the original! On the other hand, the Huffman scheme performs optimally when all symbols have a probability which is an integral power of 1/2. As this is not likely to be the case in practical systems, Huffman coding generally cannot attain theoretical performance levels.

It should be noted that although it is not theoretically optimal, the Shannon-Fano technique of code construction, which is very similar to the Huffman coding procedure, can yield lower average code lengths than the Huffman code in certain identical situations. The Shannon-Fano technique differs from Huffman coding in its binary tree construction. Source words are listed in descending order of occurrence probability, but the tree is formed by dividing remaining probabilities into halves which contain the closest composite frequency values. When the tree is completed, branches are labelled 0 and 1 appropriately and code words are read as in the Huffman technique.

## Arithmetic Coding

Unlike Huffman coding, arithmetic coding does not require that symbols be individually encoded into their minimum integral bit forms. In almost all cases, this allows a greater level of data compression, which is closer to the theoretical bound.

In arithmetic coding, relative symbol probabilities are assigned a portion of the range of probability between zero and one. If an alphabet consists of the symbols X, Y, Z, and $ with respective probabilities of 0.2, 0.2, 0.4, and 0.2, the interval breakdown of these symbols might be as shown in Table 1.

**Table 1.** Interval Breakdown of X, Y, Z, and $

| Symbol | Probability | Interval |
|---|---|---|
| X | 0.2 | [0, 0.2] |
| Y | 0.2 | [0.2, 0.4] |
| Z | 0.4 | [0.4, 0.8] |
| $ | 0.2 | [0.8, 1.0] |

As a message is received for encoding, symbol probabilities are applied to each other. This produces a range which narrows proportionally to the probability interval associated with each successive symbol. Figure 5 provides a graphic illustration of this process for encoding the message YXZZ$ using the probabilities above. The resulting interval is unique to the encoded message and any number within that interval can be used to encode it. Likewise, the decoder only requires one number from the resulting interval to recover the original message. Decoding the message from this value is accomplished in much the way that encoding occurred. For example, if 0.2280 was selected from the resulting interval in Figure 5, the decoder would first note that it was within the range of 0.2 and 0.4 and conclude that the first symbol must be Y. The decoder then determines that the only symbol which could have been encountered next and still retain 0.2280 in the resulting interval is X. This process is repeated until an end of message character is decoded. In the example, the $ represents the end of message character.

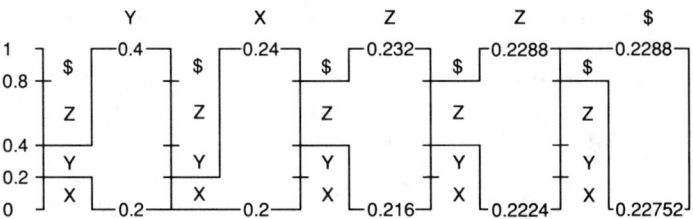

**Figure 5.** Arithmetic coding applied to the message YXZZ$.

Recalling the formula, $b_i = f(-\log_2 P_i)$, which computes the minimum number of bits, $b_i$, needed to represent a symbol with the probability, $P_i$, a comparison between the Huffman and arithmetic coding algorithms can be made for the encoding of the YXZZ$ message. Employing Huffman coding would require

$$f(-\log_2 0.2) + f(-\log_2 0.2) + f(-\log_2 0.4) +$$
$$f(-\log_2 0.4) + f(-\log_2 0.2) = 12 \text{ bits,}$$

while arithmetic coding would require only

$$f(-\log_2 0.2 - \log_2 0.2 - \log_2 0.4 - \log_2 0.4 - \log_2 0.2) = 10 \text{ bits.}$$

This number may also be obtained by subtracting the resulting upper bound from the lower bound and then applying the formula

$$f(-\log_2(0.2288 - 0.22752)) = 10 \text{ bits.}$$

From Figure 5, it becomes evident that there are a number of different ways that arithmetic coding calculations can be performed with basic arithmetic. This promotes the use of simple and cost-effective hardware and software designs which, in turn, provide faster processing times over most Huffman implementations. Also from the example, it becomes obvious that any message can be represented by a real number between zero and one. As messages become longer, the interval narrows and more bits are required for encoding. Similarly, symbols with greater probability reduce the range less and incur fewer bits than would a symbol with a small probability of occurrence.

Arithmetic coding may employ a fixed probability model, as in this example, or an adaptive one which is modified with each new symbol reception. Some systems perform preparsing of a finite-length data string to obtain the exact frequency of occurrence for each symbol. In any case, the primary concern of any data compression/decompression scheme must be satisfied—maintaining integrity at both the encoding and decoding stages.

## Ziv-Lempel Coding

The Ziv-Lempel [1978] coding algorithm is a highly effective method of noiseless data compression. This coding scheme makes use of a binary parsing tree which naturally removes data redundancy, and a code string which acts as an index into the tree as well as the compressed data stream itself.

Coding begins with the examination of a continuous stream of binary data. As a sequence of bits is encountered, it is compared to previously coded bit sequences. The incoming data string continues to lengthen until a bit is received which makes the new sequence unique to those strings already encoded. At this point, an index which points to the longest existing matching sequence is saved. In addition, the new unique bit is also saved and the new unique sequence is added to the growing list of encoded sourcewords. Saving the index and the unique bit is merely the concatenation of these values to the code string which represents the new compressed data stream. The addition of the new unique sequence is accomplished by creating a new node in the parsing tree with a label index equal to the next counter value. The counter is updated each time a new sourceword is encoded. This explanation of the Ziv-Lempel coding algorithm can be reduced to the following pseudo-code.

*Step 1*: Find the longest previously coded sourceword which matches the next sequence to be encoded.

*Step 2*: Append the index of the existing coded sourceword to the code string.

*Step 3*: Append the new unique bit to the code string.

*Step 4*: Add the new sequence to the parsing tree by creating a new appropriate node in the tree.

*Step 5*: Return to parsing incoming data.

Creation of the parsing tree and the resulting compressed code string are shown in Figure 6. The linear table and the spaces in the original data string have been included for the purpose of clarification. Note that the Ziv-Lempel coding scheme requires that the first node or the root node of the parsing tree be a null node.

From Figure 6, it can be seen that once the tree is created, coded

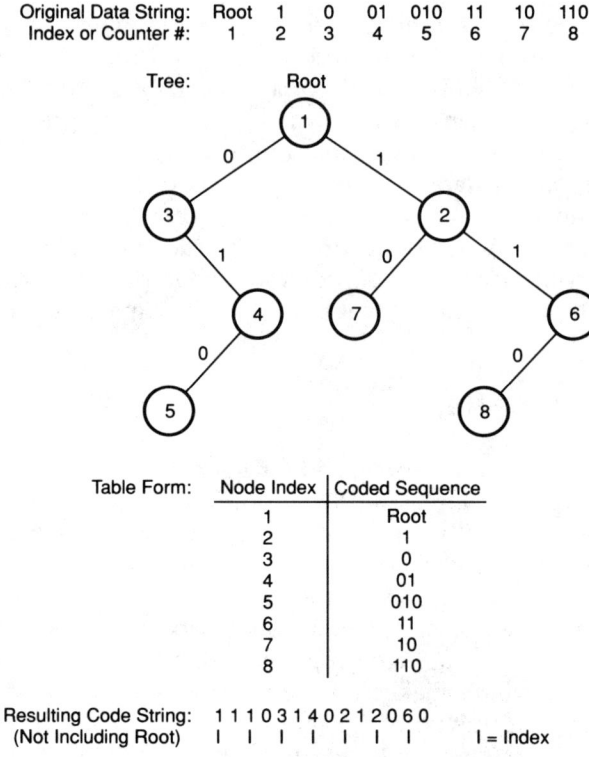

**Figure 6.** A Ziv-Lempel coding example.

sourcewords are recovered by merely tracing the path from the root to the appropriate remote node and then adding the unique bit. A small amount of overhead may be necessary for transmitting tree structure information to the decoder and for carrying out recursive tracing functions in the tree. However, this particular coding algorithm can serve as an efficient data compression method in instances where there are long, repeated sequences of data strings.

## Differential Pulse Code Modulation

As its name implies, differential pulse code modulation (DPCM) is a form of conventional PCM. The difference is that PCM systems output actual sample values while DPCM systems produce a representative measure of the difference between successive samples. If the sampling rate is fast enough, the change from one sample value to the next should be encodable with very few bits. The compression effect and its significance are obvious.

Relying on predictive methods described earlier, DPCM systems encode the difference between a predicted value and the real value for each sample. A block diagram for a basic DPCM encoding system is shown in Figure 7. The corresponding decoder regenerates the original waveform using the predicted values and the encoded differences. As in any prediction scheme, the efficiency of DPCM depends on the ability of the prediction algorithm to follow changes in the input signal.

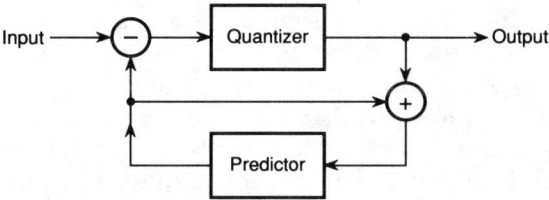

**Figure 7.** A differential pulse code modulation (DPCM) encoding system.

## Delta Modulation

Delta modulation (DM) can be viewed as a special case of DPCM. The distinction arises in that DM uses only one bit to encode the difference between the predicted and actual values. Although the use of a single bit greatly simplifies hardware design and operations, it requires significantly higher sampling rates than those used in conventional PCM systems. Also, because only two values are used to encode the difference, the size of this increment is critical. The size of the increment will directly reflect on the accuracy of the system and competes with the sampling rate as a design trade-off. Block diagrams for both DM encoder and decoder circuits are shown in Figure 8.

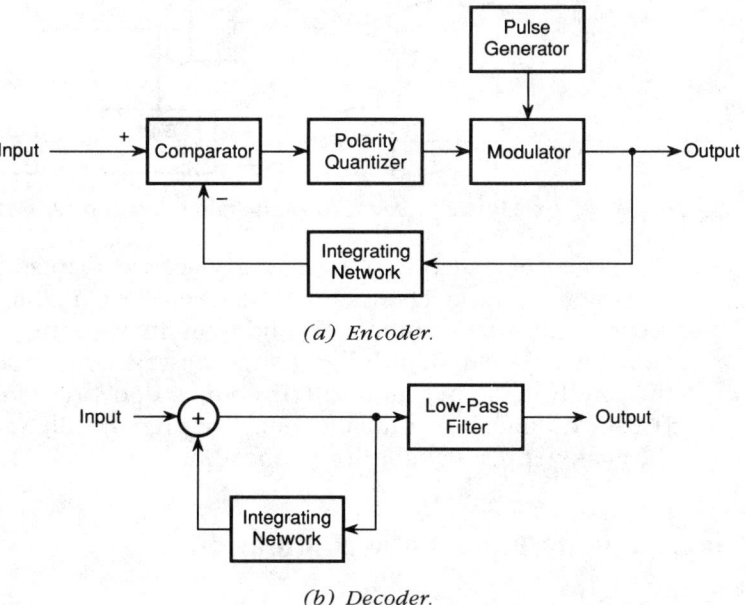

(a) Encoder.

(b) Decoder.

**Figure 8.** Delta modulation (DM) encoding and decoding processes.

In terms of audio applications, DM systems have several clear advantages and disadvantages in comparison with conventional PCM systems. Although they offer a significant reduction in encoded data, DM techniques require extreme and sometimes unobtainably high sampling rates for high-fidelity results. In addition, transient audio signals may not be tracked quickly enough by DM systems, resulting in sonic distortion. Furthermore, as the range of quantization values is only two, quantization error is always present and most evident in low-level passages. On the positive side, steep filters are not required at the output of the decoder, and A/D and D/A converters can be replaced with much less expensive integrating circuitry.

## Adaptive Delta Modulation

Adaptive delta modulation (ADM) is a modification of delta modulation which overcomes some of the drawbacks of DM. In particular, adaptive techniques are implemented to reduce transient and quantization distortion. The adaptive process employs an adjustable increment size. As the input signal changes more quickly, the increment size is increased. If the input signal remains nearly constant, the increment size is decreased to provide better tracking of the signal with less quantization error. A block diagram of a basic ADM encoder is shown in Figure 9.

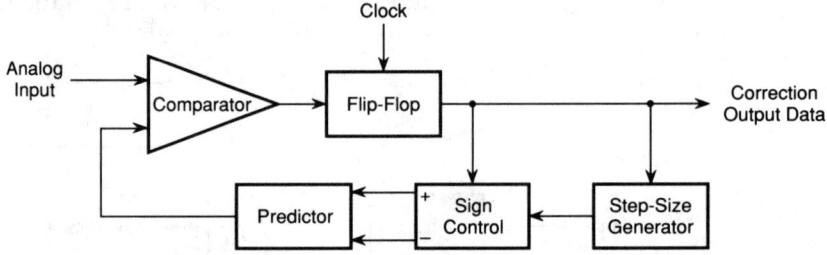

**Figure 9.** An adaptive delta modulation (ADM) encoder.

As with any adaptive system, design of the adaptation algorithm is critical to its performance. The trade-off between design complexity and prediction accuracy must be manipulated, and integrity with the decoding process must be maintained. Similar design issues must be considered with the inverse of this method, sometimes called companded predictive delta modulation (CPDM), which leaves the increment size fixed while varying the amplitude of the analog input signal prior to encoding.

## Adaptive Delta Pulse Code Modulation

Adaptive delta pulse code modulation (ADPCM), sometimes referred to as adaptive differential pulse code modulation, combines properties of both

ADM and conventional PCM encoding. Specifically, the 1-bit encoding scheme is replaced with a multibit scheme. In addition to the increase in quantization levels, a more accurate and effective method of generating variable increment sizes is achieved by assigning each quantization level its own step-size scale factor.

Because of the uniqueness of ADPCM, its encoding scheme is often referred to as a bit rate reduction (BRR) encoder. The basic block diagram of a BRR encoder is shown in Figure 10. Information pertaining to the amplitude and spectrum distribution of the signal is supplied by the range and filter inputs at the front end of the encoder. Depending on the nature of the input signal, the BRR encoder selects a delta PCM mode of operation which makes a prediction of the input value, selects an appropriate increment size, and quantizes the difference with a sufficient number of bits. Noise shaping is added to further reduce the noise floor of the encoded signals. In terms of performance, ADPCM systems have been shown to function as well as or better than conventional PCM systems with less quantization noise and a significant reduction in data.

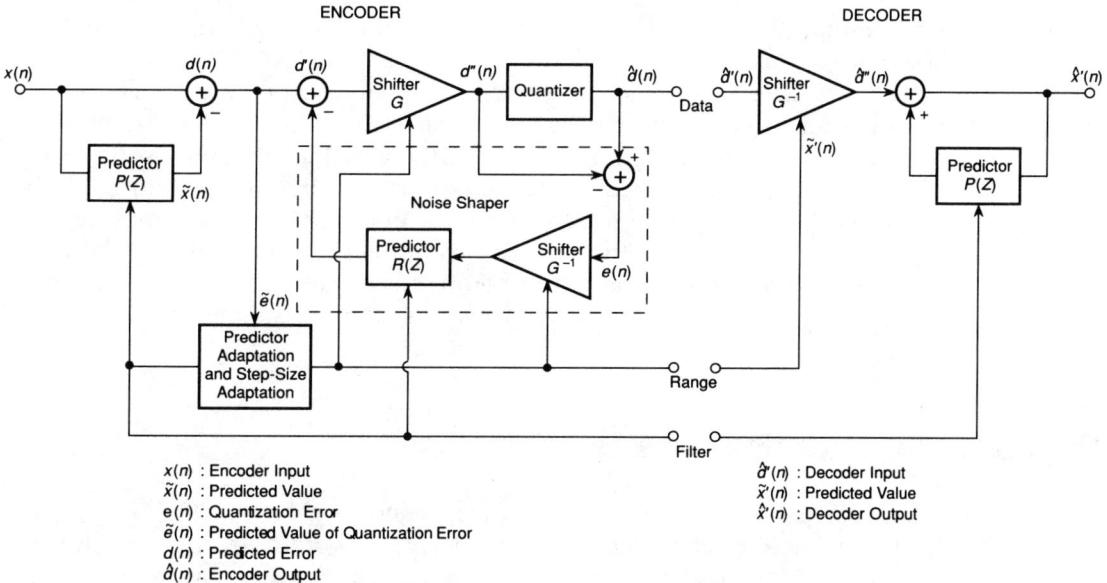

**Figure 10.** Bit rate reduction (BRR) encoding used in ADPCM systems.

## Transformations

Because of their complexity, transformations are not usually considered to be popular forms of data compression. The fact remains, however, that the same data may be represented in a more concise domain through the use of transformations. As shown in Figure 11, it is easy to conceive of the basic block diagram for a data compression system using transformations.

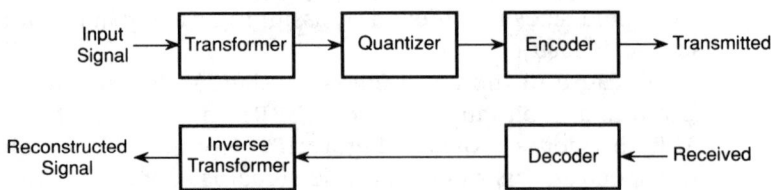

**Figure 11.** Data compression using transformations.

Orthogonal transformations such as Fourier, Haar, Hadamard, Karhunen-Loeve, and Walsh are among the most efficient of these data compression techniques. As an example, consider representing a sinewave by conventional PCM methods; several sample points are required for accurate representation. Use of a Fourier transform would yield only two frequency values, the amplitude and phase, necessary to represent the same signal. In addition, it may be possible to increase compression ratios by incorporating into transformations analogous methods of prediction/interpolation and adaptation used in PCM systems.

Digital filters, explained in Chapter 10, can also provide a form of data compression. Using Fourier and $z$ transformations, the audio signal can be represented in the frequency domain. In this domain operations such as extracting useful bands, bandlimiting signals, or even frequency shifting can be carried out. The filtered signal could now be encoded by lower sampling rates and/or fewer bits. Inverse transformations would be applied after decoding to complete the process. Digital filters could thus be used before subjecting data to one of the PCM-based data compression systems described above.

## Digital Audio Data Compression Systems

Like most major developments, data compression techniques did not originate in the technology of digital audio. Rather, they were born of research from such parents as AT&T and government agencies. For this reason, when data compression systems were first implemented in digital audio, they did not provide the audio quality demanded by listeners accustomed to 16-bit linear PCM. Even now, with concentrated effort in the digital audio field, data compression has yet to fully mature. At present, it is a design race to produce a completely transparent digital audio data compression system which is optimally efficient in data reduction. Recent methods presented here depict the evolution and current state of digital audio data compression systems. Many of these schemes have found their own place and application in the world of digital audio. However, the development of new standards lags behind.

## Design Considerations

The primary goal in designing any data compression system is to make the difference between the decoded output signal and the original input signal as small as possible. In audio applications, this goal must be extended to encompass human subjective response to varying audio errors. Specifically, the design of a digital audio data compression system must take into account the fact that certain errors will be much less audible than others. In this way, errors can be acknowledged as a consequence of a digital audio data compression system if it is known that an increase in data reduction can be achieved while the properties of psychoacoustics conceal such errors.

One subjective phenomenon relied on is masking. Masking occurs when the level and/or frequency of one sound is such that it conceals another sound. Although the masked sound is present, it is inaudible. This idea suggests that inaudible tones need not be encoded in a digital audio data compression scheme. In general, sounds of low levels are most feasibly masked by louder sounds of a similar frequency. Similarly, as sounds become farther apart in frequency, it becomes more difficult for one to mask the other. Modulation noise is an example of an error which may be overcome through masking in a digital audio data compression system.

Early digital audio data compression systems, for the most part, were wideband (or singleband) systems selected for their ease of implementation rather than their sonic quality. Later digital audio data compression systems were also wideband systems; however, they did show improvements such as equalization curves which boosted or cut certain frequency bands to condition the signal for encoding. Inverse equalization curves were used following the decoding process. Recently developed systems, and those currently being researched, are multiband systems. These systems act on individual frequency bands and therefore take advantage of the masking properties of audio signals. Bands are encoded separately and then combined during decoding. Furthermore, compression systems of this type usually employ a technique known as dynamic bit allocation. This technique allows the frequency bands with the most energy to be coded with the most bits while the lowest-energy bands utilize the fewest number of bits. Theoretically, if a particular band that is sufficiently low in energy is masked by another band, it may not be encoded at all.

## The DAT 16/12, BBC 14/10, and Video 8 10/8 Schemes

The DAT 16/12, BBC 14/10, and Video 8 10/8 digital audio data compression schemes are all wideband systems which examine the input sample value and encode it nonlinearly in a form using fewer bits. Although the concept behind their implementation is the same, their designs vary. For example, in the Video 8 system, 10-bit audio samples are compressed into 8-bit form before being sent to the heads for recording. In this format, compression is necessary because storage space is limited, as discussed in Chapter 7. In the DAT 16/12 scheme, 16-bit samples do not have to be compressed before being recorded. However, the option of compressing to twelve bits is provided to the user to reduce the required storage space and, correspondingly, the quality of the encoded audio.

## The CD-I and DVI Systems

The CD-I (*c*ompact *d*isc-*i*nteractive) and DVI (*d*igital *v*ideo *i*nteractive) formats utilize the same digital audio data compression scheme. It is a wideband ADPCM system which offers four different quality levels which may be varied according to demands of the input signal. A complete block diagram of the CD-I and DVI encoding and decoding system is shown in Figure 12.

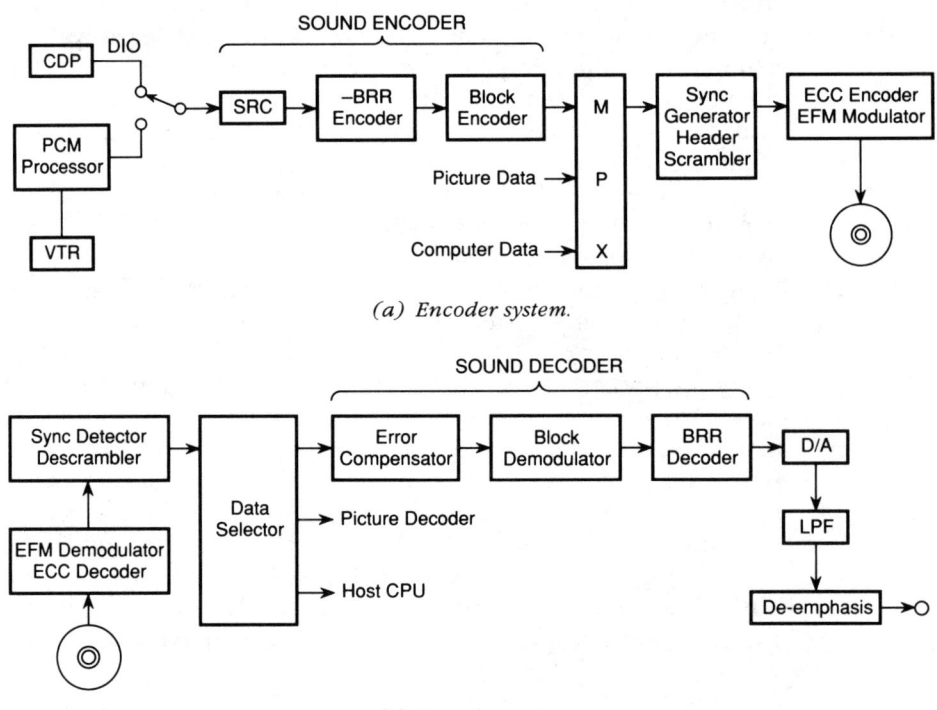

(a) Encoder system.

(b) Decoder system.

**Figure 12.** The CD-I and DVI encoding and decoding systems.

As CD-I and DVI are both multimedia digital formats which use conventional CD-Audio discs, they must significantly reduce digital audio data bit rates to permit storage of audio data in addition to video, text, and other data. Depending on the amount of nonaudio information to be encoded, a variety of ADPCM levels, or operational modes, may be used. If only audio is to be stored, the CD-Digital Audio mode may be selected. This offers the standard

CD-Audio format (44.1 kHz/16-bit/stereo) without any compression whatsoever. If only a small percentage of nonaudio information is to be stored, the Hi-Fi mode may be appropriate; it offers a 37.8 kHz sampling rate with 8-bit resolution in either a four-channel mono or two-channel stereo configuration. The Mid-Fi mode offers more capacity for video and text data, but audio quality decreases accordingly with a 37.8 kHz sampling rate and 4-bit encoding with a four-channel stereo or eight-channel mono channel option. Lastly, the Speech mode offers speech quality audio with an 18.9 kHz sampling rate and 4-bit resolution with either sixteen mono or eight stereo audio channels. ADPCM data compression is used in the latter three modes, and all modes offer a total playing time of approximately 74 minutes per channel.

Measured performance of these ADPCM modes has shown the Hi-Fi mode to yield a 17 kHz bandwidth with a 90 dB S/N ratio. Mid-Fi mode and Speech mode yield a 60 dB S/N ratio and 17 kHz and 8.5 kHz bandwidths, respectively.

## A 4:1 Compression System

One commercial data compression system compresses high-quality 16-bit PCM samples into four bits. This 4:1 compression ratio is fixed and error immunity of better than one part in 10,000 is guaranteed.

The system uses a multiband ADPCM technique together with dynamic bit allocation. Four frequency bands share the allotted four bits. Within the bands, predictive coding methods are used in conjunction with an algorithm for utilizing psychoacoustic masking. A block diagram of the encoding and decoding counterparts is shown in Figure 13.

There are many remarkable aspects of this design. The system can operate with sampling rates from zero to 50 kHz. The sampling rate in use determines the frequency response of the output signal at the decoder. By identifying the start of each data segment, the system allows encoders and decoders to be used independently of one another without a synchronizing word clock. Another synchronization mode allows a number of systems to be synchronized together for multichannel capabilities. The system also offers a multiplexing scheme which allows up to eight coded channels to be multiplexed together and transmitted via a single communications link for decoding by a single decoder. The system uses the AT&T DSP16 (described in Chapter 13), a 16-bit DSP processor with a 55 ns instruction cycle time, two 36-bit accumulators, and both serial and parallel I/O ports to achieve audio compression in real time. The processing delay is only 2.5 ms at a sampling rate of 48 kHz.

The system audio specifications include a dynamic range greater than 90 dB with frequency deviations less than 0.05 dB across the bandwidth. Total harmonic distortion through the encoding and decoding process is less than 0.05%, and S/N ratios are greater than 65 dB. In addition, the crosstalk and phase deviation between a stereo pair are zero. The Apt X 100 digital audio data compression system represents one of the most advanced systems of its kind. It was developed by Audio Processing Technology, Ltd.

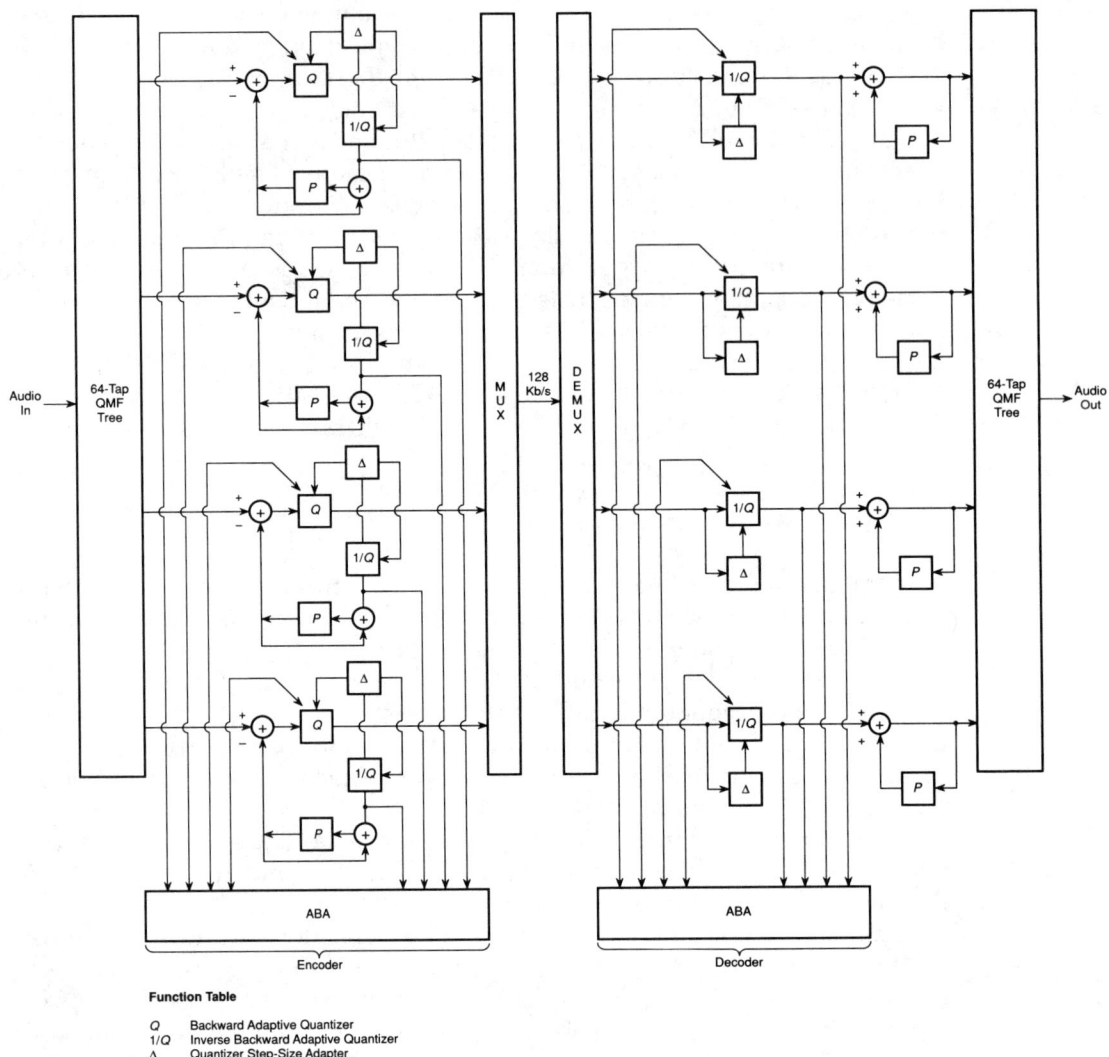

**Figure 13.** A subband ADPCM coding scheme suitable for 4:1 audio data compression, showing encoding and decoding sections. *(Courtesy Audio Processing Technology, Ltd.)*

## The Eureka Systems

Eureka digital audio data compression systems have been developed in Germany. These are multiband systems which separate the audio signal into 20 to 30 bands each approximately 1/3-octave wide. Dynamic bit allocation is employed, and very detailed models of psychoacoustic masking are used to decide how particular frequency bands are coded. As 20 to 30 different frequency bands are utilized, almost all of these bands must go without coding (allocated zero bits) to achieve significant data reduction. This implies that very good predictive methods are used within the bands themselves. Compact disc quality results have been claimed with compression factors as low as one bit per sample. The Eureka system is discussed in Chapter 9.

## DCC (Digital Compact Cassette) and the PASC (Precision Adaptive Subband Coding) System

DCC is a digital compact cassette format which utilizes a cassette shell similar to that used by conventional analog audio cassettes. DCC recorders are designed to play either digital DCC cassettes, at sampling rates of 32, 44.1, or 48 kHz, or standard analog cassettes both at the speed of $1^7/_8$ ips with equivalent playing times. Even by recording eight parallel data tracks, each with a track width of 185 $\mu$m, the need to limit tape consumption prevents DCC from recording linear PCM data as in the DAT format. Rather, a sophisticated data compression algorithm is employed to yield a dynamic range greater than 105 dB and a THD + N measurement of less than 0.0025%, with flat frequency response to the Nyquist frequency. This is accomplished with an audio bit rate of 384 Kb/s—about one fourth that of the compact disc.

The data compression algorithm is known as Precision Adaptive Subband Coding (PASC). The PASC system is based on three principles. First, the ear only hears sounds above the threshold of hearing. Second, louder sounds mask softer sounds of similar frequency, thus dynamically changing the threshold of hearing. Similarly, other masking properties such as high- and low-frequency masking may be utilized. Third, sufficient data must be allocated for precise encoding of sounds above the dynamic threshold of hearing.

Processing begins by splitting the audio signal into 32 subbands of equal width with a 24-bit fixed-point digital filter. A DSP chip which models the most sensitive human ear continuously adapts its threshold to mimic dynamic variations of the human cochlea. As the frequency and level of the audio signal varies, the real-time computations of the signal processor determine the contents of each subband relative to the threshold, and bits for encoding these signals are allocated for maximum precision, according to the dynamic model. Bits not required to encode a particular subband can be dynamically used to improve coding on other subbands. A floating-point data word is employed, with variable-length 15-bit mantissa and 6-bit exponent.

Figure 14a shows average and minimum human hearing threshold curves that were derived from critical listening tests. Figure 14b shows a tone (*A*) that exceeds the minimum threshold and thus would normally be audible. Figure 14c shows an example in which tone *A*, when in the presence of a louder nearby tone (*B*) is made inaudible and therefore does not require encoding in the PASC system. At the output of the encoder, Reed-Solomon error correction with 47% redundancy is employed along with an eight-to-ten modulation code. Data is

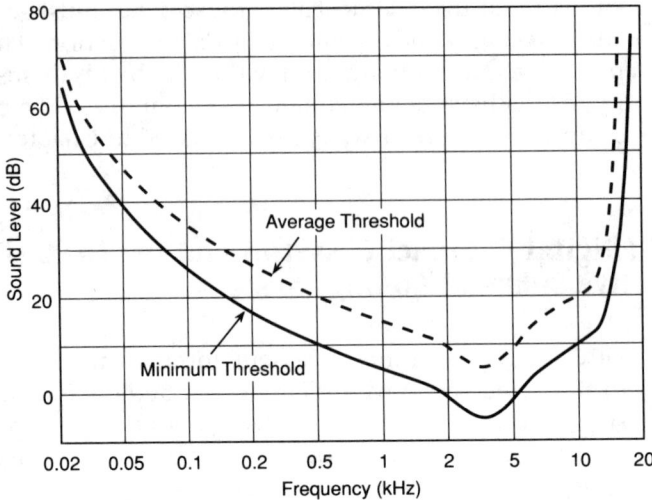

*(a) Average and minimum thresholds of hearing curves.*

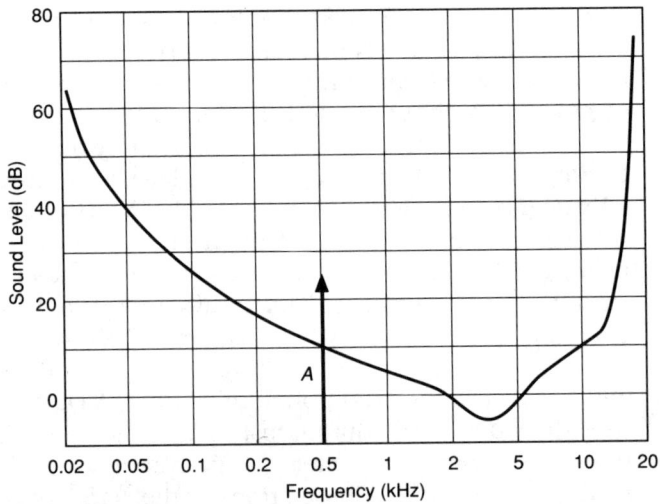

*(b) Example of a tone that would normally be audible.*

**Figure 14.** Threshold of hearing models used in the PASC compression algorithm. *(Courtesy Philips)*

distributed over eight parallel magnetic tape tracks, in two tape directions, at a rate of 95.2 Kb/s/track, yielding an overall data rate of 768 Kb/s. Even with this significant amount of overhead, data on two auxiliary tracks can be used to convey up to 400 characters per second for song titles, lyrics, artist names, and other information, while achieving high audio fidelity. The DCC system and PASC data compression method were developed by Philips. A number of data compression schemes are summarized in Table 1.

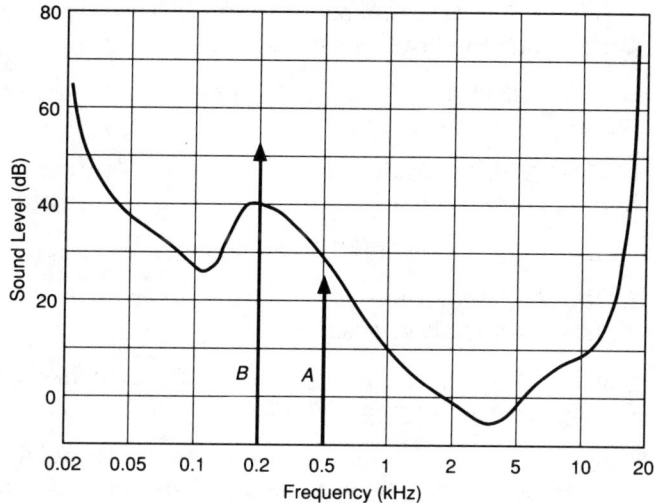

*(c) The same tone being masked by a louder tone that raises the threshold of hearing.*

**Figure 14.** (cont.)

## Conclusion

Data compression systems have been used in high-fidelity audio applications for only a very short time, thus there has been limited opportunity to test the effectiveness of such systems. Many potentially difficult questions remain unanswered.

What if several iterations of data compression encoding and decoding are performed? Will it be possible to edit a signal in compressed form? Will changes in gain, frequency, and time be possible? How will this affect the use of digital audio data compression systems in the postproduction community? In terms of spatial information, how does compression affect subtle cues like space, ambience, and stereo relationships? As processing delays are overcome in inexpensive applications such as consumer digital audio, what sonic penalty will be paid?

Clearly, data compression algorithms will be a vital part of future digital audio systems. Moreover, their implementation will challenge both the objective and subjective methods we use to assess the quality of recorded music.

## References

Audio Processing Technology Ltd, "Apt-X 100 Digital Audio Data Compression System," 1990.

———, Database News Release, Autumn, 1990.

Capellini, V., *Data Compression and Error Control Techniques with Applications*, Academic Press, 1985.

Davisson, L. D., and R. M. Gray, editors, *Data Compression*, Dowden, Hutchinson and Ross, 1976.

Fox, B., "Digital Audio Cassette Launch: Kenwood Write Once CD," *Studio Sound*, December, 1990.

Gerzon, M., "The Gentle Art of Digital Squashing," *Studio Sound*, May, 1990.

Held, G., *Data Compression: Techniques and Applications, Hardware and Software Considerations*, John Wiley & Sons, 1987.

Huffman, D. A., "A Method for the Construction of Minimum-Redundancy Codes," *Proceedings of the IRE*, 1952.

Langdon, G. G., "A Note on the Ziv-Lempel Model for Compressing Individual Sequences," *IEEE Transactions on Information Theory*, vol. IT-29, March, 1983.

Makhoul, J., Roucus, S., and H. Gish, "Vector Quantization in Speech Coding," *Proceedings of the IEEE*, November, 1985.

Moore, J., "Data Compression or Fractals Are Your Friends," unpublished, 1989.

Pohlmann, K. C., "The Philips DCC Format," *Mix*, April, 1991.

Sabin, M. J., and R. M. Gray, "Product Code Vector Quantizers for Waveform and Voice Coding," *IEEE Transactions*, ASSP, vol. ASSP-32, June, 1984.

Shannon, C. E., "A Mathematical Theory of Communication," *Bell System Technical Journal*, vol. 27, 1948.

Witten, J. H., Neal, R. M., and J. G. Cleary, "Arithmetic Coding for Data Compression," *Communication of the ACM*, June, 1987.

Ziv, J., and A. Lempel, "A Universal Algorithm for Sequential Data Compression," *IEEE Transactions on Information Theory*, vol. IT-23, May, 1977.

*Chapter 9*

# Digital Audio Satellite Broadcasting

Brent A. Karley

## Introduction

Soon after the introduction of the compact disc, many radio broadcasters began playing CDs rather than LPs or analog cartridges to take advantage of the improved sound quality of this new medium. Although digital audio provides inherent improvements over analog, many of these advantages are diminished or eliminated when the signal is converted to the analog domain for transmission. Digital broadcasting overcomes this problem by providing a digital transmission path, thus allowing consumers to hear programs reproduced over radio with compact disc quality.

Digital broadcasting is an advantageous means of transmitting audio because it permits the high fidelity of digital audio to be extended into the transmission path. Digital broadcasting is more immune to noise and distortion than conventional transmission techniques. Repeaters can detect digital signals and retransmit a new, clean, and robust noise-free signal. Accumulation of noise is thus prevented. The implementation of digital hardware provides easy serviceability, and software control systems allow transmission flexibility. Coding and interleaving yield low bit error rates (BER). The implementation of digital broadcasting thus completes the transition of major audio sources from the analog to the digital domain.

However, extensive investment is initially required by both broadcasters and distributors in new equipment, personnel, and training to build and/or refurbish broadcasting facilities. This results in increased costs to consumers. The broadcaster may also be required to purchase time on a satellite for distant broadcasting. As the number of consumers increases, the cost per consumer correspondingly decreases.

Digital audio broadcasting (DAB), a subfield of digital communications, differs from conventional analog transmission systems in signal representation and modulation technique. Amplitude modulation and frequency modulation are typically used for transmitting analog signals. Digital broadcasting is commonly transmitted by modulating the phase of a carrier signal. By performing the transmission in this manner transmission noise and distortion are reduced while the dynamic range is increased.

There are several possible digital audio sources, transmitting links, and signal paths available for digital audio broadcasting. As shown in Figure 1, this situation may involve a local studio, remote studio, or live broadcast utilizing an outside broadcasting (OB) vehicle. Audio information is digitally transmitted from remote sources to the broadcast center where it is retransmitted either by direct broadcasting to the listener (terrestrial broadcasting) or via satellite. Typically the audio signal is in digital format when received by the broadcast center and is digitally distributed from that point. A satellite may receive the signal from the broadcast center and distributes it either directly to the listener or to an earth-based distributor such as a cable television operator. Signal transmission directly from a satellite to the listener's home or business forms the basis for a direct broadcast satellite (DBS) system. Future consumer DAB systems will likely employ a combination of satellite, terrestrial, and cable transmission systems. Currently, satellite DAB is a well-developed technology and thus provides a foundation of information useful in the development of any future DAB system.

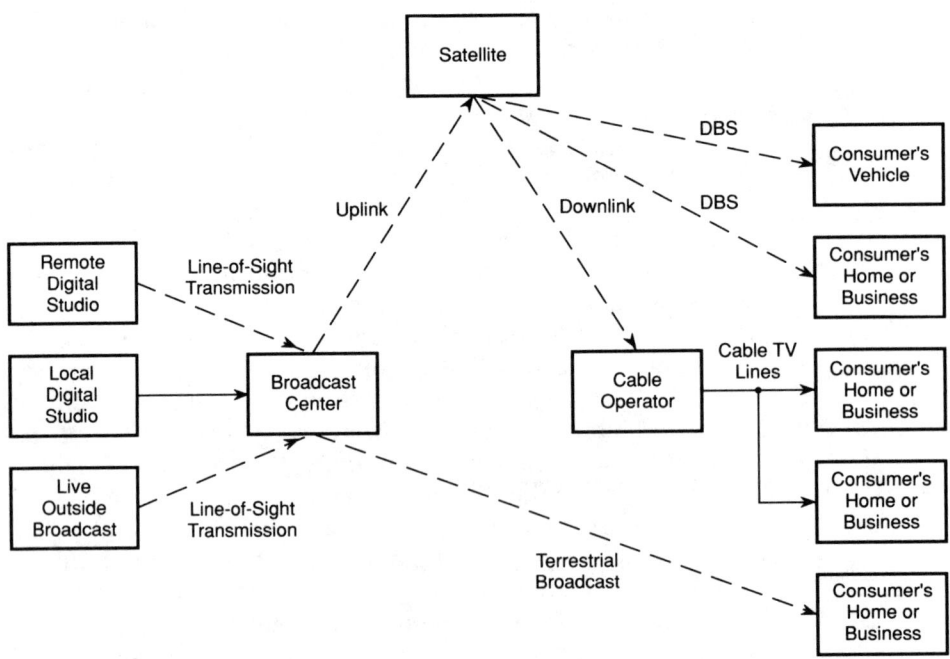

**Figure 1.** Several distribution paths are available for the transmission of digital broadcasting.

# The Digital Studio

Digital broadcasting studios differ from conventional studios in that most audio signals are generated, processed, and output in the digital domain. Several digital sources, such as compact disc players, digital audio tape players, digital audio processors, and digital video tape recorders, are used in digital broadcasting studios. The compact disc is the most common source, particularly for musical reproduction; digital audio and video tape recorders are commonly used to record and play back interviews and commercials.

Figure 2 shows a block diagram of a digital studio designed to allow for input and output of combinations of analog or digital signals. Each digital source requires interfacing to the console via an AES/EBU interface or other appropriate format. In this example there are eight inputs with two designated to receive digital signals sampled at 48 kHz, three other inputs provide for signals sampled at 44.1 kHz (such as compact discs), and one input is reserved for signals sampled at 32 kHz. Two inputs are available for analog sources such as a broadcaster's microphone or analog disc. Analog devices are introduced into the digital studio with high-resolution analog-to-digital converters. Oversampling filters may be provided following low-bit A/D converters to reduce the sampling rate through decimation and to prevent aliasing. The studio console sends each input signal to the prefader section to provide access by the studio operator and also provide level control for each input. The digital inputs are converted to a common sampling frequency of 48 kHz by sample frequency converters (SFC) to simplify the digital processing of the input signals and reduce the amount of equipment required to perform the processing. The digital input signals are summed together in a serial data stream where an optional de-emphasis circuit is provided and a master gain control maintains proper overall signal level. The de-emphasis circuit contains a digital filter with a de-emphasis response of 50/15 $\mu$s (CD standard) and can automatically be activated by an emphasis bit in the program material. A limiter provides signal conditioning to ensure proper audio signal transmission levels. Digital studios provide both an analog output and a selection of digital outputs at 32, 44.1, or 48 kHz. All digital outputs are convolved with a dither signal to reduce the effects of quantization error. Outputs are accessible for transmission via AES/EBU interfaces. An analog output is provided via a 16-bit D/A converter. Where required, sample frequency converters are used to change the 48 kHz signal to 32 kHz and 44.1 kHz outputs. The output signal is transmitted from the studio to a broadcasting center for distribution.

## Baseband Signals

The studio outputs a baseband signal in digital format. Baseband refers to any unmodulated analog or digital signal maintained at its original frequency spec-

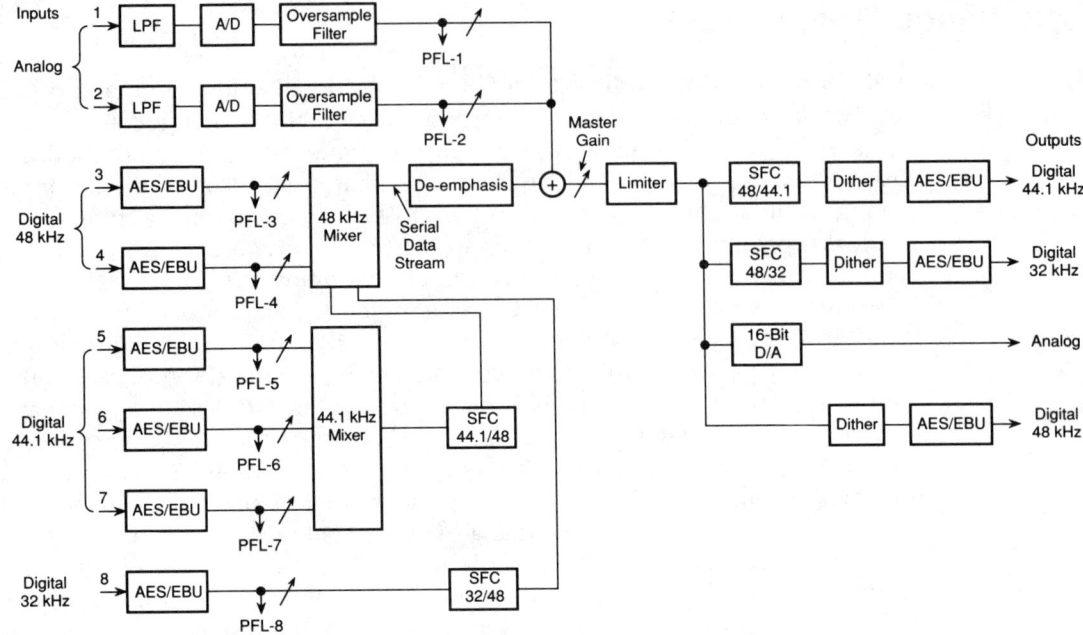

**Figure 2.** Example of a digital broadcasting studio.

trum. Digital baseband signals may be of any unipolar or bipolar format such as pulse code modulation (PCM), delta modulation (DM) or pulse width modulation (PWM). Pulse code modulation is typically used. Note that the term modulation used to describe these digital codes does not refer to modulation of a carrier signal as commonly used in communications. Modulation in this case refers to baseband coding schemes yielding baseband signals. Modulation is the alteration of a high-frequency sinusoidal carrier signal by the baseband signal. Modulation modifies a parameter (amplitude, frequency, or phase) of a carrier signal proportionally to the baseband signal. To be transmitted these digital signals must modulate a carrier signal to shift their spectra to a higher-frequency range. Baseband signals contain high amounts of power at low frequencies and cannot be efficiently transmitted over radio links with reasonably sized antennas; however, these signals are suitable for transmission over conventional optical fiber or coaxial cable. By modulating the baseband signal to a higher range of frequencies, efficient transmission of the data becomes possible.

The baseband signal output from the studio contains the audio information. Remote studios and outside broadcasts transfer the audio information via fiber optics or by modulating the digital baseband signal onto a carrier and transmitting it in the super high frequency (SHF) band to the broadcast center. Local studios transmit the signal to the transmitter of the broadcast center via optical fiber or coaxial cable directly from the AES/EBU output.

# Digital Transmission Processing

Prior to transmitting the audio information from the broadcast center the baseband signal is processed by multiplexing and modulation. This is performed at the transmitter. The processed signal is transmitted using a directional antenna aimed at a receiver located at a satellite, repeater station, home, or business.

Transmission of digital audio signals requires processing as shown in Figure 3. Analog-to-digital conversion is performed as early as possible in the processing chain to minimize noise and distortion. Multiplexing combines many digital signals into a single composite baseband signal allowing for simultaneous transmission of several signals using a single carrier frequency or cable. This eliminates the need for separate transmission paths for each signal while maximizing bandwidth capability. As noted, modulation of the baseband signal onto a carrier shifts the signal to a higher-frequency spectrum where transmission of the signal is more efficient. After the signal is received, the process is reversed to extract the original audio signal by demodulation, demultiplexing, and digital-to-analog conversion.

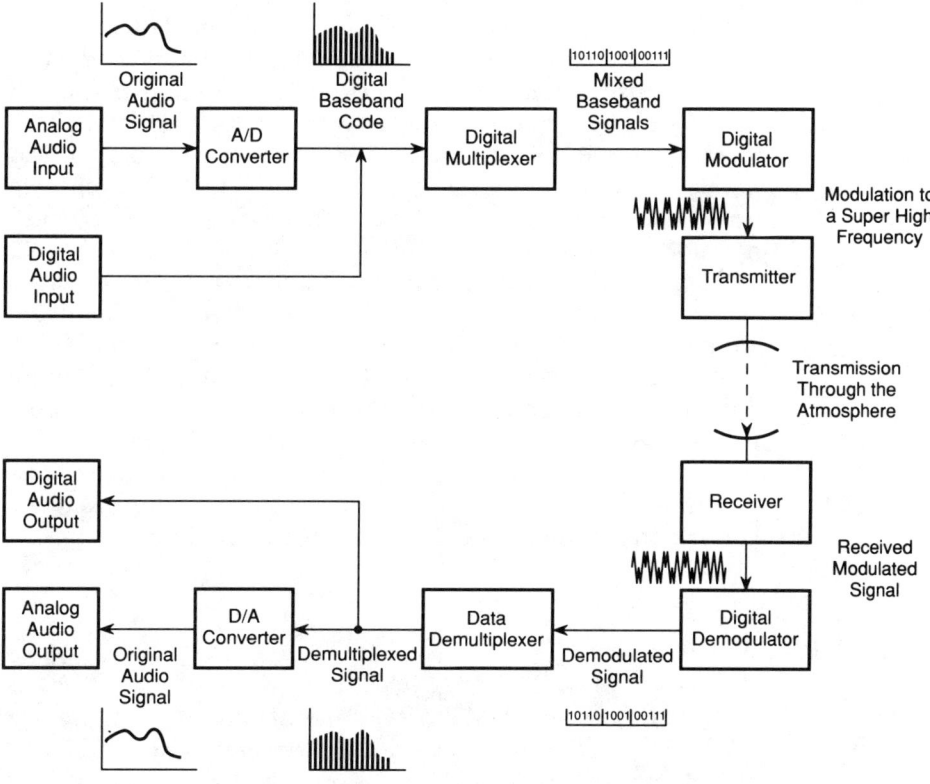

**Figure 3.** Complete transmitter/receiver signal path.

## Multiplexing

The broadcast center utilizes multiplexing to simultaneously transmit digital audio information from several different sources, such as multiple studios and outside broadcasters. The baseband signals from several digital stations are combined into a composite signal which is modulated onto a carrier for transmission. Two forms of multiplexing are used for satellite communications. Frequency division multiplexing (FDM) divides the available frequency spectrum into several smaller bands. Each input signal modulates a different carrier frequency centered within the designated bands of the available frequency spectrum. Substantial bandwidth is allotted so the input signal can be transmitted with minimal crosstalk occurring between channels. Crosstalk may result when satellite transponders are implemented at maximum power. Signals transmitted via satellite are therefore more often multiplexed by time division.

Time division multiplexing (TDM) is well suited for signals in the form of a pulse stream. It allows each user to access a satellite using the same carrier frequency at designated time intervals. Transmission time is shared by several source signals (digital audio, computer data, telemetry data, facsimile) by interleaving the data. Interleaving may be performed on a bit-by-bit basis or on a word-by-word basis. The multiplexed data stream is modulated onto a carrier prior to transmission. The receivers perform deinterleaving, separating signals to be distributed as shown in Figure 4a. The bit rates of the input signals need not be the same. Varying bit rates are accommodated by providing the higher bit rate signal with more slots in the bit stream as shown in Figure 4b. In this example, the bit rate of input $A$ is 12 bits per period ($T$) while the bit rates of inputs $B$, $C$, and $D$ are each 4 bits per period. When multiplexed, input $A$ is added to the output bit stream every other bit while inputs $B$, $C$, and $D$ are alternately added every other bit. Thus input $A$, with the higher bit rate, is accommodated by being provided more slots in the output bit stream.

The resulting bit stream is divided into frames. The minimum length of the frame is determined to be a multiple of the lowest common multiple of the input bit rates. Framing and synchronizing bits are added to identify the beginning of each frame, allowing the receiver to synchronize in time with each bit of the frame. The receiver must also be able to identify each bit in the frame by its position and distribute the bits to the appropriate output channels.

## Modulation

The composite baseband signal is modulated onto a higher-frequency carrier signal prior to transmission by one of several digital modulation techniques. This allows for efficient transmission of the signal over a radio link utilizing practical antenna sizes, and it also increases the available bandwidth over which the multiplexed signal can be transmitted.

Digital modulation refers to techniques where the carrier's amplitude, fre-

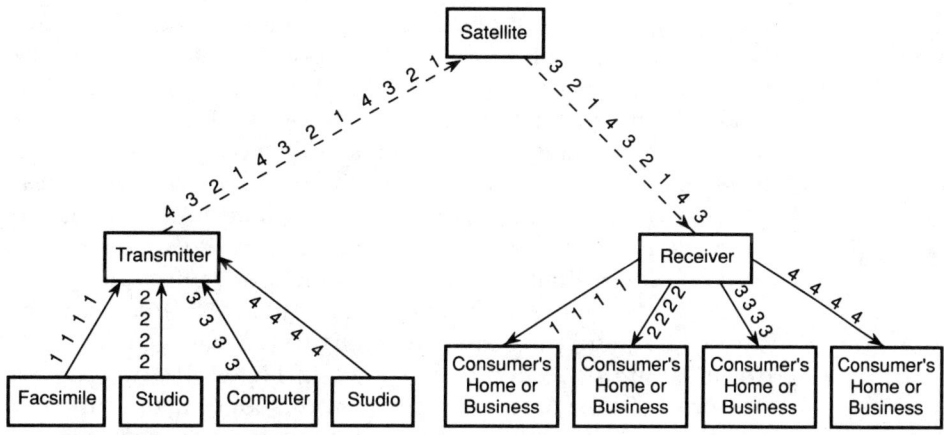

*(a) Deinterleaving is performed at the earth-based receiver.*

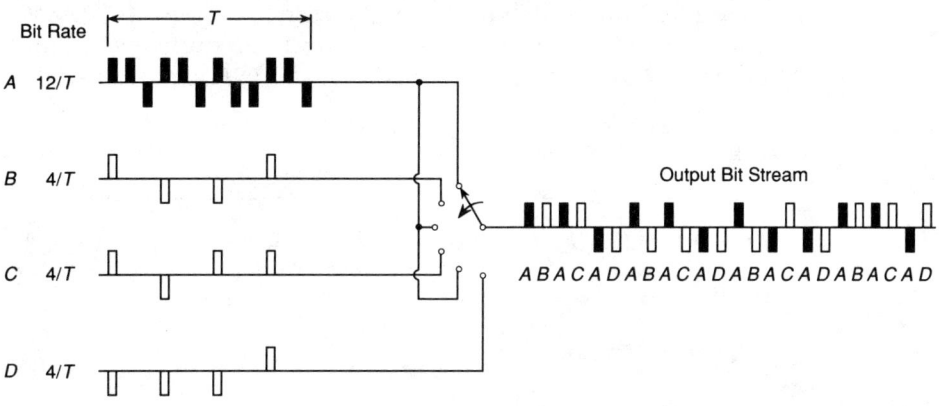

*(b) Time division multiplexing of digital signals with varying bit rates.*

**Figure 4.** Signals transmitted using time division multiplexing.

quency, and/or phase assume one of several discrete values representing the binary data of the digital baseband signal. For each of the following modulation techniques there exist two types of demodulation: coherent (synchronous demodulation) and noncoherent (envelope demodulation). Local carrier waveforms in coherent demodulation have a fixed phase relationship between the transmitted carrier and the received carrier. Noncoherent demodulation relies on detection of envelope information and is not dependent on signal phase coherence. Four common digital modulation techniques are shown in Figures 5a through 5d. Figure 5a shows amplitude modulation of a carrier signal by a baseband signal, 101101101 in this example, using amplitude shift keying (ASK). Variations of ASK exist and are listed in the figure. If the baseband signal has two levels (0 and 1), and each level is present for a period $T$, the modulation waveform of the $i$th state is equal to

$$S_i(t) = X_i(t)A_0 \cos \omega_0 t,$$

where $X_i(t)$ represents the $i$th level of a two-level waveform. Two states exist in the modulated signal. A binary 1 corresponds to an amplitude of $A_0$, and a binary 0 corresponds to an amplitude of zero. Amplitude shift keying is less common in digital signal transmission than either FSK or PSK (described below) for several reasons. The average power of the modulated signal is comparatively low, and the error probability is increased by three orders over a well-designed baseband system. Matched band-pass filters are required to demodulate the ASK signal; this adds greatly to the cost of demodulation. Relatively poor results and the expense of ASK limit it in digital modulation applications.

Figure 5b shows an example of frequency shift keying (FSK), where the carrier's frequency is varied according to the status of the input binary signal, 10110110 in this example. As shown in Figure 5b, two carriers of frequency $\omega_c + \Delta\omega$ and $\omega_c - \Delta\omega$ are generated corresponding to a binary 1 and 0, respectively. Frequency shift keying is used much more than ASK and provides several advantages over ASK such as greater noise immunity and the applicability of hard limiting, which increases the carrier-to-noise (C/N) ratio and reduces the effects of interfering signals. The carrier's constant amplitude makes it less susceptible to nonlinearities. Variations of FSK are also listed in the figure.

Amplitude Shift Keying (ASK)

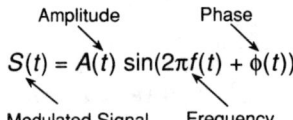
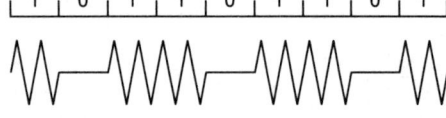

Pulse Amplitude Modulation (PAM)
M-Level Pulse Amplitude Modulation (M-LEVEL PAM)
On-Off Keying (OOK) — Coherent Detection
On-Off Keying (OOK) — Envelope Detection
Quadrature Amplitude Modulation (QAM)
M-ary Quadrature Amplitude Modulation (M-ary QAM)
Quadrature Partial Response (QAR)

*(a) Amplitude shift keying (ASK).*

Frequency Shift Keying (FSK)

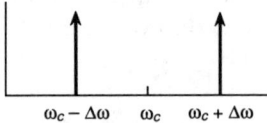
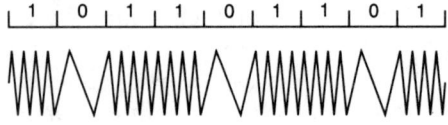

Frequency Shift Keying — Noncoherent Detection
Continuous Phase FSK — Coherent Detection
Continuous Phase FSK — Incoherent Detection
Minimum Shift Keying (MSK)
Minimum Shift Keying - Differential Encoding (MSK - Diff. Enc.)

*(b) Frequency shift keying (FSK).*

**Figure 5.** Common modulation techniques used in digital audio broadcasting.

Phase shift keying (PSK) is a form of digital modulation where the phase of the carrier is altered to one of a fixed number of states to represent binary information. Two, four, and eight states are common and referred to as binary (BPSK), quadrature (QPSK), and 8$\varphi$-PSK, respectively. Figure 5c shows the fixed states associated with each of these forms of PSK and an example of PSK modulation by a binary bit stream, 10110110 in this example. The example shows that a binary 0 is represented by a carrier signal 180° out of phase with the carrier representation of a binary 1. Quadrature PSK consists of four discrete phase states at 0°, 90°, 180°, and 270°. Eight discrete phase states exist for the 8$\varphi$-PSK modulation scheme at 0°, 45°, 90°, 135°, 180°, 225°, 270°, and 315°.

Figure 5d shows another common digital modulation technique known as quadrature amplitude modulation (QAM); it is a combination of ASK and PSK techniques. Transmission of more than 1 bit per symbol is achieved with QAM. The example shown in Figure 5d has eight available states with each corresponding to a particular sequence of three bits. Two separate carriers which differ in phase by 90° are combined to create the output waveform by amplitude modulation. There exist several types of QAM. The most common type is 4-QAM. Quadrature amplitude modulation provides an ideal combined amplitude-phase modulation technique because of its relative simplicity and high carrier-to-noise ratio.

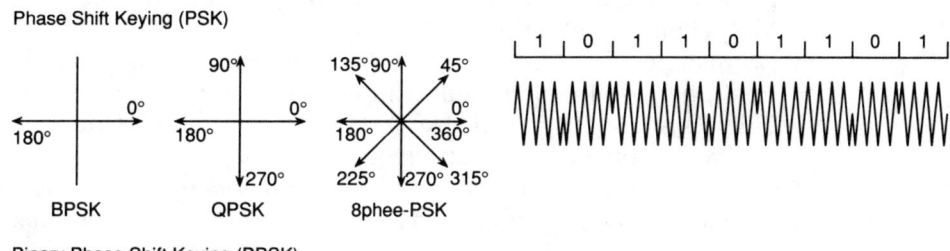

Binary Phase Shift Keying (BPSK)
Differential Encoding PSK (DE-BPSK)
Differential PSK (DPSK)
Quadrature PSK (QPSK)
Differential Quadrature PSK (DQPSK)
M-ary Phase Shift Keying (M-ary PSK)

(c) *Phase shift keying (PSK).*

Quadrature Amplitude Modulation (QAM)

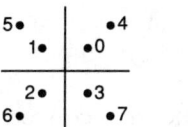
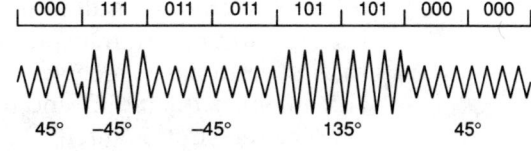

Amplitude Modulation/Phase Modulation (AM/PM)
M-ary Amplitude Phase Keyed (M-ary PSK)

(d) *Quadrature amplitude modulation (QAM).*

**Figure 5.** (cont.)

The advantages and disadvantages of each of these modulation techniques are of importance when determining the best modulating technique for a given application. Characteristics such as bit rate capabilities, modulation simplicity, use of bandwidth, carrier-to-noise ratio, and probability of error are important in designing a transmission system. The International Radio Consultative Committee (CCIR) report 378-4 provides a table of comparisons of the various modulation techniques. From this table the following observations can be made about amplitude, frequency, and phase shift modulation techniques:

Double sideband amplitude modulation with envelope detection (noncoherent) is simple but is wasteful of both bandwidth and power and thus not suitable for large-capacity applications.

Two-level frequency modulation is also simple, but, like amplitude modulation, wastes bandwidth and is only useful for small capacity applications. Three-, four-, or eight-level frequency modulation is the preferred modulation technique for digital transmission of FM radio relay systems using frequency division multiplexing.

Two-level phase modulation (BPSK) with coherent demodulation is simple, but wasteful of bandwidth. It is most suitable for small capacity systems with bit rates up to 10 Mb/s. Four-level phase modulation (QPSK) is one of the most suitable techniques for medium and high capacity systems with bit rates above 100 Mb/s. It is also adequate for small capacity systems. Eight-level phase modulation with coherent demodulation is suitable for medium capacity systems operating below 12 GHz.

Sixteen-level QAM modulation increases spectrum efficiency with only a small increase in equipment complexity, making it suitable for high capacity digital systems. High-order M-QAM and M-PSK modulation techniques function with reduced power efficiency.

Digital audio requires high bit rates. Quadrature PSK is thus the most common form of modulation used to transmit digital audio information.

## Satellite Transmitters

Earth-based transmitters are used to send (uplink) signals from specific locations on the earth to satellite receivers for retransmission (downlink) and distribution. Satellites utilize transmitters to send satellite control information back to earth or to retransmit audio, facsimile, or computer data to specifically located earth-based receivers. Transmitters consist of high-powered amplifiers and antennas to amplify and transmit information.

The transmitter antenna operates as a transducer by converting input currents and voltages into electromagnetic radiation. The audio signal is carried in the electromagnetic wave transmitted from the antenna. The electromagnetic waves that radiate from the antenna propagate through the atmosphere at slightly less than the velocity of light ($3 \times 10^8$ m/s). Antennas are designed to transmit at designated frequencies depending on the type of signal being transmitted. Figure 6 shows the frequency spectrum and locations of various broadcasting bands. Satellite transmission falls in the super high frequency (SHF) spectrum (3 to 30 GHz).

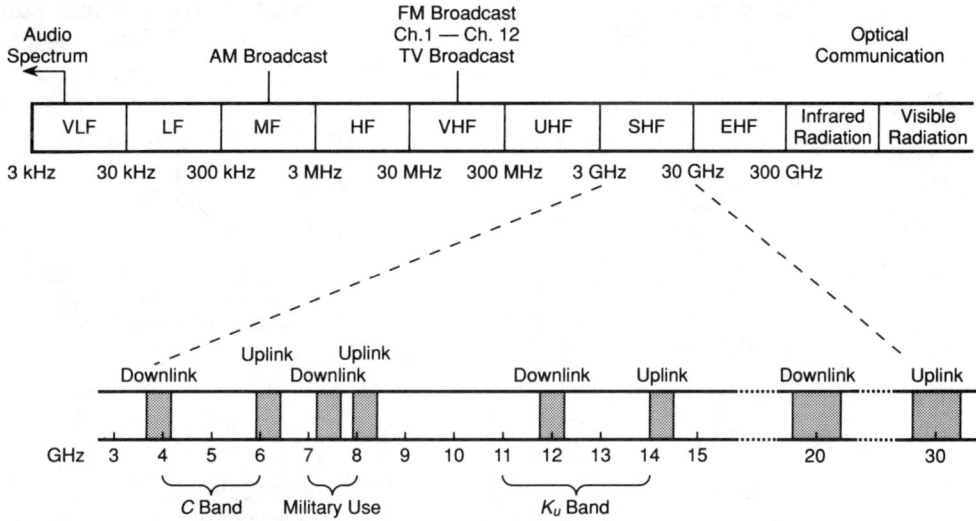

**Figure 6.** Satellite communications occur in the super high frequency band.

Electric and magnetic fields generated by the antenna are perpendicular to each other and to the direction of wave travel. These fields define the direction of polarization for a given electromagnetic wave. Polarization is described in more detail in Chapter 4. The direction of polarization is determined by the electric field and is an important electromagnetic wave characteristic controlled by the positioning of the transmitting antenna. Signals can be generated with vertical polarization, and the same frequency spectrum can be used to generate horizontally polarized signals, doubling the number of channels carried in the bandwidth. Circular polarization can also be used in signal transmission. With circular polarization the electric field rotates about the axis of propagation. Antennas can be designed to receive signals polarized with right-hand rotation or left-hand rotation. This technique also doubles the bandwidth capability of the transmission system.

Antennas utilized in satellite communications require very high gain because of the excessive losses involved in long-distance transmission. An antenna's gain is specified in decibels and is defined as a ratio of the maximum power density radiated or received by the antenna to the power density of an isotropic antenna. (An isotropic antenna transmits or receives uniformly in all directions from a point source and its gain is considered to be unity.) Antennas are typically designed to concentrate energy in a small area or beamwidth, thus increasing the gain of the antenna. Gain is measured in the direction of the maximum signal strength.

The beamwidth is defined by the angle corresponding to half the maximum power (−3 dB) as shown in Figure 7. Large gain and small beamwidth provide efficient signal transmission. Parabolic antennas exhibit these features and are therefore commonly used in satellite transmission. An antenna is placed at the focal point of the parabolic reflector. The transmitter's parabolic reflector concentrates the signal propagating from the antenna towards a receiving antenna. The gain of a parabolic reflector depends on the signal frequency and the diameter of the reflector. It is directly proportional to the area of the parabolic reflector and the square of the signal frequency:

$$G = 4\pi A v f^2,$$

where

$G$ is the directive gain,
$A$ is the effective area of the reflector in square meters,
$v$ is the velocity of the propagating wave,
$f$ is the signal frequency.

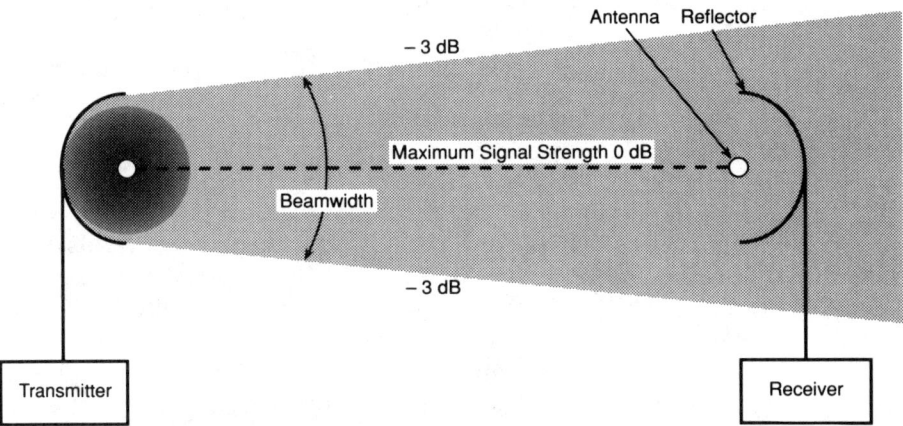

**Figure 7.** The beamwidth of an antenna lies within the angle corresponding to one half of the maximum output power.

## Satellite Design

Satellites are primarily used to receive signals from an earth-based transmitter and retransmit these signals to an earth-based receiver. The received signal is translated by satellite transponders to a lower carrier frequency prior to being retransmitted. This allows the retransmitted signal to be distinguished from the received signal. Satellites act as microwave relays with a large line-of-sight range providing long-distance transmission. Because communications satellites are located at a relatively high altitude, a large area (footprint) can be covered. Three satellites in geostationary orbit can provide global coverage, although four or more are typically used in communications systems to satisfy a large demand for communications. A delay of 270 ms exists in all satellite broadcasts because of the large distances separating the transmitter, satellite, and receiver. This is a problem only if two-way communication is required. One advantage of satellite transmission is that signals can be sent to remote locations on earth that cable networks and terrestrial broadcasting cannot reach. A satellite contains several vital systems such as a station-keeping subsystem, stabilizing subsystem, antenna subsystem, command and control telemetry subsystem, transponders, and solar panels.

## Station-Keeping Subsystem

The station-keeping subsystem maintains the satellite's geostationary orbit. Satellites are placed in geostationary orbit to maintain a fixed position relative to the earth. Following launch, satellites are placed into an elliptical orbit prior to being transferred into geostationary orbit by a station-keeping subsystem within the satellite. Once the satellite is in its geostationary orbit the earth's gravitational force and the satellite's centrifugal force maintain a geostationary orbit, although small thruster rockets are required to occasionally manipulate the satellite to overcome irregular gravitational forces due to the sun, moon, and nonuniformity of the earth's shape and density. Satellites are maintained at 22,240 miles above the equator (0° latitude) at particular longitudinal locations at least $4° \pm 0.1°$ from any other satellite. The beamwidth of earth-based receivers and transmitters is thus limited to $4° \pm 0.1°$.

The satellite's orientation with respect to the earth is called its attitude and is maintained by a stabilizing subsystem. Two common types of stabilizing subsystems are spinner-type and 3D stabilizers. The body of a spinner-type satellite rotates, acting as a gyroscope to stabilize the attitude of the satellite. The top antenna portion of the satellite is "despun" to maintain constant contact with the earth-based transmitters and receivers. A 3D stabilization satellite employs solar panels. This satellite is cube-shaped and does not spin. Three separate gyroscopes enclosed within the satellite are used to maintain stability. Each gyroscope stabilizes the satellite in one of three orthogonal directions. The satellite can thus maintain consistent contact with earth-based transmitters and receivers.

## Power Systems

The electrical power subsystem converts light energy from the sun into useful electrical energy. Solar cells are mounted on the rotating cylindrical body of spinner satellites or on the large flat panels (solar sails) of 3D satellites. Motors are activated to keep the solar cells facing the sun. DC-to-DC converters change the electrical output into the voltages required to operate the equipment on the satellite including the transponders, low-noise amplifiers, high-powered amplifiers, gyroscopes, and geostationary equipment. Storage batteries provide power during a satellite eclipse when the solar cells are shielded from the sun. The solar cells recharge the batteries between eclipses.

## Antennas

Parabolic antennas are commonly utilized on satellites to provide high energy concentration and directionality. Antennas on satellites receive control signals from the satellite command center for activating various subsystems on the satellite while other antennas receive and retransmit audio data and other information. Command signals control equipment such as antennas, solar cell panels, and positioning thrusters. Often as many as six antennas are incorporated into a single satellite. Two receiving antennas are used to receive oppositely polarized signals, while two retransmitters send oppositely polarized signals. A fifth antenna is used to receive command information for control of the various subsystems, and a sixth antenna is used to transmit data pertaining to the satellite's operation back to the satellite control center. The size of these antennas varies with a typical diameter being 1 m or less.

## Transponders

Each signal received by the satellite is passively amplified by the satellite's receiving antenna and transferred to a transponder circuit. The satellite transponder amplifies, filters, and translates the incoming signals to a lower frequency.

Satellites with up to 24 transponders are common, with some newer, larger satellites carrying 50 or more. Each transponder is capable of carrying multiple channels (typically 8 or 12), but the number of channels is limited by the power availability of each transponder. Limited bandwidth (36 to 54 MHz per transponder) also limits the number of channels possible. The total bandwidth utilized by a satellite (in the $C$ or $K_u$ band) is typically 500 MHz. Two separate transponders are required to process the vertically polarized signals and horizontally polarized signals within a common frequency band. One half of the transponders receive vertically polarized signals (typically even channels), and the other half receive horizontally polarized signals (typically odd channels) with each half utilizing the same frequency band. A vertically polarized transponder is shown in Figure 8; horizontally polarized transponders use the same architecture.

A single high-gain, low-noise amplifier (LNA) is used to amplify each set of vertically and horizontally polarized signals. Two separate LNAs are required to process vertically and horizontally polarized signals since common

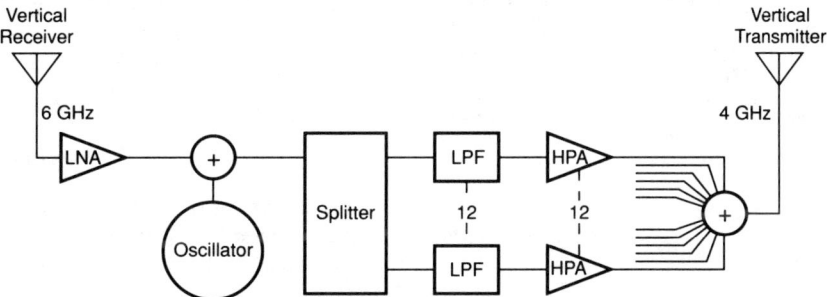

**Figure 8.** Vertically polarized transponder.

frequency bands are used. The incoming signal is very low (as much as 200 dB lower in level than the transmitted signal) and requires high gain to be processed. The signal is combined with a fixed frequency oscillation signal producing a sum and difference output. The sum signal is removed by filtering while the difference signal is passed to the high-powered amplifier (HPA). The center frequency is thus shifted down an amount equal to the oscillator frequency. A demultiplexer between the mixer and the filter uses band-pass filters to separate the signals into individual channels. Separate HPAs increase the power of the separate signals. The signals are remixed by an RF multiplexer prior to being transmitted back to earth via an output transmission antenna. The output antenna transmits a narrow, highly concentrated beam back to earth.

Most satellites have 24 active transponders with 6 to 8 inactive spares. Redundant transponders are provided to improve reliability. In the event that an active transponder malfunctions, a spare may be switched into operation in its place by the command and telemetry subsystem which is controlled by personnel stationed on earth.

## Satellite Receivers

Earth-based receivers perform the opposite function of the transmitter by receiving the signal, amplifying, and demodulating it and ultimately recovering the composite baseband signal. Figure 9a shows a block diagram of an earth-based receiver. The receiver antenna is designed to select the proper downlinked signal from the satellite. Earth station receiver antennas are commonly parabolic in shape, much like those of satellites. As noted, the receiver's beamwidth is typically limited to 4° to minimize reception from adjacent satellites that would cause interference. As the required beamwidth decreases, the receiver diameter must increase. Common diameters are 3 m for a 4 GHz downlink signal ($C$ band) and 1 m for a 12 GHz downlink signal ($K_u$ band).

The antenna feed is positioned to select either vertically or horizontally polarized signals.

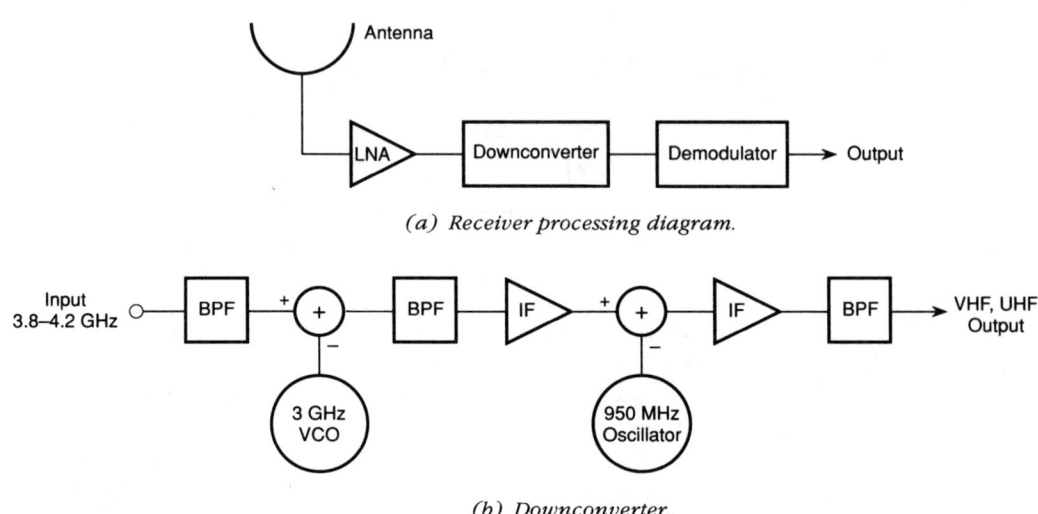

(a) Receiver processing diagram.

(b) Downconverter.

**Figure 9.** The earth-based receiver processes the received signal.

The received signal is passively amplified by the antenna and output to a low-noise amplifier (LNA). A feed horn is positioned at the focal point to receive the downlink signal from the satellite. Feed horns contain the antenna and electronics of the LNA. The LNA is often located at the focus of the antenna, where it is connected to the feed horn to limit transmission line loss between the feed horn and the LNA. Some antennas utilize subreflectors at the focal point and mount the LNAs at a different location on the antenna. The signal amplified by the LNA may contain up to 12 multiplexed channels in a 500-MHz bandwidth. After amplification by the LNA the signal is passed to the downconverter, which reduces the 4 GHz downlinked frequency to a VHF or UHF frequency. This process is shown in Figure 9b. Downconverting is performed as the signal is mixed with a VCO signal at about 3 GHz. The resulting difference signal (approximately 880 MHz) is passed to an IF amplifier and mixed with a fixed-frequency oscillator set at approximately 950 MHz. The difference signal is amplified by a second IF amplifier and passed on to the demodulator. Bandpass filters in the circuit are tuned to the appropriate signal frequency band to eliminate undesired noise. The reduction in the carrier frequency minimizes coaxial cable loss. The output of the demodulator is the original composite baseband signal. The composite baseband signal is demultiplexed at the output to separate the individual signals. The receiver's output signals correspond to the original transmitter input signals.

The quality of a receiver is based on a figure of merit as determined by evaluating two important characteristics: antenna gain ($G_a$) and noise temperature ($T_e$). The $G/T$ figure of merit is determined in dB per kelvin as

$$G/T = 10 \log(G_a/T_e).$$

The gain of an antenna depends on the antenna diameter and the frequency of the received signal. Antenna gain is determined as

$$G_a = 4\pi A/\lambda^2,$$

where $A$ is the area of the receiving antenna (effective area of the parabola) in square meters and $\lambda$ is the received signal's wavelength in meters. The noise temperature of the antenna depends on the system gain and is equal to the sum of all noise contributions by the various receiver circuits. Noise temperature can be determined by measuring the noise power of the system and dividing it by the product of the signal bandwidth and a constant:

$$T_e = P_n/kB,$$

where

$k$ is the Boltzmann constant ($1.38 \times 10^{-23}$ J/K),
$B$ is the signal bandwidth in hertz,
$P_n$ is the noise power of the receiver's electronic circuits.

## Direct Broadcast Satellites

Satellite retransmission of television and digital audio signals may be received directly by the consumer utilizing a parabolic antenna and digital tuner. This is referred to as a direct broadcasting satellite (DBS) system. The direct broadcasting system is typically supported by the consumer through a monthly fee.

In 1977 the World Administrative Radio Conference (WARC) met to plan the Broadcast Satellite Service in the frequency band from 11.7 to 12.2 GHz ($K_u$ band). Table 1 shows some of the specifications established at this conference. This frequency band was specifically allocated for DBS with 40 channels in this frequency range allotted in blocks of five to individual countries with each channel having a bandwidth of 27 MHz. Each DBS system utilized in the $K_u$ band has been designed around the WARC-1977 requirements established at this conference.

The first high-powered DBS system was established in 1976, prior to WARC-1977, as a joint venture between the United States and Canada known as the Communication Technology Satellite (CTS). It transmitted at 200 W and operated in the $K_u$ band. In 1978 the Japanese initiated a medium-powered DBS satellite system using the BS series satellites (also known as Yuri). These first direct broadcast satellites were prototypes that provided few commercial functions.

**Table 1.** WARC Specifications for 11.7 to 12.2 GHz

| Parameter | Specification |
|---|---|
| Channel Bandwidth | 27 MHz |
| Channel Spacing | 19.18 MHz |
| TV Modulation | FM |
| Power Flux Density | $-103$ dBW/m$^2$ |
| Receiving Beamwidth | 2° |
| Figure of Merit | $G/T = 6$ dB/K |
| Carrier-to-Noise Ratio | 14.0 dB |
| Receiver Diameter (Minimum) | 0.9 m |
| Receiver Noise Figure | 8.0 dB |
| Number of Transponder Channels per Country | 5 |

## DBS in Japan

A DBS satellite (BS-IIa) was launched in 1984 by the National Space Development Agency (NSDA) of Japan. It was designed to operate in the $K_u$ band for five years with three transponders. The satellite was designed to receive signals at 14 GHz and transmit at 12 GHz. The Japanese broadcasting network, NHK, began direct broadcast operations in May, 1984, using one operational transponder on BS-IIa. Unfortunately, within three months two of the three transponders failed and the third transponder failed after the fourth month. The BS-IIb satellite followed in 1986, providing two operational transponders. The BS satellites, located at 110° east at an altitude of 22,240 miles, are 3D satellites providing an output power of 100 W per channel. NHK has performed several experiments with high-definition television (HDTV) as well as multichannel digital audio broadcasting using the BS-II satellites to establish and confirm future capabilities of DBS. The BS-IIb satellite was used to broadcast the 1988 Seoul Olympics using the 1125/60 HDTV format and currently provides 24-hour service over one of the two channels, while the other channel provides 18.5 hours of broadcasting each day. Video information is transmitted at a maximum frequency of 4.5 MHz, while the audio subcarrier is 5.7272 MHz with a 2 MHz bandwidth.

Two modes of digital audio programming are provided as shown in Table 2. Mode A offers four channels, sampled at 32 kHz with 14/10-bit compression with near instantaneous companding. Mode B provides two-channel stereo sampled at 48 kHz and full 16-bit linear quantization. Dynamic range and distortion are thus improved with mode B, which is comparable to compact disc specification. The BS-II system employs a quadrature phase shift keying (QPSK) modulation scheme. Error correction and control bits are incorporated into the bit stream as shown in Figure 10.

**Table 2.** BS-II Satellite System Parameters

| Parameter | | Value | |
|---|---|---|---|
| Frequency Bandwidth | | 27 MHz | |
| TV Modulation | | FM | |
| Video Signal | | | |
|   System design | | 525 lines, M/NTSC | |
|   Maximum frequency | | 4.5 MHz | |
| Audio Signal | | | |
|   Carrier frequency | | 5.7272 MHz | |
|   Modulation | | QPSK | |
|   Bit rate | | 2048 Kb/s | |
|   Mode | A | | B |
|   Number of channels | 4 | | 2 |
|   Signal bandwidth | 15 kHz | | 20 kHz |
|   Sampling frequency | 32 kHz | | 48 kHz |
|   Quantizing | 14/10-bit | | 16-bit |
|   Companding | Near instantaneous | | Linear |
|   Additional data capacity | 480 Kb/s | | 224 Kb/s |
|   Dynamic range | 84 dB | | 95 dB |
|   Total harmonic distortion (THD) | 0.08% | | 0.003% |

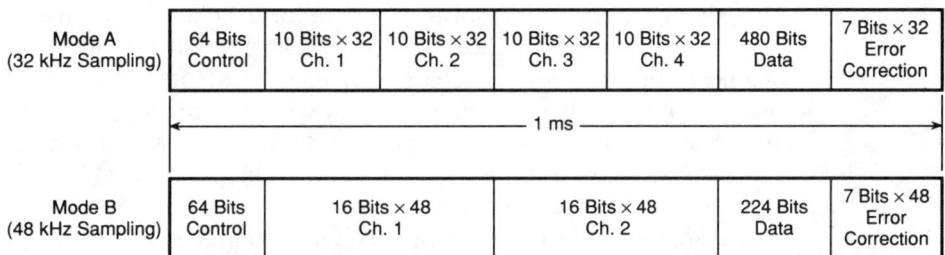

**Figure 10.** Frame specifications for the BS-IIa satellite system.

The BS-II system footprint covers all of Japan, including neighboring islands to the east. This demonstrates an advantage of DBS since terrestrial transmission to the islands would be difficult. The satellite transmission antenna is designed with a parabolic reflector with offset-feed multihorns. Receiving antennas may be as small as 75 cm in diameter on the main island and as large as 1.8 m in remote locations of Japan. Reception in South Korea can be guaranteed with a 3 m antenna. Earth-based transmitters are located at several locations in Japan to continually provide broadcasts even in locally heavy rains. Antennas having diameters of 5 and 8 m provide the main control center's contact with the satellite. Earth-based transmitting power is 350 W nominal, with a maximum capability of 1.4 kW. Several backup transmitters are available throughout Japan and are connected by land lines to secure the reliability of the uplink system. Mobile transmitters with 2.5-m parabolic antennas are used for live or remote broadcasts. The mobile transmitters transmit with a maximum power of 1.2 kW. Satellite monitoring is performed using seven stations throughout Japan measuring the level of the DBS signal as well as precipitation effects. Thus, if the satellite alters its attitude or power flux density, it will be detected and compensated by the satellite control center.

The BS-IIIa satellite is scheduled to replace the BS-IIb satellite. It is equipped with three transponders designed for television use (particularly for HDTV experimentation) with an output power greater than 120 W per channel. The added power is provided to extend transmission capabilities to Okinawa. The expected lifetime of BS-IIIa is seven years.

## DBS in Australia

A direct broadcast satellite system is available in Australia via the AUSSAT satellite system. Two satellites were launched in 1985 and a third in 1987. Six channels of this system incorporate adaptive delta modulation (ADM) for coding audio. The ADM system used is a modulation method developed by Dolby Laboratories. The system provides a 15 kHz bandwidth transmitted within 11 $\mu$s during video line blanking. Dolby ADM is cost-effective due to low-cost integrated circuits used for decoding. The audio channels are contained in a data burst with a clock period of 47 ns: One 2-bit symbol occupies three clock periods, there are 13 ADM bits per line for each audio channel, one step-size bit every two lines per channel, one bandwidth-control bit every two lines per channel, and two error-concealment bits per line per channel. Four utility data bits and a 20-bit symbol burst for clock synchronization and video clamping are contained in each line blanking interval. This was designed as a custom B-MAC system incorporating digital audio channels. (B-MAC is a transmission system designed to completely separate the audio, chrominance, and luminance signals and time multiplex them in the baseband range.)

## DBS in Europe

In Western Europe the French and Germans have agreed bilaterally to jointly develop respective satellite systems called TDF (French) and TV-SAT (German). These systems share common subsystems, modular components, and the D2-MAC transmission format while differing in frequency operation, polarization, and antenna design. The European version of digital broadcasting is referred to as digital satellite radio (DSR) and is based on the results of the WARC 1977 conference.

In 1987, Germany's TV-SAT-1 was launched, but was abandoned after mechanical failure of the solar panels caused the satellite to become inoperable. The French TDF-1 was launched in 1988 and has operated successfully. The German TV-SAT-2 was launched in 1989. A medium-powered satellite called DFS (also known as Kopernikus), another German satellite, is available in Europe for DSR. TV-SAT-2 has five transponder channels, one of which is available for DSR. It was agreed that each country would be assigned five channels with one of two possible polarization patterns. Germany's TV-SAT system received channels 2, 6, 10, 14, and 18 with left-hand circular polarization.

Table 3 provides a list of transmission parameters and specifications for DSR in Europe. Sixteen stereo or 32 mono signals can be transmitted with a 32 kHz sampling frequency and 16-bit quantization. Coding is performed by 16/14-bit floating point PCM with a 3-bit scaling factor. Error correction and interleaving is performed by 63/44 Bose-Chaudhuri-Hocquenghem (BCH) block code error correction as shown in Figure 11a. Figure 11b shows the structure of the main data frame. The BCH code is a random-error-correcting cyclic code. Simple decoding procedures make it a desirable code. For any positive integers $m$ and $t$ ($t < 2^{m-1}$), there exists a $t$-error-correcting $(n, k)$ code with $n = 2^m - 1$ and $n - k \leq mt$. The minimum distance $(d_m)$ between code words is $d_m \geq 2t + 1$. The transmission system is able to correct two errors per BCH block (63 bits) and detect blocks with up to five errors.

**Table 3.** Digital Satellite Radio System Parameters

| Parameter | Value |
| --- | --- |
| Frequency Bandwidth | 15 MHz |
| Modulation | QPSK |
| Bit Rate | 10.24 Mb/s per channel |
| Channels | 16 stereo or 32 mono |
| Quantization and Companding | 16/14-bit floating-point PCM with 3-bit scale factor |
| Sampling Frequency | 32 kHz |
| Error Correction | BCH |
| Multiplexing | TDM |

Bit rates of 10.24 Mb/s apply to both channels and quadrature phase shift keying (QPSK) is used to modulate the signal. The spectrum bandwidth is 15 MHz (−30 dB). Subcode is generated into the bit stream for auxiliary information, program indication, program type, and transmission mode. Time division multiplexing (TDM) is applied at the sampling frequency level to interleave the information between the 16 channels.

A 30 cm parabolic antenna is required to receive the signal from the satellite. The small size is due to the high output power of the satellite and the high error correction capabilities of the BCH code. Larger antennas are required outside the main service area in countries bordering Europe. The QPSK signal can be directly distributed via cable networks. A central DSR frequency for Germany has been determined to be 118 MHz.

DFS (Kopernikus) was launched in 1989 by Germany; it is designed for domestic applications such as teleconferences, telephone, voice transmission,

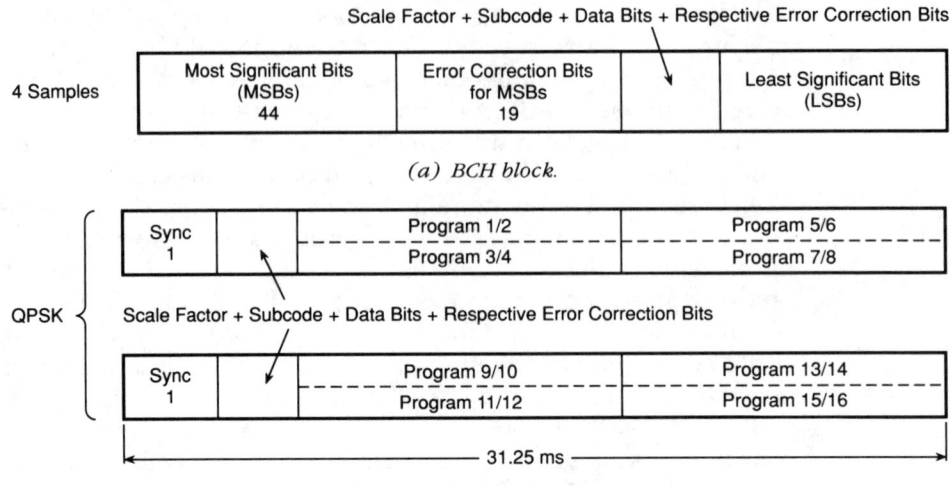

**Figure 11.** Frame specifications for the European DSR system.

facsimile, cable TV, and other data information. It has a lifetime of 10 years. DFS carries 11 transponders: 10 $K_u$ band and one $K_a$ band (20 to 30 GHz). It is located at 23.5° east and has an output power of 250 W. It is capable of providing DSR in the event TV-SAT-2 becomes inoperable.

## Terrestrial DAB Systems

Terrestrial broadcasting commonly utilizes the ultrahigh frequency (UHF) and very high frequency (VHF) bands although it is possible to use the microwave or super high frequency (SHF) bands. Bandwidths at these frequencies are reserved for television and FM broadcasting as well as air traffic control and land transportation purposes. Little space on the frequency spectrum is available to broadcast digital audio. Existing radio stations may be required to volunteer their bands for digital transmission. Although research has been dedicated to terrestrial digital audio broadcasting, no FCC standards or guidelines currently exist. Because of the competition from cable companies distributing digital radio via satellite, many FM station owners are increasingly enthusiastic about terrestrial digital audio broadcasting.

Several proposals are currently in developmental stages to bring about terrestrial broadcasting of digital audio. One of the most advanced proposals is built around a project called Eureka Project No. 147 or "EU 147/DAB." The technology developed for Eureka/147 has been a joint project of the European Broadcasting Union (EBU) and a research organization known as Eureka. This project has primarily been developed by German and French groups. The Eureka/147 project began in 1987 and is designed to provide digital audio programming throughout Europe using three distribution modes: terrestrial distribution, satellite distribution, and a hybrid system utilizing low-power terrestrial broadcasting to augment satellite distribution.

The spectrum efficiency provided by Eureka/147 is much improved over FM broadcasting, allowing up to four times as many stations in the same amount of spectrum used at present by FM stations. The spectrum requirements for digital transmission are related to bit rate; thus data compression is necessary to implement any DAB system. The Eureka/147 system implements a unique coding scheme designed to minimize the bit rate required to transmit the audio information. This coding scheme reduces the bit rate by a factor of six compared with the compact disc by reducing signal quality in areas not readily perceived, yet offering unimpaired CD quality broadcasting.

Other advantages of the Eureka/147 system involve frequency response, power efficiency, and elimination of multipath interference. FM systems normally experience interference due to multipath propagations, deteriorating broadcast quality. Eureka/147 utilizes a coding scheme called coded orthogonal division frequency division multiplexing (COFDM)/MUSICAM, which utilizes multipath echoes to reinforce the digital signal. This form of multiplexing uses both time interleaving as well as frequency interleaving. This is performed by splitting the time interleaved data among many narrow-band carriers with a certain amount of redundancy and implementing multiplexing and demultiplexing in the digital domain. The bit rate per carrier is low and thus the transmission is low and thus the transmission is insensitive to multipath echoes. Due to the redundancy, if some of the multipath signals are delayed or lost, the system can still fully reconstitute the audio signal.

The Eureka/147 system can provide for 12 to 16 stereo channels per market area depending on the amount of spectrum available. The system can transmit two to three stereo channels per megahertz of available spectrum, and all channels can be transmitted from a single transmitter, allowing each channel equivalent power outputs and thus equal coverage.

The Eureka/147 system provides an audio-frequency response up to 22 kHz, giving additional audio bandwidth compared to FM broadcasting (15 kHz). It is believed that this system would require about 1000 W of power to cover the same area as an FM station transmitting at 50,000 W.

## Cable DAB Systems

Digital audio is being introduced to millions of US households through cable networks designed for the distribution of cable television services (CATV). Both coaxial cable and optical fibers are used to transmit digital audio information to the consumer at frequencies normally assigned to a CATV channel. Coaxial cable and optical fiber are used for CATV because conventional wire pairs are not suited for high-frequency transmission (beyond 500 MHz) due to high resistance and increased loss of energy from radiation. CATV cable overcomes both problems and eliminates crosstalk between adjacent cables. Error correction and control bits are used to provide data integrity. The digital signal requires modulation (typically QPSK) at VHF or UHF frequencies. The signal is converted to analog form by a digital tuner obtained by the consumer (typically from the cable operator).

### CATV Audio Tuners

The CATV tuner is a component specially designed to demodulate the input signal, decode the digital information, remove the audio data, and convert it into an analog signal. Data associated with all channels is demodulated and made available at the tuner. The user selects which channels are to be accessed for listening. The audio signal (usually in analog form) is ultimately passed to the consumer's home stereo system to be acoustically reproduced. The digital audio data is separated from the cable television information by a directional coupler and directed to the tuner. The tuner allows the user to choose a station (typically 20 or more stereo channels are available), display related information (selected channel, channel format, and station name) and output an audio signal to the home stereo. When provided by a cable company the digital broadcast is available to paying subscribers only.

One tuner designed for European cable formats receives signals transmitted at 118 MHz in 4-PSK modulation with a bandwidth of $\pm 7$ MHz. The tuner is capable of receiving up to 16 stereo or 32 mono channels sampled at 32 kHz, utilizing a 16/14-bit floating-point transmission method. A flat frequency response is thus provided from 10 Hz to 15 kHz. A dynamic range of 92 dB and a signal-to-noise ratio of 110 dB ($A$-weighted) is reported with greater than 80 dB channel separation and a nonlinear distortion factor of less than 0.01%. The output electronics includes four-times oversampling filters and dual 16-bit D/A converters.

Other units are known to provide a flat frequency response to 20 kHz utilizing 44.1 kHz as the sampling frequency. Up to 28 stereo channels are available in some digital tuners.

# References

Angus, R., "Sound for Space, Digital Broadcasting, Digital Components," *Audio Magazine*, February, 1990.

Berfeld, B., Weiss, D., and S. Widmar, "Transmission Suite for Digital Broadcasting," *Audio in Digital Times: AES 7th Conference Proceedings*, May, 1989.

Bridle, M., "Satellite Broadcasting in Australia," *IEEE Transactions on Broadcasting*, vol. 34, no. 4, December, 1988.

Caramanilis, S., "Oscar, Amateur Radio Satellite," *Elektra*, 1976.

Chen, Z., Wang, J., and K. Feher, "Effect of HPA Non-linearities on Crosstalk and Performance of Digital Radio Systems," *IEEE Transactions on Broadcasting*, vol. 34, no. 3, September, 1988.

*Digital Audio Broadcasting: Status Report and Outlook*, National Association of Broadcasters, Washington, DC, 1990.

Douglas, R. L., *Satellite Communications Technology*, Prentice-Hall, 1988.

Feher, K., *Advanced Digital Communications Systems and Signal Processing Techniques*, Prentice-Hall, 1987.

——, *Digital Communication Satellite/Earth Station Engineering*, Prentice-Hall, 1983.

Forrest, J. R., "Commercial Broadcasting for Europe," *IEEE Transactions on Broadcasting*, vol. 34, no. 4, December, 1988.

Fujimoto, M., Grim, J. P., Guo, C. H., Haquet, G., and G. M. Maier, "Small and Lightweight DBS and FSS Converters," *IEEE Transactions on Consumer Electronics*, vol. 36, no. 3, August, 1990.

Georgy, J., "Satellite Broadcasting in France," *IEEE Transactions on Broadcasting*, vol. 34, no. 4, December, 1988.

Gusmao, A. M., and N. L. Esteves, "ENCAP-4: An OQPK-type Modulation Technique for Digital Radio," *IEEE Proceedings on Communications, Radio and Signal Processing*, February, 1988.

Heymann, R., Irvin, G., Jessner, H., Kriedt, N., and N. Ozoguz-Geibier, "A Multipurpose Four IC Satellite Concept," *IEEE Transactions on Consumer Electronics*, vol. 36, no. 3, August, 1990.

Killen, H. B., *Digital Communications with Fiber Optics and Satellite Applications*, Prentice-Hall, 1988.

Klank, O., and D. Rottman, "DSR-Receiver for Digital Sound Brodcasting via the European Satellites TV-SAT/DF," *IEEE Transactions on Consumer Electronics*, vol. 35, no. 3, August, 1989.

Konishi, Y., Special Issue on Satellite Broadcasting, *IEEE Transactions on Broadcasting*, vol. 34, no. 4, December, 1988.

Konishi, Y., and Y. Fukuoka, "Satellite Receiver Technologies," *IEEE Transactions on Broadcasting*, vol. 34, no. 4, December, 1988.

Le Floch, B., Halbert-Lassalle, R., and D. Castelain, "Digital Sound Broadcasting to Mobile Receivers," *IEEE Transactions on Consumer Electronics*, vol. 35, no. 3, August, 1989.

Maegele, M., and T. Hentrich, "The German TV-SAT Broadcasting Satellite System," *IEEE Transactions on Broadcasting*, vol. 34, no. 4, December, 1988.

Matsushita, M., and S. Yokoyama, "Experience on Operating a DBS System (DB-2) in Japan," *IEEE Transactions on Broadcasting*, vol. 34, no. 4, December, 1988.

McNally, G. W., "Digital Audio in Broadcasting," *IEEE Transactions*, ASSP, October, 1985.

Miller, J. E., "Application of Coding and Diversity to UHF Satellite Sound Broadcasting Systems," *IEEE Transactions on Broadcasting*, vol. 34, no. 4, December, 1988.

Miyasaka, E., "A Sound Reproduction and Transmission System for HDTV," *Audio in Digital Times: AES 7th Conference Proceedings*, May, 1989.

Pizzi, S., "Digital Audio Applications in Radio Broadcasting," *Audio in Digital Times: AES 7th Conference Proceedings*, May, 1989.

Stripp, D., "BBC Digital Audio—A Decade of On-Air Operation," *Digital Audio Collected Papers*, Audio Engineering Society, 1983.

Sukow, R., *et al.*, "Radio's Digital Evolution," *Broadcasting*, October 17, 1988.

Swanson, H., "Digital AM Transmitters," *IEEE Transactions on Broadcasting*, vol. 35, no. 2, June, 1989.

Townsend, A. A. R., *Digital Line-of-Sight Radio Links*, Prentice-Hall, 1988.

*Chapter 10*

# Digital Signal Processing: Theory
### Jayant Datta

## Digital Signals and Systems

One great advantage of digital signal processing, or DSP, is that the processing takes place in the digital domain, which offers both flexibility and superior sonic performance, making DSP central to digital audio. Audio signals that exist in nature are analog; therefore the signal must first be digitized. This is achieved by sampling the waveform at discrete points in time and then quantizing the amplitude of the samples to a finite number of allowed values. As a result, the original continuous time, continuous amplitude signal, $x(t)$, is lost and in its place there is a digitized, discrete time, amplitude quantized signal $x_i(nT)$, where $x_i$ is the appropriate quantized amplitude, $T$ is the sampling period, and $n$ is an integer. If the sampling time is normalized, then, in general, the digitized signal is $x(n)$.

For example, a sinewave of frequency $f$ is represented in the analog domain as

$$x(t) = \sin \omega t,$$

where the angular frequency $\omega = 2\pi f$. If this signal is sampled every $T$ seconds, the sampling frequency is $f_s$ Hz or $\omega_s$ rad/s. The relationship between the sampling period $T$ and sampling frequencies are $f_s = 1/T$ and $\omega_s = 2\pi/T$. The discrete form of the equation becomes

$$x(n) = \sin \omega n,$$

where $n$ is an integer. The continuous and discrete signals are shown in Figure 1. The values of $n$ and $t$ are not equal to each other; the signal is sampled at discrete values of $t = nT$. Looking at the discrete samples it is difficult to determine the frequency of the sinusoidal waveform. There appear to be eight samples in each period of the sinusoidal wave. Thus all that can be said is that the signal has a frequency that is one eighth the sampling rate. The reference in the digital domain is the sampling frequency; everything is taken relative to it. It would not be visually possible to tell the difference between a 2 kHz sinewave sampled at 32 kHz and a 3 kHz sinewave sampled at 48 kHz. This is because in both cases the sampled signal has a frequency that is one sixteenth the sampling frequency. Subsequently, both will show 16 samples in each period of the sinewave.

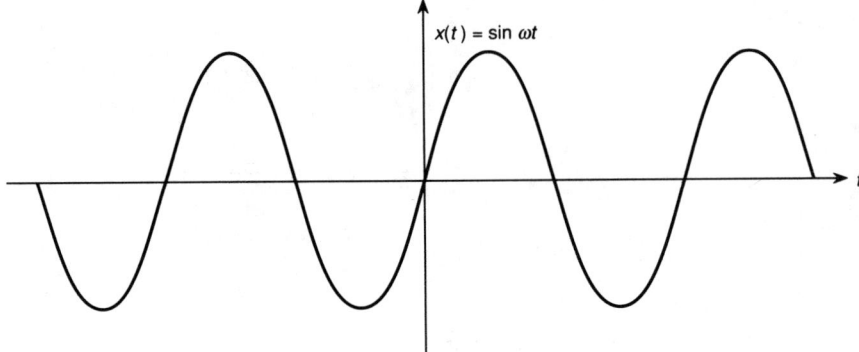

*(a) Continuous time signals are defined for all time and have continuous amplitude values.*

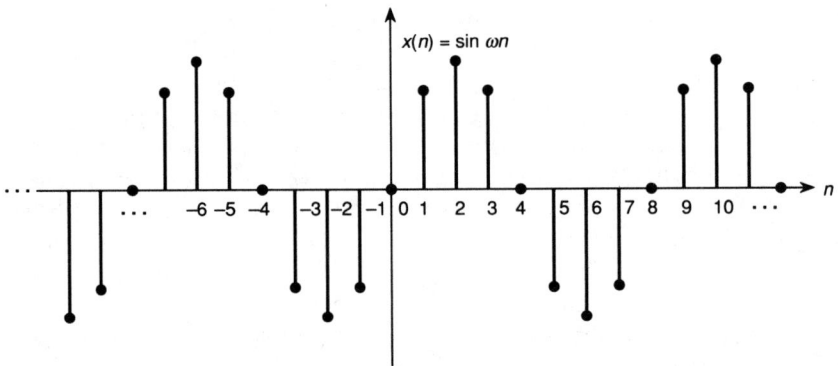

*(b) Discrete time signals are defined only at the sampled times and can take on only certain quantized amplitude values.*

**Figure 1.** Continuous and discrete time signals.

At first glance it appears that sampling results in a loss of information; the continuous signal exists for all time but only discrete points are sampled. However, discrete time sampling allows recovery of the original signal if the Nyquist criterion is satisfied. This requires the sampling frequency to be greater than twice that of the highest frequency present in the signal. Violation of this condition results in aliasing. The measurement of sample amplitudes requires quantization. The continuous amplitude signal of the analog domain would require an infinite number of bits to accurately record the value of the sample at a sample time. In reality, only a finite number of bits is used, for example, 16 bits for digital audio. Thus amplitude quantization leads to distortion, which can be particularly objectionable for low-level signals. A discussion of sampling and quantization is presented in Chapter 2.

The sampled signal $x(n)$ is a sequence of values. There are two broad kinds of sequences—two-sided sequences and one-sided sequences. These are shown in Figure 2. A two-sided sequence can have nonzero values for both positive and negative values of $n$. One-sided sequences are further divided into left-sided sequences and right-sided sequences. A left-sided sequence may have any value for $n$ less than some number $N_1$, but it is identically equal to zero for all $n$ greater than $N_1$. A right-sided sequence has the opposite definition: it is identically equal to zero for all $n$ less than some number $N_1$ and may have any value for $n$ greater than $N_1$.

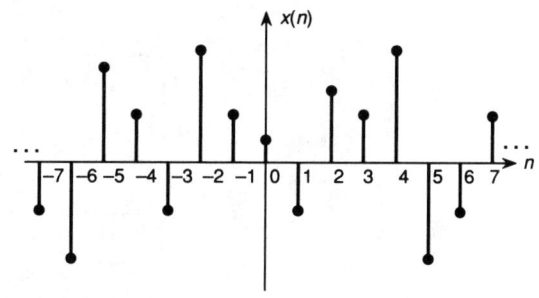

*(a) Two-sided sequence.*

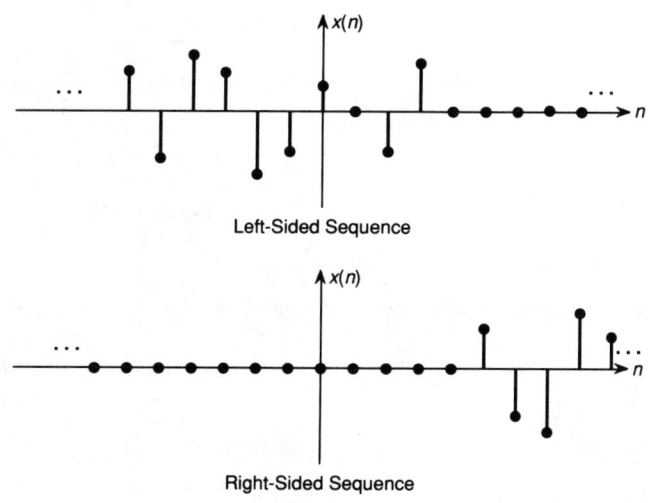

*(b) One-sided sequences.*

**Figure 2.** Two-sided and one-sided sequences.

Discrete systems that are linear, time-invariant, and causal are of great importance. A discrete system accepts one or more digitized inputs $x(n)$ and produces one or more digitized outputs $y(n)$ as shown in Figure 3a. A linear system has two properties—homogeneity and superposition. Homogeneity requires that the amplitude of the output be proportional to the input throughout the full range of amplitudes. If an input $x(n)$ produces an output $y(n)$, then a scaled input $ax(n)$ produces the scaled output $ay(n)$ as shown in Figure 3b. Superposition requires that each input signal be treated independently of the others. The input $x_1(n) + x_2(n)$ produces the output $y_1(n) + y_2(n)$ as shown in Figure 3c. Combining these two properties of a linear system, an input $a_1x_1(n) + a_2x_2(n) + \ldots + a_Nx_N(n)$ would produce an output $a_1y_1(n) + a_2y_2(n) + \ldots + a_Ny_N(n)$. The input consists of the summation of numerous scaled signals. The output is the sum of the response of the system to each individual scaled signal as shown in Figure 3d. Linearity is important because it means that no new spectral components are added to the signal.

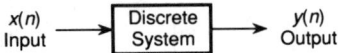

*(a) Input-output relationship for a discrete system.*

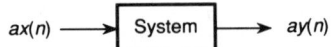

*(b) A system exhibiting homogeneity.*

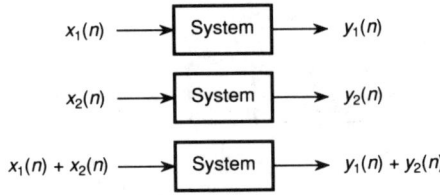

*(c) The property of superposition.*

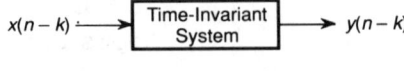

*(d) A linear system showing the properties of homogeneity and superposition.*

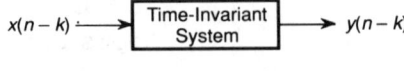

*(e) A time-invariant system.*

**Figure 3.** Properties of discrete, linear, time-invariant systems.

A time-invariant system, shown in Figure 3e, produces an output $y(n-k)$ for an input $x(n-k)$, for any integral delay $k$. In other words, delaying the input simply delays the output by the same amount of time. Thus the output is independent of the time of origin of the signal and depends only on the input waveform. A system is called causal or physical or realizable if the output does not depend on future values of the input. Causal systems, as opposed to noncausal or anticipatory systems, can be realized physically. Noncausal systems are mainly of theoretical importance and may be realized by introducing delays.

## The Impulse Response

The impulse response of a linear system, denoted by $h(n)$, may be considered to be the equivalent of its signature. It fully characterizes the system's response. Mathematically, the system's response to a delta function $\delta(n)$ is called its impulse response. The delta function, as shown in Figure 4, is an impulse at $n = 0$ and is also called an impulse function. It contains equal energy at all frequencies and is defined in the discrete time domain as

$$\delta(n) = 1, \quad \text{if } n = 0 \quad \text{and} \quad \delta(n) = 0, \quad \text{if } n \neq 0.$$

The impulse response can be used to find the time-domain response, frequency response, and phase shift characteristics of a system. A system is said to be *stable* if every bounded (finite) input produces a bounded output and if the impulse response has finite energy. A necessary and sufficient condition on the impulse response for stability is

$$\sum_{n=-\infty}^{\infty} \mid h(n) \mid < \infty,$$

that is, the sum of the absolute value of each sample of the impulse response must be a finite number. This ensures that a finite input will produce a finite output.

If the input to a causal system is the delta function, then its output will be $h(n)$. The impulse response characterizes the system; an arbitrary response (for an arbitrary system) is shown in Figure 5. Since the input occurs only at $n = 0$, $h(n)$ must be zero for all negative values of $n$, because $h(n)$ cannot anticipate the response before the input arrives at $n = 0$. As a result, causality demands that $h(n) = 0$ for $n < 0$. This alternate definition allows $h(n)$, and hence the system, to be tested directly for causality.

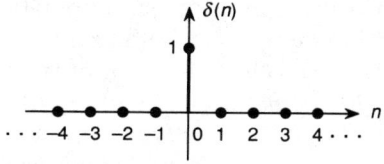

**Figure 4.** The delta or impulse function.

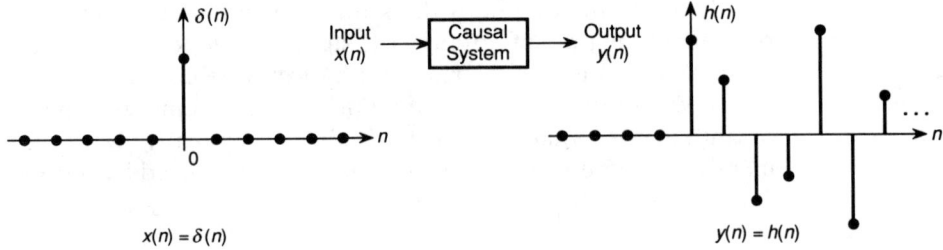

**Figure 5.** The impulse response of an arbitrary causal system.

## Convolution

The output of a linear system is the convolution of the input and the impulse response. The convolution operation is denoted by *. Mathematically, $y(n)$ is the convolved response of $x(n)$ and $h(n)$:

$$y(n) = x(n) * h(n) = h(n) * x(n)$$
$$= \sum_{k=-\infty}^{\infty} x(k)\, h(n-k) = \sum_{k=-\infty}^{\infty} h(k)\, x(n-k),$$

where $x(k)$ and $h(k)$ represent the input and impulse response sequences, $x(-k)$ and $h(-k)$ represent the same sequences reversed, and $x(n-k)$ and $h(n-k)$ represent the reversed sequences shifted to the right by $n$ units. Figure 6 shows examples of reversed and shifted sequences. The expression $x(k)\, h(n-k)$ represents the multiplication of the sequence $x(k)$ and the reversed and shifted sequence $h(n-k)$ for a particular value of $n$ and $k$. The summation of all such multiplications for a particular value of $n$ but all values of $k$ can be expressed as

$$\sum_{k=-\infty}^{\infty} x(k)\, h(n-k).$$

This gives the value of $y(n)$ for a particular value of $n$, say $n_1$. To find the next value of $y(n)$, $n=n_1+1$, the sequence $h(n-k)$ is shifted to the right by one more unit. The multiplication and summation is carried out as before to yield $y(n)$ for $n = n_1 + 1$. This method of convolving could be described by a reverse-shift-overlap-multiply-sum operation, as shown in Figure 7. By studying the figure, the method for finding the convolved output $y(n)$ for a system may be described as follows. The impulse response is reversed and $h(0)$ of the sequence $h(-k)$ is aligned with $x(n)$. This automatically makes the $h$ sequence $h(n-k)$. The overlapping regions of $x(k)$ and $h(n-k)$ are multiplied term by term, and all these products are added to produce $y(n)$. The successive values of $y(n)$ can be found by moving the reversed and shifted impulse response one step to the right each time and again repeating the procedure.

This is carried on until the two sequences no longer overlap. The same numerical result can be obtained by interchanging the roles of $x(n)$ and $h(n)$. If the two sequences being convolved have lengths $n$ and $m$, then the convolved output has length $n + m - 1$.

A second way to look at convolution is to use the fact that the system is linear and time invariant. By definition, a system's response over all time to an impulse is the impulse response. A linear system's response to a scaled impulse is a scaled version of the impulse response. A time-invariant system's response to a delayed

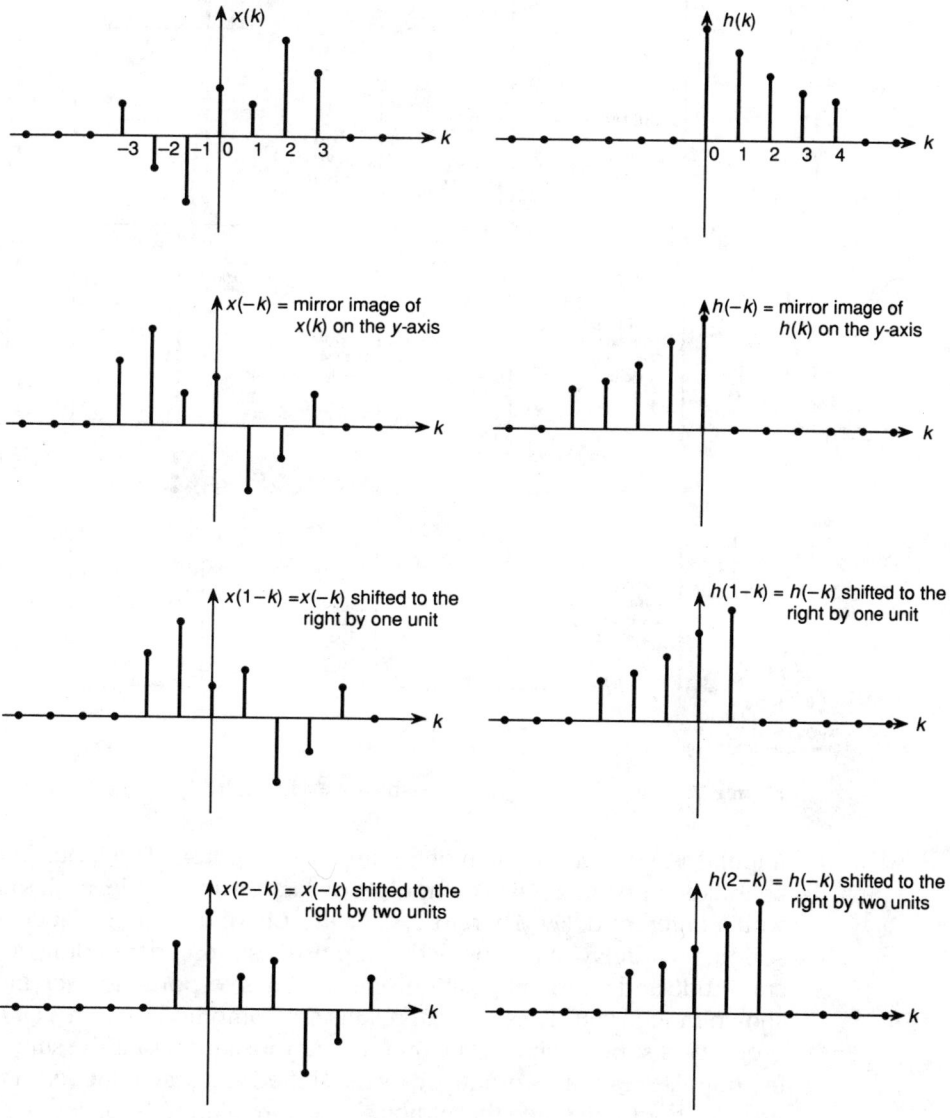

**Figure 6.** Reversing and shifting sequences.

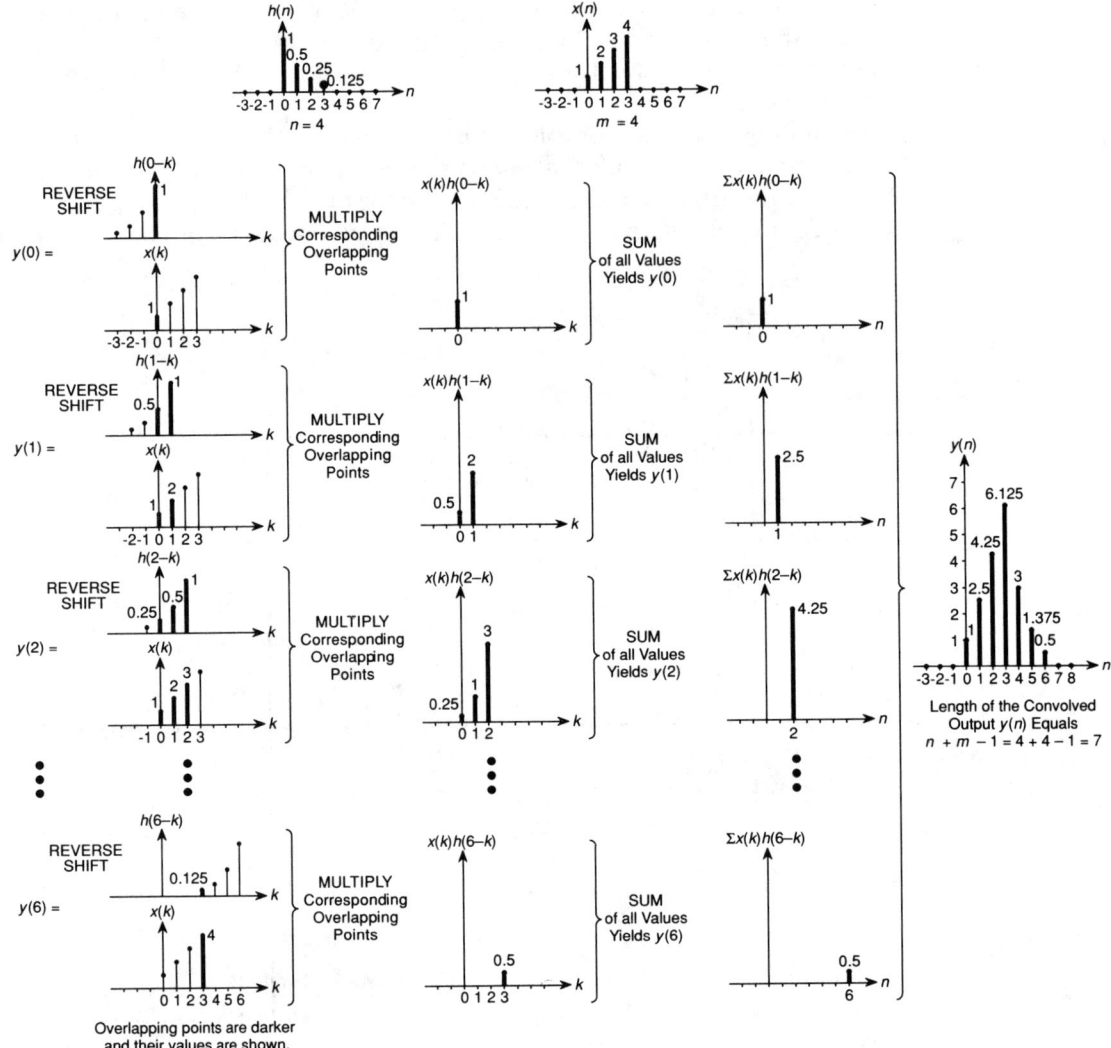

**Figure 7.** Convolution by reverse-shift-overlap-multiply-sum operator.

impulse is a delayed version of the impulse response. The input sequence could be looked upon as a train of impulses of varying amplitude, each sample arriving with a different delay. Therefore, each sample of the input sequence produces a scaled, time-delayed version of the impulse response, depending on the sample's amplitude and temporal position. The system's response to each input sample is shown in Figure 8. These partial results are combined to form $y(n)$ by using the property of superposition. The output at any instant $y(n)$ is the sum of the parts of the impulse responses produced by the shifted and scaled inputs for that instant of time. It should be noted that although the convolution is carried out by using two different methods (Figures 7 and 8), the results are the same.

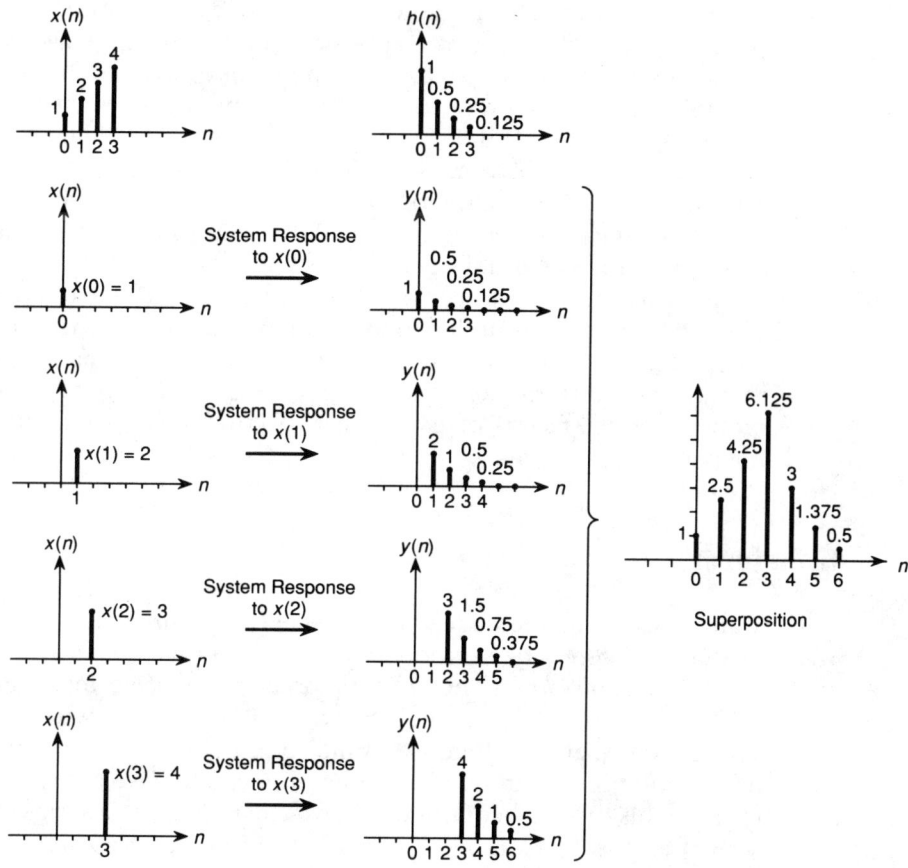

**Figure 8.** Convolution performed by superposition of the partial impulse responses of the input signal.

## Spectra

There are two domains or ways of representation that are generally used to look at signals: the time domain and the frequency domain. In the time domain, the way the amplitude of a signal changes with time is studied. In the frequency domain, the absence and presence of certain frequencies is studied. Music consists of different frequencies of varying amplitudes, and if it is passed through a bank of audio bandpass filters, their outputs would denote the presence and absence of frequencies and the relative amounts of the frequencies present. This is the display of a spectrum analyzer, where the incoming signal is broken down into its component frequencies. In other words, it is a look at the frequency domain. Generally speaking, if the amplitude variation in the time domain is slow, it signifies the presence of low frequencies. A rapid change in the amplitude means there are high frequencies present. By applying the Fourier transform to the waveform of a sound, one can determine the amplitudes and frequencies responsible for a particular waveshape.

It is easy to find the spectrum of simple tones with a definite frequency and periodically repeating waveform. In music, the spectrum changes rapidly with the passage of time. If the waveform is studied over a long duration, then the measurement of spectrum is more accurate. The spectrum obtained in this case, however, is the value for the entire period, and there is no information about the spectrum during a particular subinterval of that time period. The accuracy of the spectrum decreases as an attempt is made to measure over a smaller time period and narrow the measurement to a particular subinterval. This problem can be understood if the extreme case of the subinterval consisted of a single sample. A single sample gives no information about the spectrum of the waveform. As a result, the measurement of the spectrum is a tradeoff between the factors of accuracy of spectrum and spectral information during particular intervals.

## Transforms

Transforms are mathematical tools that allow the movement from the time domain to the generalized frequency domain and vice versa. There are a large number of transforms. They may be broadly classified into continuous transforms, series transforms, and discrete transforms. Continuous transforms are applied to signals continuous in time and frequency. If a continuous time signal has only certain frequencies, then it is said to have a discrete or line spectrum. In this case, the continuous transform reduces to a series transform. Thus series transforms are a special case of continuous transforms and are applied to continuous time, discrete frequency signals. Discrete transforms are applied to discrete time, discrete frequency signals. The relationship between the time and frequency domains, and principal transforms are shown in Figure 9. Using the transforms shown, it is possible to move between the time and frequency domains. By setting the frequency variables to particular values, it is possible to move to particular portions of the frequency domain.

## Laplace Transform

The Laplace transform is an example of a continuous transform. It is a mathematical operation that maps a time-domain function $x(t)$ into a complex frequency-domain function $X(s)$. It is defined as

$$X(s) = \int_{-\infty}^{\infty} x(t) e^{-st} dt.$$

Thus $x$, a function of time, is converted to $X$, a function of complex frequency $s = \sigma + j\omega$. The value $s$ has a real part $\sigma$ and an imaginary part $\omega$. This complex

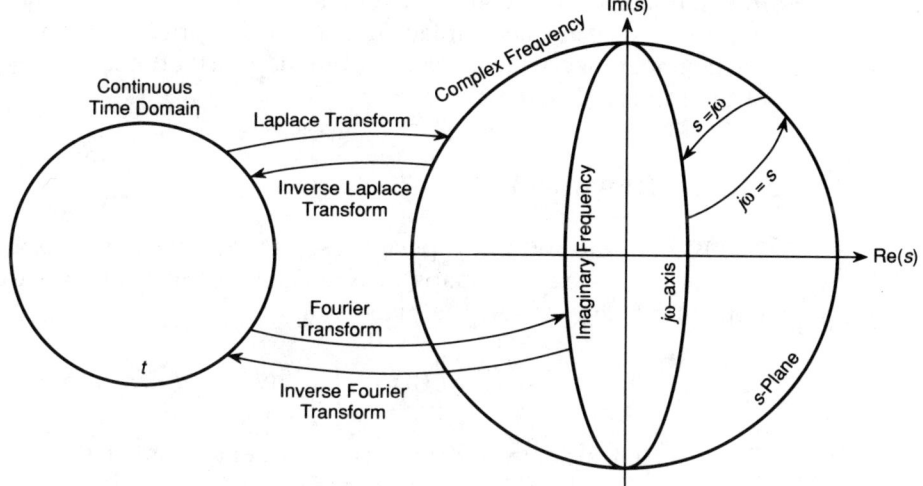

*(a) Continuous time domain and frequency domain.*

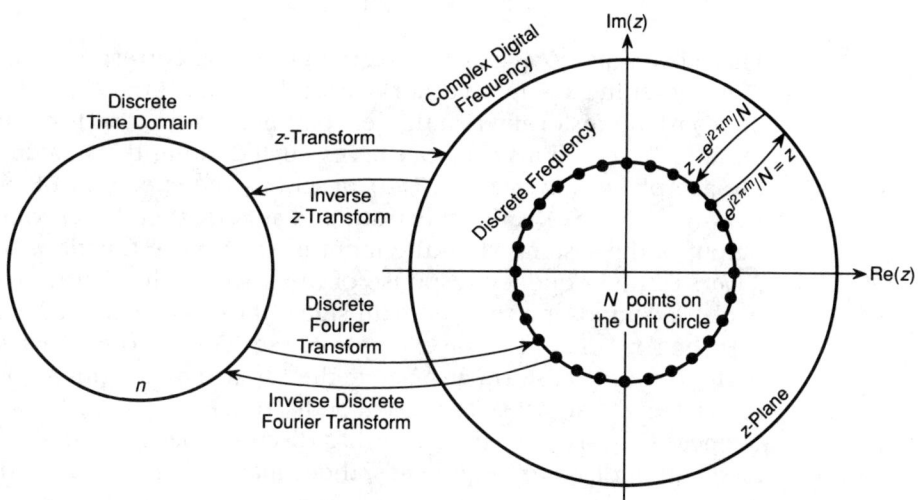

*(b) Discrete time domain and frequency domain.*

**Figure 9.** Mappings between the time and frequency domains.

frequency lies in the $s$-plane, a two-dimensional plane having $\text{Re}(s)$ and $\text{Im}(s)$ axes. The function $x(t)$ can be recovered from $X(s)$ using the inverse Laplace transform, defined as

$$x(t) = \frac{1}{j2\pi} \int_{c-j\infty}^{c+j\infty} X(s)\ e^{st}\ ds.$$

This converts $X$, a function of complex frequency to $x$, a function of time. The integration takes place along a vertical line in the s-plane, where the real part of $s = c$, a constant. The Laplace transform is of primary importance to the analog engineer, because most analog circuit transfer functions are given in terms of the Laplace variable $s$.

## The Fourier Transform

The Fourier transform is a special case of the Laplace transform. It is a mathematical operation that maps a time-domain function $x(t)$ to a frequency-domain function $X(j\omega)$. It equals

$$X(j\omega) = \int_{-\infty}^{\infty} x(t) e^{-j\omega t} dt.$$

The inverse Fourier transform maps a frequency-domain function $X(j\omega)$ onto a time-domain function $x(t)$ and is defined as

$$x(t) = \frac{1}{2\pi} \int_{-\infty}^{\infty} X(j\omega) e^{j\omega t} d\omega.$$

These two equations may be obtained from the corresponding Laplace equations by setting $s = j\omega$. In other words, the Laplace transform is identical with the Fourier transform when the real part of $s$ is zero, that is $\sigma = 0$. This implies that the Fourier transform can be evaluated along the imaginary axis of the Laplace plane. Physically, $X(j\omega)$ describes the spectrum of the signal $x(t)$. However, if $X(j\omega)$ is the spectrum of a system, then it describes the spectral output of the system when the input is an impulse function. In other words, $X(j\omega)$ is the frequency response of the system. The Fourier transform effectively decomposes a time-domain signal into its constituent frequencies.

The Fourier series is an example of a series transform. This is a special case of the Fourier transform and hence the Laplace transform. A Fourier series results when the signal is composed of discrete frequencies. This has two consequences—the Fourier transform has discrete spectra, and since the signal is made up of discrete frequencies, the signal is periodic in the time domain.

## The Discrete Fourier Transform

The discrete Fourier transform (DFT) is an example of a discrete transform. This is the Fourier transform of a sampled signal $x(n)$. If the signal $x(t)$ is uniformly sampled and converted to $x(n)$, the Fourier transform integral can be approximated by a summation:

$$X(j\omega) \cong T \sum_{n=-\infty}^{\infty} x(nT) e^{-j\omega nT},$$

where $T$ is the sampling period. This yields uniformly spaced values for the frequency $\omega$. The signal is considered over all time. In a practical situation, due to limitations of hardware, only a finite number of samples are considered. This is called a finite Fourier transform. If the sample consists of $N$ points, the $N$-point DFT and its inverse are defined respectively as

$$X(m) = \sum_{n=0}^{N-1} x(n) W_N^{nm},$$

$$x(n) = \frac{1}{N} \sum_{m=0}^{N-1} X(m) W_N^{-nm},$$

where $W_N = e^{-j2\pi/N}$ It should be noted that $W_{Nk}(=\exp[-j2\pi/N])$ is equal to $W_N^{k+pN}$. This is because

$$\begin{aligned} W_N^{k+pN} &= \exp[-j2\pi(k+pN)/N] \\ &= \exp[(-j2\pi k/N)\exp(-j2\pi p)] \\ &= \exp(-j2\pi k/N) \cdot 1 \\ &= \exp(-j2\pi k/N) = W_N^k, \end{aligned}$$

for any integer $p$. This shows that $W_N$ has period $N$. As a result, $X(m) = X(m \pm pN)$ and $x(n) = x(n \pm pN)$. Thus $X(m)$ and $x(n)$ are both periodic, and these sequences can be constructed for any $n$ or $m$ by using a periodic extension of $x(n)$ or $X(m)$ from the fundamental period.

In DFT algorithms, $X(m)$ is usually computed at $N$ equally spaced frequencies in the interval 0 to $\omega_s$. The spectral components are separated by $\omega_s/N$. The DFT is evaluated at the discrete frequencies given by $\omega = m(\omega_s/N)$, for $m = 0, 1, 2, \ldots, N-1$. The term $X(m)$ represents the level of the frequency $m(\omega_s/N)$ contained in the signal $x(n)$ and is often referred to as bin $m$. These are the only signals of which $x(n)$ can be composed, since these are the only frequencies whose periods are integrally related to $N$. The frequencies present are all observed to be harmonically related to the fundamental frequency $\omega_s/N$ (corresponding to $m=1$, or bin 1), except for $DC$ (corresponding to $m=0$, or bin 0). Thus the DFT breaks down the time-domain signal in terms of a harmonically related set of discrete frequencies. The bin number is used to specify the particular harmonic of the fundamental frequency. The amplitude of each bin gives a measure of the power spectrum since power is related to the square of the amplitude.

Properties of the DFT are given in List 1. This list shows that for a real waveform (all audio signals), the DFT is even in its magnitude response. Further, it is known to be periodic. As a result of these two properties, the DFTs for only the positive-half frequencies are usually shown, since the remainder can be inferred from this information. Due to the presence of positive and negative frequencies (except for DC and Nyquist frequency), the amplitude of the response at these bins is half of what they should be. This is because the energy at these frequencies is equally divided between the positive and negative frequencies. Properties 10 and 11 in List 1 show that convolution in the time domain is equivalent to multiplication in the frequency domain and multiplication in the time domain is equivalent to convolution in the frequency domain. They are thus called duals of each other, and this is called the property of duality.

**List 1.** Properties of the Discrete Fourier Transform

1. The DFT is linear. The DFT of $ax(n) + by(n)$ is $aX(m)+bY(m)$.
2. If $x(n)$ has even symmetry, that is, if $x(n) = x(-n)$, or $x(n) = x(N-n)$, then its DFT is also even, $X(m) = X(-m) = X(N-m)$.
3. If $x(n)$ has odd symmetry, that is, if $x(n) = -x(-n) = -x(N-n)$, then its DFT is also odd, $X(m) = -X(-m) = -X(N-m)$.
4. If $x(n)$ is real, then its DFT has even real part and odd imaginary part, that is, $X(m) = X^*(-m) = X^*(N-m)$, where $X^*(m)$ is the complex conjugate of $X(m)$.
5. If $x(n)$ is imaginary, then its DFT has odd real part and even imaginary part, $X(m) = -X^*(-m) = -X^*(N-m)$.
6. The DFT of $x(-n)$ is $X(-m)$.
7. The DFT of $x^*(n)$ is $X^*(-m)$.
8. The DFT of $x(n-k)$ is $X(m)W_N^{mk}$.
9. The DFT of $W_N^{-nk}x(n)$ is $X(m-k)$.
10. If two sequences $x(n)$ and $h(n)$ are convolved in the time domain to obtain the output $y(n)$, then the DFT $Y(m) = X(m)H(m)$. That is, the DFTs $X(m)$ and $H(m)$ are multiplied.
11. The dual to property 10 is that multiplication of $x(n)$ and $h(n)$ results in the convolution of the DFTs. That is, $Y(m) = X(m)*H(m)$.

In the evaluation of the DFT each $X(m)$ requires $N$ complex multiplications and $N-1$ complex additions. Since there are $N$ spectral components, the DFT in its raw form requires $N^2$ complex multiplications and $N(N-1)$ complex additions. The memory required is $2N$ for $x(n)$ and $X(m)$ and $N^2$ for the $W_N^{nm}$ coefficients. Implementation of the DFT in hardware shows a variation proportional to $N^2$ in both computation and memory. For example, a 1024-point DFT requires about a million complex multiplications and additions each. In short, the DFT is typically a non-real-time process.

## The Fast Fourier Transform

The fast Fourier transform (FFT) refers to a collection of algorithms that exploit both the symmetry and periodicity of the $W_N^{nm}$ coefficients to streamline the calculation of the DFT. The FFT algorithm requires $N$ to be an integral power of two. For sequences of arbitrary length, zeros are padded to make the total number of samples an integral power of two. It requires $N\log_2 N$ computation steps and memory size of 2N. Numerically, both the DFT and the FFT give the same results; however, the FFT computes the same result faster using a more dedicated configuration. For $N = 1024$ the computation steps are on the order of 10,000. This represents a 100-fold increase in speed over the DFT. This improvement in speed often allows the processing to be done in real time. The FFT could be implemented in software or by using hardware modules.

## The z-Transform

The z-transform is yet another type of transform. It is equivalent to the Laplace transform of sampled data and is the building block of digital filters. Mathematically, the z-transform is described as

$$X(z) = \sum_{n=-\infty}^{+\infty} x(n) z^{-n}.$$

The DFT is the Fourier transform for sampled signals, and the z-transform is the Laplace transform for sampled signals. The Laplace transform is a generalization of the Fourier transform in continuous time theory. Similarly, in discrete time theory, the z-transform is a generalization of the DFT. The z-transform, unlike the DFT, is not something that is calculated. It is a mathematical tool used primarily in the theory of digital filters. It is in a sense more general than the DFT, since it includes the DFT as a special case. The z-transform is of considerable interest in the general theory of DSP, because all digital filter transfer functions are described in terms of the digital variable z.

The inverse z-transform is defined as

$$x(n) = \frac{1}{2\pi j} \oint_c X(z) z^{n-1} dz.$$

This is a contour integral in the z-plane over any closed path in the region of convergence. The region of convergence is explained in more detail below. In most cases the inverse z-transform is calculated by breaking $X(z)$ into partial fractions and using a look-up table to convert the z-expression into a discrete time-domain sequence.

If $x(n)$ is a right-sided sequence starting at $n = 0$, then the summation for the z-transform exists only from 0 to $\infty$. This is called the engineering or one-sided z-transform and is represented by

$$X(z) = \sum_{n=0}^{+\infty} x(n) z^{-n}.$$

If $z = \exp(j2\pi m/N)$, then the z-transform becomes the N-point DFT. This means that the values of N equally spaced sample points on the unit circle of the z-plane directly give the N-point DFT, as shown in Figure 10.

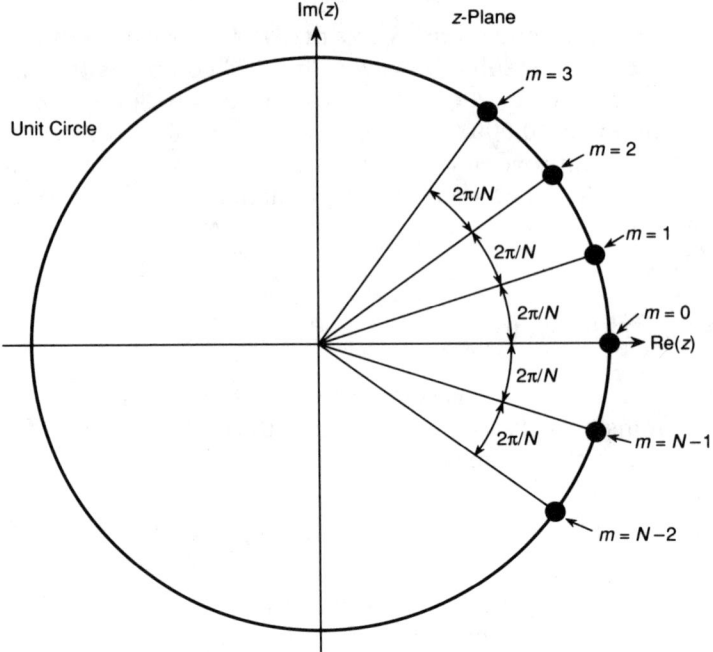

**Figure 10.** The *N* sample points on the unit circle of the *z*-plane that give the DFT are each separated by an angle of $2\pi/N$.

As noted before, the output of a linear, time-invariant system is the convolution of the input and the impulse response $[y(n) = x(n) * h(n)]$. It can be shown that convolution in the time domain is equivalent to multiplication in the *z*-domain $[Y(z) = X(z) \cdot Y(z)]$. The *z*-transform of the impulse response, $H(z)$, is called the transfer function, and it can be obtained by dividing the *z*-transform of the output $Y(z)$ by the *z*-transform of the input $X(z)$: $H(z) = Y(z) / X(z)$.

If $x(n)$ is the delta function, then $X(z) = 1$. Therefore $H(z) = Y(z)$. This means that in the *z*-domain, the transfer function is identical with the output if the input signal is the delta function. It will be recalled that a system's time-domain response to the delta function is the impulse response $[h(n) = y(n)]$. The transfer function $H(z)$ is thus the *z*-transform of the impulse response. Therefore,

$$H(z) = \sum_{n=-\infty}^{+\infty} h(n) \, z^{-n}.$$

List 2 summarizes the properties of the *z*-transform. Property 6 shows the relationship between time-delayed signals and their *z*-transforms. This could be used to find the *z*-transform of a time-domain equation directly. For continuous time systems, the input-output relationship is described by continuous time equations. These might be linear, higher-order, or differential equations. In discrete time systems, the input and output exist only at sampled times and these systems are described by means of difference equations. Thus there often arises the need to deal with signals that are shifted in the time domain and the need to find their $z$-transforms. In general, let $X_k(z)$ be the $z$-transform of $x(n-k)$. Then,

$$X_1(z) = z^{-1}X(z),$$
$$X_2(z) = z^{-2}X(z),$$
$$X_3(z) = z^{-3}X(z),$$

and so on using property 6.

**List 2.** Properties of the z-Transform

1. The $z$-transform is linear. The $z$-transform of $ax(n) + by(n)$ is $aX(z) + bY(z)$.
2. The ROC for a right-sided sequence is the exterior of some circle in the $z$-plane and must include $\infty$ for a causal sequence.
3. The ROC for an anticausal sequence is the interior of some circle.
4. Convolution in the time domain is equivalent to multiplication in the $z$-domain.
5. Multiplication in the time domain is equivalent to complex convolution in the $z$-domain.
6. The $z$-transform of $x(n-k)$, a sequence that has been delayed by $k$ units, is given by $z^{-k} X(z)$.

This shows that the expression $z^{-1}$ in the $z$-domain is the equivalent of a unit delay in the time domain. It is thus called the unit delay element. It is discussed below in the context of DSP components. Consider the difference equation

$$y(n) = x(n) - x(n-1) - y(n-1).$$

The $z$-transform for this can be found by using property 6

$$Y(z) = X(z) - z^{-1}X(z) - z^{-1}Y(z).$$

where $X(z)$ and $Y(z)$ are the $z$-transforms of the input and output $x(n)$ and $y(n)$, respectively. Grouping the $X(z)$ and $Y(z)$ terms together yields

$$Y(z)(1 + z^{-1}) = X(z)(1 - z^{-1}).$$

The transfer function is given by

$$H(z) = \frac{Y(z)}{X(z)} = \frac{1 - z^{-1}}{1 + z^{-1}}.$$

Thus, property 6 greatly simplifies the process of finding the transfer function, $H(z)$, of a discrete system that is described by means of a difference equation.

## The Region of Convergence

The $z$-transform of a function has a region associated with it called the region of convergence (ROC). In this region, the $z$-transform converges to a finite solution, whereas outside this region the $z$-transform grows in magnitude and takes on undefined values. Subsequently the latter region is excluded from its ROC.

The relationship of the delta function with respect to the ROC may be explored as follows: if $x(n)$ is the delta function, then

$$x(n) = 1, \quad \text{for } n = 0, \quad \text{and} \quad x(n) = 0, \quad \text{for } n \neq 0,$$

and the $z$-transform $X(z)$ is

$$X(z) = \sum_{n=-\infty}^{+\infty} x(n) z^{-n}.$$

Therefore

$$X(z) = 1.$$

This shows that the $z$-transform of the delta function is unity. Further, it converges to this value irrespective of the value of $z$, implying that its ROC is the entire z-plane, as shown in Figure 11a.

Similarly, the ROC and the $z$-transform of the unit step function may be derived. If $x(n)$ is the unit step function $u(n)$, then

$$x(n) = 0 \quad \text{for } n < 0, \quad \text{and} \quad x(n) = 1 \quad \text{for } n \geq 0,$$

and the $z$-transform $X(z)$ is

$$X(z) = \sum_{n=0}^{+\infty} z^{-n}.$$

This is a geometric series with ratio $z^{-1}$, and it converges to $1/(1 - z^{-1})$, if $|z^{-1}| < 1$. Therefore,

$$X(z) = \frac{1}{1 - z^{-1}}.$$

Its ROC is $|z^{-1}| < 1$, or $|z| > 1$. Since $z$ is a complex number, $z = x + jy$, where $x$ and $y$ are the real and imaginary components of $z$, respectively. Therefore, the magnitude of $z$ can be written as $|z| = (x^2 + y^2)^{1/2}$. The ROC can be written as $(x^2 + y^2)^{1/2} > 1$. A circle of radius $r$ and center at the origin is defined by the equation $x^2 + y^2 = r^2$. Therefore the ROC for $|z| > 1$ means the ROC is the entire region outside the unit circle, as shown in Figure 11b.

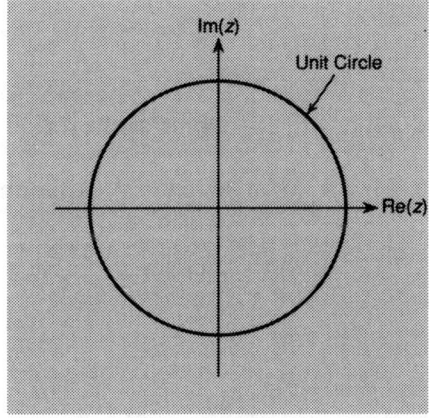
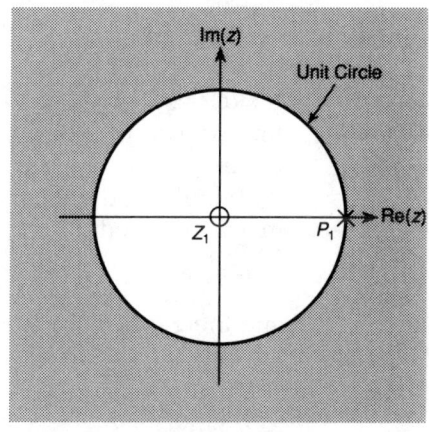

(a) The ROC for the delta function is the entire z-plane.

(b) The ROC for the unit step function is the exterior of the unit circle.

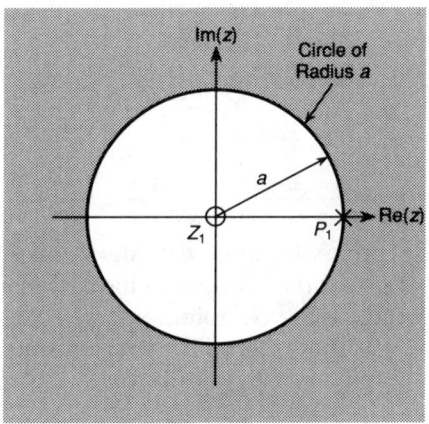
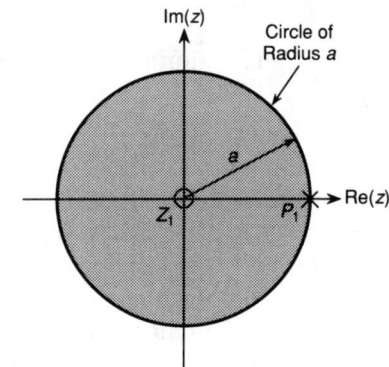

(c) The ROC for a causal exponential sequence $a^n$ is the exterior of the circle of radius a.

(d) The ROC for the anticausal sequence $-a^n$ is the interior of the circle of radius a.

**Figure 11.** The region of convergence (ROC) for various sequences.

If $x(n)$ is a causal exponential sequence, its ROC and z-transform may be derived:

$$x(n) = 0 \quad \text{for } n < 0, \quad \text{and} \quad x(n) = a^n \quad \text{for } n \geq 0,$$

then the z-transform is

$$X(z) = \sum_{n=0}^{+\infty} a^n z^{-n} = \sum_{n=0}^{+\infty} (az^{-1})^n.$$

This is a geometric series with ratio $az^{-1}$, and it converges to $1/(1 - az^{-1})$ if $|az^{-1}| < 1$. Therefore,

$$X(z) = \frac{1}{1 - az^{-1}} = \frac{z}{z - a}.$$

Its ROC is $|az^{-1}| < 1$ or $|a/z| < 1$ or $|z| > |a|$. This ROC corresponds to the exterior of a circle of radius $a$ with its center at the origin as shown in Figure 11c.

Similarly, the ROC and $z$-transform of an anticausal exponential sequence may be derived:

$$x(n) = -a^n \quad \text{for } n < 0, \quad \text{and} \quad x(n) = 0 \quad \text{for } n \geq 0,$$

and the $z$-transform is

$$X(z) = \sum_{n=-\infty}^{-1} -a^n z^{-n} = \sum_{n=-\infty}^{-1} -(az^{-1})^n.$$

Replacing $n$ by $-n$, and changing the limits of the summation to 0 and yields

$$X(z) = 1 - \sum_{n=0}^{\infty} (a^{-1}z)^n.$$

The summation is a geometric series with ratio $a^{-1}z$, and it converges to $1/(1 - a^{-1}z)$ for $|a^{-1}z| < 1$, or $|z/a| < 1$, or $|z| < |a|$. Therefore, for $|z| < |a|$,

$$X(z) = 1 - \frac{1}{1 - a^{-1}z} = \frac{z}{z - a}.$$

The ROC is the interior of the circle of radius $a$, as shown in Figure 11d.

The last two examples of exponential sequences have the same expression for their $z$-transforms, but their ROCs are complementary. This shows that it is important to know both $X(z)$ and the ROC to be able to uniquely identify a time-domain sequence.

## The Unit Circle and Digital Frequencies

If the substitution $z = e^{j\omega}$ is made in the definition of the $z$-transform, then the $z$-transform becomes the Fourier transform. The relation $z = e^{j\omega}$ is the equation of the unit circle in the complex plane. This means that the evaluation of the $z$-transform along the unit circle gives the frequency response. The $z$-transform becomes the $N$-point DFT if the substitution $z = e^{j2\pi m/N}$ is made in the definition of the $z$-transform, and the summation is taken over $m$ for $m = 0, 1, 2, \ldots, N-1$, where $m$ is the bin number. The substitution $z = e^{j2\pi m/N}$ for $m = 0, 1, 2, \ldots, N-1$, is equivalent to sampling the $z$-plane at $N$ equally spaced points along the unit circle. Since the $N$-point DFT is known to be periodic with period $N$, the values that are allotted to $m$ are often $-N/2, -(N-2)/2, \ldots, -2, -1, 0, 1, 2, \ldots, (N-2)/2, N/2$. This leads to an $N+1$-point sequence, and because of periodicity, the value of the DFT at $-N/2$ is identical with the value at $N/2$, and one of these can be arbitrarily dropped to give an $N$-point sequence.

Digital frequencies are denoted by $\Omega$ and analog frequencies by $\omega$. Using a sampling frequency of $\omega_s$, the relation between the digital and analog frequencies is

$$\Omega = 2\pi\omega/\omega_s = \omega T.$$

Since the sampling frequency $\omega_s$ is a constant, $\Omega$ and $\omega$ share a linear relationship. Using the relationship between analog and digital frequencies, $X(m)$ is found to give the contribution of the digital frequency $\Omega = m(2\pi/N)$, for bin numbers $m = -N/2, -(N-2)/2, \ldots, -1, 0, 1, \ldots, (N-2)/2, N/2$. The angular spacing between adjacent points is $2\pi/N$. For $m = 0$, the DC component is found at $\Omega = 0$. For each successive positive value taken by $m$, the frequency increases and $\Omega$ increases in the counterclockwise direction by an angle of $2\pi/N$, until $m = N/2$, where the Nyquist frequency is reached and $\Omega = \pi$. Similarly, for each successive negative value taken on by $m$, the negative frequency increases and the digital frequency $\Omega$ moves in the clockwise direction by an angle of $2\pi/N$, until $m = -N/2$, where the negative Nyquist frequency is reached and $\Omega = -\pi$. Thus the fundamental period is shifted from $[0, 2\pi]$ to $[-\pi, \pi]$. These are equivalent because of the periodic nature of the DFT.

Bin number $m = -N/2$ corresponds to $\Omega = -\pi$, which corresponds to the analog frequency of $-\omega_s/2$. This shows that the point on the unit circle at $\Omega = -\pi$ corresponds to the negative Nyquist frequency, $-\omega_s/2$. This point can be represented in rectangular coordinates as $(-1, 0)$ on the $z$-plane.

Bin number $m = 0$ corresponds to $\Omega = 0$, which corresponds to the analog frequency of zero. This shows that the point on the unit circle at $\Omega = 0$ corresponds to the DC component. This point can be represented in rectangular coordinates as $(1, 0)$ on the $z$-plane.

Bin number $m = N/2$ corresponds to $\Omega = \pi$, which corresponds to the analog frequency of $\omega_s/2$. This shows that the point on the unit circle at $\Omega = -\pi$ corresponds to the positive Nyquist frequency, $\omega_s/2$. This point can be represented in rectangular coordinates as $(-1, 0)$ on the $z$-plane. This is the same as the point for $m = -N/2$, showing the periodic nature of the DFT. Figure 12 summarizes the location of digital frequencies on the unit circle of the $z$-plane.

## Digital Filters

A general difference equation is

$$y(n) + b_1 y(n-1) + b_2 y(n-2) + \ldots + b_N y(n-N) = a_0 x(n) + a_1 x(n-1) + a_2 x(n-2) + \ldots + a_M x(n-M).$$

This can be rewritten in a compact form:

$$y(n) = \sum_{i=0}^{M} a_i x(n-i) - \sum_{i=1}^{N} b_i y(n-i).$$

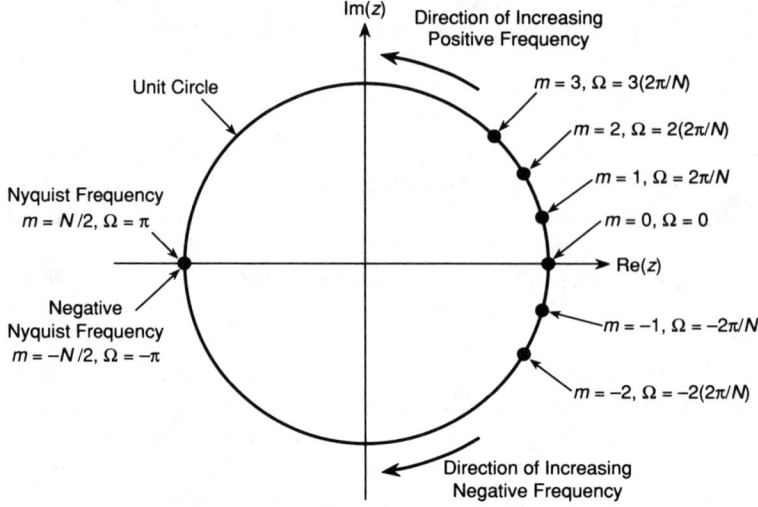

**Figure 12.** The relationship between the DFT coefficient $m$, the digital frequency $\Omega$, and its location on the unit circle.

This input-output relationship of digital signals is commonly referred to as a *digital filter*. Taking the $z$-transform of the digital filter gives

$$Y(z) = \sum_{i=0}^{M} a_i z^{-i} X(z) - \sum_{i=1}^{N} b_i z^{-i} Y(z)$$

$$= X(z) \sum_{i=0}^{M} a_i z^{-i} - Y(z) \sum_{i=1}^{N} b_i z^{-i}.$$

Grouping the $Y(z)$ and the $X(z)$ terms together,

$$Y(z) \left( 1 + \sum_{i=1}^{N} b_i z^{-i} \right) = X(z) \sum_{i=0}^{M} a_i z^{-i}.$$

The transfer function, $H(z) = Y(z) / X(z)$, from the above equation is

$$H(z) = \frac{\sum_{i=0}^{M} a_i z^{-i}}{1 + \sum_{i=1}^{N} b_i z^{-i}} = \frac{N(z)}{D(z)}.$$

$N(z)$ is the numerator of the transfer function, and it has $M$ roots, which are called the *zeros* of the system, because at these values the transfer function has a value of zero. The denominator of the transfer function is $D(z)$, and it has $N$ roots, which are called the *poles* of the system, because at these values the denominator becomes zero and the function is not defined. Perhaps the best way to visualize the magnitude of the transfer function is to imagine the $z$-plane containing extremely long and thin poles at the pole locations. A membrane is stretched across the $z$-plane and is forced to touch the $z$-plane at the location of the zeros. The resulting membranous surface is a representation of the magnitude of the transfer function. Much can be deduced from this contour. For example, the frequency response of the function can be determined by tracing the contour of the transfer function along the unit circle.

## Pole and Zero Patterns and the System Response

One of the advantages of a graphical representation of a system in terms of the locations of its poles and zeros is that the frequency and phase response can be visually estimated for simple cases. As noted, a linear system does not add any new spectral components to the output that are not present in the input. Thus, an excitation provided at a particular frequency will have an output at that same frequency. The amplitude of the output is the amplitude of the input multiplied by the system's response at that frequency. The phase shift at a particular frequency is defined to be the angle of the complex number that represents the system's response at that frequency.

The frequency response is calculated by traversing along the unit circle; it has even symmetry for real sequences. This results in a symmetrical frequency response folded about the point $\Omega = 0$ on the unit circle. The $z$-contour traced on the unit circle is like a cylinder that has an uneven top edge. If this cylinder is cut vertically at $\Omega = \pi$ and opened, it would represent the frequency response for the fundamental period $-\pi$ to $\pi$. Since this response displays even symmetry, half the information is redundant, and the response from 0 to $\pi$ is all that is really required. The same principle holds for the phase response except that the phase response exhibits odd symmetry for real sequences.

The system's response can be found by considering a point $A$ located on the unit circle that makes an angle $\Omega$ with the $x$-axis at the origin as shown in Figure 13. The system poles are denoted by an X at the pole locations $P_1$, $P_2$, ..., $P_N$, and the system zeros are denoted by O at the zero locations $Z_1$, $Z_2$, ..., $Z_M$. The system response at the digital frequency $\Omega$ is given by the product of the lengths of the vectors joining $A$ with the zeros, divided by the product of the lengths of the vectors joining $A$ with the poles. For the phase response, the angles that the poles and zeros make with the point $A$ is of importance. The phase at a particular point is the summation of the angles made by the zeros, less the summation of the angles made by the poles (as shown in Figure 13). It should be noted that the contribution of a pole or zero at the origin has no effect on the magnitude response, since the length of a vector from the origin to the unit circle always has a magnitude of one. It affects the phase response by adding a linear phase component. By the same token, a pole or zero situated near the unit circle has a large effect on the system's response.

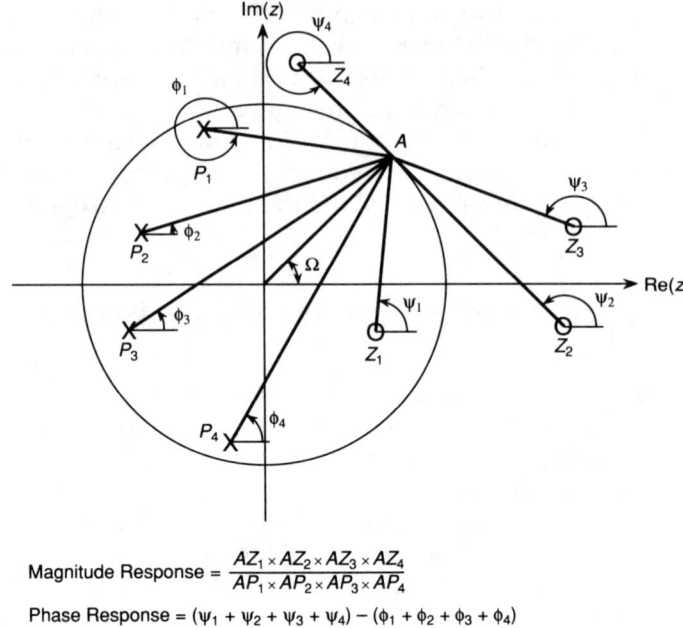

**Figure 13.** Magnitude and phase response of a system obtained from its pole and zero locations.

## Stability

Stability of a system is an important consideration. In the analog domain, the location of the poles in the $s$-plane is used to determine the stability of a system. Vertical lines on the $s$-plane map onto circles in the $z$-plane. The imaginary axis of the $s$-plane maps onto the unit circle in the $z$-plane. The right half of the $s$-plane maps onto the exterior of the unit circle, while the left half of the $s$-plane maps onto the interior of the unit circle of the $z$-plane. Stability of a causal system requires that the poles be situated in the left half of the complex $s$-plane. This in turn demands that the poles of a causal system be inside the unit circle of the $z$-plane. The unit circle therefore plays an important role in the $z$-plane. Some systems converge only on the unit circle; this is analogous to systems converging on the imaginary axis of the $s$-plane. The imaginary axis of the $s$-plane is the special case of the Fourier transform. This implies that systems that converge only on the unit circle of the $z$-plane only have a Fourier transform but no $z$-transform. In the $z$-plane, the ROC is bounded by the system poles. In the case of a right-sided sequence, it is the exterior of some circle, and if the right-sided sequence is a causal sequence, then the ROC includes $\infty$. For a left-sided sequence, the ROC is the interior of some circle. A sequence that has both a causal and an anticausal component will, in general, have an annular (doughnut-shaped) ROC.

Applying the stability condition to an example of a causal exponential sequence (Figure 11c), it can be deduced that it will be stable only if the pole lies inside the unit circle. Since the pole lies at $a$ and bounds the ROC, this requires that $a \leq 1$ for stability. This is an intuitive result that should be expected. The condition $a > 1$ makes the exponential sequence increase in magnitude without bounds and would naturally be unstable. The condition $a = 1$ is the same as the unit step function and borders on stability. The condition $a < 1$ is a decreasing sequence and is the only stable case. All these scenarios are shown in Figure 14.

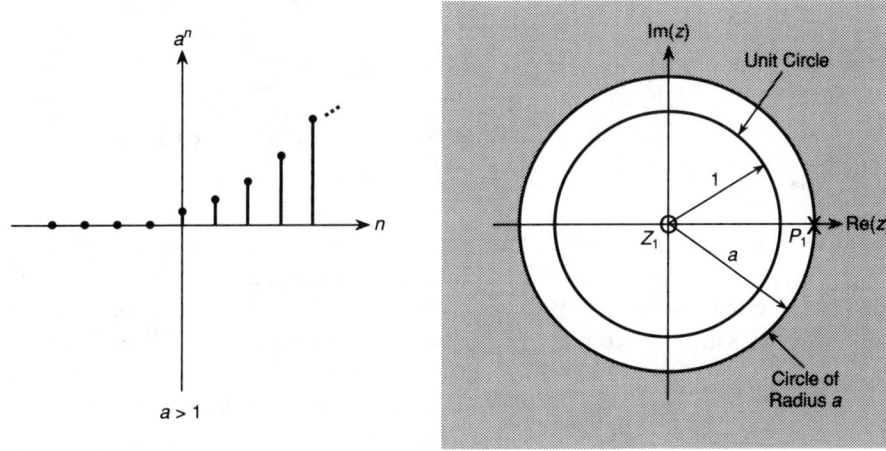

(a) *The location of the pole outside the unit circle leads to instability.*

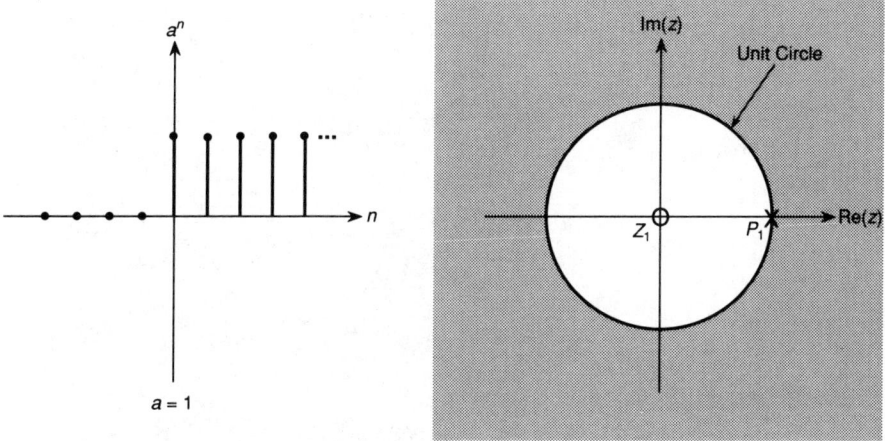

(b) *The location of the pole on the unit circle borders on stability.*

**Figure 14.** The influence of the pole location on stability.

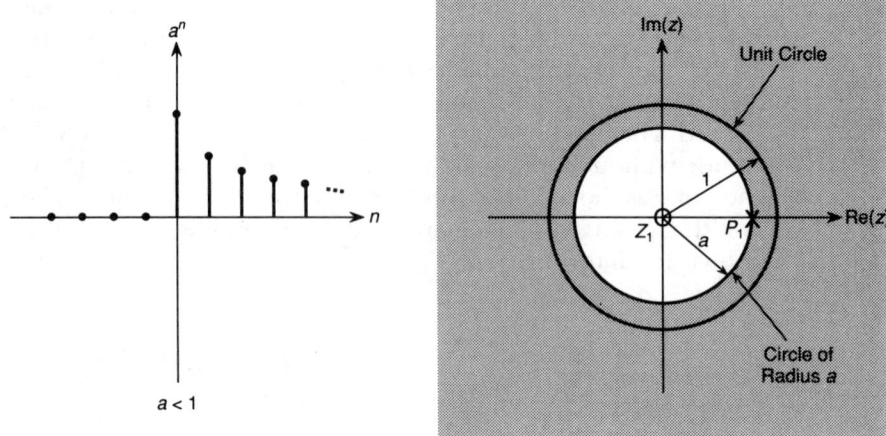

*(c) The location of the pole inside the unit circle leads to stability.*

**Figure 14.** (cont.)

The position of the poles is crucial. The poles cannot be in the ROC because the transfer function is not defined at the poles and is contradictory to what the ROC represents. The ROC is bounded by the poles. A causal system has an ROC that is the exterior of some circle of radius $R$ and must include the point $z = \infty$. Therefore, all the system poles must lie on or inside this circle of radius $R$. A necessary condition for stability is that the poles should lie inside the unit circle. Therefore, the ROC is the exterior of some circle that is smaller than the unit circle, in other words, that is $R < 1$. This leads to the conclusion that the ROC must include the unit circle and $z = \infty$ for a stable and causal filter, as shown in Figure 15.

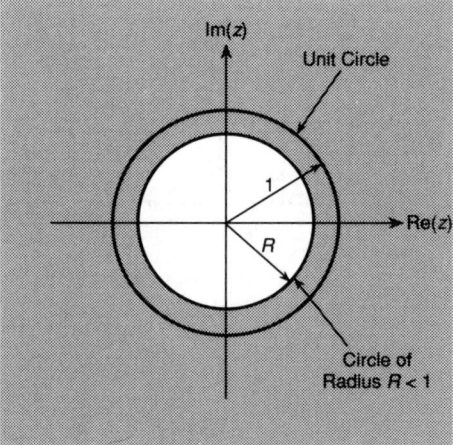

**Figure 15.** The ROC for a stable and causal filter must include the unit circle and $\infty$.

## DSP Components

The hardware implementation of the transfer function of a digital filter is the realization of the corresponding difference equation. In other words, it relates the output $y(n)$ and its delayed versions $y(n-1)$, $y(n-2)$, ... to the input $x(n)$ and its delayed versions $x(n-1)$, $x(n-2)$, ... A closer look at the general transfer function of a digital filter,

$$y(n) = -b_1 y(n-1) - b_2 y(n-2) - \ldots - b_N y(n-N)$$
$$+ a_0 x(n) + a_1 x(n-1) + a_2 x(n-2) + \ldots + a_M x(n-M),$$

reveals that each term is composed of delayed signals multiplied by some coefficients. All such terms are summed to produce the present output. The above equation can be realized in hardware using three basic elements. A unit delay when given an input $x(n)$ produces an output $x(n-1)$ as shown in Figure 16a. It is represented either as a box labelled as $z^{-1}$ or $D$. As the name suggests, this block delays the signal by one unit of time. If $L$ unit delays are cascaded, then an input of $x(n)$ produces an output $x(n-L)$. The unit delays could be ordinary shift registers through which the data is clocked, or memory locations. A multiplier provides gain coefficients $a_i$ and $b_i$. If a multiplier has a gain of $g$, then for an input of $x(n-i)$, the output would be $gx(n-i)$, as shown in Figure 16b. A summing block produce the sum of all the inputs at its output. In contrast to the other two components, which accept a single input and produce a single output, the summing block accepts multiple inputs and produces a single output, as shown in Figure 16c. The multiplier and delay elements can be used in conjunction to form any term $a_i x(n-i)$ or $b_i y(n-i)$. The summing block is used to add all such terms together to produce the present output. By using these three basic building blocks any digital filter can be realized.

## Filter Classification

Using the DSP components, the general digital filter can be realized in hardware as shown in Figure 17. The paths along the top half of the filter which feeds the input $x(n)$ to the summation block are called feedthrough paths. Similarly, the paths along the lower half which feeds the output $y(n)$ back to the summation block are called feedback paths. Therefore the coefficients $a_0, a_1, \ldots, a_m$ are referred to as the feedforward coefficients, whereas $b_1, b_2, \ldots, b_N$ are referred to as the feedback coefficients. In this case, since the output, $y(n)$, is fed back, these filters are also called recursive filters. This feedback could lead to instability. A system with poles contributes an exponential sequence to the impulse response at each of these poles, making the impulse response potentially infinite in duration. Recursive filters are usually called infinite impulse response (IIR) filters. The term IIR filter is a misnomer because the impulse response need not necessarily be infinite in duration. Generally, the terms IIR and recursive have come to be used interchangeably.

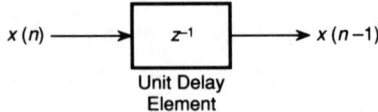

*(a) The unit delay element.*

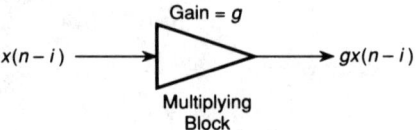

*(b) The multiplier.*

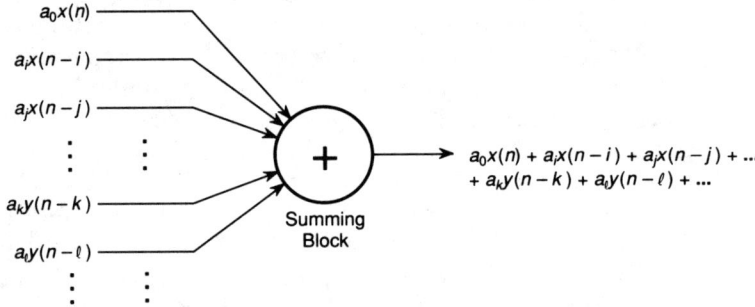

*(c) The summation block.*

**Figure 16.** The components required to realize a digital filter in hardware.

Consider the filter shown in Figure 18a. A nonzero input $x(n)$ would appear at the output, and, after a unit delay, half the previous output would appear at the output. Specifically, if $x(n) = \delta(n)$, the impulse function, then the output (the impulse response) would be an exponentially decaying sequence of infinite length, as shown in Figure 18b. The difference equation that describes the operation of the filter is

$$y(n) = x(n) + 0.5y(n-1),$$

since the present output at any instant is the summation of the present input and half of the previous output.

For the special case where all the $b_i$ coefficients of a digital filter are zero,

$$y(n) = \sum_{i=0}^{M} a_i x(n-i),$$

$$H(z) = \sum_{i=0}^{M} a_i z^{-i}.$$

The system has $M$ zeros. Its hardware realization is shown in Figure 19. Since the output is not fed back, this is called a nonrecursive filter. The response of such a

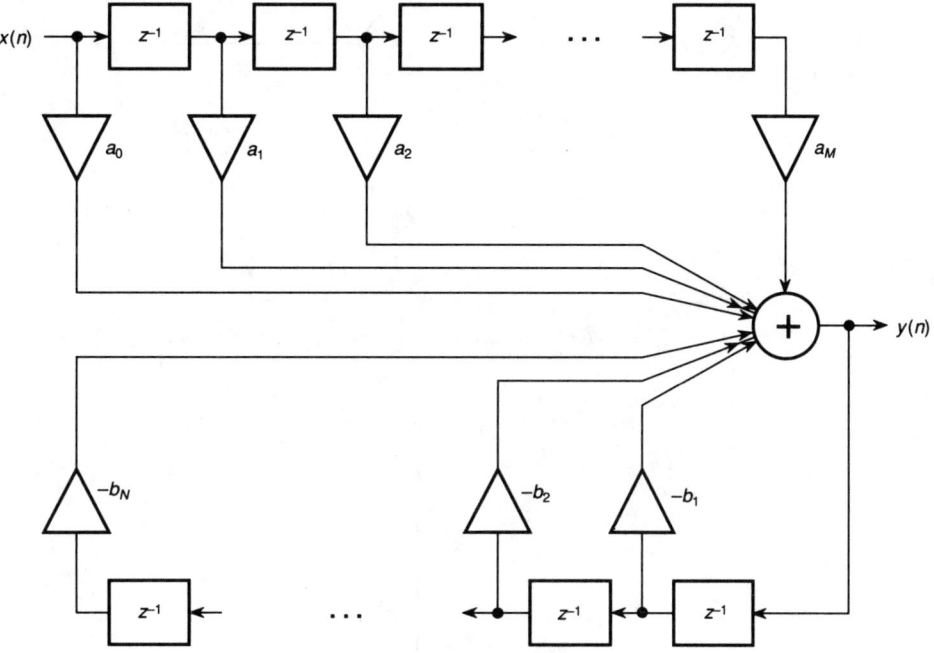

**Figure 17.** The hardware realization of the general digital filter using the three basic building blocks.

filter to an impulse function has a finite duration during which it is nonzero. Hence, it is called a finite impulse response (FIR) filter. All nonrecursive filters are FIR filters, but all FIR filters need not be nonrecursive. The terms FIR and nonrecursive, however, have come to be used interchangeably.

An FIR filter could be designed to imitate the exponential decay produced by the IIR filter. Since the output cannot be fed back, the input has to be delayed and scaled and fed to the output. Consider the FIR filter having coefficients $a_i = (0.5)^i$:

$$y(n) = x(n) + (0.5)\,x(n-1) + (0.5)^2\,x(n-2) +$$
$$(0.5)^3\,x(n-3) + \ldots + (0.5)^M x(n-M).$$

In other words, the variables $a_0$, $a_1$, $a_2$, $a_3$, and $a_M$ (see Figure 19) would now equal 1, 0.5, $(0.5)^2$, $(0.5)^3$, and $(0.5)^M$, respectively. The impulse response of this filter decays exponentially but does not go to zero because it has a finite length, as shown in Figure 20. In this case the length is $M+1$. This approximates the IIR filter impulse response and would be identical with the IIR filter if it were allowed to have an infinite number of terms. However, the hardware realization of this FIR filter requires more components than the IIR filter.

A finite impulse response and the absence of feedback ensures stability of the FIR filter. An FIR filter works by constructing the impulse response for each input and scaling the response by the amplitude of the input and shifting

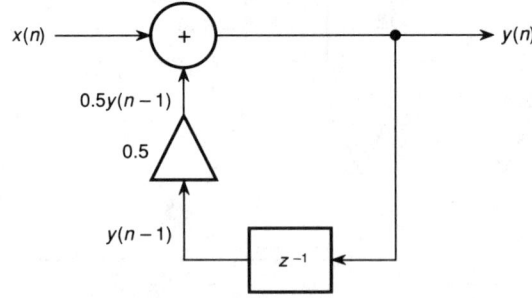

*(a) The hardware realization of an IIR filter (note the feedback path).*

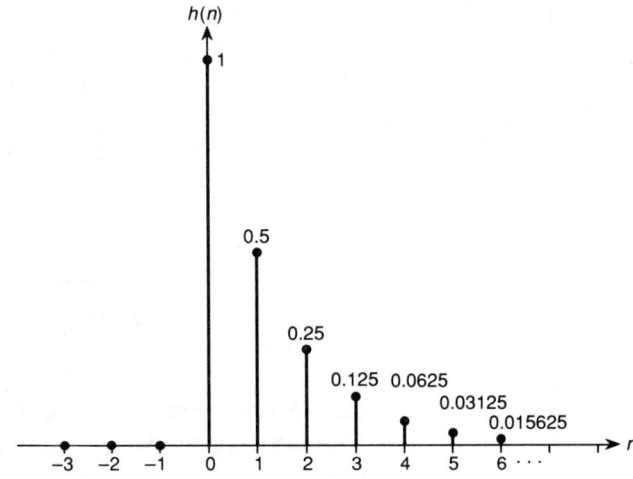

*(b) The exponentially decaying infinite impulse response of the filter.*

**Figure 18.** An example of an IIR filter.

it by the input sample's position in time. All scaled and shifted responses are added to yield the final output.

In the case of the IIR realization (shown in Figure 18a), if the multiplication coefficient is less than one, the impulse response is an exponentially decaying response. Changing the coefficient to one results in an impulse response that is one for all positive time. If the coefficient is greater than one, then the impulse response starts at one and increases exponentially leading to instability until a maximum value is reached. This demonstrates the potential instability problem of an IIR filter—a factor that is nonexistent for FIR filters. Note that this instability occurs when the pole is situated outside the unit circle.

In the case of the IIR filter considered, the impulse response is created by making the present output a fraction of the previous output. This fraction is determined by the multiplying coefficient. Since the data is repeatedly multiplied by this coefficient, a slight inaccuracy in this coefficient would result in a large difference after many recursions. As a result, there could be a difference in performance between the filter's theoretical design and actual realization.

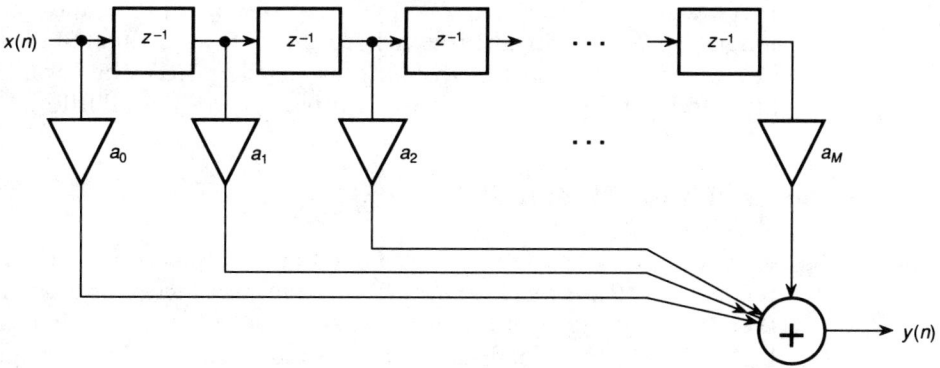

**Figure 19.** Hardware realization of the general nonrecursive (FIR) filter.

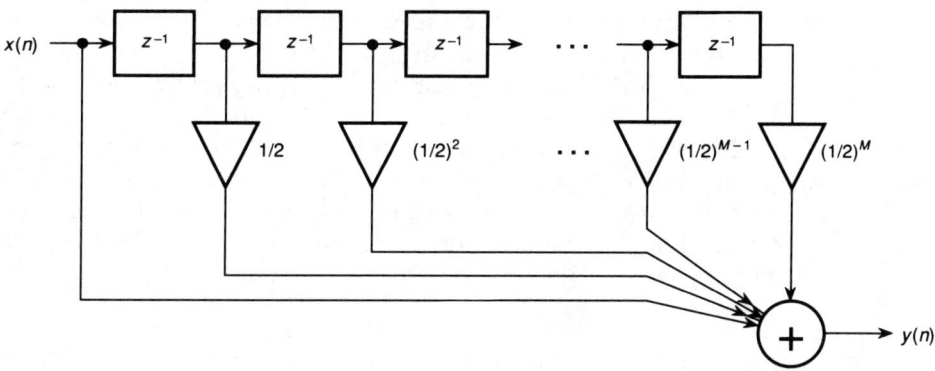

*(a) The hardware realization of an FIR filter.*

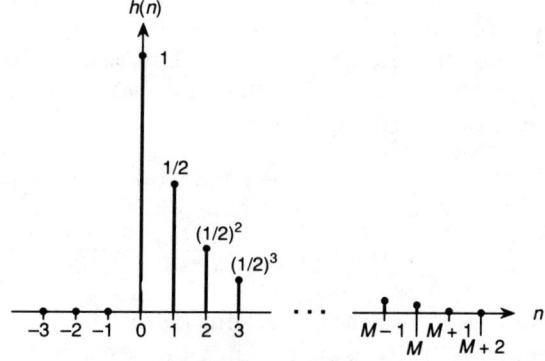

*(b) The exponentially decaying impulse response of the filter (note finite length).*

**Figure 20.** Impulse response of the FIR filter that approximates an exponentially decaying impulse response (note the finite length).

In a practical situation, the coefficient would be represented by a finite wordlength. In some critical cases, where the poles are situated close to the unit circle, the finite wordlength might cause the pole to be located outside the unit circle; this could lead to instability in the designed filter.

## Comparison of IIR and FIR Filters

If the transfer function is the ratio of two polynomials in $z^{-1}$, then it results in a recursive or IIR filter. However, if the transfer function is a polynomial in $z^{-1}$, then the resulting filter is a nonrecursive or FIR filter. FIR filters are always stable, and they can be made causal by introducing a delay. The design methods for both filters are well defined. If a method called analog-to-digital transforms (discussed below) is used, then IIR filters are easier to design. Otherwise, FIR filters are easier to design. An IIR filter will have higher selectivity and sharper cutoff than an FIR filter of the same order. The FIR requires greater computation, resulting in greater hardware and software complexity, making it a more expensive filter to use. The operation of FIR filters is easier to understand; the operation of IIR filters is made more complicated by feedback. For most audio applications linear phase is an important criterion. Since FIR filters can have exactly linear phase they are preferred to IIR filters, which can be made to have linear phase with phase equalization. Therefore, most audio applications use the more expensive FIR filters because of the advantage of inherently linear phase. Infinite impulse response filters are used for such applications as reverberation, where phase nonlinearity is part of the effect desired.

## Examples of Digital Filters

Consider the difference equation $y(n) = [x(n) + x(n-1)]/2$. The present output is seen to be the mean value of the present input and the past input. This serves as an averaging or smoothing filter, and the expected response would be that of a low-pass filter. This assumption could be verified by finding the pole/zero pattern, which in turn would require the filter to be transformed to the $z$-domain, as follows:

$$Y(z) = [X(z) + z^{-1}X(z)]/2$$

$$H(z) = \frac{Y(z)}{X(z)} = \frac{1 + z^{-1}}{2} = \frac{z+1}{2z}.$$

This gives rise to the pole/zero pattern shown in Figure 21a (zero at $z = -1$, pole at $z = 0$). By using the method described above, it is possible to estimate the frequency response. In this case the denominator, $AP_1$, remains 1. The numerator, $AZ_1$, starts with a value of $2 (\Omega = 0)$ and with increasing frequency gradually decreases to a value of zero ($\Omega = \pi$). The resulting frequency response is shown in Figure 21b—a low-pass filter as expected. The difference equation may be realized in hardware by adding the present and delayed input and multiplying the result by 0.5 to obtain the output as shown in Figure 21c.

Now consider the difference equation $y(n) = [x(n) - x(n-1)]/2$. The

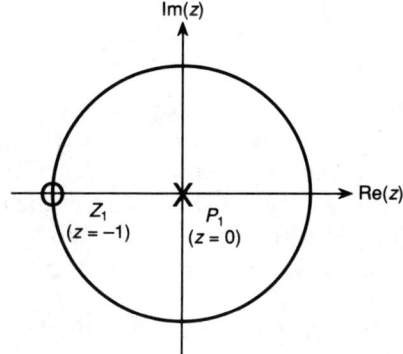

*(a) The pole and zero locations of the filter.*

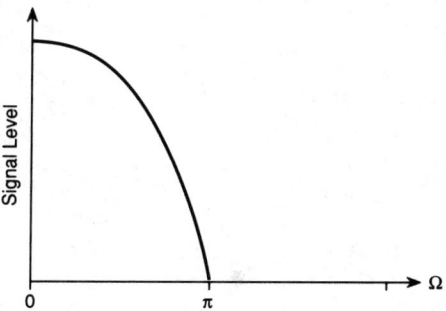

*(b) The frequency response of the filter.*

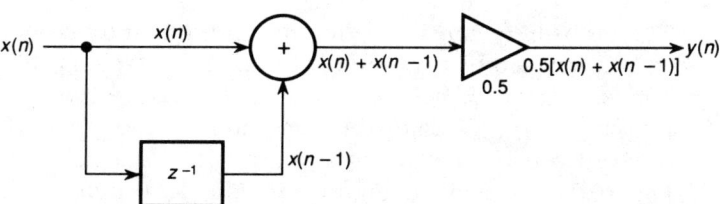

*(c) The realization of the digital filter using the three DSP building blocks.*

**Figure 21.** An example of a digital low-pass filter.

present output is half the difference between the present and past input. Since this filter emphasizes the difference in adjacent values of the input signal, it would seem to be a high-pass filter. Transforming the difference equation into the z-domain yields

$$Y(z) = [X(z) - z^{-1}X(z)]/2$$

$$H(z) = \frac{Y(z)}{X(z)} = \frac{1-z^{-1}}{2} = \frac{z-1}{2z}.$$

This gives rise to the pole/zero pattern in Figure 22a. (zero at $z = 1$, pole at $z = 0$). When compared with Figure 21a the zero location has changed from $z = -1$ to $z = 1$. In this case the frequency response is that of a high-pass filter as shown in Figure 22b. The hardware realization is shown in Figure 22c. When this is compared with Figure 21c, it will be noticed that an extra multiplication coefficient of $-1$ was all that was required to change the low-pass filter into a high-pass filter.

The pole/zero pattern in Figure 23a shows two zeros at $z = \pm 1$ and two conjugate poles situated on the imaginary axis at $z = \pm 0.5j$. The frequency response in Figure 23b shows that this is a bandpass filter response.

Once again, the transfer function can be written in terms of the system zeros and poles:

$$H(z) = \frac{Y(z)}{X(z)} = \frac{(z+1)(z-1)}{(z+0.5j)(z-0.5j)}$$

$$= \frac{z^2 - 1}{z^2 + 0.25} = \frac{1 - z^2}{1 + 0.25z^{-2}},$$

$$Y(z) + 0.25\, z^{-2} Y(z) = X(z) - z^{-2} X(z).$$

Taking the inverse z-transform yields the difference equation:

$$y(n) + 0.25\, y(n-2) = x(n) - x(n-2),$$
$$y(n) = x(n) - x(n-2) - 0.25\, y(n-2).$$

The terms here consist of the present input and output and also the input and output delayed by two units of time. Thus two delay elements are required to convert $x(n)$ to $x(n-2)$ and two more to convert $y(n)$ to $y(n-2)$. The present output $[y(n)]$ is given by the summation of the input $[x(n)]$, the negative of its delayed version $[-x(n-2)]$ and a quarter of a delayed version of the output $[-y(n-2)]$, as shown in Figure 23c. The delay branches of $x(n-2)$ and $y(n-2)$ are combined to obtain the realization shown in Figure 23d.

The plot in Figure 24a also contains two poles and zeros, but this time the poles are on the real axis at $z = \pm 0.5$, and the two zeros are situated at $z = \pm j$. The frequency response of Figure 24b is that of a bandstop filter. As before, the transfer function can be written in terms of the system poles and zeros:

$$H(z) = \frac{Y(z)}{X(z)} = \frac{(z+j)(z-j)}{(z+0.5)(z-0.5)}$$

$$= \frac{z^2+1}{z^2-0.25} = \frac{1+z^{-2}}{1-0.25z^{-2}},$$

$$Y(z) - 0.25\, z^{-2}Y(z) = X(z) + z^{-2}X(z).$$

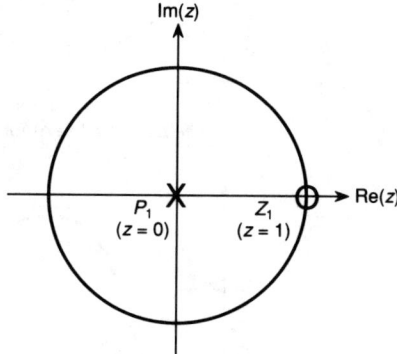

*(a) The pole and zero locations of the filter.*

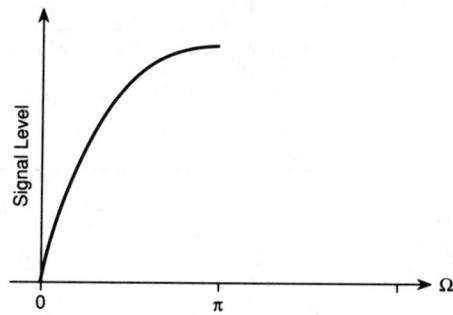

*(b) The frequency response of the filter.*

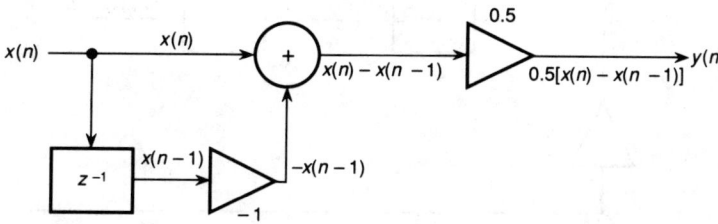

*(c) The realization of the digital filter using the three DSP building blocks.*

**Figure 22.** An example of a digital high-pass filter.

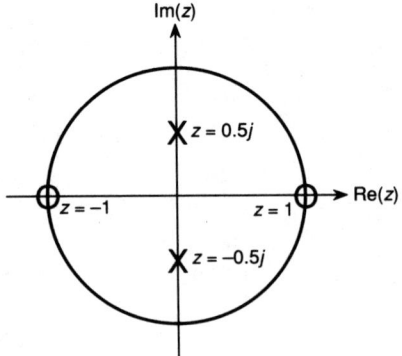

*(a) The pole and zero locations of the filter.*

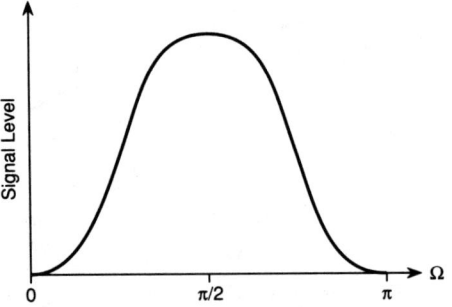

*(b) The frequency response of the filter.*

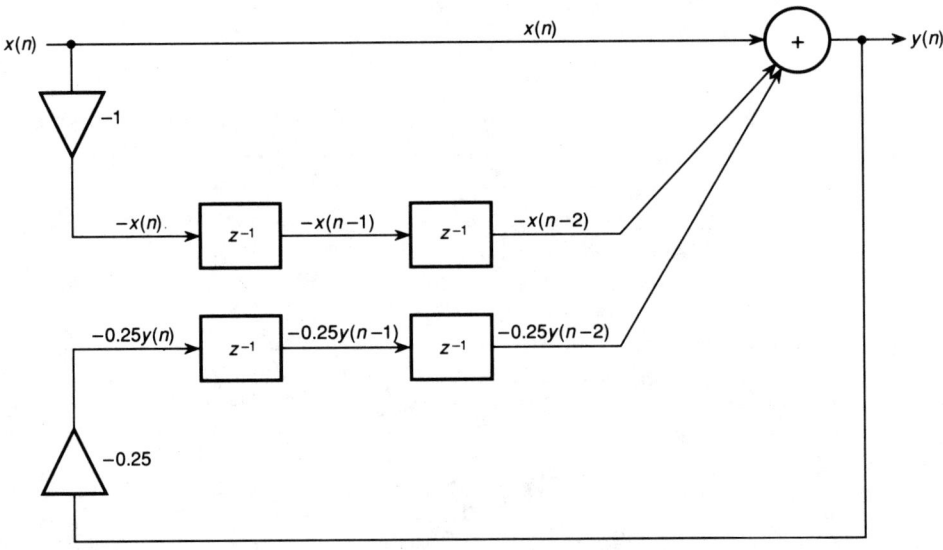

*(c) The realization of the digital filter using the three DSP building blocks.*

**Figure 23.** An example of a digital bandpass filter.

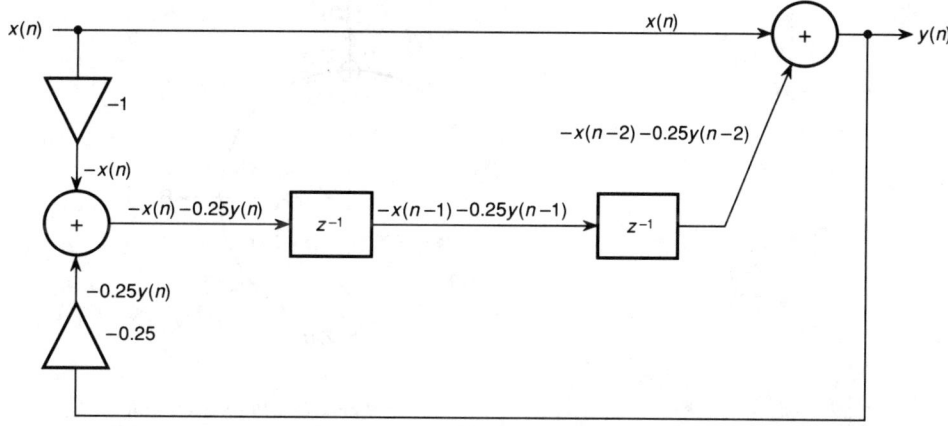

*(d) Modifying (c) by combining the delay elements together.*

**Figure 23.** (cont.)

The difference equation can be obtained by taking the inverse $z$-transform, and this is used to obtain the hardware realization:

$$y(n) - 0.25\, y(n-2) = x(n) + x(n-2).$$
$$y(n) = x(n) + x(n-2) + 0.25 y(n-2).$$

Its hardware realization is shown in Figure 24c. Once again, on comparing the last two hardware realizations, it is seen that the two schemes are almost identical except for the multiplication coefficients. Therefore, it is observed that the type of filter can be changed merely by changing the value of the multiplying coefficient, making a digital filter much more flexible than an analog filter.

## Errors

Digital filters do have their drawbacks. There are errors that exist. It is probable that the multiplying coefficients cannot be represented exactly in binary. As a result the signal passing through the multiplying block emerges slightly different from what is desired. If the number of bits used to represent the coefficient is increased, the error will be lower, and it will take a larger number of operations to produce a significant error. As a result of this, there are many processors that have coefficient words of 24 bits or more.

If the result of the addition of two numbers is larger than the number of bits can handle, then the value wraps around, or starts at the lowest value. This might be viewed as erroneously obtaining the negative peak instead of the positive peak. Sonically, this would result in an undesirable pop. It would be better to use some kind of saturating arithmetic to ensure that overflow values

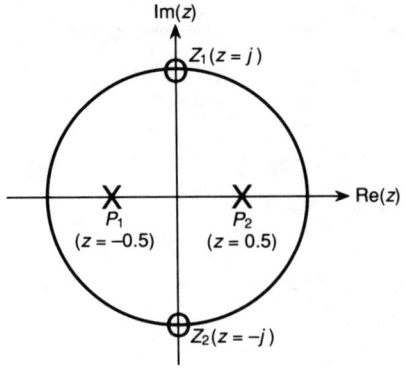

(a) The pole and zero locations of the filter.

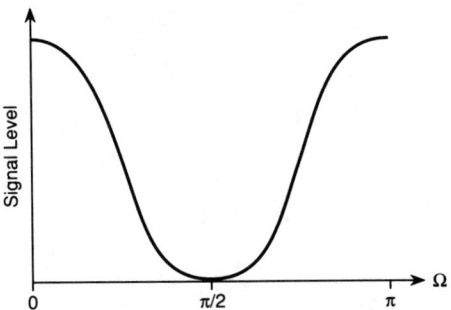

(b) The frequency response of the filter.

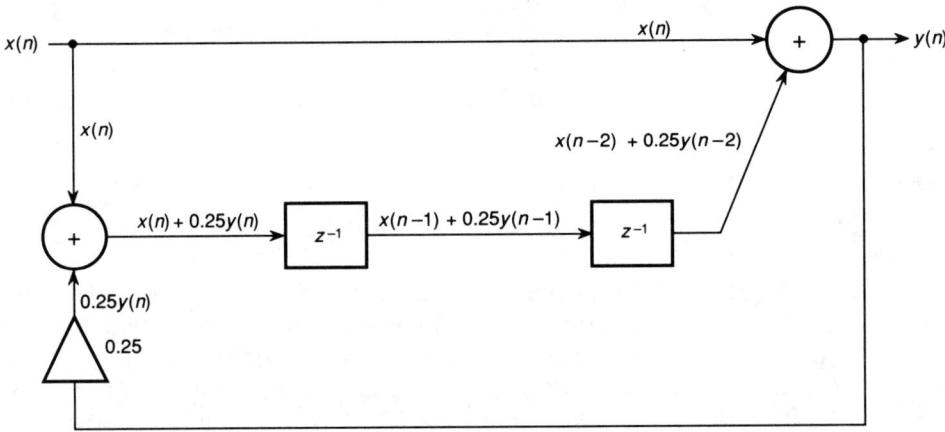

(c) The realization of the digital filter using the three DSP building blocks.

**Figure 24.** An example of a digital bandstop filter.

are set equal to the maximum value that can be handled. This is akin to clipping, which, although undesirable, is sonically a better solution.

The addition of two $n$-bit numbers could result in a $(n+1)$-bit number, whereas the multiplication of two $n$-bit numbers could result in a $2n$-bit number. This would imply that in order to maintain accuracy, the internal processing of the 16-bit samples might require 56-bit representation or more. After the processing is done, the wordlength has to be reduced to the original length. Truncation and rounding-off cause errors. In order to minimize its effect, all truncation or rounding-off should be accompanied by digital dithering, so that, statistically speaking, the error on the average is reduced.

Coefficient errors cause a change in the pole and zero locations and result in a slightly altered response and, in an extreme case, could push a pole out of the unit circle, resulting in an unstable filter. The sensitivity of the system to any coefficient can be mathematically determined by partially differentiating the system's transfer function with respect to the coefficient of interest. The question of error is a trade-off between cost on one hand and performance on the other.

## Types of Digital Filter Realizations

Without any loss of generality, the transfer function of a digital filter can be assumed to have the same number of poles and zeros, that is, $M=N$. The transfer function in that case can be written as

$$H(z) = \frac{\sum_{i=0}^{N} a_i z^{-i}}{1 + \sum_{i=1}^{N} b_i z^{-i}},$$

where the number of poles and zeros are $N$. A digital filter can be realized in hardware in numerous ways. The departure of the implemented filter's performance from the one that was designed on paper is affected not only by the finite wordlength but also by the particular form of realization chosen. Some configurations give rise to more errors than others.

Figure 25 shows a direct form I realization of a digital filter. This structure is very simple and is obtained easily and directly from the equation of the digital filter. It requires $2N$ delay elements and a single summation block. According to Dattoro [1988], this topology is usable for audio purposes.

The transfer function of the digital filter, $H(z)$, can be written as the product of its numerator and denominator as follows:

$$H(z) = \left( \sum_{i=0}^{N} a_i z^{-i} \right) \left( \frac{1}{1 + \sum_{i=1}^{N} b_i z^{-i}} \right) = \frac{Y(z)}{W(z)} \frac{W(z)}{X(z)},$$

where

$$\frac{Y(z)}{W(z)} = \sum_{i=0}^{N} a_i z^{-i} \quad \text{and} \quad \frac{W(z)}{X(z)} = \left( 1 + \sum_{i=1}^{N} b_i z^{-i} \right)^{-1}.$$

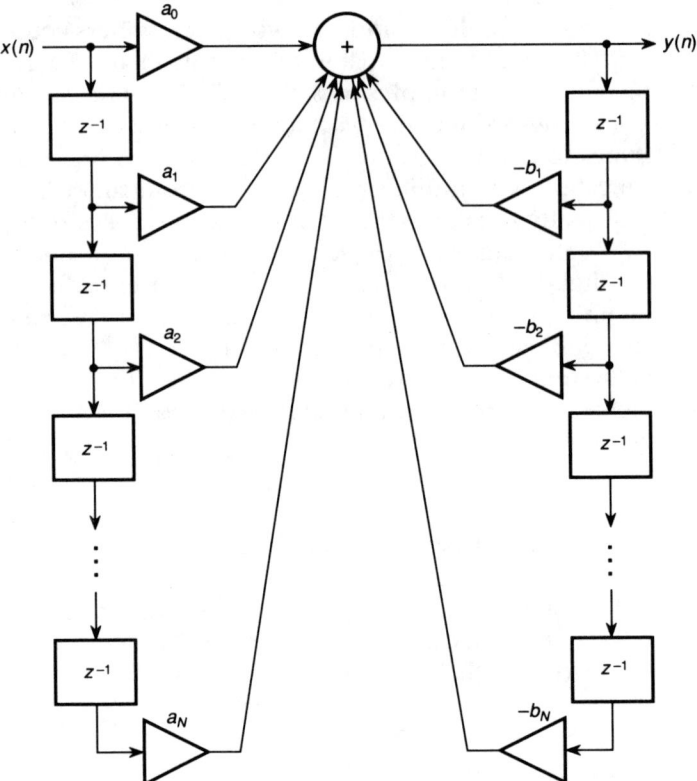

**Figure 25.** Direct form I realization of a digital filter.

In the discrete time domain these yield

$$w(n) = x(n) - \sum_{i=1}^{N} b_i\, w(n-i),$$

$$y(n) = \sum_{i=0}^{N} a_i\, w(n-i).$$

The realization of the above equations, called the direct form II realization, is shown in Figure 26a. The delays in the two columns can be combined to form a single column of $N$ delays, as in Figure 26b. This is also called a canonic realization because the number of delay elements equals the order of the digital filter equation. This configuration has two summing blocks, hence two sources of truncation error, which are in the feedback and the feedforward paths. This arrangement is not suitable for audio purposes because of the circulation of these errors.

The biquadratic realization is used to implement higher-order filters. Biquadratic filters (commonly known as biquads) are made up of first- or second-order sections. The finite wordlength of the coefficient and error recirculation problems rise geometrically with higher-order filters. Therefore,

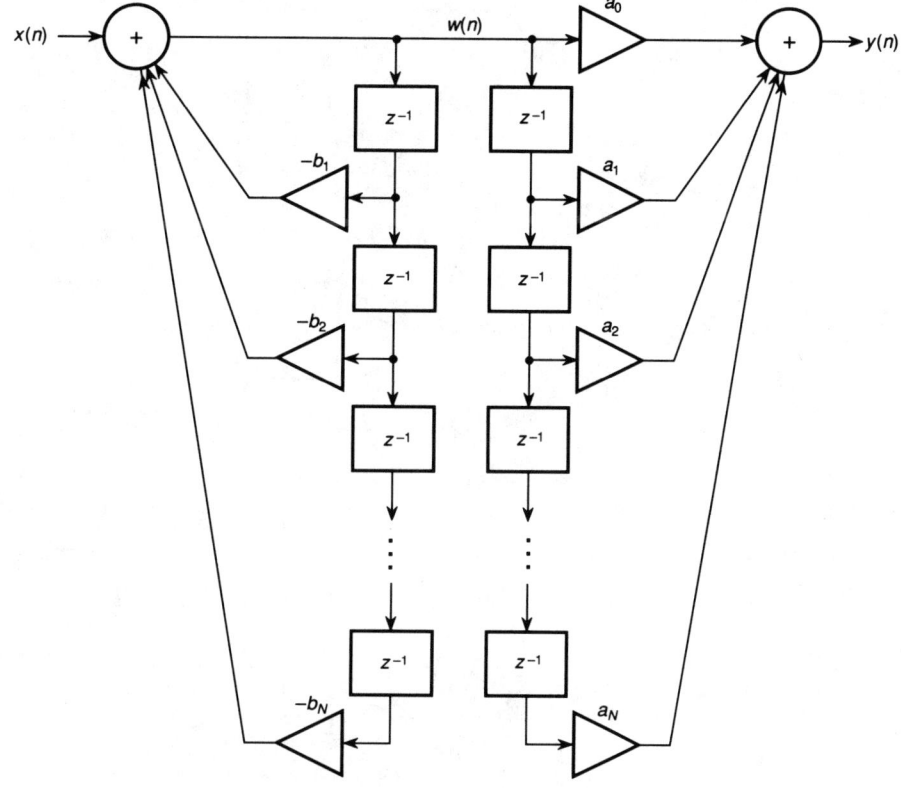

*(a) The noncanonic form realization.*

**Figure 26.** Direct form II realizations. *(cont. p. 334)*

smaller sections (biquads) are used in either a cascade or parallel form to implement the higher-order filter in audio.

The transfer function $H(z)$ can be written as a product of first- and second-order sections. The cascade realization is shown in Figure 27a. The filter performance depends on the ordering of the biquad sections in cascade and also on the poles and zeros for each section. First poles and zeros that are close together are paired, and then sections are ordered from smallest pole magnitude to largest pole magnitude. This gives more dynamic range. The function $H(z)$ can also be written as a partial fraction expansion of first- and second-order sections. The parallel realization is shown in Figure 27b. Since the sections are in parallel, the question of the order of the sections does not arise in this case.

# Digital Filter Design Techniques

The design of digital filters involves determining a transfer function in the $z$-domain. It has to be either a rational function in $z^{-1}$ for an IIR filter or a

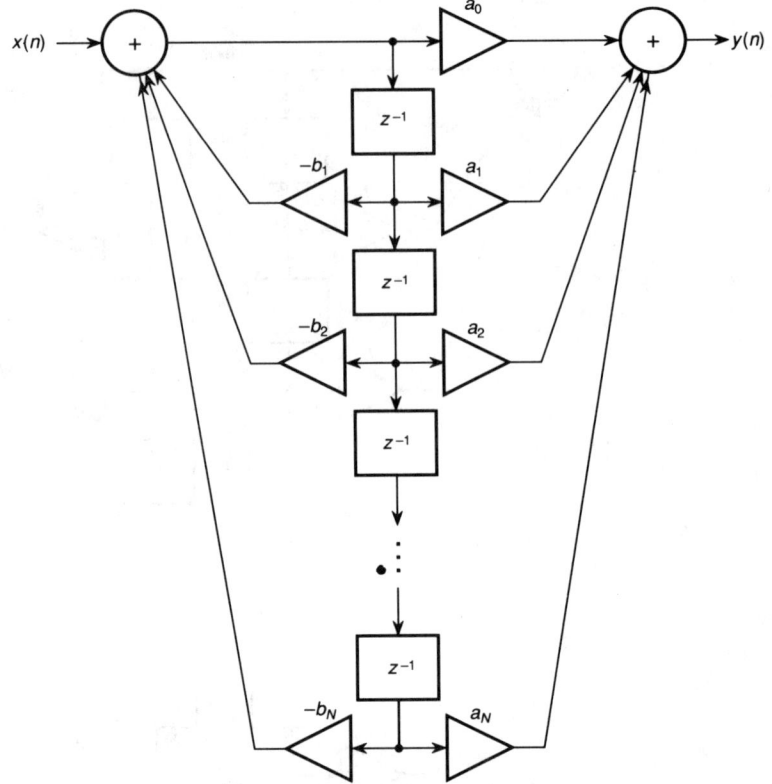

*(b) The canonic form realization; here the number of delay elements used is equal to the order of the filter.*

**Figure 26.** (cont.)

polynomial in $z^{-1}$ for a FIR filter. The transfer function is determined by keeping the filter's performance specification in mind. This specification could be defined either in the frequency or time domains. Filters allow certain frequencies to pass through without any modification, collectively called the passband, while blocking others, called the stopband. In audio applications, filters are usually specified by their amplitude response in the frequency domain. The performance of the theoretical result and the actual implementation differ, mainly because of finite wordlength considerations. It is advisable to simulate a filter, using finite wordlength precision, and then compare this with the desired response. The choice of a particular design method depends on whether an IIR or a FIR filter is desired. For many audio applications phase linearity is important. This places more importance on FIR filters.

## General Analog-to-Digital Transformations

Digital transformations allow the design of digital filters directly from the analog expression of the filter. The analog filters are usually specified by the Laplace variable $s$. Digital filters are specified by the variable $z$. Digital

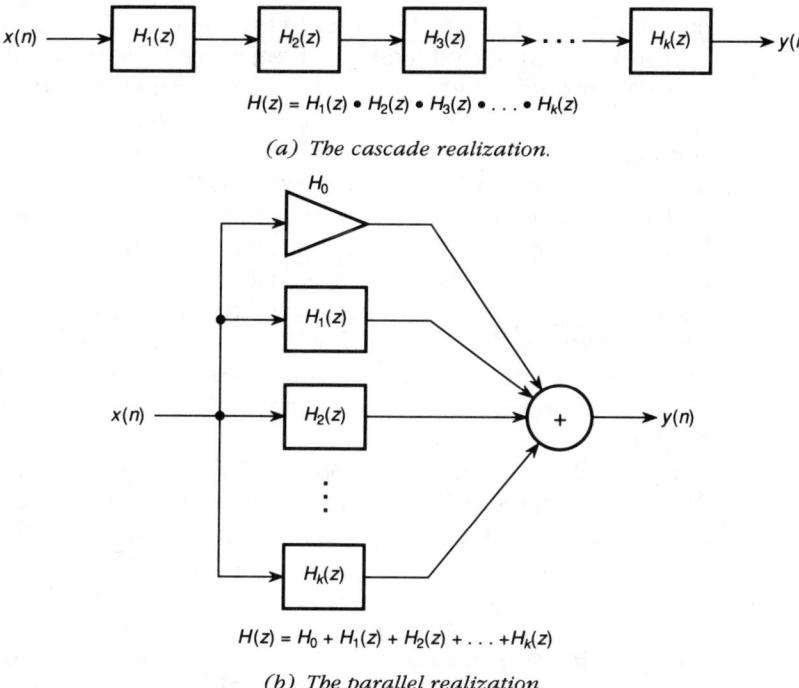

*(a) The cascade realization.*

*(b) The parallel realization.*

**Figure 27.** The biquadratic realizations of a digital filter.

transformations or *s*-to-*z* transforms provide the designer with a means to obtain the digital transfer function that is equivalent to a given analog transfer function. The *s*-to-*z* transform provides the result directly and is akin to algebraic substitutions in its relative simplicity.

Frequency transformations allow a given type of filter to be changed to another type of filter by means of a substitution. Usually, the starting point is an analog low-pass filter. Using the frequency transformations, this can be changed into a high-pass, bandpass, or bandstop filter. Digital transformations and frequency transformations could be coupled together and called general analog-to-digital transformations. This allows the designer to move directly from an analog low-pass filter to any type of a digital filter.

There are numerous analog-to-digital transformations, each with its own peculiarities. These transformations do not necessarily ensure that a stable analog filter will result in a stable digital filter. One of these transformations, the bilinear transform, is more suitable because of certain properties. It ensures that a stable analog filter results in a stable digital filter. Moreover, it exactly preserves the frequency-domain characteristics. However, the bilinear transform introduces frequency compression (explained below).

There are four kinds of bilinear transformations generally used to realize IIR filters. In each case the starting point is an analog low-pass filter that has been normalized in the frequency domain such that its pass-band frequency $\omega_p$ corresponds to unity. It is possible to transform this to realize a low-pass, high-pass, bandpass, or bandstop filter in the digital domain by using the appropriate bilinear analog-to-digital transformation. The transforms are shown in Figures 28a through 28d along with the relationship between the analog and digital frequencies. This relationship is used to plot the digital filter responses starting in each case with the analog low-pass filter as shown.

As noted, analog and digital frequencies share a linear relationship, with the digital frequency interval $[-\pi, \pi]$ corresponding to the analog frequency range $[-\omega_s/2, \omega_s/2]$. Following the bilinear transformation, however, Figure 28a shows that the entire range of positive analog frequencies $[0, \infty]$ is mapped onto the digital frequency range $[0, \pi]$. This means that analog and digital frequencies no longer share the linear relationship. The bilinear transform has compressed the entire range of positive analog frequencies into the digital frequency interval $[0, \pi]$, resulting in what is called frequency compression or warping. One effect of frequency warping is that digital filters have a sharper rolloff than analog filters. In the case of the high-pass filter also, the analog frequencies $[0, \infty]$ are mapped onto the digital frequencies $[0, \pi]$. The bandpass and bandstop filters map the entire analog frequency range $[-\infty, \infty]$ on the digital frequencies from zero to $\pi$.

There are four commonly used families of analog filters. Butterworth filters have a flat amplitude response in their passband, and their poles lie uniformly spaced on the unit circle in the $s$-plane. They have medium characteristics with respect to both frequency selectivity and phase. Chebyshev filters have an equiripple amplitude response in either the passband or the stopband, and their poles are located on an ellipse in the $s$-plane. They are more frequency selective but less phase sensitive than Butterworth filters. Elliptic or Cauer filters have an equiripple amplitude response in both the passband and stopband. They are the most frequency selective but the lowest in phase sensitivity. Bessel filters have very poor frequency selectivity but are linear phase filters. The order of a Bessel filter would have to be much higher than those of the other filters for a particular frequency specification.

In summary, if the application in question requires high selectivity and is phase insensitive, then Chebyshev and elliptic filters can be used, since they would require lower-order filters (in general, the lower the order, the less complex is the implementation). If the application does not require high selectivity but is phase sensitive, then Bessel filters can be used. These two groups describe the two extreme cases. Very often, phase-sensitive applications require high selectivity, which results in a compromising situation; Butterworth filters could be used for this purpose. Generally the order and family of the analog filter for a particular application has to be determined, its analog transfer function calculated using a filter handbook, and finally the appropriate analog-to-digital transformation can be used to obtain an equivalent digital filter.

## The Design of FIR Filters with Arbitrary Responses

The digital filters obtained in the manner described in the previous section mimic an equivalent analog filter. Their performance is by no means superior

Chapter 10: Digital Signal Processing: Theory

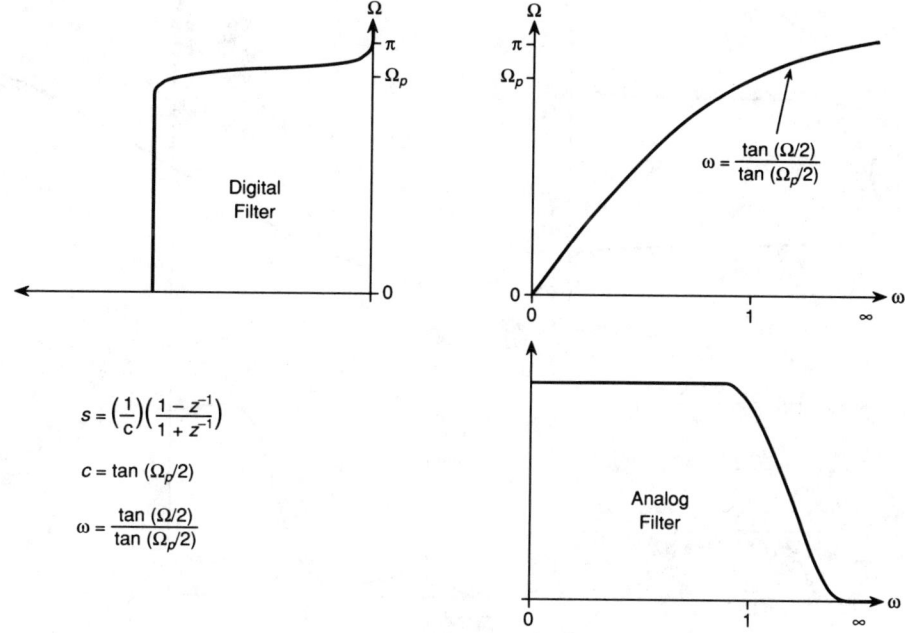

(a) A digital low-pass filter obtained from an analog low-pass filter.

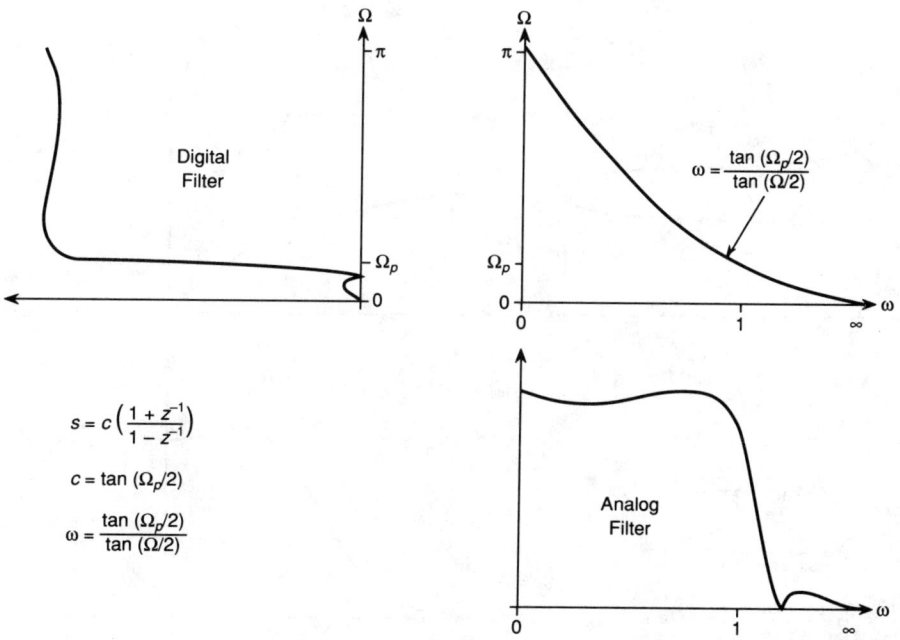

(b) A digital high-pass filter obtained from an analog low-pass filter.

**Figure 28.** The mathematical equations involved in the bilinear transform and a graphical representation of the construction of the four types of filters.

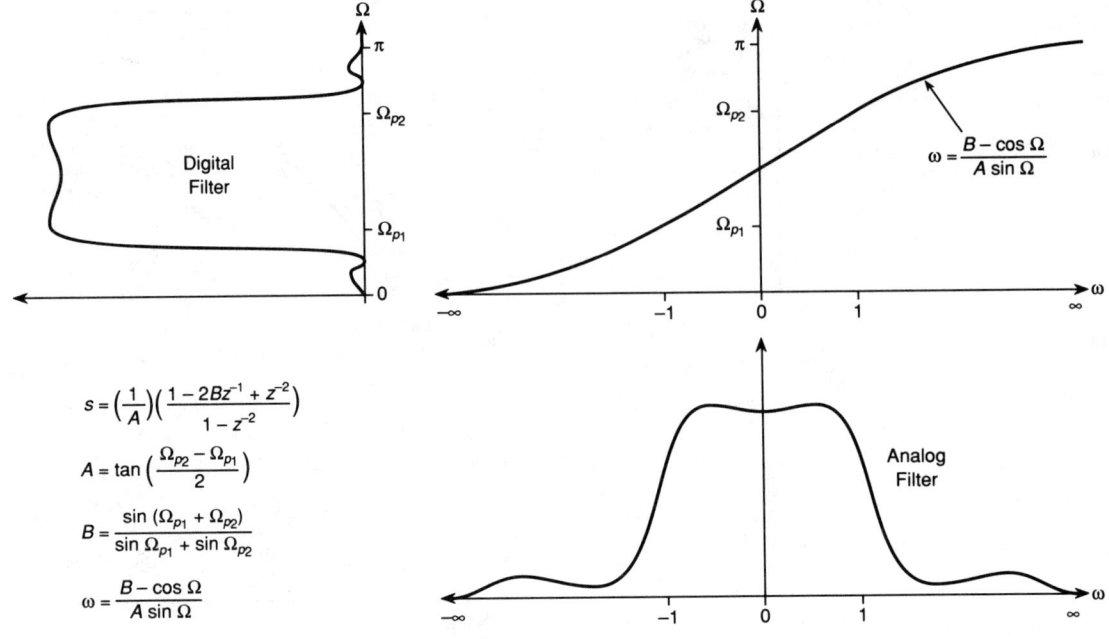

(c) A digital bandpass filter obtained from an analog low-pass filter.

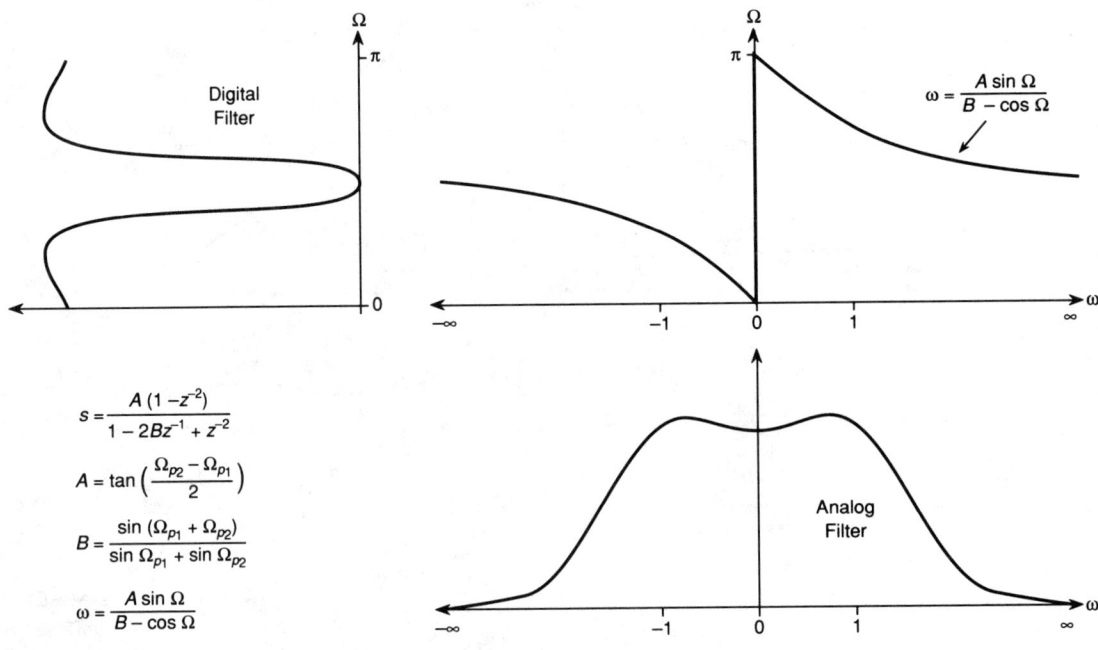

(d) A digital bandstop filter obtained from an analog low-pass filter.

**Figure 28.** (cont.)

to analog filters. Analog filters are generally rational functions in $s$, and applying general analog-to-digital transformations would in general lead to rational functions in $z^{-1}$. In other words, these digital filters are IIR filters, and they can make use of the well-established theory of analog filters. That is, things that can be done in the analog domain can also be done in the digital domain. However, aping analog filters underutilizes the capabilities of DSP. An application that requires a frequency response that cannot be met by standard analog filters would demand the power of DSP. For these more general cases, numerous methods have been developed that require computers because the procedures are algorithmic, computation-intensive and require a few iterations to converge to a solution. Computer-aided designs can be used for the realization of both IIR and FIR filters. The following section will concentrate only on the FIR filter because it has linear phase.

## Windowing

Often, the desired frequency response, $H_d(e^{j\Omega})$, is piecewise constant with sharp transitions at the boundaries. This gives rise to impulse responses that are noncausal and infinite in duration—an IIR filter. The easiest way to convert this into an FIR filter is to truncate the IIR impulse response to $N$ samples. This can be made causal by applying a delay to the impulse response. Direct truncation leads to the well-known Gibbs phenomenon—there is about a 9% overshoot and ripple before and after a discontinuity. Increasing $N$ does not decrease the amplitude of the overshoot; instead, the frequency response shows rippling over a smaller frequency range. Rather than have a window that cuts off the impulse response suddenly, it is more desirable to have a window that gradually tapers off the impulse response at the extremes. Thus windowing allows the conversion of an infinite impulse response to an approximate finite impulse response.

An FIR filter is uniquely specified either by $N$ samples of its frequency response or the $N$ coefficients of its impulse response. The desired frequency response is denoted by $H_d(e^{j\Omega})$, the corresponding impulse response is $h_d(n)$, and the window function by $w(n)$. In general, $h(n)$, the windowed finite impulse response, can be written as the product of the desired impulse response $h_d(n)$ and a finite duration window function $w(n)$. This can be expressed as

$$h(n) = h_d(n)\ w(n).$$

The time-domain multiplication results in convolution in the frequency domain. For the case of direct truncation, the window is called the rectangular window. Figures 29a through 29d show the rectangular window $w(n)$, its spectrum $W(e^{j\Omega})$, a desired frequency response $H_d(e^{j\Omega})$, and the frequency response $H(e^{j\Omega})$ obtained as a result of convolving $H_d(e^{j\Omega})$ and $W(e^{j\Omega})$, which shows characteristics that are typical of the windowing method. As a result of convolution, the sharp discontinuities of $H_d(e^{j\Omega})$ become transition bands in $H(e^{j\Omega})$. The width of the transition region depends on the width of the main lobe of $W(e^{j\Omega})$. The smaller the main lobe, the sharper the transition between the passband and the stopband. The ripples in the filter's response can be decreased if the side lobe levels of the window are low. The ideal window requires a small main lobe which contains most of the energy and very small side lobe levels that rapidly decrease in energy. These are conflicting requirements, and the windows that are used make a trade-off between the transition width and the ripple. Some of the popular windows that are used are described in Table 1, along with the width of main lobes and peak amplitude of the side lobes. Their responses are shown in Figure 30. The table clearly shows that the ripple in the stopband is attenuated at the cost of an increased transition width. Usually, large attenuation is desired in the stopband, but in this case the price that is paid is a larger transition width. A large transition width implies that it is difficult to know the position of the passband and stopband in advance. This is a direct result of convolving the desired frequency response with that of the window and cannot be avoided.

## Computer-aided Design of FIR filters

The windowing method works satisfactorily, but it does not optimize any parameter. In effect, for a given value of $N$, the windowing method does not pro-

**Table 1.** Some Popular Windows

| Window | Equation* | Transition Width | Side Lobe (dB) |
|---|---|---|---|
| Rectangular | $w(n) = 1$ | $\frac{4\pi}{N}$ | $-13$ |
| Triangular (Bartlett) | $w(n) = \frac{2n}{H-1}$ for $0 \leq n \leq \frac{N-1}{2}$ <br> $w(n) = 2 - \frac{2n}{N-1}$ for $\frac{N-1}{2} < n < N$ | $\frac{8\pi}{N}$ | $-25$ |
| Hanning | $w(n) = \frac{1}{2}\left[1 - \cos\left(\frac{2\pi n}{N-1}\right)\right]$ | $\frac{8\pi}{N}$ | $-31$ |
| Hamming | $w(n) = 0.54 - 0.46 \cos\left(\frac{2\pi n}{N-1}\right)$ | $\frac{8\pi}{N}$ | $-41$ |
| Kaiser | $w(n) = \frac{I_0\left[\pi a \sqrt{1 - \left(\frac{2n}{N-1} - 1\right)^2}\right]}{I_0(\pi a)}$ <br> $a = 2.0$ | $\frac{12\pi}{N}$ | $-46$ |
| Blackman | $w(n) = 0.42 - 0.5 \cos\left(\frac{2\pi n}{N-1}\right) + 0.08 \cos\left(\frac{4\pi n}{N-1}\right)$ | $\frac{12\pi}{N}$ | $-58$ |

*All windows $w(n)$ are defined only for $0 \leq n < N$.

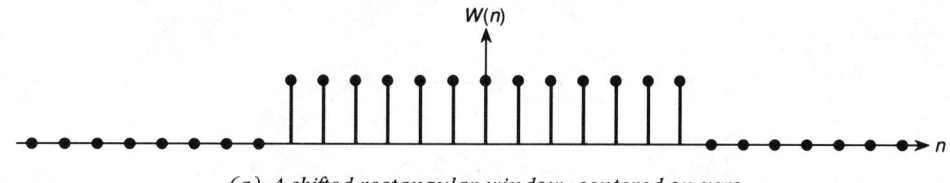
*(a) A shifted rectangular window, centered on zero.*

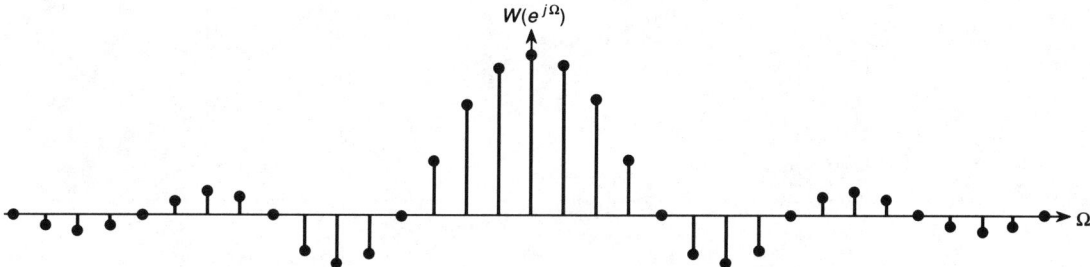
*(b) The spectrum of the window is described by a (sin x / x) function.*

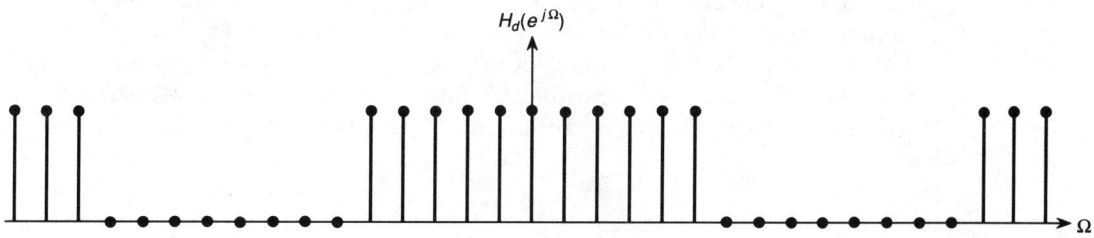
*(c) The desired frequency response is that of a low-pass filter with sharp transitions.*

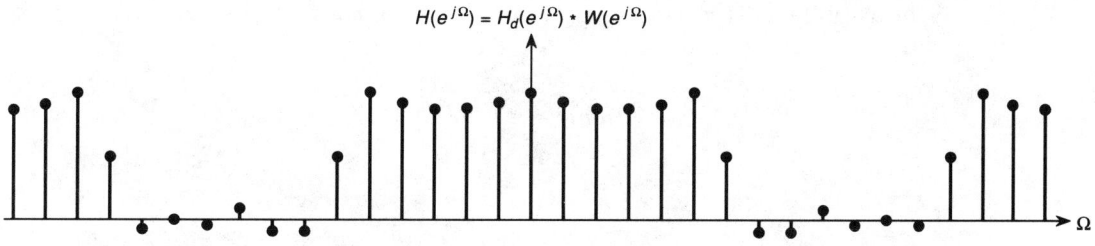
*(d) The windowed frequency response is the convolution of the spectrum of the window and the desired frequency response. This introduces the transition regions and ripples in the filter's response.*

**Figure 29.** The windowing technique to approximate a brickwall low-pass filter.

vide an optimal solution. Alternatively, the design procedure may be geared to provide the desired response keeping in mind some criterion, for example, the minimization of the maximum error in some frequency band. Thus a better

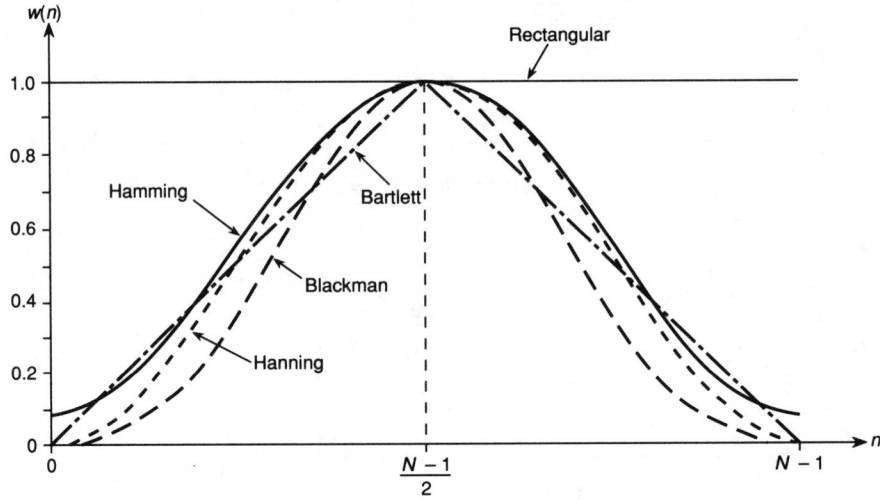

**Figure 30.** The responses of some commonly used windows.

filter may be designed, but the procedure is algorithmic and iterative in nature and may require the solution to a set of linear inequalities. All this requires the computation power of a computer for its solution. Different designers often use different criteria. Thus the algorithms used could be different and often the methods used are proprietary programs. Two popular techniques are outlined below.

## Frequency Sampling

The $N$-point DFT relations between the impulse response and the transfer function are

$$H(m) = \sum_{n=0}^{N-1} h(n)\ W_N^{nm},$$

$$h(n) = \sum_{m=0}^{N-1} H(m)\ W_N^{-nm}.$$

The second equation shows that if the frequency response is specified at $N$ equally spaced points around the unit circle, the points can be used to determine the $N$-point impulse response. Direct application of this method results in an unsatisfactory frequency response, as shown in Figure 31a. The response obtained is exact at the $N$ specified points, but there is a finite error between them where the response is obtained by means of interpolation. In general, the smoother the desired frequency response, the smaller the error between the sample points. One method that could be used to reduce the error is to allow a value between the passband and stopband to take a value between zero and one, as shown in Figure 31b. This introduces a transition region, but the value of this sample can be optimized, such that the maximum gain in the stopband is reduced and the ripple in the passband is eliminated. The intermediate value is found in an algorithmic fashion using a computer. Further improvement can be obtained by making two of the points variables, as shown in Figure 31c. This results in a transition width that is twice as wide, but the gain in the stopband is reduced even further. If $N$ is doubled, the transition width would remain the same, but the amount of computation would increase. This method gives excellent results in most cases. However, there is a limitation on the frequencies at which the response can be specified; in particular, they must be integral multiples of $2\pi/N$. If the passband and stopband frequencies are not one of the integral multiple frequencies, then control is lost over the amplitudes of the frequency response at these important frequencies. Once again, if $N$ is increased such that the samples can come close to these passband or stopband frequencies, then the amplitudes of these frequencies can be specified. This is not an efficient method because of the increase in computational load.

## Equiripple FIR Filters

In the frequency sampling method, the error is concentrated in the transition region, and the error is lower in regions away from the transition region. It is found that if the error is spread out over the entire frequency range, then the required order of the filter is lowered. As a result of spreading the error into the passband and the stopband, the filter's frequency response displays ripples. For the case of the low-pass filter shown in Figure 32 the ripples in the passband and the stopband are $\delta_1$ and $\delta_2$, respectively. These filters are designed iteratively using a computer. There are five parameters under consideration; these are $M$ (which is related to the number of extrema there are in the response of the desired filter and to the length of the impulse response), $\delta_1$ (the passband ripple), $\delta_2$ (the stopband ripple), $\Omega_p$ (the passband frequency) and $\Omega_s$ (the stopband frequency). It is not possible to specify each of these parameters independently of the others. There are two broad categories of algorithms that allow certain parameters to be optimized while keeping others fixed. In one case $M$, $\delta_1$, and $\delta_2$ are fixed, while $\Omega_p$ and $\Omega_s$ are allowed to vary. In the other case $M$, $\Omega_p$, and $\Omega_s$ are fixed while $\delta_1$ and $\delta_2$ are optimized.

# Conclusion

In conclusion, it may be useful to quickly itemize some of the key points in the study of DSP. Analog signals are converted to digital signals $x(n)$. The signals may be further processed using linear, time-invariant, causal systems that are characterised by an impulse response, $h(n)$. The processed signal, $y(n)$, is the convolution of $x(n)$ and $h(n)$.

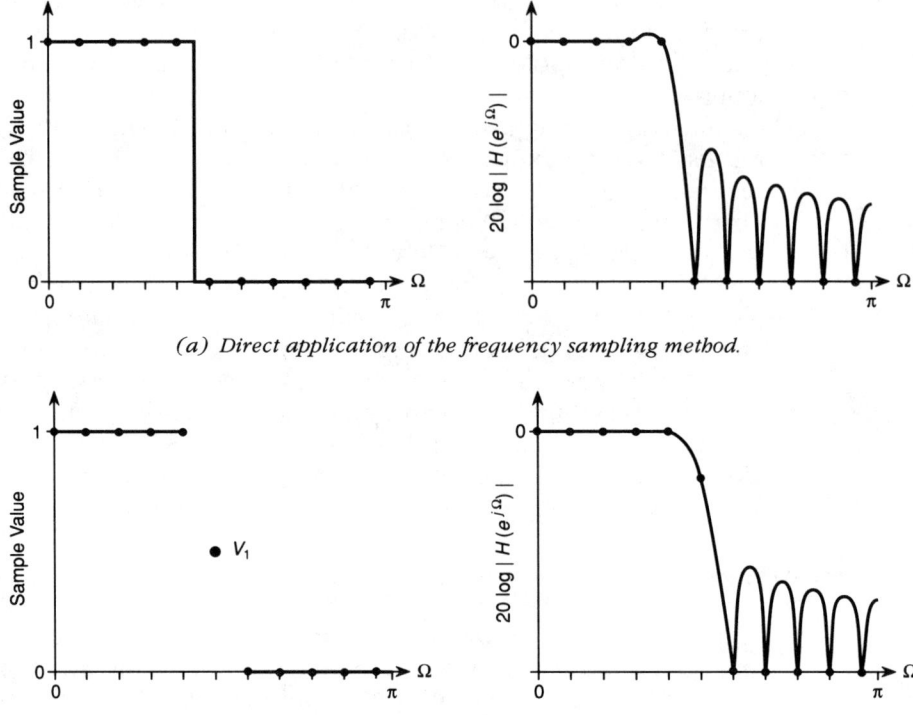

*(a) Direct application of the frequency sampling method.*

*(b) Allowing the value of one point in the transition band to be varied results in a larger attenuation in the stopband and a flat passband.*

*(c) Allowing two points to be optimized results in a still larger attenuation in the passband at the expense of a wider transition region.*

**Figure 31.** Results of the frequency sampling method.

Transforms allow a time-domain signal to be transported to the frequency domain and vice versa, without any loss of information. The DFT determines the spectra of a digital signal in terms of a harmonically related set of discrete frequencies. The FFT uses algorithms to compute the DFT orders of magnitude faster, often enabling the processes to be run in real time. Convolution in

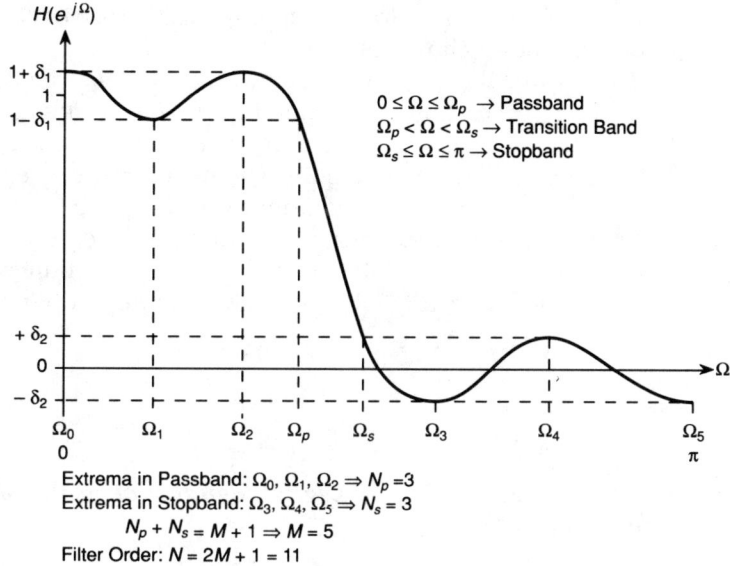

**Figure 32.** An equiripple FIR filter showing extrema in the frequency response at $\Omega_0$, $\Omega_1$, $\Omega_2$, etc., and the relationships among the various parameters.

the time domain is equivalent to multiplication in the frequency domain, and vice versa.

The $z$-transform converts a discrete time signal into a generalized digital frequency function. It includes the DFT as a special case. The $z$-transform is a powerful tool in DSP and is the building block of digital filters. In general, the digital filter transfer function, $H(z)$, has poles and zeros. The location of the poles and zeros can be used in conjunction with the unit circle to find the frequency response. The poles must be located inside the unit circle for a stable filter. The transfer function determines whether the filter is an IIR or FIR filter.

FIR filters can have linear phase and are preferred in many audio applications. Filters can be realized in hardware using numerous configurations. They must be analyzed to find the location of the error producing sources, to study the flow pattern of the error within the hardware and then estimate its overall effect at the output.

There are quite a few techniques to design digital FIR filters. In the windowing technique, an infinite impulse response is multiplied by a finite length window to create a finite impulse response. This straightforward method yields satisfactory results, but has a wide transition region, and the passband and stopband frequencies cannot be specified.

Better results are obtained by using frequency sampling. Here the values of the points in the transition region are optimized under the constraint that the maximum error is minimized. Lower- order filters can be obtained by using equiripple FIR filters. In this case, some of the parameters of the filters are fixed, while others are varied or optimized. The use of computers is imperative in these design techniques.

Digital signal processing is a powerful tool that can do virtually everything that analog processing can do, and more. One advantage of DSP is flexibility. By accessing new coefficients (at the press of a button), a person could create a sharp notch filter to remove hum. The next minute, the filter's second button could provide the loudness contour to compensate for a low listening level. A third button could emulate the acoustics of Carnegie Hall in the user's living room—it's all done by DSP. The possibilities are endless; moreover, processing takes place in the digital domain, which offers superior sonic performance.

## References

Bateman, A., and W. Yates, *Digital Signal Processing Design*, Computer Science Press, 1989.

Dattorro, J., "The Implementation of Recursive Digital Filters for High Fidelity Audio," vol. 36, no. 11, pp. 851–878, November, 1988.

Lindquist, C.S., *Adaptive and Digital Signal Processing*, Volume 2, Steward & Sons, 1989.

Oppenheim, A.V., and R.W. Schafer, *Digital Signal Processing*, Prentice Hall, 1975.

Peled, A., and B. Liu, *Digital Signal Processing: Theory, Design and Applications*, Robert E. Krieger Publishing Co., 1985.

Pohlmann, K.C., *Principles of Digital Audio*, 2nd edition, Howard Sams & Co., 1989.

Rabiner, L.R., and B. Gold, *Theory and Application of Digital Signal Processing*, Prentice Hall, Inc., 1975.

Robinson, E.A., and M.T. Silvia, *Digital Signal Processing and Time Series Analysis*, Holden Day, Inc., 1978.

Strawn, J., editor, *Digital Audio Signal Processing*, William Kaufmann, Inc., 1985.

Watkinson, J., *The Art of Digital Audio*, Focal Press, 1988.

*Chapter 11*

# Digital Signal Processing: Applications

Jayant Datta

## Introduction

In the 1980s digital audio established itself as a reliable and mature technology with very diverse applications, even in secondary technologies. This is supported by the fact that areas in which audio is not of primary importance (e.g. motion pictures, television, etc.) are shifting to digital audio. Not only is digital audio superior to analog audio with regard to dynamic range, channel separation, imaging, and overall fidelity, but it also lends itself to the power of digital signal processing (DSP). The DSP chip has become the workhorse of the audio industry as designers realize its full potential. It combines the versatility of a complete effects package with the sheer processing power of a digital computer.

All processing in a DSP chip takes place in the digital domain. As a result, the signal does not suffer from the adverse effects of analog processing. In analog technology, each function requires a different, dedicated processor—a parametric equalizer section is different from a compressor or a reverberation unit. A single DSP chip is capable of performing all these functions and many more. The processing possibilities that appear as a result of its flexibility are considerable. Digital signal processing found early applications in soundfield processing. This technology has recently been modified for the car as well. Consumers are beginning to find more and more sophisticated DSP products appearing in the market. Digital signal processing is used in state of the art digital mixing consoles, where it helps to create new recording techniques. It makes the digital audio workstation environment more powerful in its signal processing and editing capabilities. Even loudspeaker crossover network characteristics can be improved by DSP. At the same time, DSP may also be used to improve the sonics of a noisy recording from the early 1900s.

In technology there are no free lunches, and DSP is no exception. DSP chips are costly, though the prices continue to fall. A computer engineer must program the chip, and, subsequently, the end user has to bear the high cost of programming. Digital signal processing is inherently a very computation-intensive technology; therefore, some of the more complicated and specialized processes cannot be executed in real time. In favor of DSP, it must be mentioned that these specialized operations sometimes cannot be carried out in the analog domain, and, even if they could, the results would be vastly inferior. When all things are considered and opinions taken, there is no doubt that DSP will play a dominant role in future audio products.

# Digital Soundfield Processing

When a person hears live music, as opposed to music reproduced by means of loudspeakers in the listening room, he or she generally hears much more than just the direct sound from the musical source. Each real-world acoustical venue, from a small room to a large cathedral, has its unique acoustical characteristic that it imparts to the music that is played in it. The sound, which is in effect processed by the room, depends on a number of things, such as the shape of the room, the volume of the room, the absorptivity of the materials in the room, and the position of the listener in the room. All these factors play a vital role in the final perception of the musical experience by the listener. The artificial creation of these effects by digital means is called digital soundfield processing.

## Soundfields

A soundfield is composed of direct sound and delayed reflections and echoes of the direct sound. The auditory impression of the soundfield is primarily composed of the perception of four elements—loudness, reverberation, extension, and equalization. Loudness refers not only to the sound pressure level (SPL) of the direct sound but also to the SPL contributed by the reflected sound. The perception of reverberation depends on two factors. One is the reverberation time (the time it takes for the SPL to decrease by 60 dB). For example, the reverberation time of a typical consumer residential listening room is about 0.2 s, while that of a large auditorium could be over 2.5 s. The second factor is related to the average absorption ratio of the room (the ratio of the absorbed acoustic power to the incident acoustic power). The lower the ratio (e.g. 0.2), the more live the room is said to be, and the higher the ratio (e.g. 0.4), the more dead it is. The perception of extension is related to the time base of sound reflection. Extension gives the auditory illusion of being in a larger space. The materials that make up any venue also equalize the reflected sound waves. High-frequency signals are absorbed by room furnishings and boundaries, while the low-frequency energy is lost by transmission through the walls of the venue.

Consider the nature of the soundfield in a typical concert hall, as shown in Figure 1. First the direct sound reaches the listener, shortly after it is emitted. The listener is able to locate the sound source using this sound. This is followed by reflections from the ceiling or walls or from behind the stage. Primary reflections (one bounce) usually precede other reflections (two bounces or more). Horizontal reflections that arrive from the walls are called lateral reflections, whereas the vertical reflections from the floor and the ceiling are called nonlateral reflections.

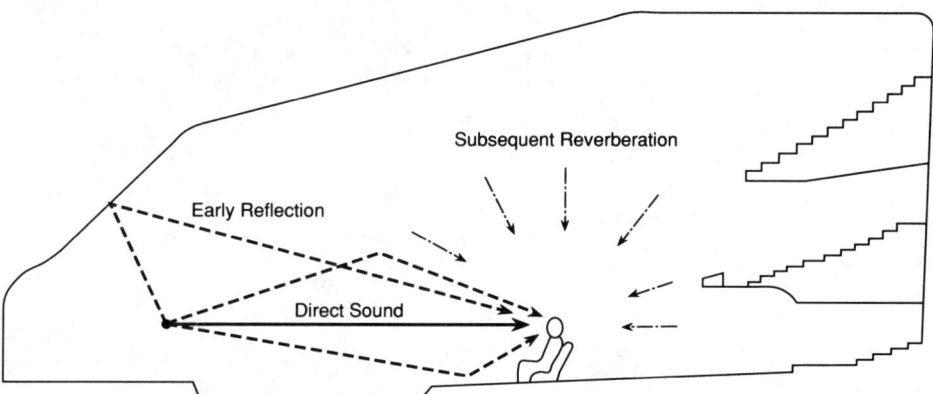

**Figure 1.** The soundfield in a concert hall. *(Courtesy Fujitsu Ten)*

In general, reflections that arrive within 50 to 80 ms of the direct sound are called early reflections, while subsequent reflections are called late reverberation. The early reflections are usually primary and secondary (two bounces) reflections. They provide a sense of depth, enhance the direct sound and strengthen the perception of loudness and clarity. The delay time between the arrival of the direct sound and the beginning of early reflections is called the initial delay gap (IDG). The loudness of lateral reflections and a large IDG contribute to the perception of extension. Early reflections are followed by subsequent reverberation resulting from complex and repetitive reflection and attenuation of the original sound as it reflects from walls, ceilings, floor and other objects. This sound reaches the listener from all directions and provides a sense of depth or volume, and therefore extension. Figure 2 shows the reverberant soundfield created in the time domain. The amplitude variation is seen to be similar to an exponential decay. The subsequent reverberation region is dense, which means that there are a larger number of reflections present per unit time.

Figures 3a through 3d show the direction of arrival of sound with the passage of time. Each diagram is the equivalent of a snapshot of the intensity and direction of the sound arriving at a particular instant of time. The radial lines show the direction of the arriving sound, and their length is a measure of the power or loudness of the sound. As can be seen, during the first 50 ms, sound mainly comes from the front—these are the direct sound and the early reflections. Subsequent reverberation sounds reach the listener from all directions.

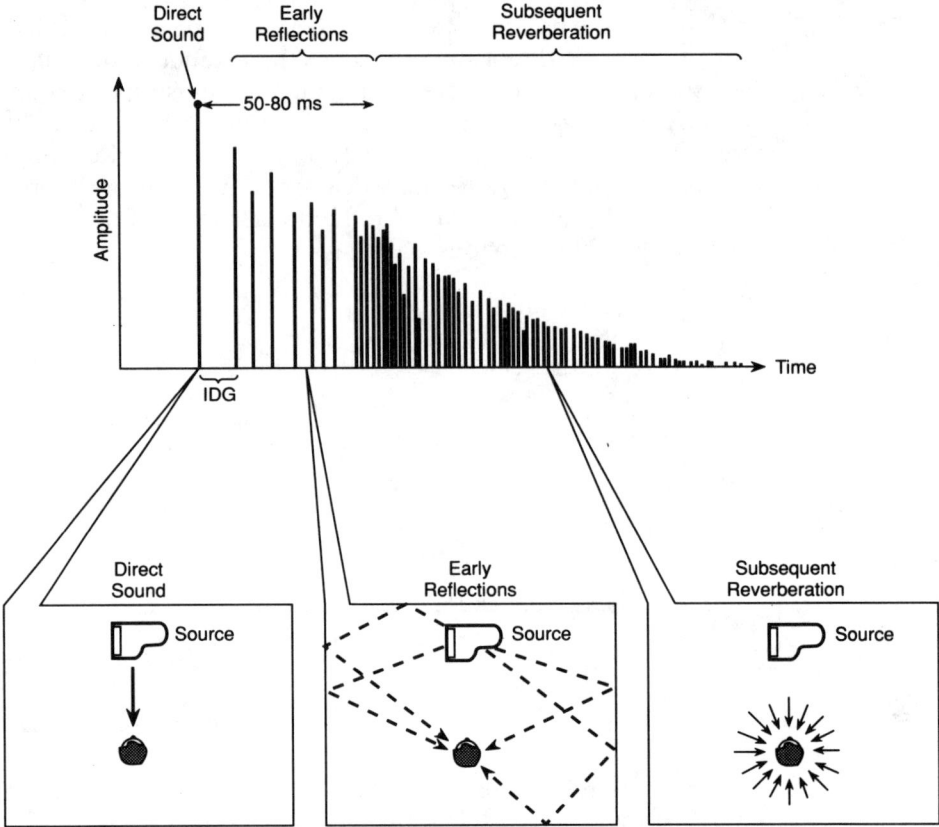

**Figure 2.** Time base of a reverberant soundfield.

## Soundfield Processing in the Home

Figure 4a shows the impulse response of a typical concert hall, while Figure 4b shows the impulse response of a typical consumer's listening room. The greatest differences that should be expected are a shorter IDG in the listening room (because the reflecting surfaces are so close to the listener) and a large reverberation time for the concert hall.

In the case of stereo reproduction in a typical home, there are only two sources of sound (loudspeakers placed in front of the listener), and the impulse response of the room has neither the large IDG nor a reverberation characteristic like that of a concert hall. Subsequently, the listener is not fooled into believing that he or she is in a concert hall. The feeling of a live musical event could be created if the impulse response of the room could be modified such that it had a larger IDG and a longer and denser subsequent reverberation. In addition, the listener needs to be immersed in a soundfield surrounded by a plethora of sources which simulate the early and higher-order reflections both

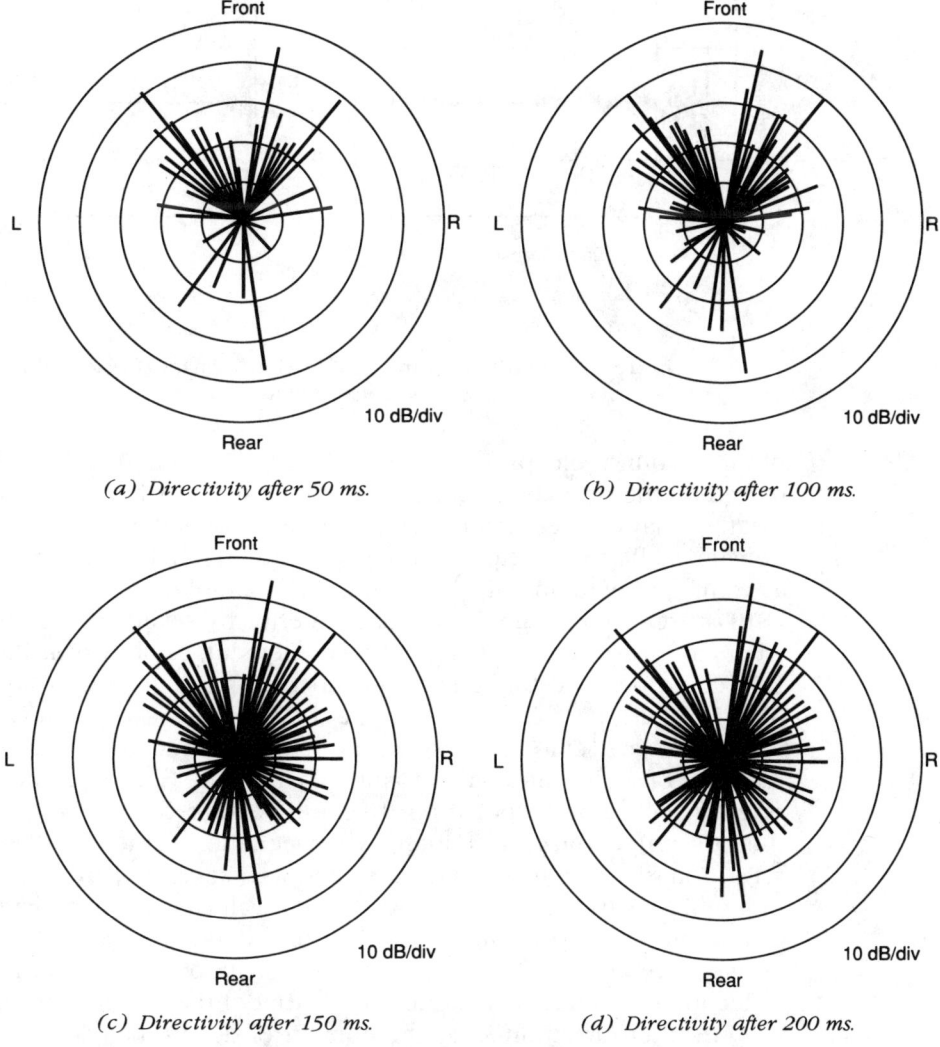

**Figure 3.** Directivity diagrams of reflected sound. *(Courtesy Fujitsu Ten)*

in their direction of arrival and their attenuation in amplitude and high-frequency content.

In reality, there are limitations to the hypothetical setup described above. Some of the reverberation times can be very long and complicated, so these have to be approximated by shorter and less complicated ones, and practical considerations generally limit the number of loudspeakers to about six, including the original stereo pair. However, the acoustics of a larger soundfield can be reproduced in a smaller soundfield by electronically changing the initial delay times and the nature of the reverberation. The reverberation characteristics of a larger space can be imitated by delaying and attenuating the input signal by appropriate amounts such that the result is an electrical analog to the

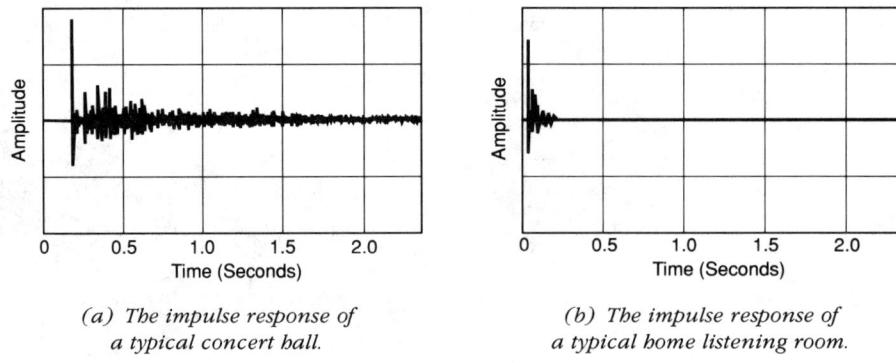

(a) The impulse response of a typical concert hall.

(b) The impulse response of a typical home listening room.

**Figure 4.** The differing impulse response characteristics of rooms. *(Courtesy Fujitsu Ten)*

physical soundfield present in the simulated room. The basic element required for the simulation is the delay element. It has been seen in Chapter 10 that the delay element is an integral part of the digital filter. Once in the digital domain, audio signals can be processed without much deterioration in quality, as compared with analog processing. For example, the bucket brigade delay (BBD) used in analog processing is inferior by several orders of magnitude when compared with the digital delay element, as shown in Figure 5. Thus DSP could be used for soundfield processing. Referring to the general FIR filter in Figure 19 of Chapter 10, if the coefficient $a_0$ is set to unity and all other coefficients are set to zero, except say $a_{10}$, which is set to 0.9, then the impulse response of such a filter will be as shown in Figure 6. The output at $n = 0$ is the original signal, the output at $n = 10$ could be looked upon as the beginning of the early reflection, and the interval between them is the IDG. The entire early reflection region consists of discrete and well-separated arrivals of sound. Setting other coefficients after $a_{10}$ to nonzero values could simulate this region. This is often done by using a long electronic delay line that is tapped at various points, as required. Therefore, an FIR filter can be used to generate the early reflection characteristic and feed the output of the reverberator directly.

FIR filters are limited by the number of delays that can be summed in a practical processor, thus it is not possible to use an FIR filter with a very large number of coefficients to generate every sample required in the reverberation. The reverberation region could be obtained by using feedback or recirculation as shown in Figure 7a (refer also to Figures 18a and 18b in Chapter 10). This IIR filter has an exponentially decaying impulse response corresponding to the envelope of the subsequent reverberation, and this simple circuit has the potential of generating many reflections. The frequency response of this filter is shown in Figure 7b; it is the familiar comb filter. The sharpness of the teeth depends on the attenuation $g$. For values of $g$ near unity, the reverberation time becomes very long, and the frequency response becomes so sharp that only frequencies that fall on the peaks reverberate, while the others die out quickly. The nonuniform frequency response obviously rules out using this circuit. An alternative circuit is shown in Figure 8a, and its impulse response is shown in

Chapter 11: Digital Signal Processing: Applications

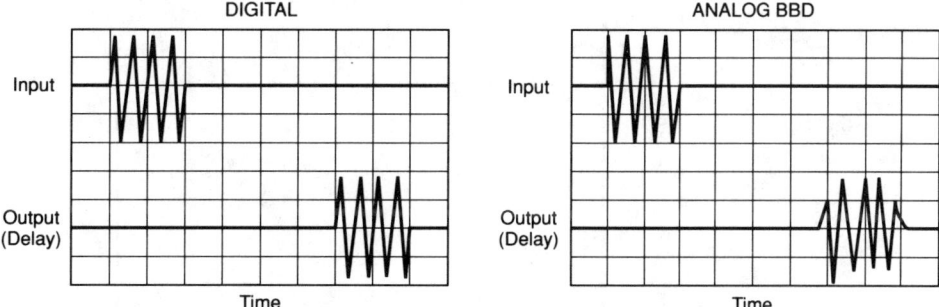

*(a) The input and output transient responses are identical in a digital delay, but the output is misaligned in an analog delay.*

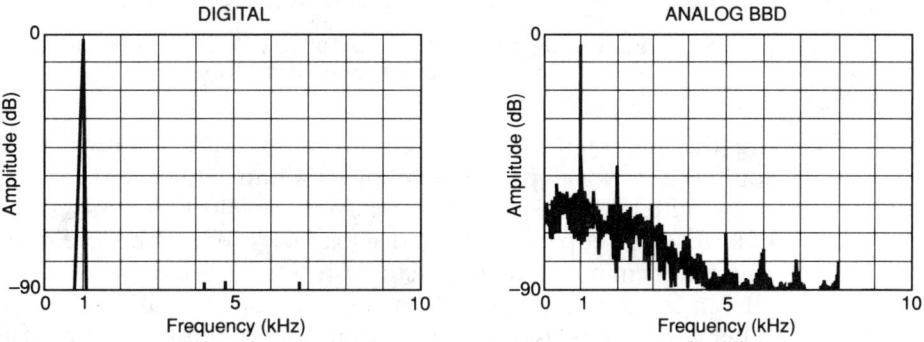

*(b) A spectral analysis of a 1 kHz signal reveals a larger amount of noise in the analog delay.*

**Figure 5.** The performance characteristics of an analog bucket brigade delay versus a digital delay. *(Courtesy Sony Corp.)*

Figure 8b. It acts as an all-pass filter with a flat frequency response, while providing a reverberating circuit. However, in practice it is found that an all-pass reverberation circuit sounds unnatural because in nature reverberation is more frequency dependent. The solution is to use a number of frequency dependent comb filters in parallel and to pass the output of these comb filters to the all-pass reverberation circuit.

The final configuration is shown in Figure 9. The output consists of the direct sound, to which are added the early reflections generated by the FIR section, and to this is added the frequency dependent subsequent reverberation generated by the comb filters ($C_1$, $C_2$, $C_3$) in parallel with the all-pass reverberation generator (AP) placed in series. This general configuration can be used to simulate the reverberation of various kinds of acoustic environments by changing the impulse response of the filter system. In addition, the FIR filter taps can be varied to change the IDG, thus changing the apparent room size. The choice of gain and delay in the circuit has a strong bearing on the frequency dependency and the reverberation time. These change the listener's perception of the volume and other characteristics of the venue. Thus, chang-

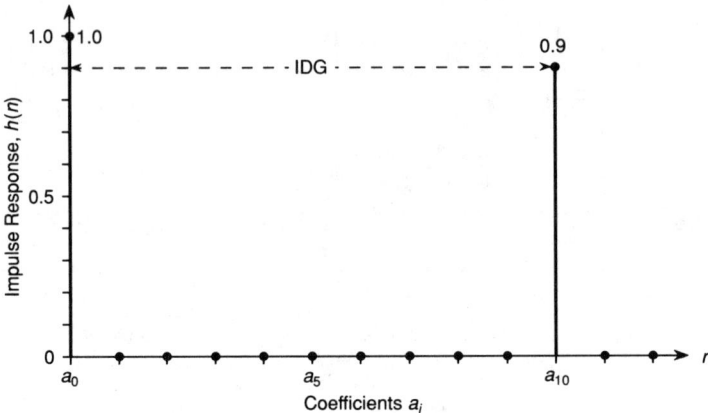

**Figure 6.** Impulse response of an FIR filter showing the IDG and the first reflection.

ing the parameters of the circuit allows the creation of entirely new soundfields. Not all parameter choices sound natural, and as a result the design of soundfield processors still remains an art. Most processors come with a host of factory preset parameters and also allow the user to store a large number of parameters in memory. The factory presets are good starting points and generally create the ambience of an average concert hall, cathedral, stadium, or other venue. The user has the option of fine tuning the dimensions, liveness, equalization, and other parameters of the particular environment according to his or her taste. If the channels are processed separately, then the user could select a position in the chosen venue that is nearer one wall than another. The user does not have to know DSP to specify parameters. He or she could specify the dimensions in meters, or the average absorption ratio as a number, and the processor is able to convert these values into corresponding coefficients in the circuit. The acoustical venues can be changed by pushing buttons on the front display of the processor or remote controller, from the comfort of the user's favorite armchair. Changing algorithms causes a change in the coefficients, which may destabilize the generated audio waveform. Thus many processors use an audio muting circuit during the changeover. The output fades out and fades in with the new coefficients. This takes place so quickly that the effect is inaudible.

It was noted at the outset that natural reverberation is very complicated. In preparation for the design for an ambience processor, the impulse response of a venue is often measured by using either four or six closely spaced microphones as shown in Figure 10. This allows spatial information to be captured and the particular venue to be characterized. The timing differences between the microphones are used to calculate the location of the source of the sound, whether direct or reflected. The average of many similar locations can be studied. Then these reverberation patterns are simplified. This is the first step where there is a loss of information. The choice has to be judicious. After that,

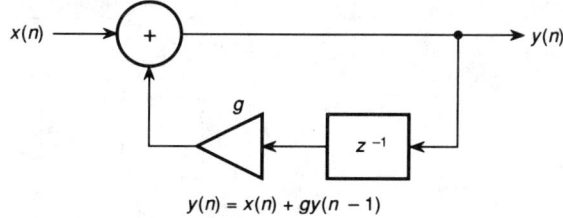

$$y(n) = x(n) + gy(n-1)$$

*(a) A simple IIR filter with feedback path with gain g.*

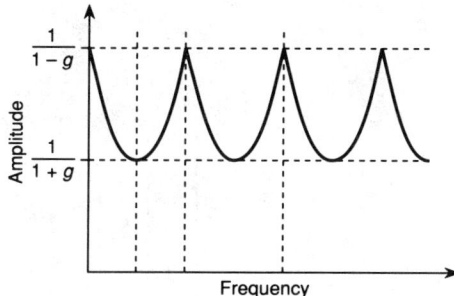

*(b) The frequency response is that of a comb filter.*

**Figure 7.** An IIR recirculation block with gain g.

DSP is used to simulate the necessary pattern. Even the result of this approximation is a great improvement over stereo. When the processor is shut off, the user feels that the sound stage has collapsed to the region between the main loudspeakers.

## Soundfield Processing in the Car

Home systems have benefited from DSP in soundfield processing since 1986, but the car environment has been enhanced by DSP ambient processing techniques only since 1990. There are some advantages in the car audio setup from the point of view of soundfield processing. Unlike the home system, cars have multiple speakers and are a natural candidate for soundfield processing. The listening positions in a car are also fixed. Among the chief drawbacks are the very small volume of the car, resulting in a very small reverberation time, and the wide variability in the materials that make up the interior of the car—from reflecting glass surfaces to absorbing, plush upholstery. Another problem peculiar to the car is that the listener is closer to one of the speakers and hears the sound from that first. This one-sided sound results in a loss of true stereo imaging unless a time alignment is applied.

As might be imagined, the position of the loudspeakers is of prime importance. A smaller loudspeaker unit offers more choices in placement but lacks sufficient bass response. The problem is resolved by using many smaller speak-

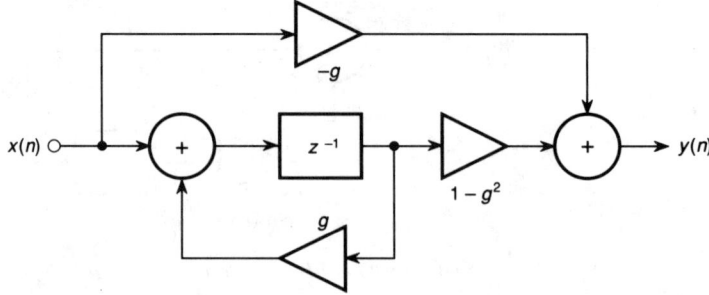

*(a) The hardware realization of the all-pass filter reverberation block.*

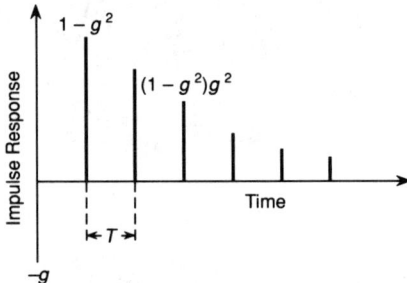

*(b) The impulse response of the all-pass filter reverberation has the dense impulse response required to simulate a real-world reverberation.*

**Figure 8.** Reverberation generation using an all-pass filter building block.

ers spread out in the interior and placed close to ear level and larger speakers not critically positioned serving as the low-frequency channel. When speakers are mounted on the front doors or in the underdash panel, the image often falls below the listener. To raise the level for a more realistic sound stage, small speakers can be mounted on the side mirror positions. Rear speakers can be positioned on the rear package tray so that sound is reflected by the glass and diffuses throughout the car to provide a filling sound. Speakers in the rear door expand the auditory room size sideways and backwards. There is often a center speaker in the front that serves two purposes. It can be used to balance the one-sided sound that the listener usually hears. At the same time, it expands the sound stage a little higher and forward.

The impulse response of a car interior is shown in Figure 11. It can be seen that, compared with a room (Figure 4b), the car interior IDG is smaller, the amplitude of the subsequent reverberation is larger, but the reverberation time is smaller. These effects can be attributed to the smaller listening space of the car. The basic approach to soundfield processing in the home and in the car are much the same. In both cases, the impulse response must be modified from the existing response to one that would be present at the desired venue. Of the large number of surround channels present in the car, in most cases only the primary pairs carry stereo information, and the others are used to create the

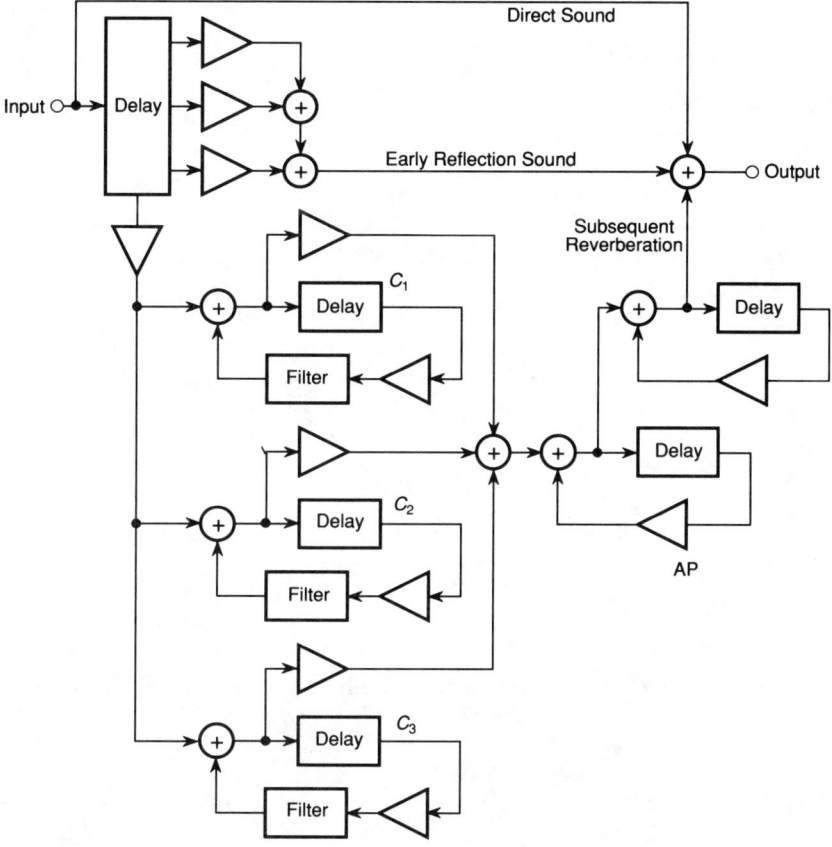

**Figure 9.** Schematic diagram of a reverberation processor.
*(Courtesy of Sony Corp.)*

impression of a larger space, as shown in the block diagram in Figure 12. The difference in the distribution of sound arrival patterns with and without soundfield processing is shown in Figures 13a and 13b. In the first case, asymmetry due to the proximity of the listener to one of the speakers is readily observed. This is no longer true in the second case, which results in an improved stereo image. The number of signals coming from the rear has also increased, providing a more natural setting, in accordance with the experience of live music.

In summary, the generation of the soundfield mainly concerns delaying the direct sound and modifying it. Digital filtering is particularly adept at generating delays, so DSP can be used to advantage in this application. Soundfield processing gives the listener the impression of being surrounded by music and a sense of being at the live performance. It typically requires two to four extra speakers that may be placed at the side, behind the listener, or behind the main speakers. The suggested placement varies from manufacturer to manufacturer, depending on the particular algorithm used to generate these extra

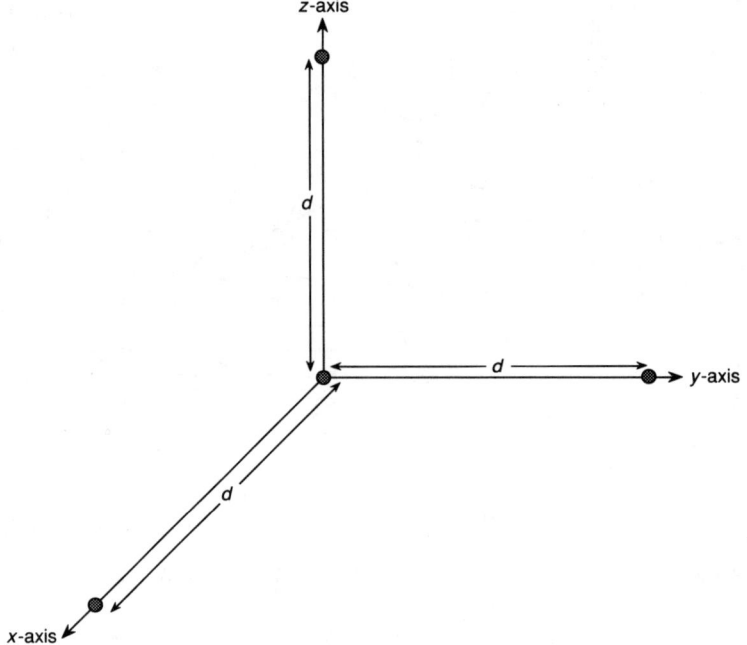

*(a) The four microphones technique: One is at the origin and the other three are at equal distances away from it along the* x-, y-, *and* z-*axes.*

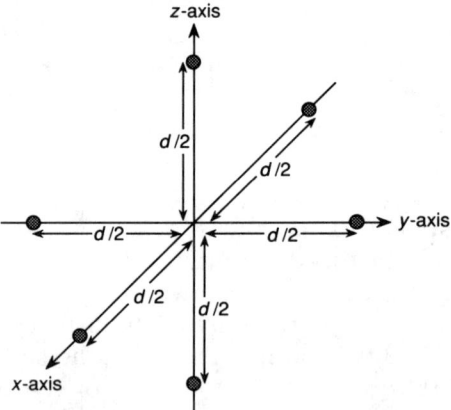

*(b) The six microphones technique: The configuration consists of three equidistant pairs that are placed along the* x-, y-, *and* z-*axes symmetrical to the origin.*

**Figure 10.** The orientation of the microphones used to compute the sound sources in an acoustic location.

signals from the original left and right channel information. Many processors use the sum or the L + R signal to generate the reverberation. As the processing power of the DSP chip improves, the left and right channels are being processed separately, and these are combined at a later stage with the appropriate

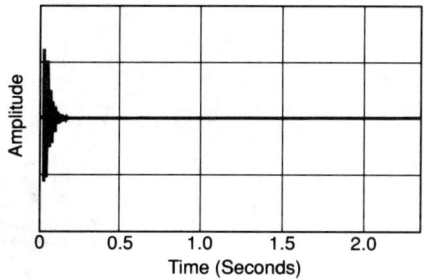

**Figure 11.** Impulse response of the interior of a car.

amount of delay and attenuation and fed to a particular surround channel for an even more realistic effect. Digital signal processing can also be used to create delay blocks in audio channels, to compensate for asymmetric speaker placement.

## Other DSP Applications in Consumer Audio

A digital sound source such as the compact disc (CD) or digital audio tape (DAT) is commonplace in the consumer market. Most of these sources are digital only with respect to the way in which they store data. Once the data is retrieved from the storage medium, the bulk of these "digital" components communicate with the outside world through their analog outputs. This is because the majority of consumer signal processing components, such as the pre-amplifier and equalizer, have traditionally been and still are largely analog. If the signal from the digital sources requires processing, it is subjected to various stages of analog processing. These analog stages add noise, introduce

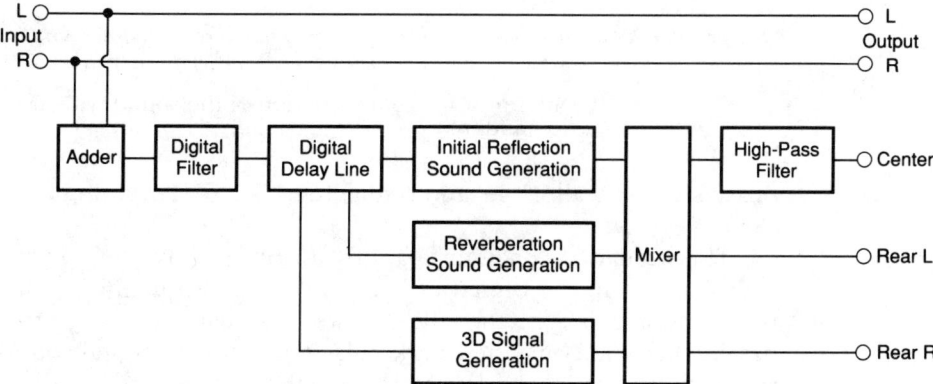

**Figure 12.** Block diagram of the soundfield processing employed in the car.

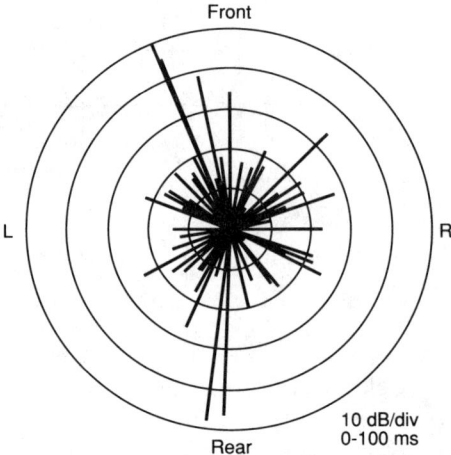

*(a) Without soundfield processing, sound comes mainly from the front and more from the left, due to the proximity of the driver to the left front speaker.*

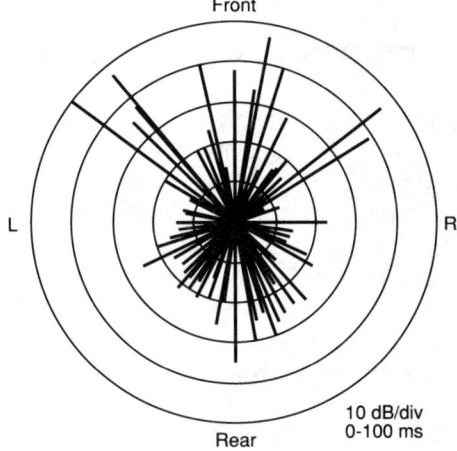

*(b) Sound after being processed has more energy in the rear channels and also compensates for the proximity of the listener to the left front speaker.*

**Figure 13.** Difference in the distribution of the sound with and without soundfield processing. *(Courtesy Fujitsu Ten)*

phase shift, and alter the stereo imaging, thus deteriorating the quality of the signal.

This scenario is changing rapidly. Until recently, the power of processing audio signals in the digital domain was confined mainly to the professional recording studio. With the falling cost of DSP hardware, this processing power has reached the consumer marketplace. Consumer digital soundfield processors have recently been joined by digital equalizers, receivers, preamplifiers, and even portable stereos that feature the flexibility and fidelity of DSP. An audio component with a DSP chip in it becomes a versatile tool that can be programmed to do various kinds of processing stemming from the flexible nature of the DSP circuit architecture. Providing an extra function is sometimes as simple as providing one more push button, and a few lines of code. Indeed, the number of push buttons and space available for a display on the front panel of a component could very well be the limiting factor to the number of functions available to the user.

Analog sources must be digitized prior to processing. Only after all the necessary modifications have been carried out is the digital signal passed to the outside world either through a digital output to another digital component or through a D/A converter to the analog world. During the intermediate processing stage, the processor offers total control over the signal. For example, graphic and parametric equalization may be provided. The center frequencies of the ten or twelve equalizer bands may be selected from a hundred choices in the audio range. The $Q$ of the filter can be varied from very wide to very narrow settings, and the boost/cut incremental resolution may be 0.1 dB, for extremely precise applications. Not only is this digital equalizer sonically superior, but the control offered is much greater than that provided by analog equalization. Dynamic range compression/expansion is another helpful feature that allows the user to make copies of wide dynamic range CDs onto tape for playback in the car. It is also possible to tailor the sound of a CD according to personal taste and make a copy of the same on DAT, without leaving the digital domain.

Digital signal processing has many applications in the car, both for audio processing and noise abatement. The interior of a car has very uneven frequency response characteristics and can benefit from equalization. Since the processed ambience signals are in the digital domain (analog inputs such as the tuner or the cassette player are digitized using an A/D converter), DSP could be used to provide equalization using digital filters, in addition to providing extremely flexible filtering characteristics.

Unlike the situation in the home, the ambient noise conditions in a car depend on many uncontrollable factors. The sources of noise are numerous. The engine and wind noise change with the speed of the car. The nature of the surface of the road and tires are factors affecting tire noise. The traffic density around the car and the kind of road (an open interstate highway as opposed to a tunnel) also plays a role in determining noise. The changing ambient noise level would require the user to continuously change the audio system gain setting to hear the softest passages yet avoid making the louder passages too loud. To avoid this problem, DSP could be used to provide dynamic compression as shown in Figure 14. In this example the original signal has soft passages that are buried in the background noise. When the signal is compressed, the softer passages become louder and can be heard above the background noise, while, at the same time, the louder passages are made softer. This results in the signal being comfortably audible at the expense of some dynamic range. The level of compression could be varied dynamically by taking into account the changing ambient noise level and the level of the softer passages being played. This allows the wide dynamic range of the digital sources to be fitted into the window created by the ambient noise level at the lower limit and the threshold of pain at the upper limit.

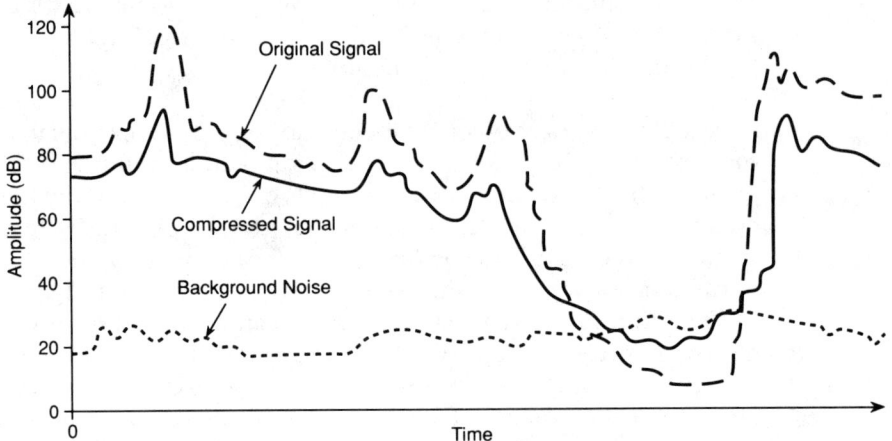

**Figure 14.** Compression of a highly dynamic audio signal to fit into a smaller dynamic window.

The DSP circuit could also be programmed to cancel the ambient noise in the car. It could read the speed of the car from a sensor, and microphones could monitor the relative road noise level. Using these inputs, it could approximate the noise level in the car and generate an inverse acoustical signal to cancel it. This would work well on the average; however, it is not possible to predict the exact nature of traffic and road conditions, which are dynamic and unpredictable quantities. Whereas a DSP muffler can easily cancel noise at the source (manifold), it is relatively more difficult to cancel noise in the passenger cabin itself.

These DSP applications bring consumers close to the dream of realizing an all-digital audio system (except for the loudspeaker, which has to interface with the analog world). An audio system with soundfield processing and all the features described above costs less today than a single soundfield processor did when it was first introduced. This is a clear indication of the decreasing costs of DSP products and the undeniable entry of DSP into the consumer market.

## Loudspeaker Crossover Networks

Loudspeaker drivers cover different frequency ranges—like the woofer (low frequencies), midrange (midfrequencies), and tweeter (high frequencies). Therefore, the audio signal that is output from the source must be split into frequency bands before being applied to the respective driver. The crossover network is primarily used to do this.

In the case of idealized drivers, the crossover network should aim to provide a flat frequency response when the combined acoustical outputs of all driver units are considered. The cutoff slope of the crossover filters should be steep enough to protect the drivers from out-of-band signals that could damage them. The polar response should be as uniform as possible. This means that ideally the combined output should not be a function of the angle measured from the axis of the loudspeaker. In practice, the combination of sounds from different drivers causes interference and results in a nonideal polar response. Theoretically, if the crossovers had an infinite cutoff rate such that any frequency was output only from one driver, there would be no interference—the ideal polar response could be attained. However, an infinite cutoff rate or a brickwall filter exists only in the pages of a textbook.

A crossover network must attempt to provide linear phase of the combined output. The subjective importance of linear phase is not yet established. However, it is safe to say that the ear is relatively insensitive to phase nonlinearities present in a monaural soundfield. But when dissimilar phase nonlinearities are present in a multichannel soundfield, the image localization cues may be severely damaged, resulting in shifts in perceived sound source position. Some trained listeners are particularly sensitive to "phasiness," but others are apparently not endowed with the power to hear phasiness. Still, linear phase is a desirable specification—an acceptable polar response would require the outputs from the different drivers to be in phase with each other. As noted in Chapter 10, FIR filters can have linear phase. Thus FIR filters are more suitable for digital crossover networks.

FIR filters can be designed with sharp cutoff characteristics to approximate the desired response. The digital filter has certain desirable properties that are unattainable by analog means. Digital filters with steep cutoff slopes can be designed while maintaining linear phase. With extra effort additional improvements can also be made. The driver units generally do not have a flat frequency response. The digital filter can be designed to compensate for this. Ideally, all the drivers in a loudspeaker should be in one plane, but the woofers are larger and the cones often have their drive units recessed deeper into the cabinet than the smaller and shallower tweeters. As a result of this, the high frequencies reach the listener slightly before the low frequencies do. There is a frequency dependent time lag introduced that was not present in the original signal. Some speaker manufacturers have recognized this problem and use a slanted face for the speaker enclosure so that all the drivers are aligned. With a digital crossover, the signals to the different drivers can easily be delayed by appropriate amounts to precisely correct for any misalignment in the position of the drivers.

Using the power of DSP, it is possible to equalize the response of the combination of audio system and listening room. It is assumed that the loudspeaker driver units have been frequency compensated and that the amplifier used has a flat frequency response. If such a system is fed pink noise and this is analyzed at the listening position, it can be concluded that the resulting frequency aberrations are due to the effects of the room. The frequency error function can be applied to an adaptive filter to adjust the filter coefficients for a more uniform response. This method attempts to equalize the response at a single point in the room. Perfect equalization cannot be achieved because the acoustic path has delays in addition to a skewed frequency response. It has been found that a more uniform response over a larger listening area can be obtained if the filter coefficients are adjusted to minimize the sum of square errors between the equalized responses at multiple points in the room and delayed versions of the original signal.

The more drivers used, the more complicated the design of the crossover becomes. A two-way system is the easiest case—the crossover network consists of a low-pass section feeding the woofer and a high-pass section feeding the tweeter. The stringent linear phase and flat magnitude response conditions leads to a relationship between the time domain outputs of the two drivers. The output of the tweeter section is obtained by subtracting the woofer section output from the midpoint of the current input. This means that only one of the outputs needs to be calculated; the other one can be derived from it. Normally, two convolutions (explained in Chapter 10) would be required to find the output of the two drivers, but in this case a single convolution suffices for both outputs.

If steep filters are used, the low-pass filter output feeding the woofer will be sufficiently bandlimited to enable the sampling rate for that section to be reduced by a factor of about six or eight (a process known as decimation). As a result, the time between samples is increased by the same factor. The extra time can be used to do more processing. The frequency range of this section is also reduced by the same factor. The end result is that the resolution for equalizing the woofer section increases by a factor of 36 to 64. Implementation would require two DSP chips. One chip would generate the outputs for both sections and the equalization curve for the tweeter, and the second one would generate the equalization curve for the woofer. The block diagram for the implementation of the loudspeaker crossover network is shown in Figure 15.

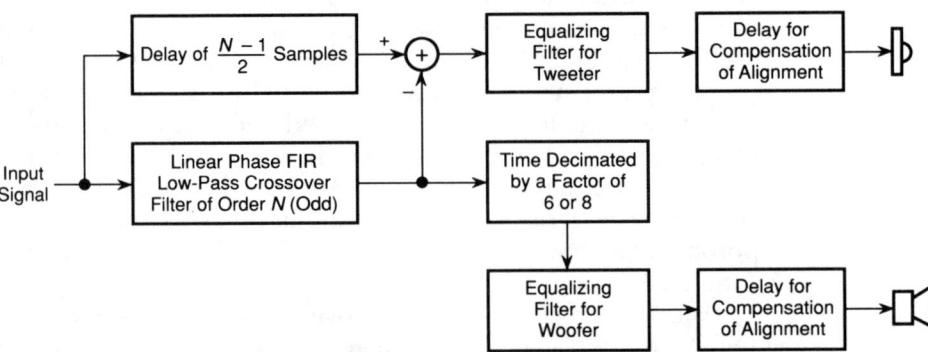

**Figure 15.** The implementation of the digital loudspeaker crossover and frequency compensation network.

# Digital Mixing Consoles

There is no denying the fact that digital audio has exposed the inadequacies of traditional analog equipment. The presence of digital-ready amplifiers and speakers only serves to lay bare these weaker links in the chain. It does not make sense to have a digital signal source and a final digital storage device, yet pass the signal through an analog mixing console, where processing takes place in the analog domain.

A true digital console would process everything in the digital domain. Figures 16a through 16c show the differences in approach between the analog and digital setup for oft-performed mixer functions. In analog, gain control can be realized by a simple potentiometer. In digital, an A/D converter is required to convert the position of the variable resistor, and this value is used to multiply the audio data. Mixing in the analog domain requires a simple adder configuration consisting of an op amp and resistors, but the digital counterpart requires a digital adder and a multiplexer. The task of routing requires a multipole switch in analog, but it requires a demultiplexer and an encoder in the digital implementation. All this processing must take place in real time (about 20 $\mu$s). Since the mixing console deals with a large number of inputs, the digital mixing console is in fact a full-fledged digital computer, processing audio and control data.

Digital mixing consoles have the capacity to accept either digital or analog inputs; the latter are converted to digital at the front end. To make a complete transition to the digital domain, each microphone input must be followed by an A/D converter. In some applications these digital signals are converted to light pulses and connected to the mixer using fiber optics. This eliminates contamination of the signal by external fields and loss of high frequencies—common problems associated with long runs of analog signals in cables. This application is described in detail in Chapter 5. Phase cancellation is a common problem when microphones are used. This happens whenever two or more microphones pick up sound from the same source but are at different distances from it. As a result, there is a phase difference in the outputs of the microphones, which, when combined, cancel each other at some frequencies and reinforce each other at other frequencies. This problem can be eliminated by providing a delay unit in each input channel, such that the microphone path lengths can be electronically made the same.

Digital consoles are often ergonomically designed using the concept of assigned controls. This can be provided in two configurations. In the first case, instead of each input module carrying its own set of knobs and switches, only one control module is used. This works in combination with a group of push buttons, each corresponding to an input channel. Touching any button assigns the control module to that channel. This is similar to sliding a window that is only one module wide along the top of a conventional console to expose only the relevant input module. Each channel could be assigned a four-character abbreviation for identification purposes, such as BASS, OBOE, TUBA, SNRE,

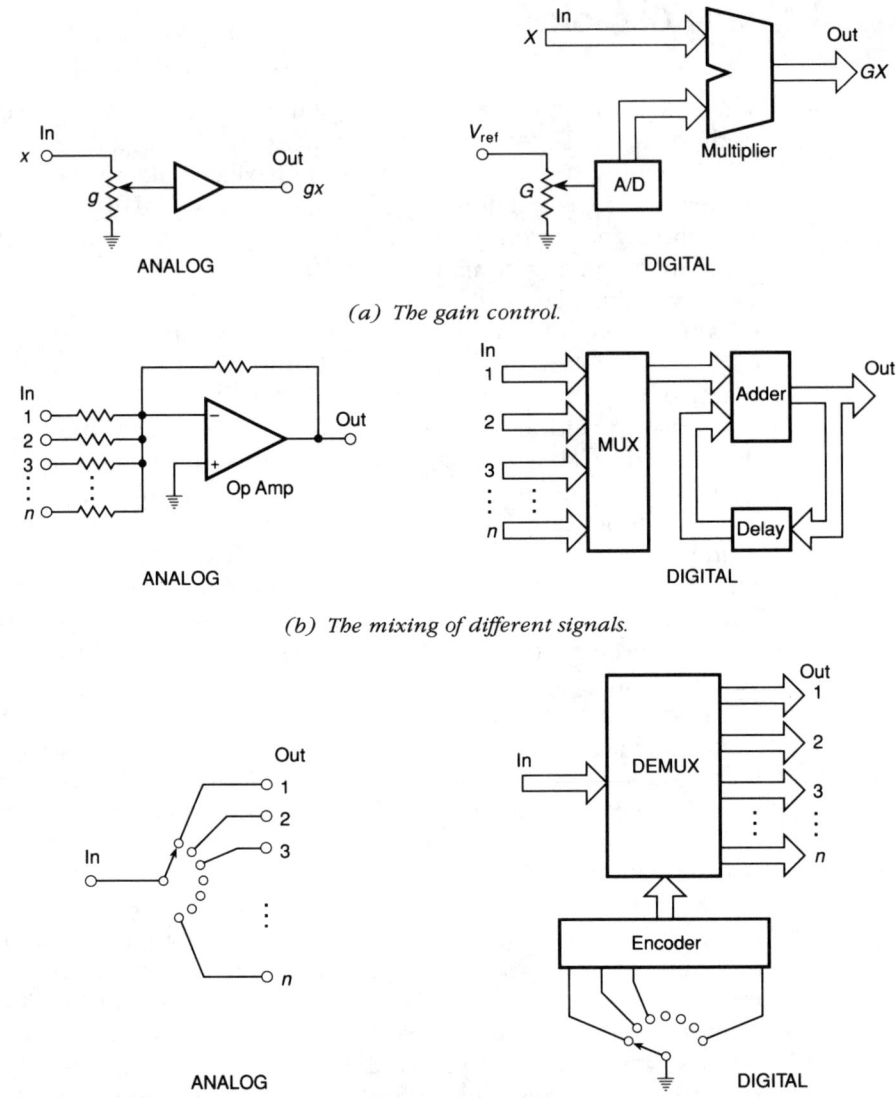

(a) The gain control.

(b) The mixing of different signals.

(c) The routing of signals.

**Figure 16.** Analog versus digital implementations of some mixing console functions.

etc. When a channel is selected, the appropriate mnemonic appears. The entire console takes on an uncluttered appearance that is easier and quicker to work with. The balance provided by the engineer has the potential of sounding better due to an additional reason—he or she does not have to reach out to adjust some parameter and move away from the sweet spot, since everything is near at hand.

The second control assignment method uses as many modules as there are input channels. However, each module has only a single control fader on it which works in conjunction with a group of function buttons. The function buttons have all the effects required among them. Choosing a module and touching a function button assigns that function to the control fader of that particular module. This system is better suited for a smaller number of input channels; otherwise, it will become as wide as some of the large analog consoles that exist. These two methods show that a given control set may be used to perform either the same set of functions for a variety of channels or different functions for a single channel. A hybrid configuration could allow a combination of these two control assignment configurations.

In all cases the control knobs operate encoders. These controls are motor driven and quickly move to the last set position for that particular operation; this is because many engineers work by feel. The positions of the controls are stored in memory and they are updated as soon as a change is made. In effect, the memory content at any instant is analogous to a snapshot of the console setup. This control information can be saved and written to a diskette. The console can then be set up in exactly the same fashion at a later date when a session is resumed, without the delay associated with manually setting and tweaking the controls to duplicate the original sound. Further, the diskette can be used to set up a similar console at another studio location. Automation is as easy as adding time code and varying the contents of the control memory accordingly.

The control memory retains the positions of the controls in binary code. This raises the question of how many bits are required for control. Note that this digital word is entirely different from the audio data signal. The number of control bits would depend on the particular function. Psychoacoustically speaking, there are no absolute thresholds for human auditory perception as they vary from individual to individual. For example, it is certain that using only two bits for panning would be inadequate. This would split the soundstage into merely four discrete regions. Panning under these circumstances from left to right, would result in an image that would jump from one location to the next. In the case of panning, there are two separate issues that need to be considered—static versus dynamic panning. In the case of static panning, once the position of the source is established in the soundstage, it is not altered. Dynamic panning is used to move the source across the soundstage to add the effect of movement, etc. One can use relatively fewer bits for static panning, but for a dynamic pan to sound smooth, a large number of bits is required. The minimum number of bits required for a smooth transition must be found experimentally and then a safety margin has to be provided. It is found that a smooth pan may require eight bits. The volume control on the other hand, to which the listener is more sensitive, may require 16 bits.

It is known that adding two 16-bit numbers could result in a 17-bit number. If the internal processing of audio data was confined to only 16 bits, this could lead to problems where the signals are mixed (or added) together—there would be no headroom. However, the internal processing uses many more bits and has a higher dynamic range and hence eliminates the problem of internal overloading. Consider a 56-input mixer that uses 32 bits or more for internal processing (called the internal wordlength). A single parallel data bus as wide as the internal word length can accommodate all 56 signals by staggering each individual channel's signal slightly in time with respect to each other. The 56 samples are arranged in sequence between the sample clock sync pulses. This method, called time division multiplexing, permits the use of internal data paths that are no wider than the internal data wordlength for multiple input systems. Since the DSP chip has to perform a large number of computations for each channel, it is not usually used in a time division multiplexed mode. The processing is distributed amongst as many DSPs as there are inputs. This makes the computation burden more manageable and leads to a modular construction.

## The Tapeless Studio

DSP technology can be used to form a high-quality digital audio recording environment with the capability to edit and process sounds digitally. The hard drive of the computer becomes the storage medium, doing away with magnetic tape. Hardware interfaces are used to capture sound. Digital sources, such as DAT, can be used to dump data into the hard drive. Digital audio requires a large amount of memory. For example, using a sampling rate of 44.1 kHz, stereo 16-bit audio requires about 10.5 megabytes (MB) of memory for every minute of music. Therefore, large hard drives of about 650 MB capacity are used to store about an hour's worth of music.

In one system, software sets up the computer so that up to four digital audio tracks can be created. It is possible to monitor up to three tracks while recording, to record on two channels simultaneously, to mix tracks, use sound-on-sound, and use tracks in a loop; in other words, the software emulates a four-track recorder. Once the sound is in the computer, DSP is used to provide digital equalization and effects. Any region of any track can be selected by marking off the region in a visual waveform display. Edit points can be selected by scrolling the waveform horizontally. Noisy points can be eliminated by using drawing tools that allow screen editing. The results may be previewed to hear and see how well it works.

DSP tools include graphic and parametric equalization. Any section of a track can be equalized. The section is continuously played in a loop and the sound can be monitored as the parameters are adjusted until the desired effect is achieved. The graphic equalizer has variable center frequency and bandwidth and is comparable to a parametric equalizer. The parametric section offers the normal peak/notch filters in addition to shelving filters. The parameters have wide ranges, and virtually any equalizing filter can be created by entering the parameters. For example, a sharp notch filter may be created to eliminate hum from a recording. The spectrum of signals can be determined by using FFTs. The FFTs are displayed in three dimensions, with the third dimension being time. In addition to showing the frequency content of a signal at any instant of time, they also show the variation of frequency with respect to time. Conversion of sampling rates of signals is another possibility offered by DSP. Data compression techniques can be used to save disk space. A 4:1 compression ratio would require about one fourth the disk space occupied by the uncompressed signal.

For mixing down, snapshots of the fader/pan/equalization states can be retained. The system is compatible with SMPTE and supports MIDI files. The mix can be automated by specifying the beginning and end-point states and the transition time. The mixed versions are all saved as separate files, making it possible to compare many different mixes of a recording. The final mix can be written from the hard drive to a DAT. The system described is an extremely versatile digital audio production tool that can perform a wide array of tasks, such as editing, mixing, equalization, and effects. Using the tools, a four-track project could be done entirely in the digital domain. This system is available through DigiDesign.

# Early Restoration Processes

The recording characteristics in the early days of recording were very noisy and fall in the category of lo-fi. Not so the performers, as there can be found a wealth of superb performances. The primary source of music today is the CD. Today's customer expects a certain high quality of sound from the CD. This means that the recording companies cannot release these old recordings into the CD format directly. This results in a loss for both sides—the recording company loses revenue because it cannot provide items for which there is a demand, and a generation of music is lost to the consumer. Therefore, devising a way to salvage these recordings is well worth the effort.

In the 1970s, Stockham and his colleagues used a deconvolution technique to restore and commercially release a number of recordings of the operatic tenor Enrico Caruso that were made in the early 1900s. One of the assumptions made was that the recording process in the old recordings acted as a filter. In that case, these old recordings can be viewed as the convolution of the music with the impulse response of the recording process in the time domain (see Chapter 10 for more details on convolution in linear systems). Alternately, the spectrum of the old Caruso recordings could be looked upon as the product of the spectrum of the music and the spectrum of the recording process. Further, it was assumed that the spectral contents of a modern and old performance of the same piece are similar.

FFT methods were then used to find the average long-term spectrum of an old and modern recording. For such a large number of data, the FFTs of successively overlapping segments were found and the resulting outputs were summed and averaged. The difference between the two calculated spectra was attributed to the old recording process. The inverse of this transfer function would be the filter required to restore the recording. An alternative method is to use a sound that requires a very large bandwidth and is present in the recording. If only this segment of the performance is used to determine the transfer function of the correcting filter, then much the same results can be obtained as in the more comprehensive method described above.

## A Restoration Workstation

A collection of computer software has been introduced which permits the removal of frequently encountered noises on recordings and also provides means for restoring old recordings with a noisy background. A personal computer augmented with DSP circuitry forms the hardware basis. In the case of noise in the form of pops and clicks, the traditional approach has been to use some form of editing. This manual technique is not only time consuming but it also destroys part of the original recording. Using a workstation, the audio data from the original recording is stored in the hard drive as a soundfile. It is possible to visually examine the soundfile on the monitor. Clicks and pops can be clearly distinguished from the surrounding signal as shown in the upper signal in Figure 17. To manually declick, the problem area is marked off, and the interpolation mode is selected from the menu. Rather than leaving a gap at the site of the click, the system analyzes the signal before and after the click and replaces the click with what it estimates the program material should have been, as shown in the lower signal in Figure 17. This method is sonically superior to manual editing because it does not shorten the program length and maintains continuity on both sides of the click by interpolating over the gap.

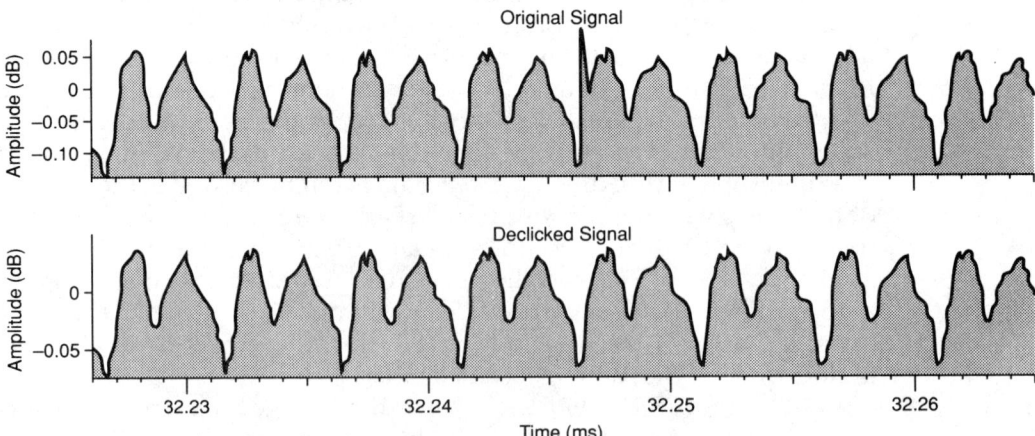

**Figure 17.** Removing the click and pops from a signal. *(Courtesy Sonic Solutions)*

Declicking every click in this manner could take a long time. Most frequently encountered noises fall into categories with certain characteristics that have been predefined in the system. The operator listens to the soundfile and determines the kinds of noises present. The system then automatically searches for these problem areas in the soundfile and corrects them. The modified and unmodified versions are compared, to find if any transients were lost in the process, in which case it would be better to leave that section untouched. Further, there could be some noise present that does not fit any of the predefined categories, and this can be manually removed.

The real power of the system comes to the fore in reducing the background noise in noisy recordings. As much as 20 dB of noise reduction is claimed. If the original recording is not very tightly edited, then there would be some blank spaces between pieces of music. These regions are representative of the combination of the recording process, recording venue, and noise in the recording format. This segment is important in the restoration process. The audio range is split into 2000 spectral bands or bins. The noise sample's content at each of these bins is examined as shown in Figure 18a. The level of each bin is noted and acts as the threshold point. If the noise sample is truly representative of the recording process, then the threshold levels in the 2000 bins could be considered its signature in the frequency domain. The soundfile, which typically consists of both signal and noise components, is similarly split into 2000 bins as shown in Figure 18b. The content of each bin is compared with the threshold values. If they match, it means there is no signal in that bin, only noise at that frequency. Subsequently, the output of that bin is attenuated. The degree of attenuation is a complex function of various factors evaluated by the system. The high resolution of 2000 bins is one of the keys to the superior results of this system, but it makes the process computation-intensive (over 53 million computations per second of program material). Therefore, this is currently not a real-time process. The NoNoise system is commercially available from Sonic Solutions.

Restoration methods such as these are possible because of the superior powers of DSP. The algorithms require considerable computation and hence are generally non–real-time processes. Advances in computer technology may allow them to be real time in the near future.

# Conclusion

In summary, it seems that digital audio and DSP audio applications are moving along the same path. After its introduction, digital audio proceeded to rapidly proliferate throughout the field of audio. Digital signal processing entered the field of audio later, but it has surely made its presence felt by introducing radical changes in the way signals are processed. The success story should come as no surprise to anybody familiar with the power, flexibility, and superior sonic quality of DSP.

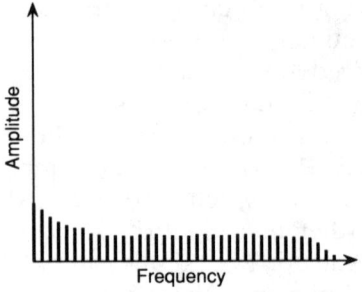

*(a) The spectral threshold values extracted from a sample of the background noise.*

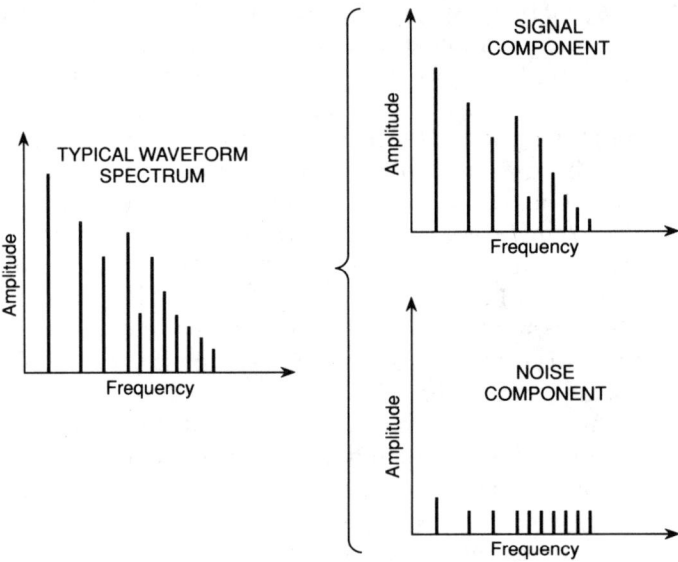

*(b) Typical waveform consisting of a signal and the background noise, and separation of the signal and noise components of this waveform.*

**Figure 18.** Suppressing the background noise in a recording.

Soundfield processing was the first application of DSP to audio. Different acoustic venues could be simulated by simplifying the complex natural reverberation characteristics of a venue. As a result of price reductions, it is now possible to find a stand-alone consumer unit with DSP that offers many functions—filtering, equalization, expansion/compression, soundfield processing, etc., at a low cost. This signifies that the gestation period for DSP is over and that consumers are beginning to receive the full benefit of DSP products. Besides being used at home, soundfield processing has also hit the road, where it is ideal for increasing the perceived acoustic space of a car. In addition, DSP can provide equalization correction for the interior of the car and detect the varying noise level in the car and dynamically compress the music so that soft passages are not buried in the ambient noise, and loud passages are not too loud.

The weakest link in the audio chain is the loudspeaker. Driver units have a limited frequency range, and even within this range the response is not flat. Digital FIR filters that provide steep filtering while still maintaining linear phase are used in crossover networks. These filters can be used for some additional functions as well—the nonuniform frequency response of the driver units can be compensated for, the signal to the driver units can be delayed so that they are time aligned, and the entire audio system and room combination can be equalized.

Digital signal processing has also been widely employed in professional audio applications. For example, some digital mixing consoles have been introduced. During mixing, a signal goes through many stages of processing. Each analog stage deteriorates the quality of the signal, but digital processing can eliminate this problem to a large extent. The concept of control assignability is used to greatly reduce the size of consoles. This leads to an ergonomically superior product that has an uncluttered appearance and more working space.

Some systems have been developed which provide a studio-quality digital recording environment in a single package. This is possible by using software to simulate a digital recorder. Hardware interfaces allow it to communicate with external sources and storage devices. A DSP chip is used to provide precise and powerful editing and processing of signals, hence bringing more power to the tapeless studio.

Digital signal processing is also used to salvage old, noisy recordings. By using FFT measurements, the transfer function of the recording process can be found and then equalized. Modern recordings can also have a noisy background or extraneous noises. These are edited on workstations by using software packages that automatically detect the presence of clicks and pops and eliminate them. A noisy background can be reduced by about 20 dB by using DSP to recognize the characteristics of the noise present and then suppress it.

# References

Berger, I., "QED: DSP," *Audio Magazine*, August, 1990.

Berkovitz, R., "Digital Equalization of Audio Signals," *Digital Audio Collected papers*, Audio Engineering Society, 1983.

De Lancie, P., "NoNoise Sound Restoration from Sonic Solutions," *Mix*, September, 1987.

Dean, R., "Digital Mixing: A Reality at Last," *Mix*, April, 1985.

Elliot, S. J., and P. A. Nelson, "Multiple-Point Equalization in a Room Using

Adaptive Digital Filters," *Journal of the Audio Engineering Society*, vol. 37, no. 11, November, 1989.

Griesinger, D., "Practical Processors and Programs for Digital Reverberation," *Audio in Digital Times: AES 7th Conference Proceedings*, May, 1989.

Griesinger, D., "Theory and Design of a Digital Audio Signal Processor for Home Use," *Journal of the Audio Engineering Society*, vol. 37, no. 1/2, January/February, 1989.

Monforte, J., "Sonic Restoration through NoNoise," *Recording Engineer/Producer*, September, 1989.

Pohlmann, K. C., "Digital Mixing Consoles," *Mix*, September, 1985.

Poyten, P., "Digidesign's Sound Tools," *Mix*, December, 1989.

———, "Playing with a Full Deck at Digidesign," *Mix*, July, 1990.

Schuck, P. L., "Digital FIR Filters for Loudspeaker Crossover Networks II: Implementation Example," *Audio in Digital Times: AES 7th Conference Proceedings*, May, 1989.

Shapton, D., and M. Mattingley-Scott, "True Digital Audio Mixers," *Recording Engineer/Producer*, February, 1989.

Sutton, M., "Yamaha DMP7 Digital Mixing Processor," *Mix*, October, 1987.

Wilson, R., Adams, G., and J. Scott, "Application of Digital Filters to Loudspeaker Crossover Networks," *Journal of the Audio Engineering Society*, vol. 37, no. 6, June, 1989.

Yamamoto, K., "A Newly Developed Digital Signal Processing Technology for Consumer Product Applications," *Audio in Digital Times: AES 7th Conference Proceedings*, May, 1989.

Zucker, I., "Reproducing Architectural Acoustical Effects Using Digital Soundfield Processing," *Audio in Digital Times: AES 7th Conference Proceedings*, May, 1989.

*Chapter 12*

# Low-Bit Conversion and Noise Shaping

Ken C. Pohlmann

## Introduction

Although multibit linear PCM converters can be manipulated in a variety of ways to decrease zero-cross error and glitches and improve low-level linearity, it can be argued that classical PCM architecture has reached the limit of its performance. This has stimulated development of an important digital signal processing technique: low-bit conversion. These systems are characterized by very high oversampling rates, noise shaping, and conversion of wordlengths of one or a few number of bits. These systems demonstrate that a waveform can be represented with many bits at a low sampling rate (as in PCM) or with a few bits or even one bit at a high sampling rate (as in low-bit). Modern low-bit converter systems are highly complex and use both digital signal processing and code to voltage (current) conversion as parts of an overall technology. If either part uses a single bit as its output, the technolgy itself is colloquially called a 1-bit system. In many designs, sigma-delta modulators are employed. In any case, the systems share the goal of translating nonideal converter behavior into uncorrelated, benign noise and largely shifting it out of the range of human hearing. However, as we shall see, low-bit systems differ one from the other; in fact, low-bit architecture offers more design latitude than multibit PCM conversion.

## Low-Bit Conversion

Parallel conversion of PCM audio data words, whether 16 or 18 bits, has been the mainstay of virtually all digital audio products. This architecture has proven itself to be highly successful. Careful design and calibration permit accurate conversion of the audio signal. However, as we observed in Chapter 3, expensive A/D and D/A converters, labor-intensive calibration procedures during manufacture, and sophisticated circuit design are required to achieve this performance and maintain it over the life of the converter. Understandably, some manufacturers have sought to develop alternative conversion systems.

Low-bit conversion systems cost less than high-quality 18-bit conversion yet provide high linearity. For example, the Philips "Bitstream" method has been designed to convert sampled data to an analog waveform using only one bit. This system uses a PDM (pulse density modulation) signal, thus the system must boost the bit rate by an extreme amount. In particular, it uses 256-times oversampling to produce a data rate of 11.2 MHz; this high oversampling rate is needed because a 1-bit D/A converter is used to convert the signal. Specifically, a capacitor is charged and discharged according to the "1" or "0" value of the data; the result is an analog waveform which reflects the encoded waveform through time averaging of the output bit. Errors in the reference values will generate a gain offset error but, in theory, not a linearity error. In practice, nonlinearities could result from idle patterns in the noise shaping circuitry.

Low-bit conversion is a radical departure from multibit PCM conversion. A conventional PCM converter divides the signal in multiple amplitude steps; as we have seen, this is prone to errors. Low-bit conversion divides the signal in terms of time, keeping amplitude changes constant; errors can be more tightly controlled. Figure 1 shows the concept behind low-bit conversion and its relation to multibit conversion. When the amplitude representations in Figure 1a are placed on their side, as in Figure 1b, the signal is now expressed in terms of time. When these signals are low-pass filtered, the result is an analog waveform, just as in the multibit method. However, since the weighted bits of a binary sample are transformed into the low-bit data stream using digital methods, the major sources of nonlinearity in conversion are avoided. Unlike multibit converters, the conversion accuracy of low-bit systems can be increased without trimming or regulation. Importantly, not only is the quantization error produced by low-bit converters lower than in multibit converters, it is also uncorrelated to the audio signal, and hence benign. Multibit PCM conversion is an analog process, whereas low-bit conversion is a digital process; for the same reasons that digital processing is superior to analog, low-bit conversion can be superior to PCM conversion. To not appear ungrateful, it is important to note that many of the techniques which make low-bit conversion a success are a result of the many refinements undertaken to improve multibit conversion.

Low-bit conversion methods have become increasingly useful in digital audio products, particularly in applications where conventional conversion methods present limitations. For example, sigma-delta techniques have shown themselves to be extremely competitive in A/D applications because they obviate the need for brickwall anti-aliasing filters. Although low-bit methods use familiar techniques such as oversampling, they also employ wholly new and sophisticated kinds of DSP algorithms, such as noise shaping and decimation. In particular, considerable processing is required to implement noise shaping and decrease the high in-band noise levels otherwise present in low-bit conversion. A variety of low-bit A/D and D/A architectures has been devised, each

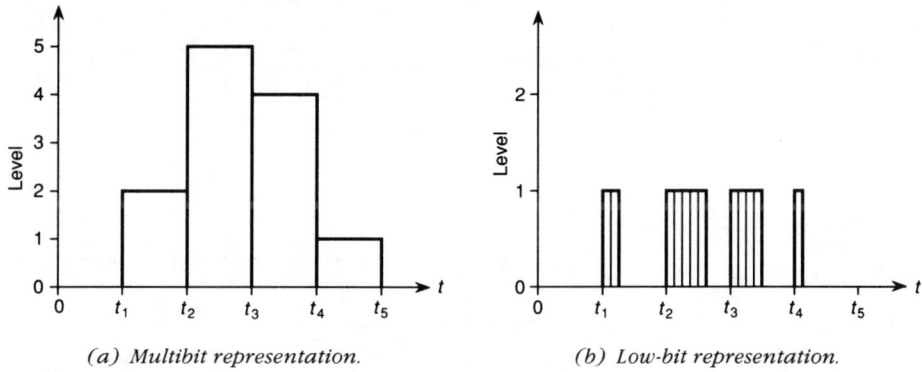

**Figure 1.** A comparison of multibit signal representation versus low-bit representation.

with different algorithms and orders of noise shaping. Together, they demonstrate that the evolution of digital audio technology, even for the most fundamental tasks, such as conversion, continues at a rapid pace. Although some low-bit converters operate with one bit whereas others employ a few bits, the terms low-bit and 1-bit are typically used interchangeably and will be used thusly here.

## Sigma-Delta Modulation

As we observed in Chapter 2, in pulse code modulation a signal is sampled and quantized into discrete steps; the maximum signal amplitude determines the maximum quantizer range. A PCM quantizer is represented in Figure 2a. Quantization error is present uniformly across the Nyquist frequency band $f_s/2$ and cannot be removed from the signal. If quantization is performed at a higher sampling rate $R \times f_s$, where $R$ is the oversampling rate, the error is spread across the $R \times f_s/2$ band; hence the noise in the audio band is reduced by 3 dB for every factor of two oversampling.

As Hauser [1991] has pointed out, oversampling will increase the signal-to-noise ratio as follows:

$$S/N \text{ (dB)} = 6.02(k + 0.5l) + 1.76,$$

where $k$ is the number of quantization bits and $l$ is the number of octaves of oversampling. Thus an oversampling A/D converter performs as well as a longer wordlength A/D converter, yielding a benefit of 0.5 bit per oversampling octave. While useful, the benefit is limited; for example, a 10-bit improvement would require an $l$ of 20 octaves—an oversampling factor of 1 million. It is the aim of noise shaping to introduce a high-pass function in the noise spectrum and thus improve oversampling performance.

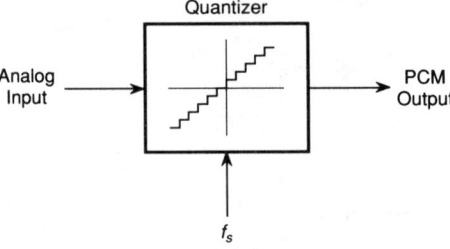

*(a) Pulse code modulation (PCM) converter.*

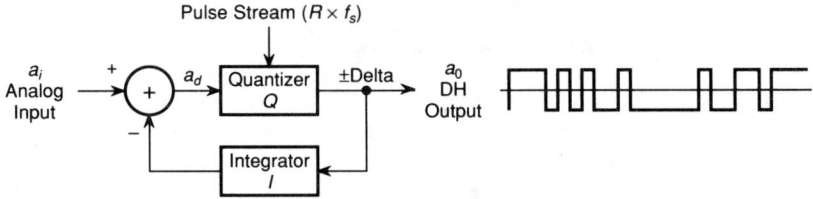

*(b) Delta modulation (DM) encoder.*

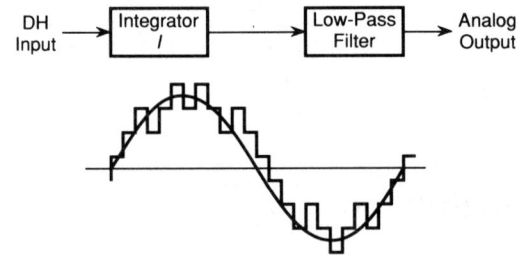

*(c) Delta modulation (DM) decoder.*

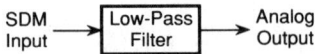

*(d) Sigma-delta modulation (SDM) encoder.*

*(e) Sigma-delta modulation (SDM) decoder.*

**Figure 2.** A comparison of modulation methods used in A/D and D/A conversion. *(Courtesy Philips)*

Delta modulation and sigma-delta modulation (sometimes called delta sigma) were developed in the 1940s and 1960s, respectively, and used for voice telephony applications. Limitations prohibited their use in high-quality music applications until the emergence of high-speed digital signal processing techniques in the 1980s. Differential pulse code modulation (DPCM) is a technique in which the derivative of the signal is quantized. When signal changes between samples are small, the quantizer's range can be reduced. With very high oversampling rates, the changes between sample periods are made very small, thus the quantizer can be reduced to 1-bit. A 1-bit DPCM coder is known as a delta modulator (DM). In other words, the DM codes the differences in the signal amplitude instead of the signal amplitude itself.

A delta modulation encoder is shown in Figure 2b; it is known as a single integration modulator. The analog input signal is compared with the integrated output pulses and the delta (difference) signal is applied to the quantizer. The quantizer generates a positive pulse when the difference signal $a_d$ is negative, and a negative pulse when the difference signal is positive. When the difference signal is zero (no input or DC) the quantizer toggles between positive and negative pulses. The system's sampling rate is determined by the rate at which bipolar output pulses are generated.

A delta modulation decoder is shown in Figure 2c. This system consists of an integrator and a low-pass filter. When the 1-bit pulses are integrated using a time constant that is long compared with the sample period, a step waveform is produced. An output analog waveform is produced by low-pass filtering this step waveform. Quantization error at the output of the integrator is white. As in PCM, oversampling decreases the error level by 3 dB for every factor of two oversampling. The dynamic range can be improved, that is, the quantization error made smaller, by making the delta (difference or step size) value smaller. The limit to which the delta value can be reduced is given by the maximum derivative of the signal. Assuming a signal of $A_m \sin 2\pi ft$, its derivative is

$$|2\pi f A_m \cos 2\pi ft| < Q/T,$$

where

$A_m$ is the maximum signal amplitude,
$f$ is the signal frequency,
$Q$ is the step size,
$T$ is the sampling period.

The maximum derivative occurs at maximum signal frequency and maximum signal amplitude. Exceeding this limit causes slope overload distortion. For music signals, the delta value must be high, resulting in high quantization error level. When the signal has lower high-frequency content, as in speech, delta can be reduced. In any case, the success of DM hinges on assumptions about the nature of the encoded signal. This dependence of the dynamic range on the signal spectrum, and good performance only when the signal has low-pass characteristics, as well as correlated patterns at low signal levels, limits single integration DM applications.

Sigma-delta modulation (SDM) was developed to overcome the limitations in delta modulation. Sigma-delta systems quantize the difference (delta) between the current signal and the sum (sigma) of previous differences. Like PCM, SDM quantizes the signal directly, and not its derivative as in DM. Thus the maximum quantizer range is determined by the maximum signal amplitude and is not dependent on signal spectrum. However, as in DM, very high oversampling rates are used to produce a 1-bit quantized code. A first-order (single integration) sigma-delta modulation encoder is shown in Figure 2d. The input to the quantizer is the integral of the difference between the input and the quantized output. This integrator forms a low-pass filter on the difference signal, thus providing low-frequency feedback around the quantizer. This feedback results in a reduction of quantization noise at low (in-band) frequencies. Unlike PCM and DM, the noise is not white but shaped by a first-order high-pass characteristic as analyzed below.

An SDM decoder is shown in Figure 2e; only a low-pass filter is required to decode the signal, to remove high (out-of-band) frequency noise. In other words, it averages the 1-bit signal to produce an analog waveform. To achieve high resolution, high oversampling rates (e.g., $256 \times f_s$) are required. However, quantization noise is highly correlated in a first-order SDM. More effective noise shaping can be achieved with high-order (multiple integration) SDM coders. Coders with orders higher than two are potentially unstable and require special design techniques, such as limiting, to prevent overflow and ensure stability.

The signal and noise transfer functions for a first-order sigma-delta noise modulator are analyzed in the $S$-domain in Figure 3. For a zero-noise source, the transfer function shows a low-pass characteristic. For a zero signal, the transfer function shows a high-pass characteristic. In other words, as the loop integrates the difference between the input signal and the sampled signal, it low-pass filters the signal and high-pass filters the noise. If the system is designed so that the signal's frequency content is less than the filter's cutoff frequency, the signal will not be affected. Conversely, oversampling spreads quantization noise over a wider bandwidth; hence noise is diminished in the audio band. If a $k$-bit A/D converter were placed in the loop, the maximum signal-to-noise ratio would be

$$S/N \text{ (dB)} = 6.02(k + 1.5l) - 3.41,$$

where $k$ is the number of bits in the converter and $l$ is the number of octaves of oversampling. This first-order circuit provides a benefit of 1.5 bits per octave; higher-order loops would further decrease quantization noise in-band, with a penalty of increased total noise power. For example, a second-order loop would yield

$$S/N \text{ (dB)} = 6.02(k + 2.5l) - 11.14.$$

## Noise Shaping

All low-bit conversion systems require noise shaping processing to reduce in-band noise levels. Through noise shaping, converter quantization errors are greatly reduced. The converter may be viewed as a wideband, spectrally weighted noise source in which the noise is more benign than multibit converter nonlinear-

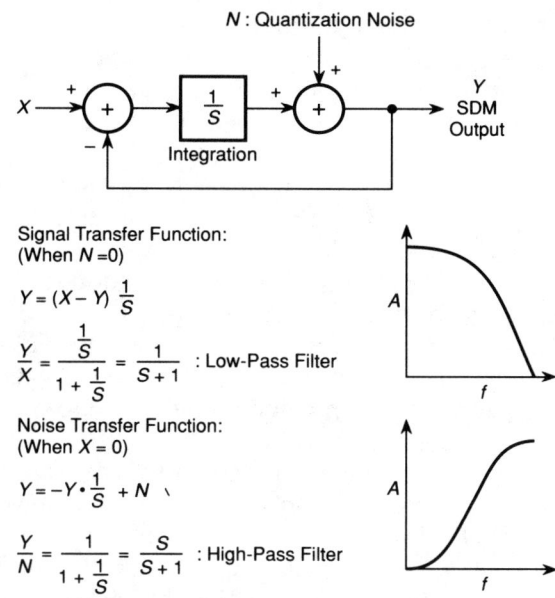

**Figure 3.** S-domain analysis of a sigma-delta modulator. *(From Park)*

ity. Thus noise shaping requires a reduced number of bits for conversion, while maintaining dynamic range with good in-band signal-to-noise ratio. As its name implies, noise shaping shifts noise away from the audio band (0 to 20 kHz) thus lowering audio band noise. The results are shown in Figure 4. Line *A* represents the requantization noise floor with 1-bit conversion. Line *B* shows requantization noise when a noise shaping algorithm is employed. Overall, both noise levels are the same; however, with noise shaping, in-band noise is decreased and out-of-band noise is increased. With noise shaping, a 16-bit signal can be represented with 16-bit performance (or much more) with a single bit.

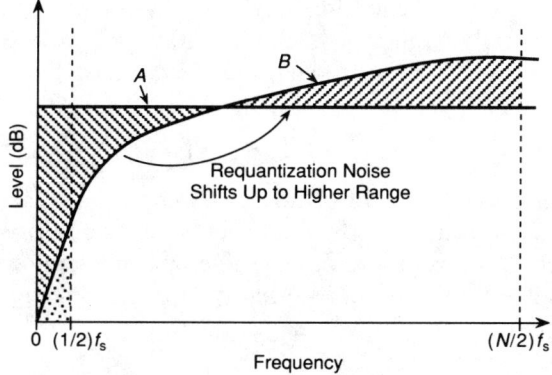

**Figure 4.** With noise shaping, requantization noise can be shifted away from the audio band. Line *A* shows the noise floor without noise shaping. Line *B* shows the noise floor with noise shaping.
*(Courtesy Matsushita Electric)*

Noise shaping reduces quantization error and improves differential linearity. The method uses the known characteristics of quantization error to modify the error and decrease its effect. Efficient noise shaping algorithms use recursion to spread the noise power across a wider spectrum and to shape the noise spectrum. In particular, the method places the error information back into the signal, much as negative feedback is used to reduce distortion in analog amplifiers. In effect, noise shaping attempts to cancel error by subtracting it from the input signal. Feedback in this discretely sampled system can operate only if a delay element a sample or more in duration is placed in the feedback loop. By placing the quantization error signal in a feedback loop, the frequency response of the error signal is altered, shifting much of its energy outside the audio band. Moreover, the noise is shaped by the approximate inverse of the loop transfer function; when a low-pass filter is placed in the loop, the noise spectrum rises with frequency.

A simple example demonstrating the workings of a first-order noise shaper is shown in Figure 5a. This noise shaper can output only one bit, with a logical 0 or 1 level; a logical 0 is output if the input signal (16 bits in this example) is less than 0.5, and a logical 1 is output if the input signal level is greater than 0.5. The remaining error signal is fed back and subtracted from the input; this error signal is always present when the input signal has a value other than 0 or 1. Figure 5b shows operation when the input signal is a constant 0.5. The output code stream is 101010, which when averaged gives the desired output value of 0.5. Figure 5c shows the operation when the input signal is a constant 0.8. The resulting code stream is 11011, which when averaged gives the desired output value of 0.8. These examples use a constant input level, but high-speed averaging applies equally to varying input music signals. In either case, the output signal from the noise shaper is applied to a 1-bit D/A converter which generates the appropriate analog voltage.

The real innovation, and diversity, associated with 1-bit systems, as we shall see, comes in the design of noise shaping algorithms. Simply put, the more complex the algorithm, the lower the noise in the audio band. More specifically, higher orders of integration in noise shaping decrease the in-band noise level. The noise shaping characteristic is based on the sigma-delta formula

$$\mathrm{NS}(z) = [(z-1)/z]^n,$$

where $n$ is the order of noise shaping.

This characteristic can be effectively comprised of n cascaded digital differentiators. As $n$ increases, the slope in frequency of the shaping function increases; thus it is more effective in suppressing low-frequency noise. In the frequency domain, we may substitute

$$z = e^{j2\pi f/f_a},$$

where $f_a$ is the noise shaping sampling frequency. Thus we may write

$$|\,\mathrm{NS}(f)\,| = [2\sin(\pi f/f_a)]^n.$$

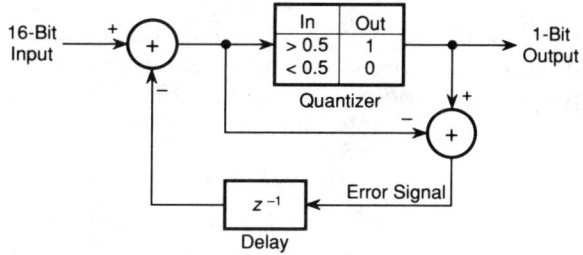

(a) First-order noise shaper circuit.

| Input | Output | Error Signal |
|---|---|---|
| 0.5 | 1 | 0.5 |
| 0.5 − 0.5 = 0 | 0 | 0 |
| 0.5 − 0 = 0.5 | 1 | 0.5 |
| 0.5 − 0.5 = 0 | 0 | 0 |

(b) Sequence of events with constant 0.5 V input level.

| Input | Output | Error Signal |
|---|---|---|
| 0.8 | 1 | 0.2 |
| 0.8 − 0.2 = 0.6 | 1 | 0.4 |
| 0.8 − 0.4 = 0.4 | 0 | − 0.4 |
| 0.8 + 0.4 = 1.2 | 1 | − 0.2 |
| 0.8 + 0.2 = 1 | 1 | 0 |

(c) Sequence of events with constant 0.8 V input level.

**Figure 5.** Principle of operation of a first-order noise shaper.
*(Courtesy Philips)*

Thus, in turn,

$$|\,\mathrm{NS}(f_a/6)\,| = 1,$$

and

$$|\,\mathrm{NS}(f_a/2)\,| = 2^n.$$

We note that the curves described by this equation equal unity value at $f_a/6$ Hz, noise is reduced only for $f < f_a/6$ Hz, and for $f_a/6 < f < f_a/2$ Hz. Thus the noise level is increased, reaching a maximum at $f_a/2$ Hz.

Figure 6 summarizes the mathematical basis of first-order noise shaping. Quantization noise is assumed to be random, and the quantizer is shown as an additive noise source. Note that the $1 - z^{-1}$ factor doubles the quantized noise power; however, the $1 - z^{-1}$ factor also shifts the noise to high frequencies. This sigma-delta modulator forms the basis for the A/D and D/A low-bit conversion systems decribed in this chapter. Using first- and higher-order noise shaping algorithms, a series of noise shaping curves can be generated, as shown in Figure 7. Clearly, as higher orders of noise shaping are employed, the in-band noise level is decreased. Dynamic range can be increased through a combination of higher oversampling rates and higher-order noise shaping filters. As the oversampling rate is increased, the portion of the curve in the audio band is relatively reduced, that is, moved toward 0 Hz. In other words, although the shape of the noise curve remains the same, high oversampling rates relatively decrease in-band noise. The input/output characteristic of a basic sigma-delta noise shaper of $n$th order is

$$Y = X + (1 - z^{-1})^n N,$$

where

$Y$ is the noise shaped output and a function of $z$,
$X$ is the input signal and a function of $z$,
$n$ is the order of differentiation of the factor $1 - z^{-1}$,
$N$ is the quantization noise (assumed to be white) and a function of $z$.

For example, a third-order noise shaper is shown in Figure 8; although theoretically correct, if used at third order or higher, this circuit would suffer operational instability due to overloading of the integrator. Thus in practice, a multistage circuit is often employed, as described below.

As the order is increased, the noise shaping characteristic changes, and in-band requantization noise is decreased. However, out-of-band noise is greatly increased; this could overly burden subsequent analog filters. A successful noise shaping circuit thus seeks to balance a high oversampling rate with noise shaping order to reduce in-band noise and shift it away from the audible range.

Noise shaping is complicated, but we can get a feel for its workings

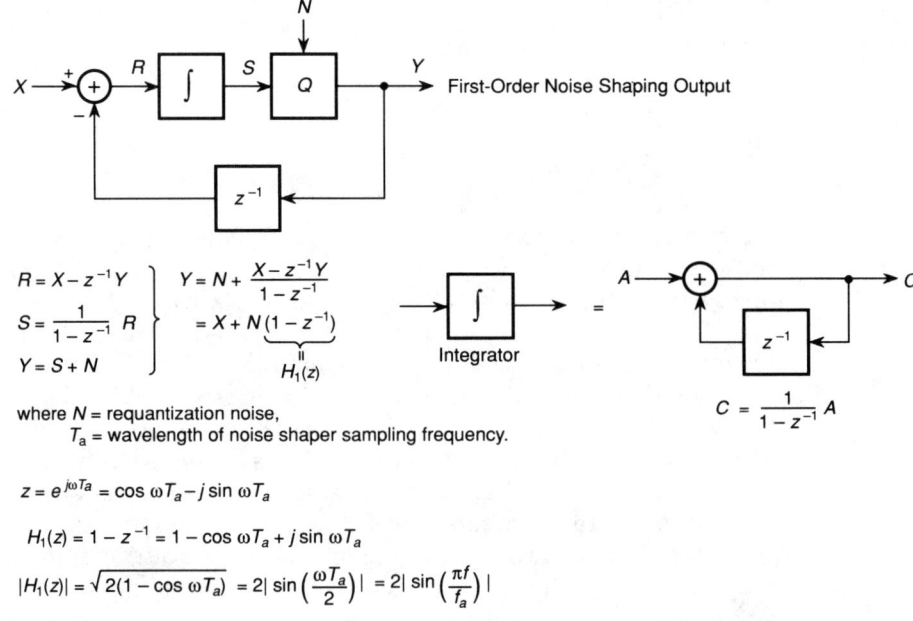

**Figure 6.** Analysis of a first-order noise shaper. *(Courtesy Philips)*

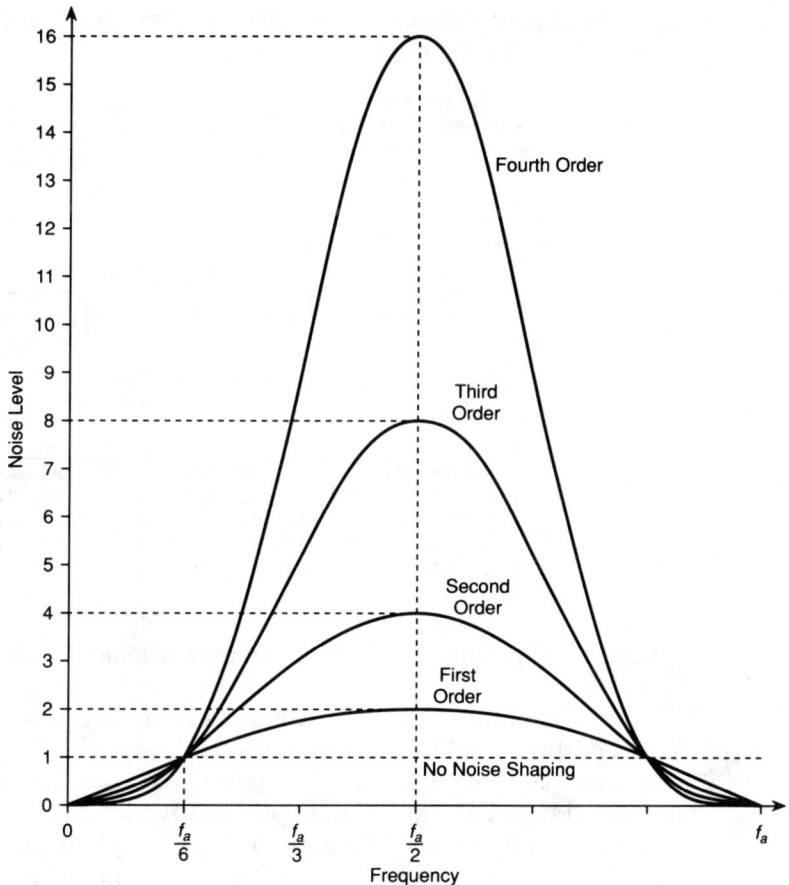

**Figure 7.** Higher orders of noise shaping result in more pronounced shifts in requantization noise.

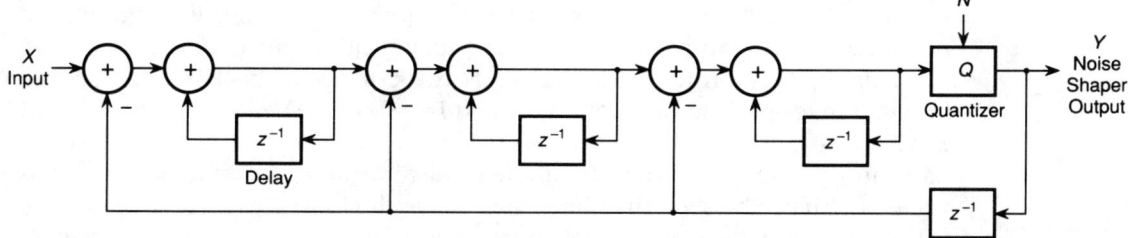

**Figure 8.** $N$th-order noise shaping could employ cascaded sections as in this third-order noise shaper. In practice, multistage designs are often employed.

through Figure 9a. This circuit performs noise shaping through feedback loops and generates a 1-bit signal for conversion into analog. In particular, this noise shaper consists of two integration (filter) loops to reduce in-band quantization noise. Although of different design from a sigma-delta modulator, it performs second-order noise shaping.

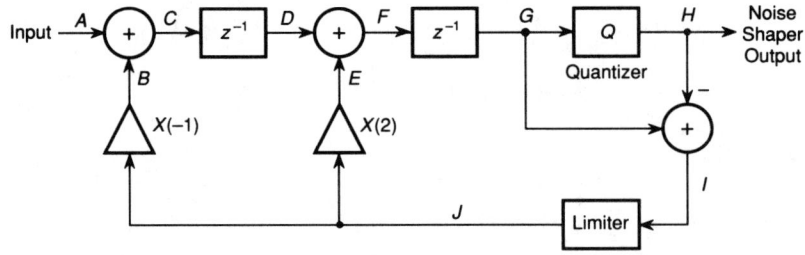

(a) Second-order noise shaper.

|   | A | B | C | D | E | F | G | H | I | J |
|---|---|---|---|---|---|---|---|---|---|---|
| $t_0$ | 0.5 | −0.4 | 0.1 | 0.5 | 0.8 | 1.3 | −0.6 | −1 | 0.4 | 0.4 |
| $t_1$ | 0.5 | −0.3 | 0.2 | 0.1 | 0.6 | 0.7 | 1.3 | +1 | 0.3 | 0.3 |
| $t_2$ | 0.5 | 0.3 | 0.8 | 0.2 | −0.6 | −0.4 | 0.7 | +1 | −0.3 | −0.3 |
| $t_3$ | 0.5 | −0.6 | −0.1 | 0.8 | 1.2 | 2.0 | −0.4 | −1 | 0.6 | 0.6 |
| $t_4$ | 0.5 | −1.0 | −0.5 | −0.1 | 2.0 | 1.9 | 2.0 | +1 | 1.0 | 1.0 |

(b) Table showing operation of circuit.

**Figure 9.** Operation of second-order noise shaping circuit. *(Courtesy Philips)*

Operation of the circuit can be explained by following the values of audio samples during several sample periods, at various points in the noise shaper. These values are shown in Figure 9b. In this example, a maximum 0 dB value is represented by a 1 (in practice, to prevent overload, an input must be scaled down in the digital filter). The input in this example is a constant 0.5 (−6 dB); for correct operation, the output of the circuit should thus be a 1-bit output with an average value of 0.5. Arbitrarily, initial values for the contents of the delay registers are selected as $D = 0.5$ and $G = -0.6$. Initially, the input is 0.5 and the value at $G$ is negative, thus the 1-bit output at $H$ is −1. The output of the quantizer is +1 if its input is positive (MSB = 0), and −1 if its input is negative (MSB = 1). The 1-bit code output from the quantizer is simply a sign bit; the remainder of the sample is fed back as a quantization error signal.

This signal is formed by taking $G - H$; this error is fed back into the double integration loops. The value of $I$ is 0.4 and $J = -0.6 + (-) -1 = 0.4$. The value of $E$ is $2 \times J = 0.8$ and $C = A - J = 0.1$. The value of $F = D + E = 1.3$. Similarly, values pass through the circuit at each sample time; for example, $D$ takes the previous value of $C$. It can be seen that the values inside the loop are larger than the unit value; in other words, wider data buses are required. Also, if large values are input to the circuit, the limiter would be needed to prevent overloading of the loops. It is the accumulation and delay elements in the feedback loops which enable noise shapers to tailor the noise floor. Ideally, with no input signal, the coder should output only a tone at $(R \times f_s)/2$, where $R$ is the oversampling rate. However, noise shaping processing also outputs signals at additional frequencies when idling. To overcome this, dither can be added to the input data so the circuit always operates with a changing signal even when the audio signal is zero or DC.

The frequency response of the shaped noise can be tailored by placing a digital filter in the feedback loop, as shown in Figure 10a. There is great flexibility in the theoretical design of the filter; for example, its parameters could be dynamically adapted so the error noise is always optimally masked by the audio signal. It is important to note that such a configuration alters the frequency response of the error signal but not that of the audio signal; as shown in Figure 10b, the configuration has the effect of passing the noise through the filter, not the signal.

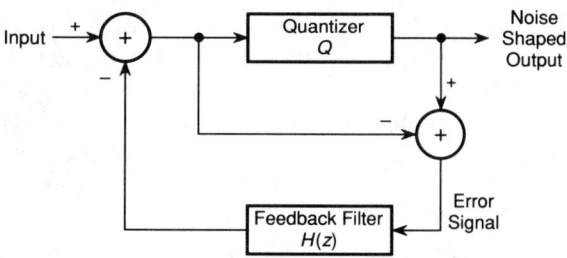

*(a) Circuit showing placement of digital filter.*

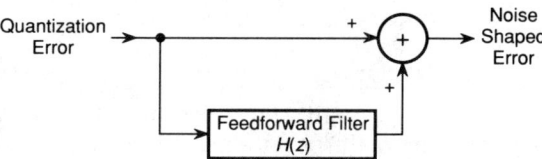

*(b) Circuit showing that error response is altered but not signal response.*

**Figure 10.** Noise response is shaped by placing a digital filter in the feedback path.

The architecture operates as a high-order noise shaper with a limiter in a feedback loop, as shown in Figure 11. Multistage noise shaping is also possible with lower-order filtering in each stage. In addition, multibit quantizing is possible, rather than 1-bit coding; this can increase dynamic range.

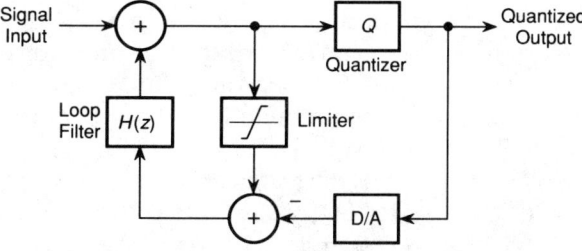

**Figure 11.** A high-order noise shaping circuit with a limiter in the feedback loop. *(From Dijkmans)*

The low-level linearity of low-order noise shaping circuits can be degraded by two problems known as idle patterns and thresholding. Given a zero input signal, a noise shaper may output a regular 1010 pattern. A very low level input may result in a similar pattern, disturbed by double 1s and 0s. If the period of the repetition of such patterns is long enough, they may be audible as a deterministic or oscillatory tone, rather than as noise. Because they occur when the channel is idling, these nonlinear patterns are called idling patterns and result in idle channel noise. The double codes will be generated, or not, depending on the duration of the input signal; the phenomenon is especially characteristic of low-amplitude, high-frequency sinewaves. Because the phenomenon has a frequency-dependent threshold level, below which the signal is not coded, the effect is sometimes called thresholding.

First-order noise shapers exhibit these effects because of their stable 1010 patterns. Higher-order noise shapers are much less prone to the problem because their output patterns are less stable. However, in many multistage designs, the effect can occur in each of the cascaded low-order stages. Thus it is important to add a dither signal in the first stage to disturb any fixed patterns.

In a traditional noise shaper design, the poles of the loop filter are at 0 Hz, as in an ideal integrator; this results in zeros in the audio band. In some noise shaper designs, a technique called zero shifting is used to modify the rising noise spectrum by shifting one or more zeros to the edge of the audio band (e.g. 18 kHz). For example, when two zeros are shifted in a third-order noise filter, noise in the range from 13 kHz to 20 kHz may be reduced, but increased below 13 kHz. Overall, the noise measurement is enhanced. However, suppression of idle patterns and thresholding effects can be reduced; thus the zero shifting technique must be used with care.

As Vanderkooy and Lipshitz [1989] have pointed out, to remove signal distortion, a noise shaping circuit must employ dithering. Figure 12 shows a noise shaping requantizer in which the $H(z)$ block contains an FIR filter with coefficients selected to perform noise shaping. In this case, $H(z)$ could be a single-sample delay $\tau$ in the feedback loop. The term $H(z)$ represents a loop error which is subtracted from the input at each next sample. This corrects for any such errors on average and gives a high-pass shape to all requantization and dither signals present inside the loop. A digital dither signal applied as shown (inside the shaping loop) is identical with a high-pass filtered dither signal applied at a point outside the loop prior to the quantizer. Figure 13a shows the spectrum of the quantized output of an undithered noise shaper when a 937.5 Hz signal of 1 LSB peak amplitude (approximately −90.3 dB) is passed though an undithered requantizer. The spectrum shows many correlated errors with this low-level input signal. When triangular digital dither is applied, a highly uncorrelated spectrum results, as shown in Figure 13b. The noise spectrum of total power $6Q^2/12$ is shaped by this expression:

$$P(\omega) = Q^2\tau/6\pi(1 - \cos \omega\tau), \quad 0 \leq \omega \leq \pi/\tau,$$

where $\tau$ is the sampling interval and $Q$ is the LSB step size.

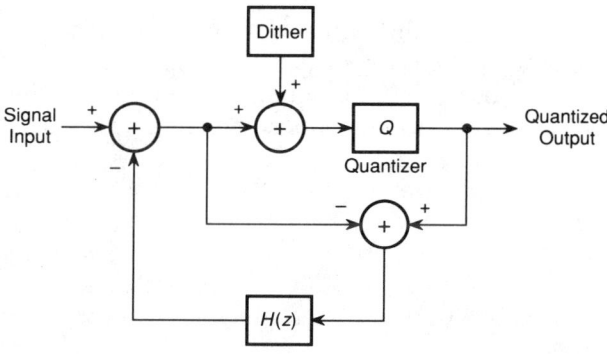

**Figure 12.** A noise shaping circuit must employ dithering to minimize distortion. *(From Vanderkooy and Lipshitz)*

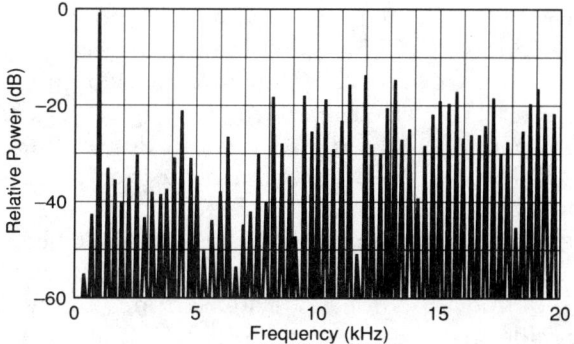

*(a) Spectrum of signal with undithered noise shaper.*

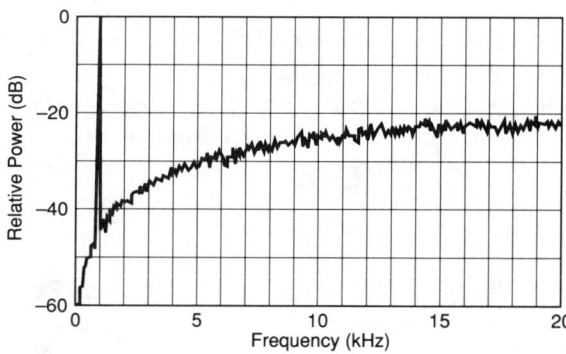

*(b) Spectrum of signal with triangular dithered noise shaper.*

**Figure 13.** Dither profoundly affects the spectrum of the signal output from a noise shaping circuit. *(From Vanderkooy and Lipshitz)*

The quantizer and the dither signal noise are both shaped by the loop. A rectangular dither signal could be applied but could result in noise modulation and limit cycle oscillation. Alternatively, a high-pass triangular dither could be applied; Requantization noise is shaped as before, but the higher frequency dither signal is shaped to even higher frequencies. Correlation, however, may result in higher overall noise. In this example, triangular dither with a white spectrum appears to yield the best results.

Noise shaping is advantageous because a simple shaper can remove quantization noise from the audio band. Higher-order noise shaping algorithms can remove even more noise overall, but relatively more noise is present near the Nyquist frequency. Hence these algorithms are more effective at high oversampling rates so there is more spectral space between the highest audio frequency and the Nyquist frequency; this allows very simple analog low-pass filters to be employed.

Another possible objective of noise shaping is reduction in the number of bits required to represent the signal. Oversampling noise shaping converters may use a 1-bit quantizer to code the signal over two quantization levels through time averaging; this avoids the degradation of differential and absolute nonlinearity which amplitude coding is prone to. Noise shaping is prerequisite in any 1-bit system. To convert a maximum-amplitude 16-bit word, a 1-bit system would have to perform $2^{16}$ toggles per conversion period; with a sampling rate of 44.1 kHz, this would demand a toggle rate of approximately 2.9 GHz. This is too fast for current technology. As the rate is slowed to accommodate hardware limitations, 16-bit performance is lost and noise levels increase to an intolerable level. Hence noise shaping is necessitated in a 1-bit coder. Looked at in another way, bit reduction at a higher sampling frequency is required to output a 1-bit signal from a 16-bit source. This, however, greatly degrades the signal's dynamic range. Noise shaping uses the error generated in the bit reduction process, returning it back to the input through negative feedback to reduce noise in the audible spectrum and hence improve dynamic range.

With any 1-bit system, because of the noise shaping employed, it is difficult to quote a meaningful figure for signal-to-noise (S/N) ratio because the noise level varies with respect to frequency. However, in general, a 1-bit system can provide an audio-band noise floor lower than that encountered in 16- or 18-bit conversion.

## Second-Order Noise Shaping in D/A Conversion

A true 1-bit D/A conversion method, known as "Bitstream" technology, is composed of three elements: oversampling, single-stage noise shaping, and pulse density modulation circuits. In the second-order implementation of the Bitstream system, the sampling rate is increased from 44.1 kHz to 11.2896 MHz—an increase of 256 times. At the same time, the 16-bit signal is converted to a 1-bit signal. It is this fast, 1-bit output that reproduces the audio waveform. The requantization error of the output signal is corrected by feedback. Instead of outputting a signal with conventional requantization error, the error undergoes further processing in the form of cancellation to attenuate its in-band level.

The processor's operation is very accurate, and hence the error of the signal is low. There are only positive and negative full-scale reference points; intermediate points are determined by time averaging. Any errors in the two points would produce an offset error, but not a linearity error. The offset error can easily be removed.

Nonintuitively, this high- or low-level pulse signal reconstructs the audio signal. The output section requires only an on and off state. All 1s would be a full positive signal, all 0s would be a full negative signal, evenly alternating 1s and 0s would be a zero-level signal, and other variations would create intermediate levels. More specifically, the method uses PDM (pulse density modulation). Figure 14 shows how a single bit, with either a high or low level, can output, in this example, a smooth sinewave. Proof of performance can be illustrated by the distortion of a PCM converter versus that of a PDM converter; the converter has lower distortion across the audio band.

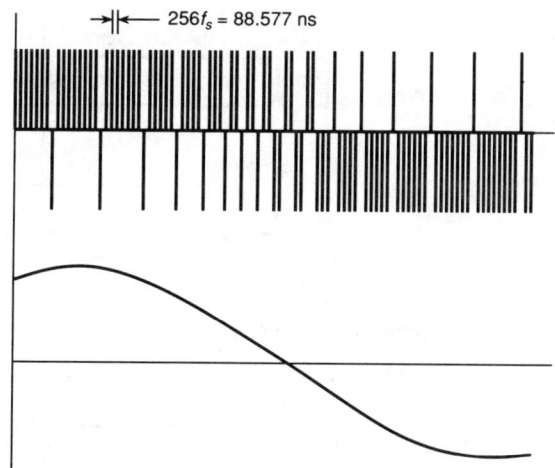

**Figure 14.** Pulse width modulation can be used to reconstruct an analog waveform.

Figure 15 shows the complete Bitstream system. The first of the three oversampling stages performs four-times oversampling to attenuate image spectra; in addition, first-order noise shaping in performed in the filter. The second stage performs 32-times oversampling. A dither signal (−20 dB at 352 kHz) is added to prevent idle patterns from causing nonlinearity. Two-times oversampling is performed in the third stage. This 17-bit signal (dither adds 1 bit to the original 16-bit signal) undergoes second-order noise shaping as described above, and a single bit is output from the quantizer. Finally, D/A conversion is accomplished at a 1-bit D/A converter via pulse density modulation, outputting 1-bit data at 256-times oversampling. A third-order analog low-pass filter removes out-of-band high-frequency components.

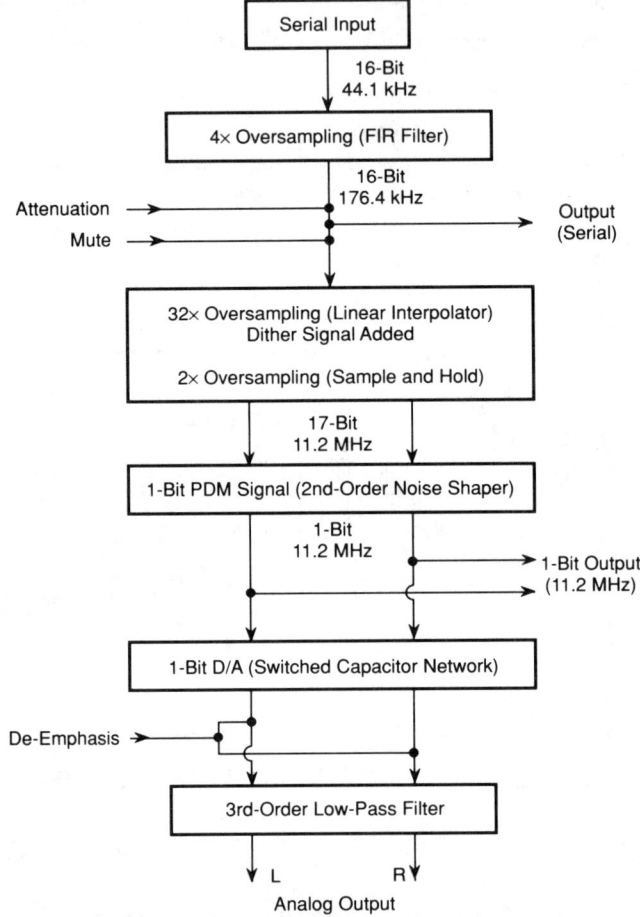

**Figure 15.** Processing elements in a second-order, pulse width modulation D/A conversion system. *(Courtesy Philips)*

The first of the three oversampling stages uses a nonrecursive FIR interpolation filter as shown in Figure 16. First-order noise shaping is performed in the accumulator of the multiplier. The signal is downscaled to prevent clipping during any signal overshooting. The gain of the filter is selected to compensate for the effects of other stages in the conversion path. In particular, a high-frequency rise is used to compensate for the aperture error (−3 dB at 60 kHz) present in the output analog filter. Thirty-two–times oversampling is performed in the second stage filter through linear interpolation. In the third stage, two-times oversampling is performed through a sample-and-hold operation.

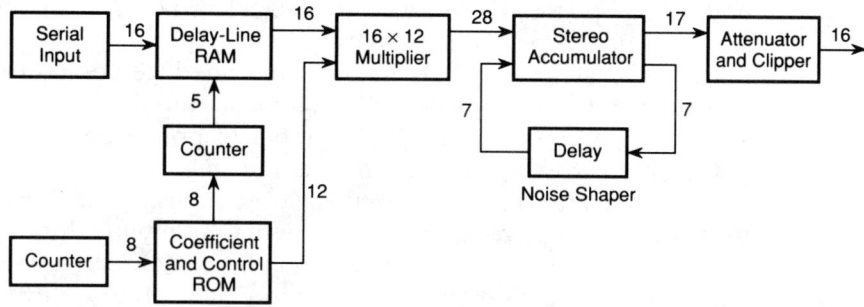

**Figure 16.** An example of a nonrecursive filter with noise shaping.

As noted, noise shaping coding can be used in conjunction with oversampling to reduce wordlength. In the case of the second-order implementation of Bitstream, the quantization noise introduced by the wordlength reduction is spectrally shaped by a low-pass feedback loop around the quantizer as shown in Figure 17. Signals are processed in two's complement format; the 1-bit code from the quantizer is the output sign bit. The remainder of each sample is fed back as a quantization error, after a limiting operation designed to prevent overflow. A 21-bit data bus is used within the loop. Second-order noise shaping is performed by adding double the error and the negative value of the error to the previous two samples:

$$Y = X + (1 - z^{-1})^2 N,$$

where

$Y$ is the noise shaped output,
$X$ is the input signal,
$N$ is the quantization error.

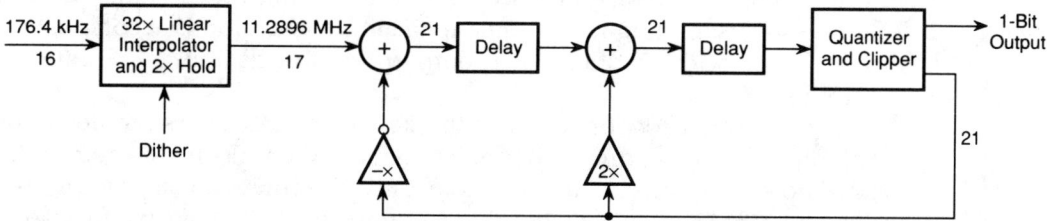

**Figure 17.** A second-order noise shaping circuit using a low-pass filter feedback loop and clipper.

Two values (±1) are output at 11.2896 MHz. This pulse density modulation signal is converted into an analog signal using a switched dual-capacitor network. Two control signals representing the data stream's logic 0 and logic 1 values control the switching of the capacitors, subject to a clock pulse. During the negative half of the clock, the first capacitor discharges while the second capacitor charges. During the positive half, if the data is a logic 1, the first capacitor is charged by taking a fixed amount of charge from the summing node of an op amp. If the data is logic 0, a fixed charge is transferred into the summing node from the second capacitor. In this way, there are only plus and minus full-scale reference points, and intermediate points are determined by time averaging. There is no MSB change around zero, for example, because zero is represented by an equal number of positive and negative full scale pulses. Zero-cross distortion is thus eliminated.

The converter retains its accuracy over environmental extremes. Because the process is strictly computational, there is negligible drift with temperature, humidity, or age. A stereo converter can be contained on one chip. The chip may be used in a differential configuration; two stereo converters can yield a theoretical 6 dB gain in signal level; in practice there is a 4 dB improvement in S/N ratio, a 2 dB improvement in total harmonic distortion (THD), and a much greater increase in channel separation. Bitstream was developed by Philips N.V. Figure 18 summarizes the mathematical basis of second-order noise shaping. Philips has also developed a single-stage noise shaper which yields third-order noise shaping.

## Third-Order Noise Shaping in D/A Conversion

Several manufacturers have developed D/A conversion methods which employ third-order noise shaping. However, their design details differ somewhat, particularly in the noise shaping algorithms and the low-bit output signal. Two third-order noise shaping conversion methods are discussed below, both using pulse width modulation.

The MASH system is a multistage third-order noise shaping method. One implementation of this design accepts 16-bit words at a nominal sampling frequency, and a digital filter stage performs 8-times oversampling and outputs 24-bit words. Noise shaping circuits output data as an 11-value signal, at a 32-times oversampling rate. Digital-to-analog conversion is accomplished via PWM (pulse width modulation), outputting 1-bit data at a 768-times oversampling rate.

As with other low-bit systems, the key to the MASH system lies in noise shaping; in this case, third-order noise shaping is employed. Generally, if cascaded noise shaping circuits exceed second order they can be prone to oscillation; the MASH circuit avoids this through its multistage configuration. A simplified schematic for the MASH noise shaper is shown in Figure 19; it contains a first-order noise shaper in parallel with a second-order noise shaper. The input signal is applied to the quantizer $Q_1$, after the error signal through the delay block is subtracted from the input. The signal output from the first loop is also applied to the second loop. The output of the quantizer $Q_2$ is differentiated and added to the output of the first loop to form the final output sig-

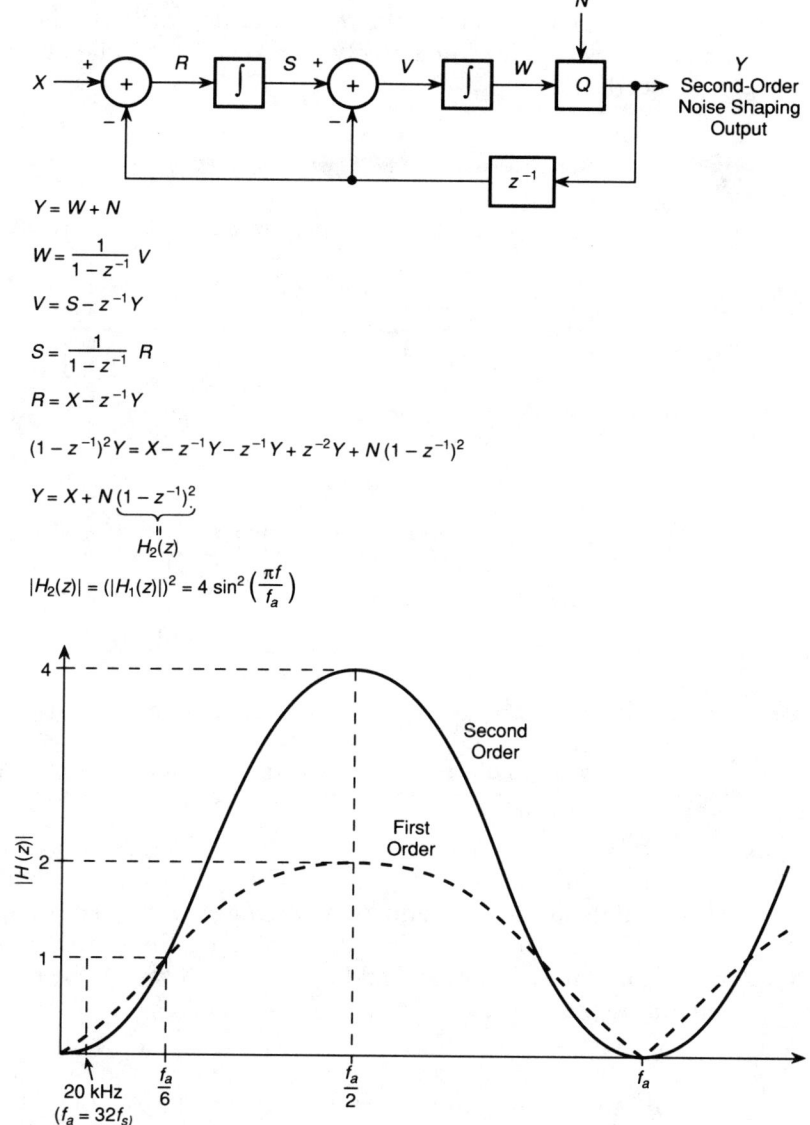

**Figure 18.** Analysis of a second-order noise shaper.

nal. Thus the requantization error of the first loop is requantized by the second and canceled by adding the requantized noise to the first loop's signal. The outputs of each stage can be characterized as follows:

$$Y_1 = X + (1 - z^{-1})N_1,$$
$$Y_2 = -N_1 + (1 - z^{-1})^2 N_2,$$

where $X$ is the input signal, the requantization errors of the local quantizers $Q_1$ and $Q_2$ are $N_1$ and $N_2$, respectively, and $Y_1$ and $Y_2$ are the outputs of stages 1 and 2, respectively.

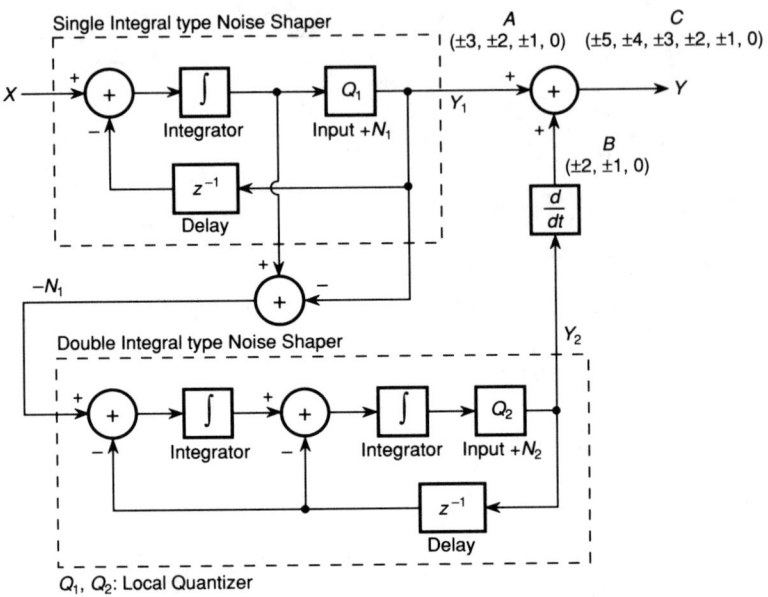

**Figure 19.** A multistage third-order noise shaping circuit outputting eleven data values prior to PWM reconstruction. *(Courtesy Matsushita Electric)*

When both sides of the second equation are multiplied by $(1 - z^{-1})$ and this is added to the first equation, we observe that the requantization error of the first stage can be canceled. By passing the output of the second stage through a differentiator and adding it to the output of the first stage, the overall circuit output $Y$ is

$$Y = X + (1 - z^{-1})^3 N_2.$$

In other words, the requantization error $N_2$ is output with a third-order differential characteristic (18 dB/octave), achieving reduced in-band noise compared to first- and second-order characteristics.

Input data is linearly requantized into a digital signal with seven values ($\pm 3, \pm 2, \pm 1$, and 0) at the main loop, and at the subloop the requantization error is requantized into five values ($\pm 2, \pm 1$, and 0). When these output values are added together, the 4-bit digital signal is output from the circuit as eleven values ($\pm 5, \pm 4, \pm 3, \pm 2, \pm 1, 0$). These different data values are shown graphically in Figure 20. Using a vertical scale to represent amplitude of the values, it can be seen that the main loop outputs seven values of rough accuracy, and the subloop outputs five values with high-frequency content, used to eliminate the requantization error of the main loop. When summed, eleven values are output. These values represent the audio signal; for example, Figure 21 shows the spectrum of 20 kHz input data, using a computer simulation.

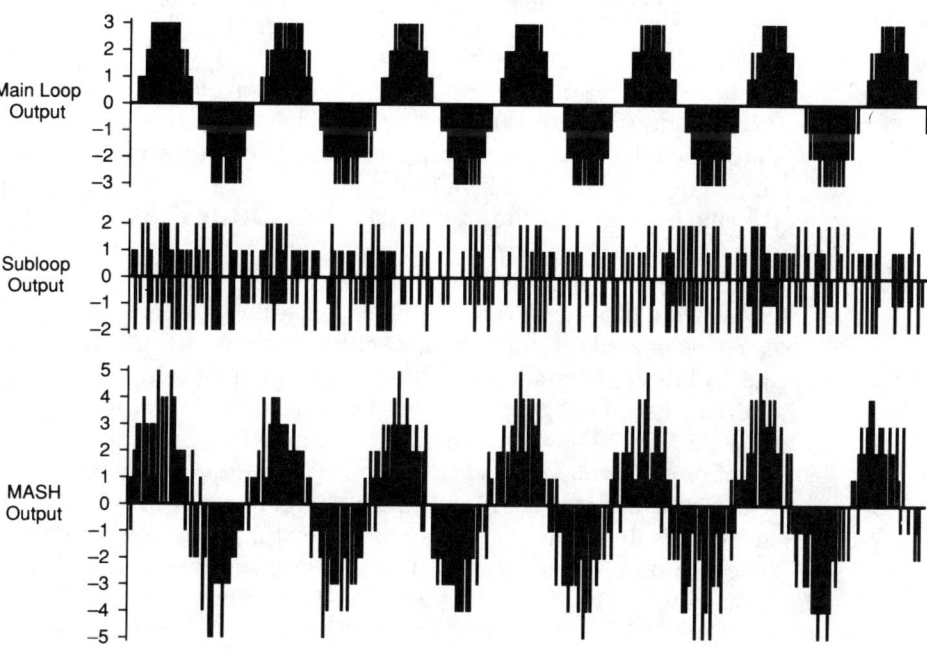

**Figure 20.** A graphical representation of the data values in a multistage noise shaper. *(Courtesy Matsushita Electric)*

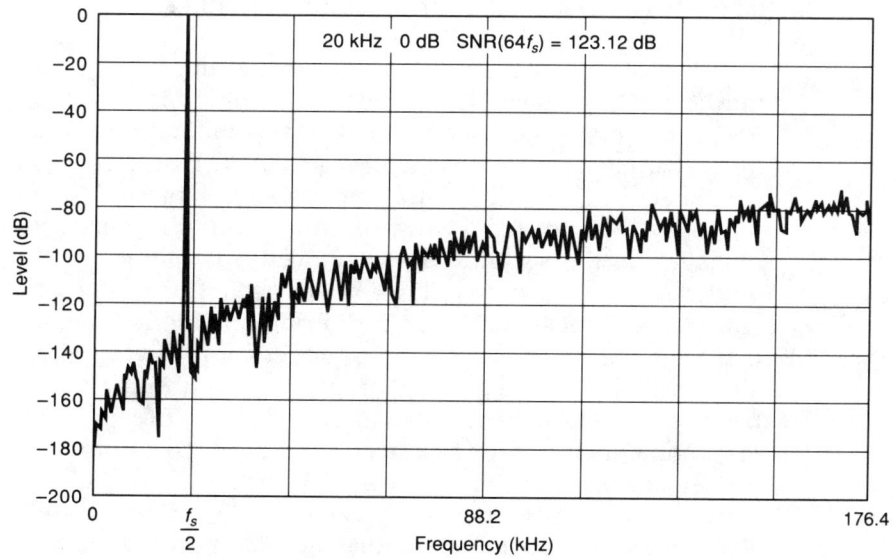

**Figure 21.** Reproduction of a 20 kHz waveform showing the effect of third-order noise shaping. *(Courtesy Matsushita Electric)*

The final element in the system is D/A conversion. The eleven-value signal is converted into pulses, each with a width corresponding to one value, as shown in Figure 22a. This can be acomplished by applying the 4-bit output of the DSP to a ROM to map 11 amplitude values into 22 time values with constant amplitude. For example, the figure shows the PWM waveforms resulting from the 0, +3, and −3 output levels. In actuality, waveforms representing ±5, ±4, ±3, ±2, ±1, and 0 are all output. The widest pulses translate into a large positive output, while the narrowest pulses translate into a large negative output, as shown in Figure 22b. It is important to note that the width of the pulses carries the vital information; the amplitude of this signal can only be high or low. At this point the signal has the form of 1-bit binary data. Because the signal is represented by a pulse width modulation waveform, conversion is performed by simply passing the signal through a low-pass filter. Because great timing accuracy can be achieved through crystal oscillators (jitter specification less than 200 ps), the widths are very accurate, and hence the error of the signal is low. Positive- and negative-going pulses may be output, to cancel common noise. This 33.8688 MHz ($768 \times f_s$) 1-bit data forms a PWM representation of the waveform. Proof of performance can be illustrated by observing the linearity of the system; it is linear through −110 dB. Such a method achieves 20-bit resolution, but greater resolution is possible. Moreover, these systems are applicable to both output and input stages as well. MASH was codeveloped by Matsushita Electric Industrial and Nippon Telegraph and Telephone Corporation.

HDLC (high-density linear converter) is another example of a third-order noise shaping architecture which uses pulse width modulation to perform conversion. This method operates similarly to other low-bit architectures in that oversampling and noise shaping are employed to provide satisfactory dynamic range. In this case, with third-order noise shaping and 64-times oversampling, a dynamic range of 118 dB is achieved.

This conversion method in similar to that employed in MASH in that it uses third-order noise shaping. However, a modified feedback path is employed. Figure 23 shows the configuration of this multistage noise shaper. Primary noise shaping is performed on the signal in the first stage; a five-value output is generated. The resulting noise component is subjected to further noise shaping in the second stage. However, under certain signal conditions, the first integrator in the second stage may overload. One solution to this is an extended feedback path (as shown in Figure 23 from point 1 to point 2); it provides an additional correction path from the second stage to the input of the first-stage quantizer. Since the first integrator receives ordinary feedback data along with data which passes through the quantizer in the first stage, the feedback path helps increase operation speed and prevent overload. Since the error components of the first stage are reduced, the output of the quantizer in the second stage can be reduced to two values. In this way, the proportion of signal to error components is greater than for the unmodified noise shaper. The overall dynamic range is improved by 6 dB while maintaining good third-order noise shaping characteristics with a noise distribution of 18 dB/octave.

In this design, the output values of the noise shaper code stream are used

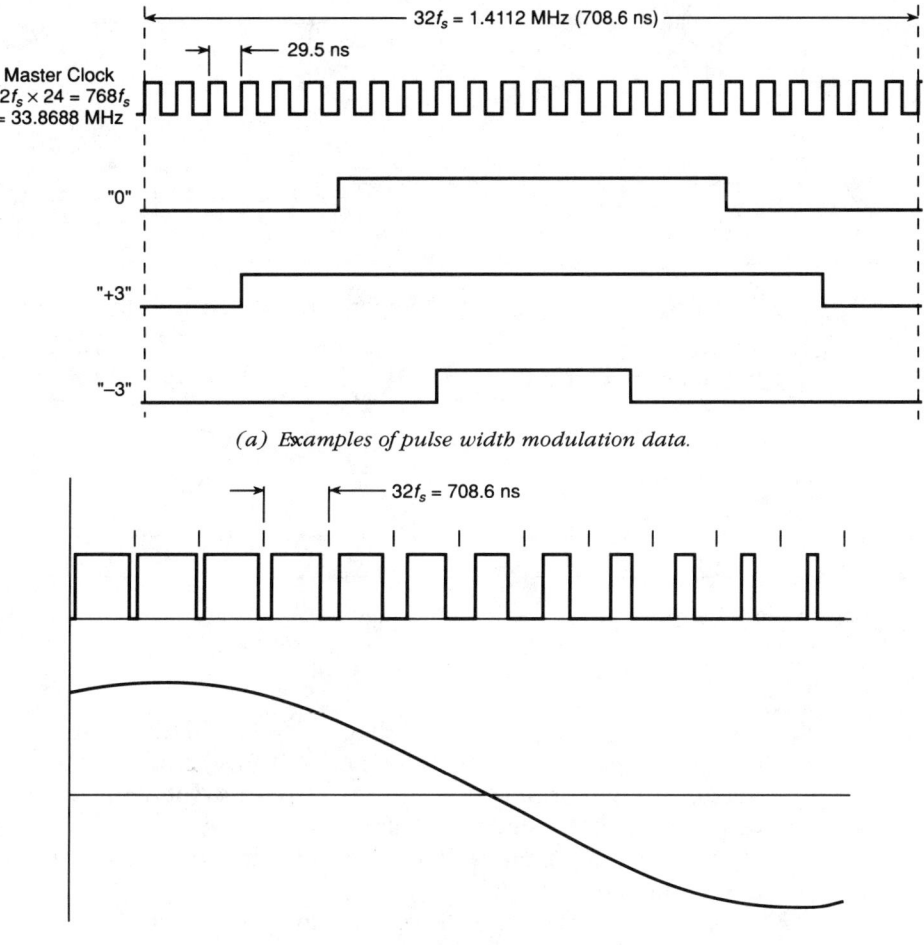

(a) Examples of pulse width modulation data.

(b) Reconstruction of analog waveform.

**Figure 22.** Pulse width modulation data is output from a MASH converter.

to generate pulses of varying width. In general, the accuracy of any pulse-type converter is proportional to the number of pulses it can generate per sample. When the pulse rate is doubled, the dynamic range is theoretically increased by a factor of ten. Hence a fast pulse rate is desirable. In this case, an output pulse generator operates at 45.2 MHz ($1024 \times f_s$). It uses pulse width modulation to output seven data values. As shown in Figure 24a, the length between logical 1 values is modulated according to input data. Modulation in pulse width, however, can create harmonic distortion caused by the asymmetrical modulation structure: While a logical 1 is modulated directly by noise shaping, a logical 0 is modulated by a complement of the maximum modulation value, with the result split into two halves, each being added to the two adjoining data blocks equally. Thus asymmetry is generated by this aperture effect, which is

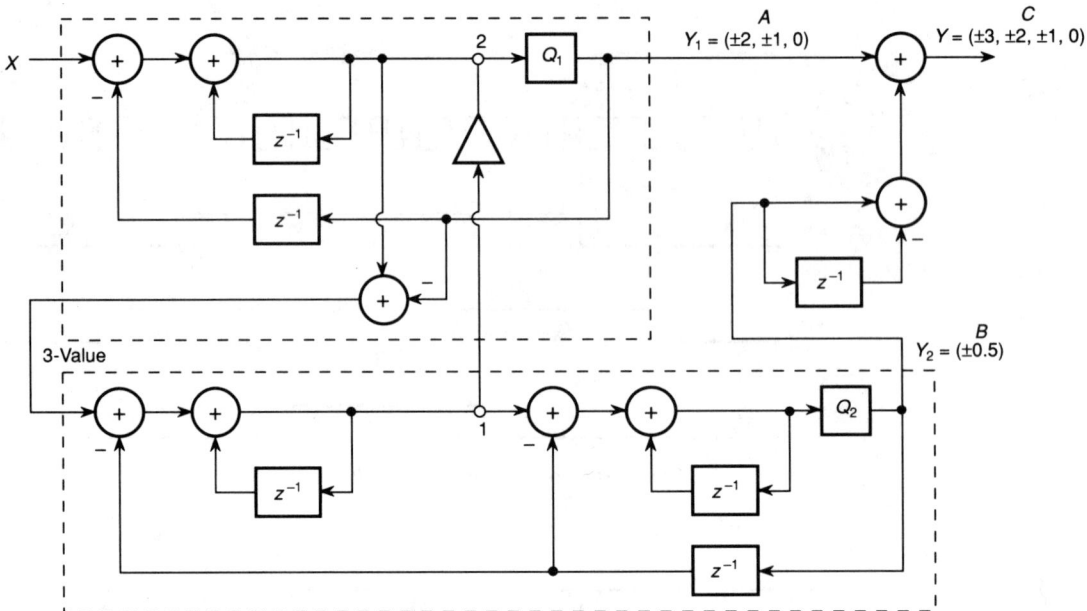

**Figure 23.** An example of a third-order multistage noise shaper with an extended feedback path from point A to point B. (*Courtesy Sony Corp.*)

caused by the ratio between the signal cycle and the pulse width. The result is even-order harmonic distortion, and it is more apparent when the signal frequency is high, or the pulse rate is low because of the relatively small number of pulses in each signal waveform. (Actually, it varies in proportion to the square of the signal frequency and in inverse proportion to the square of the operating rate.)

Although slight, this distortion can be eliminated entirely with a complementary architecture, as shown in Figure 24b. The distortion cannot be canceled by the common method of pulse inversion and differential composition. In a complementary configuration, however, the distortion can be canceled by modulating the pulses using data complementary to each other and then composing the results differentially using two pulse converters. In this case, one converter outputs a standard waveform, while the other outputs a complementary waveform. The difference between these waveforms (taken before pulse conversion in this example) forms the output signal. The 0 value generates a true null, and positive and negative pulses are symmetrical about the zero reference. As a result, spectral analysis shows the output signal to be free from even-numbered harmonic distortion components. The output analog waveform is generated through time averaging.

To help maintain a high timing accuracy, this design also employs a synchronization circuit before the D/A converter. This circuit minimizes jitter which could disrupt the time axis accuracy of the pulse width modulation output. This HDLC architecture was developed by Sony Corporation in con-

# Chapter 12: Low-Bit Conversion and Noise Shaping

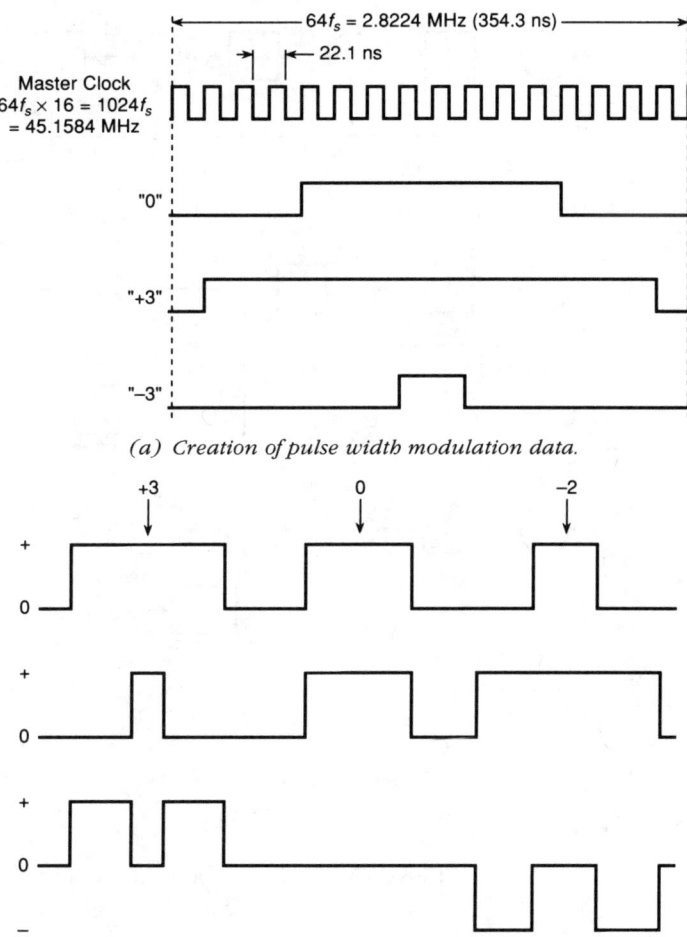

(a) Creation of pulse width modulation data.

(b) Use of complementary architecture to eliminate asymmetry distortion.

**Figure 24.** Pulse width modulation data is output from an HDLC converter.

junction with Nippon Telegraph and Telephone Corporation. Figure 25 summarizes the mathematical basis of third-order noise shaping.

## Quasi–Fourth-Order Noise Shaping in D/A Conversion

VANS (Victor advanced noise shaping) is an example of a 1-bit D/A converter architecture using eight-times oversampling, a quasi–fourth-order noise shaper, and a pulse edge modulation (PEM) converter. The noise shaper employs four loop filters in a configuration which yields in-band performance equivalent to fourth-order noise shaping. The VANS circuit is designed to oper-

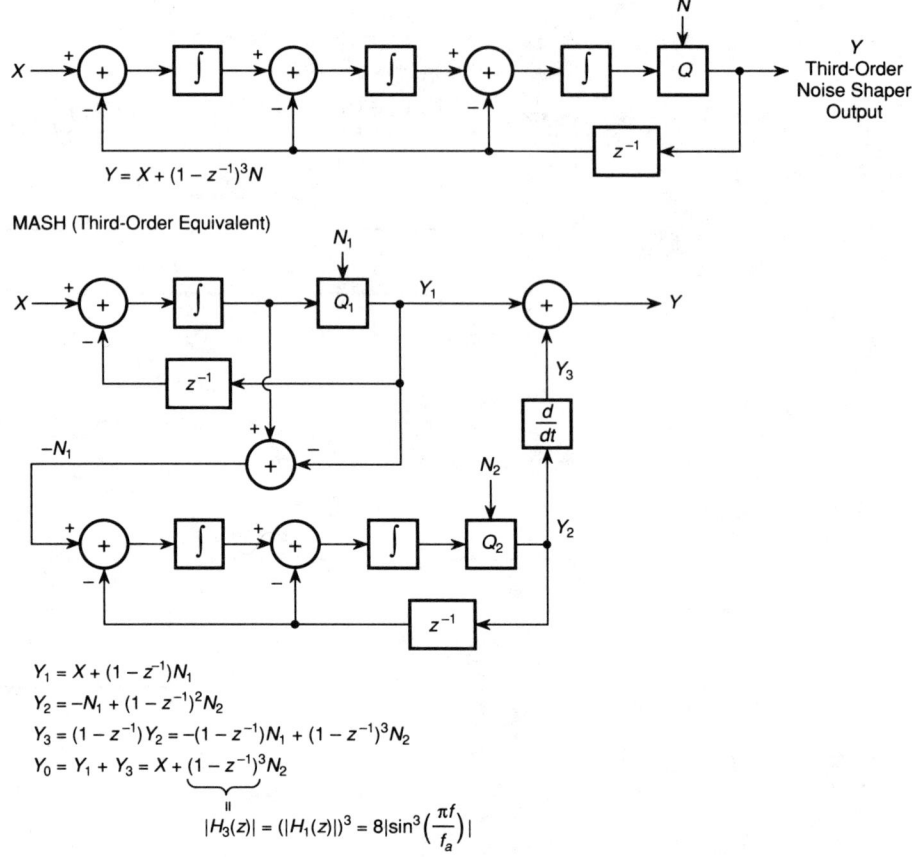

**Figure 25.** Analysis of a third-order noise shaper.

ate like a fourth-order noise shaper at audible frequencies but gradually shifts toward a second-order noise shaper at higher frequencies, as shown in Figure 26. This provides stability yet improves performance in the audio band.

Thirty-two–times oversampling is performed, and the output clock rate is 16.9344 MHz ($384 \times f_s$). With pulse edge modulation, input data is converted into a 4-bit binary pulse train with fifteen discrete values ($\pm 7$, $\pm 6$, $\pm 5$, $\pm 4$, $\pm 3$, $\pm 2$, $\pm 1$, 0). A differential configuration is used in which the rise of the leading edge of a pulse and the fall of the trailing edge of a pulse are output by two independent pulse edge modulation converters, as shown in Figure 27a. This determines the width of the pulse. Two converters output pulse trains $A$ and $C$ based on the input signal, and an analog subtractor generates a composite ($A$–$C$) signal determined by the leading and trailing edges of the pulse trains. This signal can take either a positive or negative value, as shown in Figure 27b. For example, data representing a $-1$ value would generate a short negative-going pulse, whereas data representing a $+5$ value would generate a longer positive-going pulse. When time averaged, these values create the analog out-

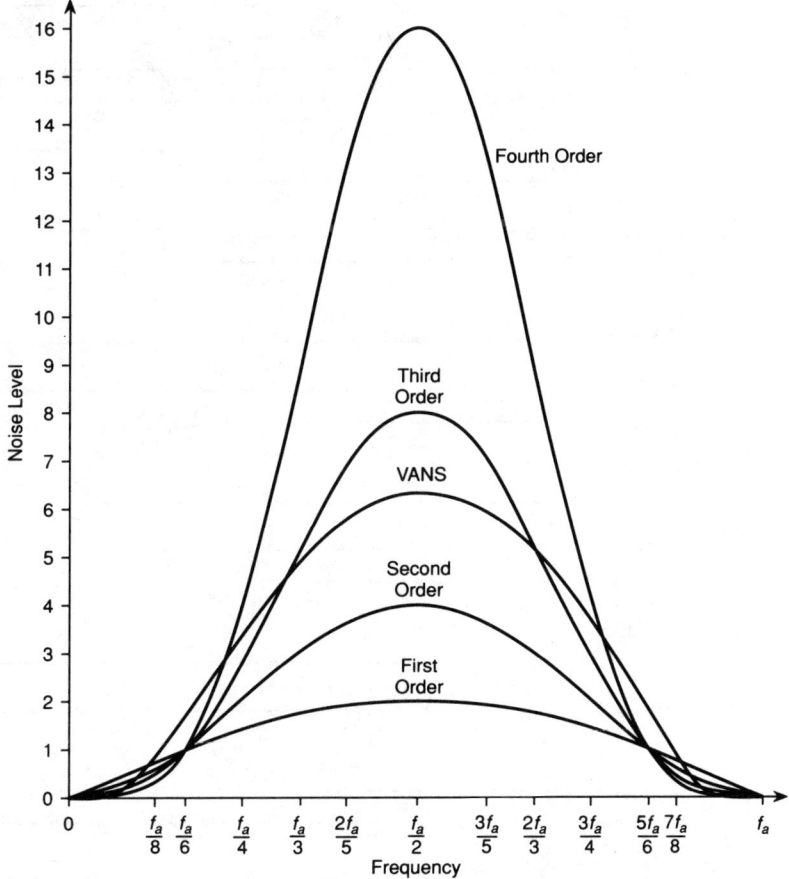

**Figure 26.** The VANS architecture provides fourth-order noise shaping in the audio band, then gradually shifts toward second-order noise shaping at higher frequencies. *(Courtesy Victor Company of Japan).*

put waveform, as in pulse width modulation conversion. VANS pulse edge conversion method was devised by Victor Company of Japan. Figure 28 summarizes the mathematical basis of quasi–fourth-order noise shaping.

## Low-Bit A/D Conversion

In the past, A/D conversion has relied on three methods: successive approximation, integrating, and flash. Successive approximation A/D converters compare the unknown input with accurately known fractions of a reference voltage. Starting with the largest fraction and rejecting any fraction that causes the sum to be larger than the unknown input, $k$ iterations are required for a $k$-bit word conversion. An integrating converter counts pulses for a period proportional to

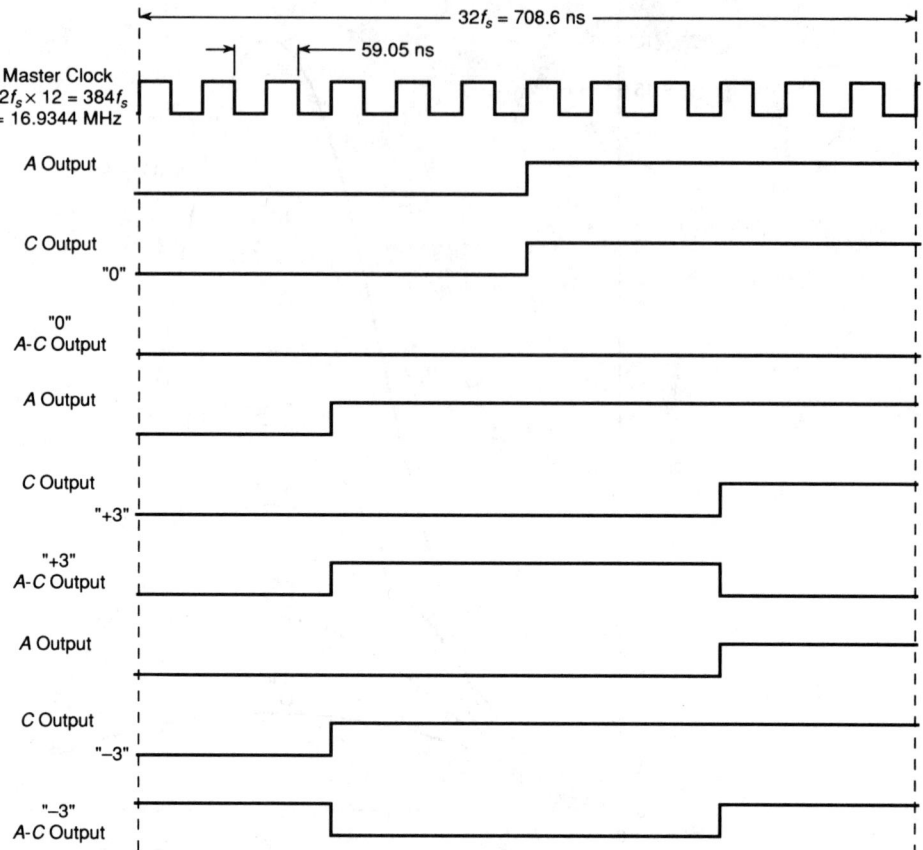

*(a) Creation of the pulse edge modulation data.*

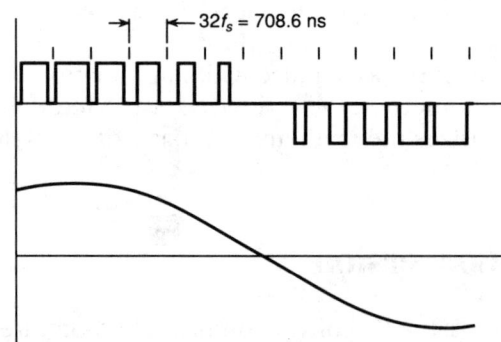

*(b) Reconstruction of the analog waveform.*

**Figure 27.** The VANS architecture uses a differential configuration to determine rising and falling edges of the output pulse.
*(Courtesy Victor Company of Japan)*

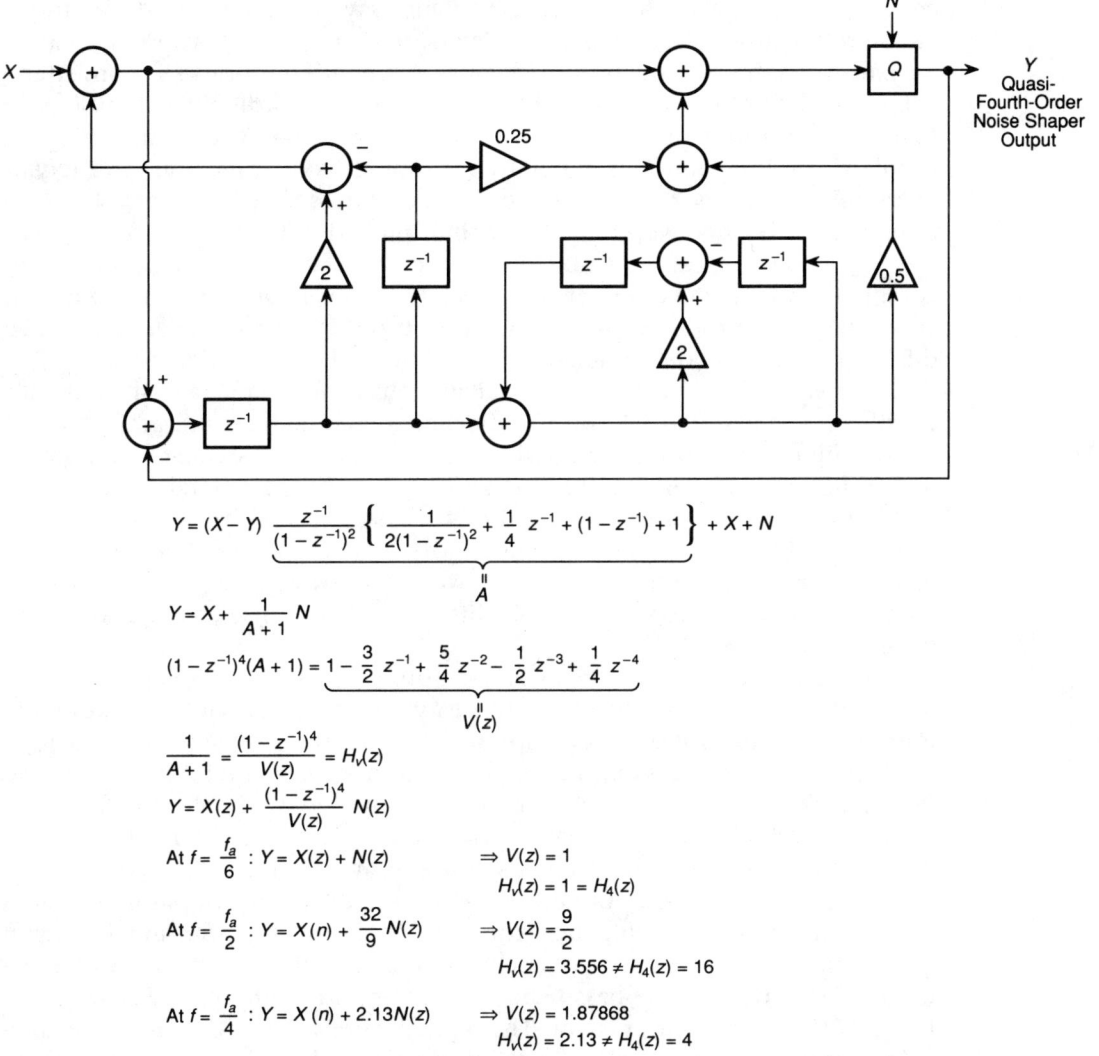

**Figure 28.** Analysis of a quasi–fourth-order noise shaper.

the input signal. Dual-slope converters are often used, to count the period required for the integral of the reference to equal the average value of the input over a fixed period. A flash converter requires only one step to form a low-resolution conversion. A series of comparators tests the input signal against a set of voltage thresholds established by a ladder network. The comparator outputs are transformed into a binary output by encoding logic. However, $2^k - 1$ comparators are required for a $k$-bit word, as well as encoding logic.

Conventional A/D converters, either directly or through associated circuitry such as brickwall filters, can contribute a substantial amount of distortion to the signal. To decrease conversion distortion, the linearity of the conversion process must be increased. One way to accomplish this is to increase the relative accuracy of the conversion by increasing wordlength; thus we have seen the introduction of longer-wordlength multibit A/D converters, using a variety of conversion architectures. Longer-wordlength multibit A/D converters improve performance, but the brickwall filters still present problems, and resolution is generally constrained to 16 bits. Thus a variety of oversampling chip sets, using low-bit architectures, have been introduced to remedy the ills of conventional input filtering and A/D conversion.

Some converter architectures use several first- or second-order coders in combination to achieve higher-order, stable noise shaping. Instead of a 1-bit code, they may produce a low-bit word of three or four bits. For example, a differential pulse code modulator differs from a delta modulator only in that the error signal is quantized to more than 1 bit. For example, a flash converter could be used inside the loop. The dynamic range increases in proportion to the resolution of the quantizer. Other designs may use a sigma-delta modulator modified to contain a multibit quantizer. Again, dynamic range can be increased according to the resolution of the quantizer. Because the output is proportional to the signal's amplitude rather than slope, it is like a PCM converter. Unlike a conventional PCM converter, however, the noise floor rises with increasing frequency. In short, this is a low-bit noise shaping A/D converter.

As noted, the real potential in oversampling converters lies in the opportunity to provide improved converter linearity specifications, yielding extremely low distortion, even at very low amplitudes. This is the special province of low-bit systems. These oversampling A/D designs use a very high initial sampling rate and take advantage of that high rate by using low-bit intermediate coding of the audio signal. Moreover, a sample-and-hold circuit is not needed. Sigma-delta modulators are increasingly used in this application.

As Adams [1986] has pointed out, oversampling converters provide high resolution not by decreasing the error between the analog input and the digital output, but by making the error occur more often. In this way the error spectrum extends beyond the pass-band of interest. Although total noise power is high, in-band noise power is low. The high bit rate is reduced to more manageable rates through decimation. Decimation may be described through a simple example. Sixteen 1-bit values could be reduced through a 16:1 decimation to a single low-bit value; for example, values 1, 0, 1, 0, 0, 1, 0, 1, 1, 0, 1, 1, 1, 1, 0, 0 would be decimated to 9/16, or 0.5625. Because there is only one (low-bit) output value for every sixteen input values, the decimator has decreased the sample rate by 16:1. As Park [1990] has pointed out, it is also important to note that decimation has increased resolution; in this example the input signal is only one bit but the decimation (averaging) process yields 4-bit resolution ($2^4 = 16$) while reducing the sampling rate. Thus oversampling followed by decimation demonstrates how speed can be exchanged for resolution.

A diagram of the A/D process is shown in Figure 29. In theory, an oversampling A/D converter is simple: The input signal is first passed through a simple analog anti-aliasing filter, and the input signal is sampled at a very fast rate of $1/T$ to extend the Nyquist frequency. The signal is applied to a coarse quantizer, which adds noise to the signal. The digital data is low-pass filtered with a cutoff at the Nyquist frequency; this removes out-of-band noise components and serves as the converter's anti-aliasing filter. Finally, the signal is resampled at a rate lower than $1/T$ (e.g. 44.1 or 48 kHz) without aliasing, for storage or processing using normal methods. A sample-and-hold circuit is not needed because an input sample can be taken during every internal clock cycle, whereas in successive approximation converters the sampled analog value must be held for the number of clock cycles equal to the number of bits being converted.

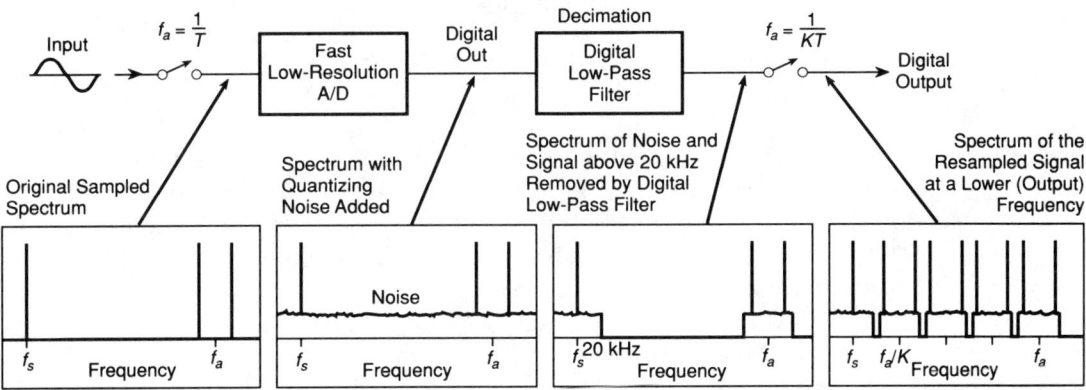

**Figure 29.** Diagram showing the theory of oversampling A/D conversion. *(From Adams)*

In oversampling A/D conversion the digital filter is a decimating filter, operating as a low-pass filter. It removes the frequency components outside the audio band to prevent aliasing between the audio signals and the resulting sampling rate. This would occur when the output of the digital filter is resampled (undersampled) at the system's sampling rate. An analog low-pass filter is required at the input to remove the frequency components which cannot be removed by the digital filter; however, because the preliminary sampling rate is high, the analog low-pass filter is low-order.

As noted, a number of methods have been devised to perform low-bit coding including delta modulation and differential pulse code modulation. Several sigma-delta methods have also been applied to A/D conversion, all using a high input sample rate and noise shaping to increase coder resolution: single and dual integrator loops, cascading of first-order sigma-delta loops, and low-bit quantizers with loop filters. The first two methods use true 1-bit coders with inherent linearity. The third method uses several bits, and noise is reduced in proportion to the number of quantizer levels used. The converter's linearity, however, depends on the linearity of the quantizer. In any case, noise performance hinges on the oversampling rate and order of noise shaping employed.

A second-order sigma-delta modulator may be shown to have the following transfer function:

$$Y = X + (1 - z^{-1})^2 N,$$

where

> $Y$ is the modulator output,
> $X$ is the input signal,
> $N$ is the quantization noise.

Then, as demonstrated by Thompson [1989], $k$-bit resolution requires an oversampling rate:

$$R = [(\pi^2/\sqrt{5}) \times 10^{6k/20}]^{2/5},$$

where $R$ is the oversampling rate defined by

$$R = f_a/f_s,$$

where $f_s$ is the output sampling frequency and $f_a$ is the oversampling frequency.

Thus 16-bit resolution would require an oversampling rate of 150. A 100 kHz output sampling rate would thus necessitate a filter sampling rate of 15 MHz. If the order of noise shaping is raised to third order, the signal delta transfer function is

$$Y = X + (1 - z^{-1})^3 N,$$

and the required oversampling rate is described by

$$R = [(\pi^3/\sqrt{7}) \times 10^{6k/20}]^{2/7},$$

Thus the required oversampling rate is 48; this is well within practical design limits.

A sigma-delta modulator can be used to create the low-bit coding from the low-pass filtered analog signal. A first-order sigma-delta modulator is shown in Figure 30. A low-resolution (1-bit quantizer) D/A converter operating at a high sampling rate is placed in a feedback loop. The input to the loop filter is the difference between the input signal and the quantized output converted back to an analog signal; this difference is theoretically equal to the quantization error. Because a coarse (1-bit) quantizer is used, quantization error at sampling time is large; however, the loop acts to produce an output bit that when averaged (through decimation) is very precise. High resolution (manifested as dynamic range) is achieved through noise shaping. The noise shaping in a sigma-delta circuit is the inverse of the transfer function of the $(1 - z^{-1})^{-1}$ filter. A filter with higher gain at low frequencies is thus desired to attenuate audio-band noise. This transfer function is essentially a high-pass filter; noise is thus shifted to a higher frequency. Because the input sampling rate is high, an anti-aliasing filter is not needed, or a simple one-pole $RC$ filter would suffice. Sigma-delta converters have recently gained prominence because of the availability of low-cost, high-speed binary multipliers and adders—a benefit of DSP development.

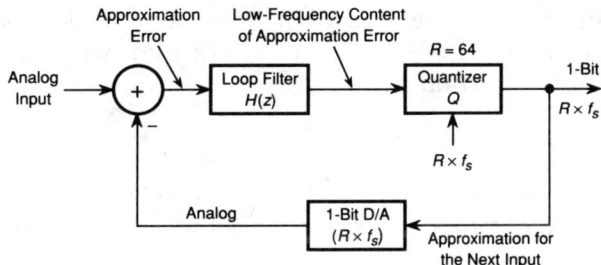

**Figure 30.** A first-order sigma-delta modulation circuit showing 1-bit D/A converter in feedback loop.

Depending on its order and design, a sigma-delta feedback loop generally consists of the following operations: subtraction of output from input to find the approximation error, filtering to extract the low-frequency content of the approximation error, 1-bit D/A conversion of the output code into a signal to subtract it from input analog signal, and quantization to output a 1-bit approximation for the next input sample. In practice, a third-order loop can be used to shape the noise toward higher frequencies, where it is removed by the subsequent decimation (undersampling) filter. As previously described in the discussion of noise shaping, the quantization noise at low (in-band) frequencies is attenuated. The feedback loop thus shapes the noise from the quantizer. The high-frequency (out-of-band) noise is suppressed by the digital decimation filter following the noise shaping.

The decimation process low-pass filters the signal and noise in the 1-bit code, bandlimiting the 1-bit code prior to sample rate reduction to remove alias components. Decimation also replaces the 1-bit coding with 16-bit coding, for example, provides a lower sampling rate, and generates a PCM output. However, the computation rate of the filter is not trivial; output samples cannot be discarded (providing decimation) until the filtering computation is complete. For example, in some designs a recursive filter may be used to reduce computation requirements. The Nyquist criterion must be observed to prevent the rising signal spectrum from aliasing with the input spectrum. However, only certain portions of the spectrum would alias into the audio band; thus the decimation filter need only attenuate those frequency bands. In particular, these frequencies are described by

$$f_{\text{alias}} = I \times f_i \pm 20 \text{ kHz},$$

where $I$ is any integer and $f_i$ is the decimation filter's intermediate resampling frequency.

The decimation filter can be designed so that its frequencies of maximum attenuation will coincide with the potentially aliasing frequencies. As Park [1990] points out, a comb filter is an expedient choice because its design does not require a multiplier. However, in many designs a comb filter cannot wholly remove out-of-band quantization noise; thus it must be followed by additional filter stages of other design. These additional stages may also be needed to compensate for high-frequency drooping caused by the comb filter. A comb filter of length $R$ is an FIR filter with coefficients equal to unity; its transfer function is

$$H(z) = \sum_{n=0}^{R-1} z^{-n}.$$

In other words, this expression shows a moving average. For example, if $R = 4$,

$$y(n) = x(n) + x(n-1) + x(n-2) + x(n-3).$$

In recursive form the transfer function can be written as

$$H(z) = \frac{1 - z^{-R}}{1 - z^{-1}}.$$

This can be expressed in terms of integration followed by differentiation:

$$Y = (1 - z^{-1})^{-1} (1 - z^{-R}) X.$$

This single-stage comb filter can easily be realized, as shown in Figure 31a. Not only is storage not required for the filter coefficients, but the burden of intermediate computations is decreased owing to the low sampling rate at the differentiator. In addition, the same topology can be used for higher orders of rate change. Indeed, in practice a single comb filter stage does not provde sufficient stop-band attenuation to prevent aliasing; thus cascaded stages are often employed, as shown in Figure 31b. In this example, four sections are cascaded, requiring eight data filters and $4(R+1)$ additions per input sample. The comb filter is designed for maximum atenuation at higher-frequency components which would alias after rate decimation. Figure 31c shows the spectrum with first-, second-, third-, and fourth-order cascaded comb filter sections.

In some decimator designs, the cascaded comb filter is followed by an FIR filter; the intermediate rate output from the comb filter is further decimated, and the FIR section provides sharp filtering when the sampling rate is reduced to nominal values (e.g. 44.1 kHz). The decimation factor is typically lower in the FIR section as compared with that in the comb filter section; however, the FIR must provide extreme stop-band attenuation. In addition, the FIR section may provide compensation for audio-band droop caused by the comb filter. FIR computation also provides linear phase response.

Consider an example in which 1-bit coding takes place at $64 \times 44.1$ kHz = 2.8224 MHz. The decimating filter may have two stages. With a $64 \times f_s$ input bit stream, the first stage may generate a multibit output sample at a sample rate of $2 \times f_s$. The second stage of the decimation filter may use a multibit multiplier with convolution performed at the output sample rate of $f_s$. In all, the decimating filter provides a stopband from 20 kHz to the half-sampling frequency of 1.4112 MHz. The analog filter at the system's input is modest, perhaps first or second order, ensuring phase linearity in the audio band. Figure 32 summarizes the operation of a 1-bit A/D converter in the frequency domain.

The use of 1-bit coding as the intermediate phase of A/D conversion simplifies the filter design. For example, a new output sample is not required for

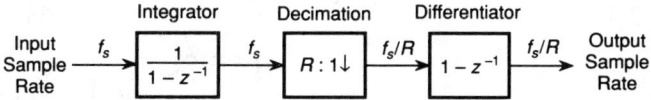

*(a) Block diagram of a one-stage comb filter.*

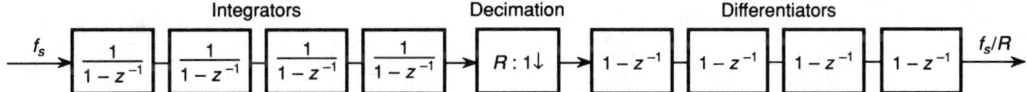

*(b) Block diagram of a cascaded four-stage comb filter.*

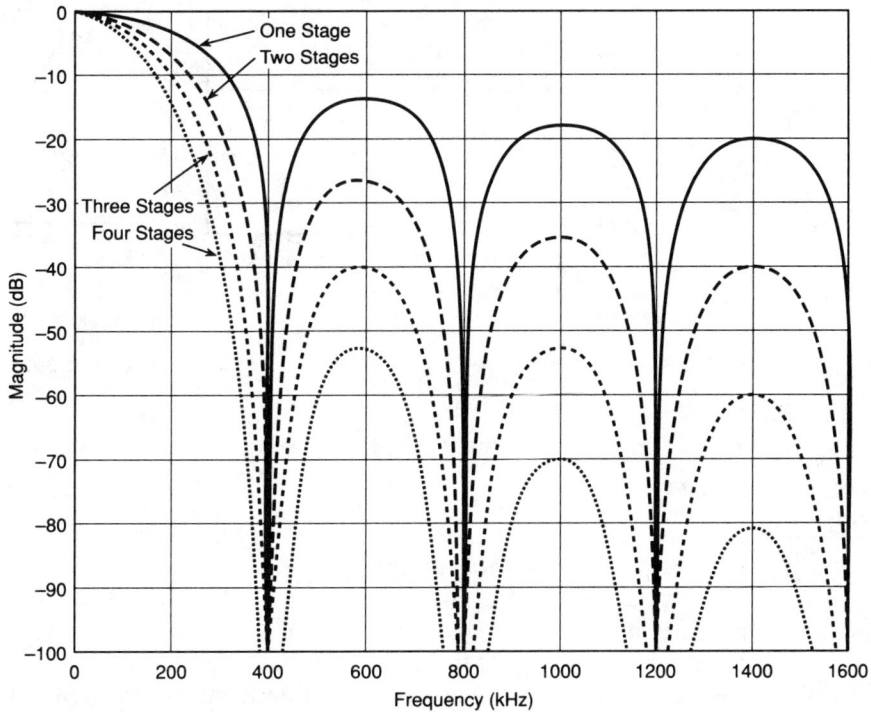

*(c) Spectrum showing response of first-, second-, third-, and fourth-order cascaded comb filter sections.*

**Figure 31.** Use of comb filters for decimation. *(From Park)*

every input bit. Because the decimation factor is 64 (in this example), an output is required only for every 64 input bits. In practice, the decimation filtering might be carried out in two stages. For example, in the Philips PCF 5022 oversampling A/D chip the first stage performs 192 multiplications per single low-bit output sample at a rate of $2f_s$. The decimation filter's second stage performs low-bit multiplication with full convolution, with an output sample rate of $f_s$. An FIR filter would commonly be used for down-sampling, because its non-recursive operation would simplify computation to one sample every $1/f_s$

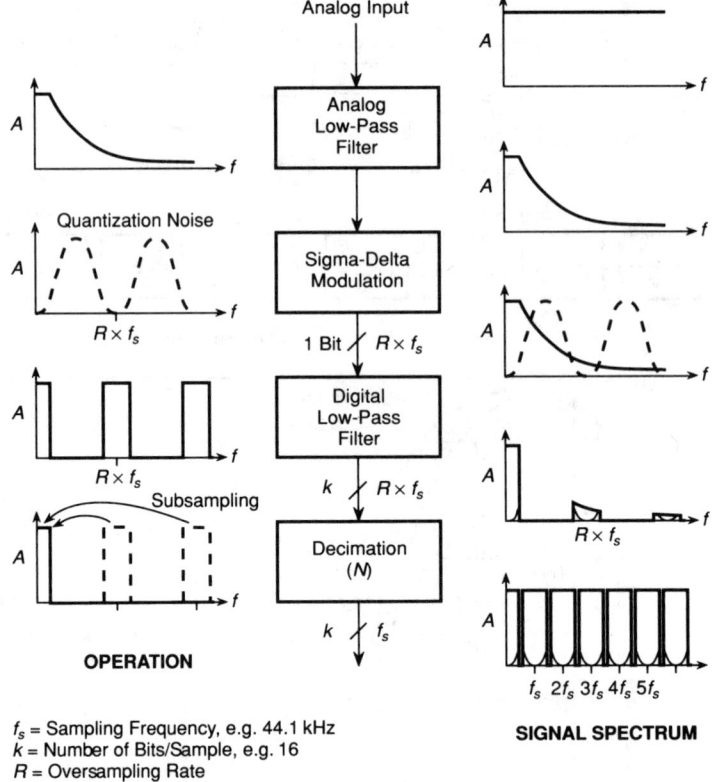

**Figure 32.** Summary of spectral characteristics of a 1-bit A/D converter.
*(Courtesy Philips)*

second. Following decimation, the result may be rounded to 16 bits, and output at a 44.1 kHz sampling rate.

Digital audio equipment containing A/D (and D/A) converters must have a stable sampling clock which in turn is frequency locked to a distributed master clock. The individual clocks must have decreasing dynamic range and very low jitter levels to prevent generated sidebands from rising to audibility. For example, a 16-bit A/D converter with an input frequency of 10 kHz would require peak jitter of less than 250 ps. In the case of sigma-delta A/D converters, as a proportion of clock period, jitter is more significant to the sigma-delta converter than a multibit converter. Amplitude errors attributable to jitter increase as the input signal frequency increases; however, because the slew rate of the input signal is equal in either type of converter, the amplitude error resulting from sinusoidal jitter is also equal in both cases. In the case of noise-induced jitter, added noise is distributed over the sigma-delta converter's increased Nyquist frequency range and low-pass filtered by the decimation circuit. Hence overall in-band jitter-induced noise is less than in a conventional converter. Thus analysis would show that oversampling sigma-delta A/D converters are no more sensitive to sinusoidal jitter than a conventional converter and indeed are less susceptible to random noise clock jitter.

# A Low-Bit A/D Converter Chip

The block diagram of a sigma-delta A/D converter chip is shown in Figure 33. It is a linear 16-bit converter, using 64-times oversampling, providing output sample rates up to 100 kHz, operating up to 6.4 MHz. The use of sigma-delta processing obviates the need for an anti-aliasing filter and sample-and-hold circuit. As with other sigma-delta A/D converters, the input signal is oversampled to extend the noise spectrum well beyond the audio band. Noise shaping reduces noise in the audio band, and the low-pass filtering is used to remove out-of-band quantization noise. Finally, the signal is decimated to reduce the sample rate commensurate with the audio band and to increase resolution.

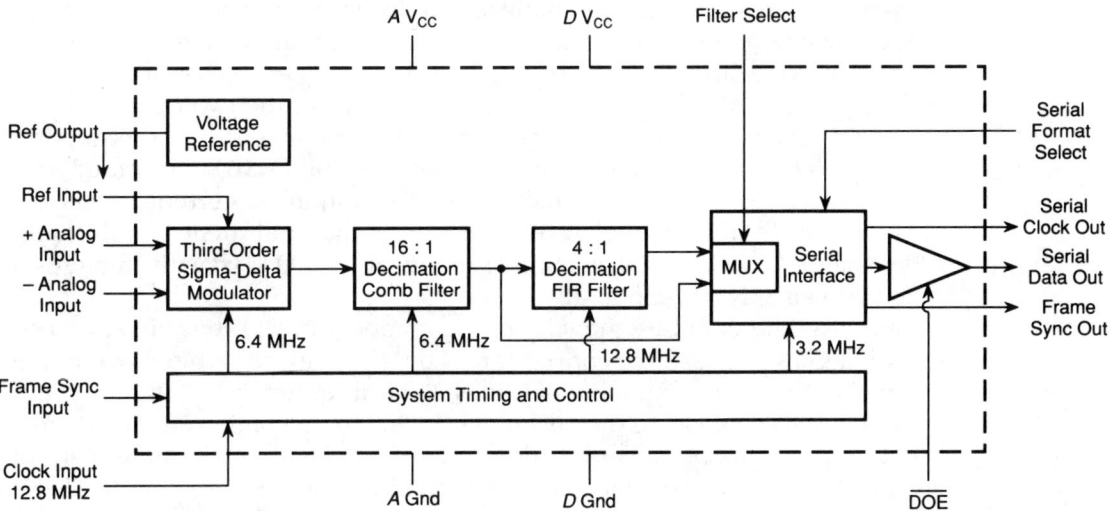

**Figure 33.** Internal block diagram of a DSP56ADC16 sigma-delta A/D converter. *(From Kloker et al.)*

The converter is designed around four major subcircuits: third-order sigma-delta modulator and noise shaper, 16:1 decimating comb filter, 4:1 decimating FIR filter, and serial interface. The third-order noise shaper places an 18 dB/octave characteristic on the quantization noise. The analog front end to the converter consists of three differential, switch-capacitor, linear integrators. Filtering and decimation are performed in two steps to reduce the complexity of the digital filter. For example, to achieve the desired stopband attenuation and filter steepness, a single-stage FIR filter with over 2800 taps would be required. Use of a multirate decimation filter system also allows a dual-mode application, as described below. The output of the modulator is filtered by a fourth-order comb filter and decimated; the sample rate is decreased by a factor of 16:1. A comb filter is used because it contains only adders and delay, thus eliminating the need for high-speed multiplication. The first-stage comb filter accomplishes initial filtering as well as decimation of the input sample rate by a factor of 16:1. Its $z$-domain transfer function may be expressed as

$$H(z) = (1 - z^{-16})^4/(1 - z^{-1})^4.$$

The equivalent frequency domain transfer function is

$$H(f) = \left[\frac{1}{16} \frac{\sin(16\pi f/f_s)}{\sin(\pi f/f_s)}\right]^4,$$

where $f_s$ is the filter sample rate.

An FIR filter is used to decimate the signal by a 4:1 factor with a low-pass response. Overall, a 64:1 decimation ratio is achieved; in other words, 63 of every 64 output samples are discarded. A stopband attenuation of −96 dB is achieved. To compensate for the response (passband droop) of the fourth-order comb filter, the FIR filter uses an inverse equalization response to achieve an overall flat response. Finite impulse response filter images occur at multiples of the comb filter output sampling rate; these are also zeros in the fourth-order comb response. The FIR filter stopband attenuates the comb response, leaving a negligible alias component at the overlap of the two responses.

A serial interface provides serial communication to a host processor; data format is 16-bit, two's complement serial, with sign bit (MSB) transmitted first. A serial interface, rather than a parallel interface, minimizes current drivers and hence noise. In all, this digital filter section is the equivalent of a thirtieth-order analog Bessel filter. The output sample rate is 100 kHz, with 16-bit resolution and an S/N ratio of 90 dB.

Because the cutoff frequencies of the comb and FIR filters are scaled by the input sampling rate, the converter may be used with any arbitrary sampling rate without changing component values. For further flexibility, this A/D converter chip is designed so the 16:1 comb filter can be connected directly to a serial output. This permits operation at faster speed (output sample rate of 400 kHz) at the expense of lower resolution (12 bit, and S/N of 72 dB). This is useful for ultrasonic applications and where lower resolution is tolerable. A general application for this chip using its full resolution is shown in Figure 34; the A/D converter is connected to a DSP processor. The DSP56ADC16 A/D converter was developed by Motorola.

## A Multilevel Quantizer A/D Converter

The sigma-delta method can be modified so that a low-bit PCM quantizer is employed. In this way, the signal's amplitude rather than slope is encoded. This can reduce the complexity of the decimation circuit. Noise shaping is still mandated; while stable loop filters of order higher than two are difficult to achieve in a sigma-delta converter, they are easy to achieve in a noise shaping PCM converter.

One such A/D chip set employs 128-times oversampling, noise shaping, and 4-bit flash conversion to achieve conversion with 20-bit resolution. The set uses an analog front end and decimator, as shown in Figure 35. The front-end module acts as a 4-bit noise shaping A/D converter; it contains 15 comparators, and a 15-

Chapter 12: Low-Bit Conversion and Noise Shaping

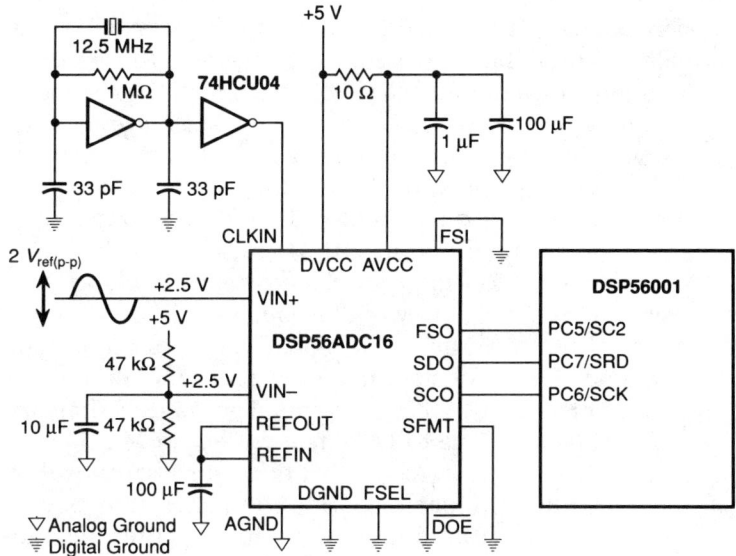

**Figure 34.** Application circuit showing interconnection of sigma-delta A/D converter (single-ended mode) and DSP processor.
*(Courtesy Motorola)*

to-4 priority encoder. Differential linearity is high because the output code transition never corresponds to an actual change in the analog input voltage. A change in the analog input causes the idle pattern at the output of the flash converter to shift by increasing the frequency of occurrence of one code, while decreasing the frequency of occurrence of another. Since the energy of the idle pattern is slightly larger than the flash converter's quantization noise, smooth transitions are ensured. Because two adjacent comparators are simultaneously active, an averaging effect is produced. This smoothes the staircase transfer function much like dither. Because the flash converter is within the feedback loop, its error is subject to negative feedback and thus has little effect on linearity. The precision resistor

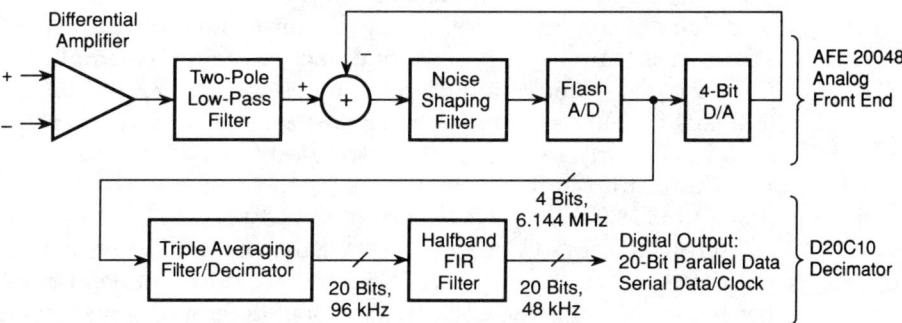

**Figure 35.** Block diagram showing architecture of an A/D converter chip employing 128-times oversampling and a 4-bit flash converter.
*(Courtesy Ultra Analog, Inc.)*

network consists of fifteen 20 kΩ resistors matched to ±0.02% tolerance. Along with two octal flip-flops, the array forms the D/A portion of the A/D quantizer. A 4-bit D/A converter is used to convert the interim output word for feedback to the summing point before integration by the noise shaping filter. The A/D converter's integral linearity is limited only by this internal D/A converter.

Three high-speed op amps with good open-loop linearity are used to implement the noise shaping loop filter; one op amp is unity-gain stable, and the other two are stable for inverting gains greater than one. Together with other components, a fourth-order characteristic is provided so that the noise floor rises with the fourth power of frequency; in addition the noise is random and uncorrelated with the audio signal. The loop filter employs four poles at low frequencies and, for stability, three zeros at high frequencies. The loop is stable because the rolloff rate is less than 12 dB/octave prior to a loop gain of unity at 500 kHz. The rolloff rate increases to 24 dB/octave at the passive pole. Filter values are selected so that three quantizer levels are used by the idling pattern with no input signal; this eliminates zero-cross distortion.

The decimator chip is a multirate filter designed to decimate the 6.144 MHz input sampling rate by a factor of 128. Two stages are employed, as shown in Figure 36. The first decimator is a simple high-speed filter comprised of three cascaded moving-average filters each with a window of sixty-four 4-bit samples; the filters output the average of the last 64 samples. The frequency response is $\sin^3 x/x^3$ and contains periodic zeros at multiples of twice the output sampling rate. The frequency response is given by

$$H(f) = \left[ \frac{1}{64} \frac{\sin(64\pi f/f_s)}{\sin(\pi f/f_s)} \right]^3,$$

where $f_s$ is the filter sample rate. Although the response of the filter is relatively flat (−1.9 dB at 20 kHz), compensating equalization must be applied at the input in the form of a two-pole filter. In addition, this filter removes components above 3 MHz.

A halfband standard FIR low-pass filter performs final decimation; it attenuates signals above the audio band so the signal may be downsampled by 2:1. Symmetry of the filter's frequency response causes half the coefficients to be zero (linear amplitude is 0.5 at one-quarter the input sampling rate). The first filter provides attenuation of frequencies above $3f_s/2$ and the second filter attenuates frequencies between $f_s/2$ and $3f_s/2$. The reduction in sample rate is accompanied by an increase in resolution from 4 bits to 20.

The 20-bit output is provided in both serial and parallel formats. As with other 1-bit converters, it is important to use a crystal or high-$Q$ $LC$ oscillator; an $RC$ multivibrator must not be used because of the large jitter it exhibits. An S/N ratio of 108 dB can be achieved with the A/D chip set. Performance can be enhanced by lowering the noise in the first integrator op amp, operating two parallel front-end circuits, and adding their outputs to form a 5-bit output; this yields an S/N ratio of 114 dB using the 20-bit output. When truncating 20 bits to 16, the effective requantization must be dithered; the LSB of the decimator chip may be

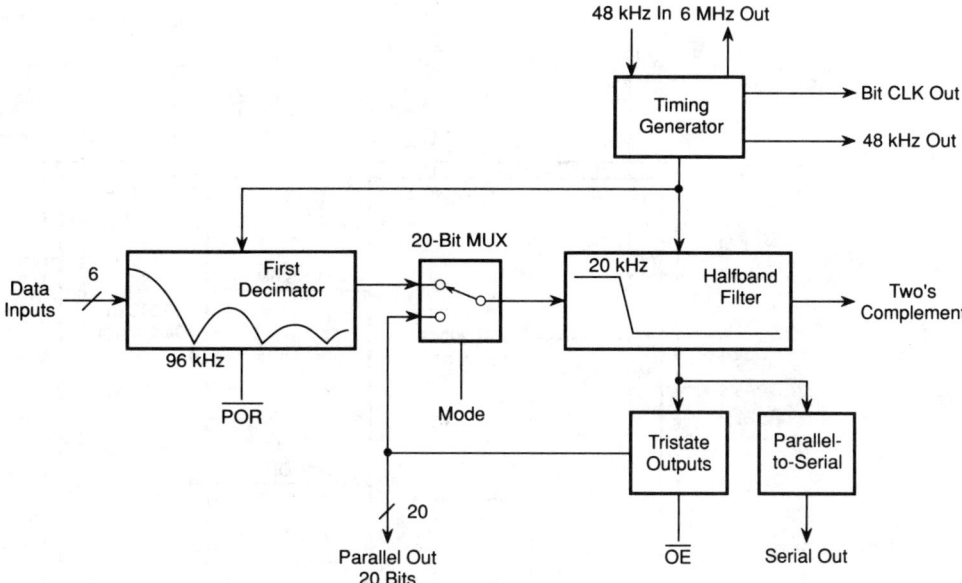

**Figure 36.** Block diagram showing architecture of a decimation circuit used in an A/D converter. *(From Adams)*

used for this purpose. An interconnection diagram of the two-package converter is shown in Figure 37. The AFE20048 analog front end and D20C10 decimator were developed by dbx and are available from UltraAnalog, Inc.

## Low-Bit A/D-D/A Converter

Because of the high degree of integration permitted by the sigma-delta conversion methods, it is possible to place a linear, 16-bit sigma-delta analog-to-digital and digital-to-analog converter (codec) on a single chip. One such chip permits input-output sampling rates up to 50 kHz with 16-bit resolution, and rates of 100 kHz with 12-bit resolution. Third-order noise shaping is used on the A/D side, and fourth-order noise shaping is used on the D/A side. The A/D section employs 64-times oversampling and 64-times decimation. A digital compression circuit is used to equalize the response to within ±0.025 dB ripple in the pass-band, with phase linearity. The D/A section employs two digital anti-imaging interpolation filters, along with an FIR compensation filter for flat pass-band response. The D/A section provides a 1-bit output signal. An analog sixth-order Bessel low-pass filter is provided on-chip, as is a temperature-compensated voltage reference for stable coding and clocking. This reference can operate in a master/slave configuration to ensure gain matching/tracking between multiple devices. Likewise, sampling coherency can be preserved between multiple converter chips to ensure interchannel phase accuracy.

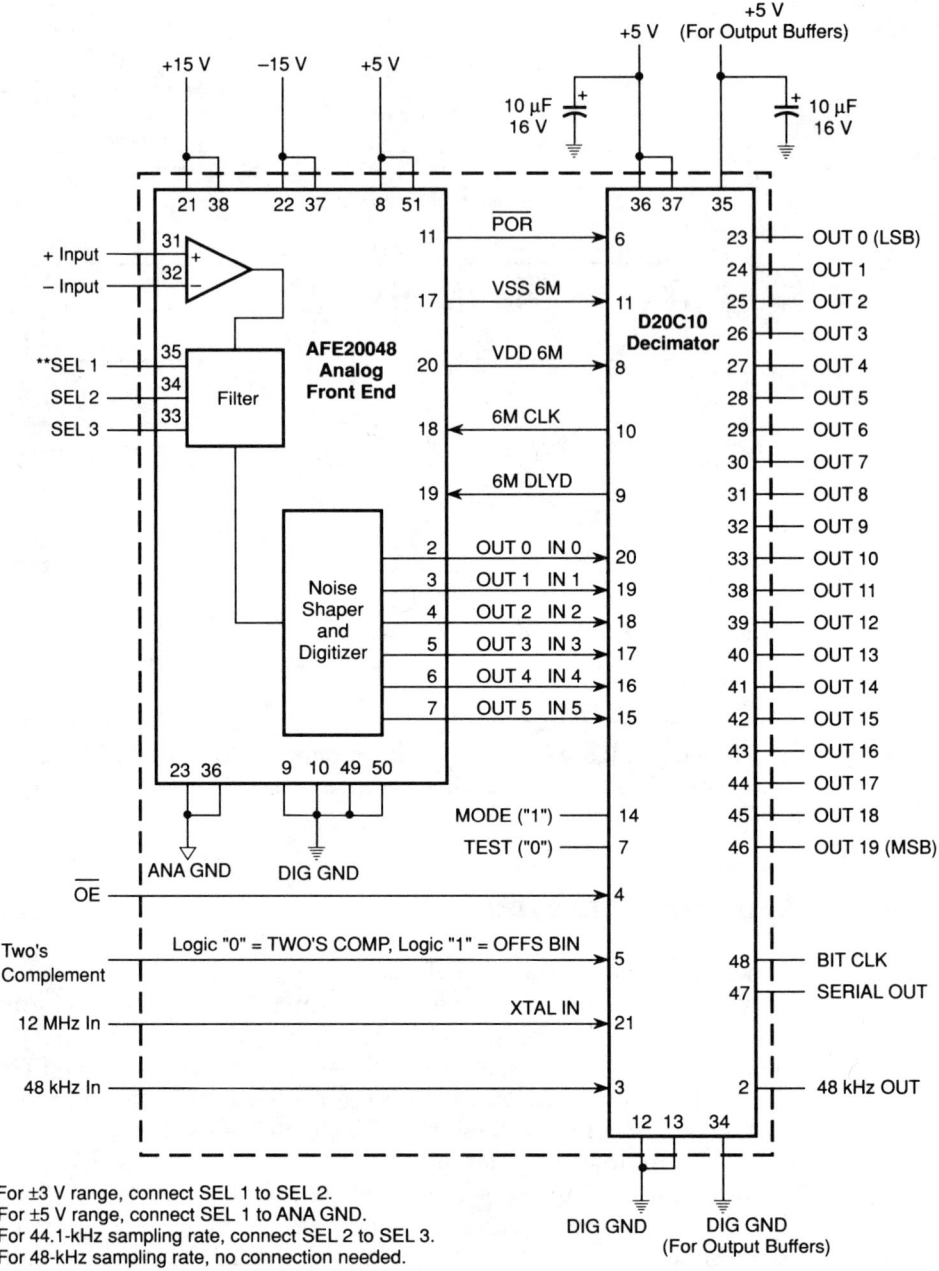

**Figure 37.** Interconnection circuit of AFE20048 analog front end and D20C10 decimation chips used as A/D converter. *(Courtesy Ultra Analog, Inc.)*

Digital data can be shifted into and out of the converters with either MSB or LSB first. An SSI bus can be implemented in several different modes.

The DSP56ADA16 provides a dynamic range of 96 dB and a signal-to-noise ratio of 90 dB. A block diagram of the DSP56ADA16 is shown in Figure 38. As with all sigma-delta converters, this converter pair is based on digital filtering techniques, thus approximately 90% of the chip is given to digital circuitry; this promotes compatibility, reliability, increased functionality, and reduced chip cost. Two of these chips form a complete conversion circuit for a stereo signal and, together with a DSP56001 chip, form a complete digital signal processing system. The DSP56ADA16 was developed by Motorola.

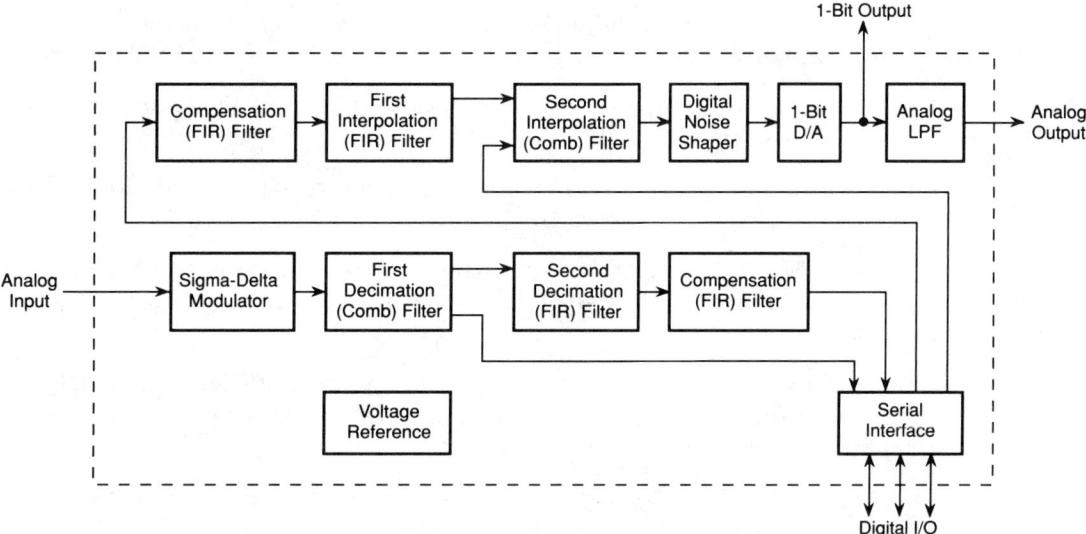

**Figure 38.** Block diagram of a DSP56ADA16 sigma-delta analog-to-digital and digital-to-analog converter chip.

## Conclusion

In addition to successfully obsoleting brickwall analog filters, low-bit A/D converters offer other improvements over conventional multibit conversion methods. As we have seen, low-bit converters can achieve increased resolution. For example, they extend the spectrum of the error between analog input and digital output far beyond the audio band. Thus the in-band noise can be made quite small. As noted, this benefit is provided by the SDM; the same circuit that codes the signal into a 1-bit stream also shifts the out-of-band noise components. In any case, in terms of phase linearity, amplitude linearity, noise, long-term stability, and other parameters, A/D and D/A converters using low-bit architectures offer significant advantages over conventional multibit conversion methods and convincingly demonstrate the utility of digital signal processing techniques.

# References

Adams, R. W., "An IC Chip Set for 20-Bit A/D Conversion," *Journal of the Audio Engineering Society*, vol. 38, no. 6, June, 1990.

———, "Companded Predictive Delta Modulation: A Low-Cost Conversion Technique for Digital Recording," *Journal of the Audio Engineering Society*, vol. 32, no. 9, September, 1984.

———, "Design and Implementation of an Audio 18-Bit Analog-to-Digital Converter Using Oversampling Techniques," *Journal of the Audio Engineering Society*, vol. 34, no. 3, March, 1986.

Agnello, A., "16-Bit Conversion Paves the Way to High-Quality Audio for PC's," *Electronic Design*, July 26, 1990.

Ardalan, S. H., "Analysis of Delta-Sigma Modulators with Bandlimited Gaussian Inputs," *Proceedings of IEEE Conference*, ASSP, vol. III, 1988.

Ardalan, S. H., and J. J. Paulos, "An Analysis of Non-Linear Behavior in Delta-Sigma Modulators," *IEEE Transactions on Circuits and Systems*, vol. CAS-34, June, 1987.

Brandenburg, K., and D. Seitzer, "Low Bit Rate Coding of High-Quality Digital Audio Algorithms and Evaluation of Quality," *Audio in Digital Times: AES 7th Conference Proceedings*, May, 1989.

Boser, B. E., and B. A. Wooley, "The Design of Sigma-Delta Modulation Analog-to-Digital Converters," *IEEE Journal of Solid State Circuits*, vol. 23, no. 6, December, 1988.

Cabot, R. C., "Testing Digital Audio Devices in the Digital Domain," *Audio in Digital Times: AES 7th Conference Proceedings*, May, 1989.

Carley, L. R., "An Oversampling Analog-to-Digital Converter Topology for High-Resolution Signal Acquisition Systems," *IEEE Transactions on Circuits and Systems*, vol. CAS-34, January, 1987.

Crochiere, R. E., and L.R. Rabiner, "Interpolation and Decimation of Digital Signals—A Tutorial Review," *Proceedings of the IEEE*, vol. 69, March, 1981.

Darling, T. F., and M. O. J. Hawksford, "Oversampled Analog-To-Digital Conversion for Digital Audio Systems," *Journal of the Audio Engineering Society*, vol. 38, no. 12, December, 1990.

DBX/CTI Research, "Application Notes, F410 Front-End/D20C10 Decimator, High-Resolution A/D Converter IC Set."

Dijkmans, E. C., and P. J. A. Naus, "The Next Step Towards Ideal A/D and D/A Converters," *Audio in Digital Times: AES 7th Conference Proceedings*, May, 1989.

Fourre, R., Schwarzenbach, S., and R. Powers, "20-Bit Evolution," *Studio Sound*, May, 1990.

Harris, S., "The Effects of Sampling Clock Jitter on Nyquist Sampling Analog-to-Digital Converters, and on Oversampling Delta-Sigma ADCs," *Journal of the Audio Engineering Society*, vol. 38, no. 7/8, July/August, 1990.

Hauser, M. W., "Principles of Oversampling A/D Conversion," *Journal of the Audio Engineering Society*, vol. 39, no. 1/2, January/February, 1991.

Hauser, M. W., and R. W. Brodersen, "Circuit and Technology Considerations for MOS Delta-Sigma A/D Converter," *ISCAS Digest*, 1986.

——, "Monolithic Decimation Filtering for Custom Delta-Sigma A/D Converters," *Proceedings of IEEE Conference*, ASSP, vol. III, 1988.

Hawksford, M. O. J., "Chaos, Oversampling, and Noise Shaping in Digital-to-Analog Conversion," *Journal of the Audio Engineering Society*, vol. 37, no. 12, December, 1989.

Hawksford, M. O. J. and W. Wingerter, "Oversampling Filter Design in Noise-Shaping Digital-to-Analog Conversion," *Journal of the Audio Engineering Society*, vol. 38, no. 11, November, 1990.

Jayant, N. S., and P. Noll, *Digital Coding of Waveforms: Principles and Applications to Speech and Video*, Prentice-Hall, 1984.

Kam, J. J. van der, "A Digital 'Decimating' Filter for Analog-to-Digital Conversion of Hi-Fi Audio Signals," *Philips Technical Review*, vol. 42, no. 6/7, April, 1986.

Kloker, K. L., Lindsley, B. L., and C. D. Thompson, "VLSI Architectures for Digital Audio Signal Processing," *Audio in Digital Times: AES 7th Conference Proceedings*, May, 1989.

Kuroda, N., JVC Corporation, personal correspondence, November, 1989.

Lee, W. L., and C. G. Sodini, "A Topology for Higher Order Interpolative Coders," *ISCAS Digest*, 1987.

Matsuya, Y., et al., "A 16-Bit Oversampling A/D Conversion Technology Using Triple-Integration Noise Shaping," *IEEE Journal of Solid-State Circuits*, vol. SC-22, no. 6, December, 1987.

Motorola, Inc., "DSP56ADC16 16-Bit Sigma-Delta Analog-to-Digital Converter," Motorola Semiconductor Technical Data, Ref. DSP56ADC16/D, 1989.

——, "Principles of Sigma-Delta Modulation for Analog-to-Digital Converters," 1989.

Naus, P. J. A., et al., " Signal-Level Distortion in Sigma-Delta Modulators," *AES Preprint 2584*, March, 1988.

Ning, H., Buzo, A., and F. Kuhlmann, "Multi-Loop Sigma-Delta Quantization: Spectral Analysis," *Proceedings of IEEE Conference*, ASSP, vol. III, 1988.

Park, S., "Multistage Halfband Filter Design for Improving Effective Resolution from Sigma-Delta Analog-to-Digital Converters," *AES Preprint 3008*, September, 1990.

Park, S., "Principles of Sigma-Delta Modulation for Analog-to-Digital Converters," Motorola Applications Note APR8/D, 1990.

Richards, M., "Improvements in Oversampling Analogue to Digital Converters," *AES Preprint 2588*, March, 1988.

Seitzer, D., et al., "Bit Rate Codecs for Audio Signals Implementation in Real Time," *AES Preprint 2707*, November, 1988.

Shah, P., "Music of the Bitstream," *Audio Magazine*, January, 1991.

Spang, H. A., and P. M. Schultheiss, "Reduction of Quantizing Noise by Use of Feedback," *IRE Transactions on Communications Systems*, vol. CS-10, December, 1962.

Spinnler, W., et al., "Implementation Aspects of Low Bit Rate Codecs for High Quality Audio Channels," *AES Preprint 2750*, March, 1989.

Stikvoort, E. F., "Higher Order One Bit Coder for Audio Applications," *AES Preprint 2583*, March, 1988.

———, "Some Remarks on the Stability and Performance of the Noise Shaper or Sigma-Delta Modulator, *IEEE Transactions on Communications*, vol. 36, no. 10, October, 1988.

Theile, G., Link, M., and G. Stoll, "Low Bit Rate Coding of High Quality Audio Signals," *AES Preprint 2432*, March, 1987.

Thompson, C. D., "A Monolithic 100 kHz 16-Bit A/D D/A Converter Using Sigma-Delta Modulation," Motorola, Austin, Texas.

———, "A VLSI Sigma Delta A/D Converter for Audio and Signal Processing Applications," *Proceedings of IEEE Conference*, ASSP, May, 1989.

Uchimura, K., Hayashi, T., Kimura, T., and A. Iwata, "Oversampling A-to-D and D-to-A Converters with Multistage Noise Shaping Modulators," *IEEE Transactions*, ASSP, vol. 36, December, 1988.

Vanderkooy, J., and S. P. Lipshitz, "Digital Dither: Signal Processing with Resolution Far Below the Least Significant Bit," *Audio in Digital Times: AES 7th Conference Proceedings*, May, 1989.

Welland, D. R., et al., "A Stereo 16-Bit Delta-Sigma A/D Converter for Digital Audio," *Journal of the Audio Engineering Society*, vol. 37, no. 6, June, 1989.

Wong, P. W., and R. M. Gray, "FIR Filters with Sigma-Delta Modulation Encoding," *IEEE Transactions*, ASSP, vol. 38, no. 6, June, 1990.

*Chapter 13*

# Digital Signal Processing: Architectures

**Matt Booty**

## Introduction

The general term digital signal processing (DSP) describes any device or algorithm operating on a signal in the digital domain. The input is in digital rather than analog form and is some type of signal. The signal is processed, or changed, in some way. This general definition helps identify the broad range of devices and algorithms that can be considered in digital signal processing.

In common practice, the term digital signal processor refers to a specific combination of general-purpose or custom microprocessor hardware and software which has been optimized to process signals. The hardware usually consists of high-speed control and arithmetic logic units combined with digital storage. The software takes the form of a control program which executes the numeric algorithm responsible for processing the signal. There is thus a distinction between digital signal processors and other forms of microprocessor hardware and software.

Perhaps the most important aspect of DSP systems is that they are designed to operate with signals, not data. A signal conveys information based on its variation over time, while data usually exists independent of time. The information in a signal depends not only on the present value, but also on past values. For example, a number in the phone book is not dependent on the number above it or below it, but an audio recording must be played back in the correct order, and our interpretation of it depends on what we have already heard.

A digital signal processor, therefore, must be able to accept a sequence of information and process it without disrupting the relationship of each item in the sequence to the others. In practical terms, a DSP system must be able to accept a series of digital samples at a certain sample rate, process the samples, and output the altered samples in the same order received and at the same sample rate. Because the processing may involve previous values of the signal and many intermediate operations, it must be able to store and retrieve several of the last values received. In many applications, processing must be performed within the constraints of the sampling period; that is, in real time. However, in other cases, off-line processing may be performed in nonreal time.

The components of any DSP system are designed to meet these requirements. The high-speed arithmetic logic unit (ALU) is required because of the high sample rates used to represent real-time audio, video, and control signals in digital form. For example, individual channel rates of 44.1 and 48 kHz are common in digital audio applications. The ALU must perform many operations, such as addition or multiplication, and must reference many values, for each sample received. In real-time applications, it must complete those calculations before the next sample is received. Obviously, the control unit, or central processing unit (CPU), must be equally fast. The instructions which determine how the individual samples are to be processed, or how the algorithm is to be implemented, must be executed many times faster than the samples are received. For each operation carried out by the ALU, the CPU must transfer a sample from the input to the ALU, tell the ALU which operation to carry out, transfer the sample to the output and possibly also to a storage medium for later reference.

With this in mind, it becomes apparent that DSP chips are microprocessors which have been optimized to perform rapid numeric calculations on a series of digital samples. The gain in speed is partly obtained by limiting the number of functions the processor can perform. But it can perform them very fast. Since most digital signal processing algorithms involve repeated multiplications, the ALU found in a DSP chip may be designed to perform multiplication in one step, rather than through a series of additions. However, it may not be able to do subtraction. Since samples must be moved quickly from input to output, the CPU found in a DSP chip will have many specialized provisions for transferring data but may lack other common microprocessor features. A digital signal processor is thus a hardware microprocessor capable of rapidly executing software algorithms which call for fast, repeated arithmetic operations on a series of numbers.

A list of digital signal processing applications is shown in Chart 1. The most common of these may be digital audio, which often uses DSP chips to implement the digital forms of familiar analog operations such as filtering and gain change. Computer generated imagery (CGI) and digital video applications use DSP chips to perform tasks such as tracing the path of a ray of light from one point to another in an animated image, or for changing the brightness of a digital video recording as it is played back. Cellular phone networks and satellite transmission systems use DSP chips to perform error correction and noise reduction on communications signals carrying massive amounts of information. Even everyday products such as compact disc players and automobile antilock braking systems contain DSP elements.

Obviously, the kinds of devices listed above contain more than just a digital signal processing chip. The DSP chip is almost always subsidiary to a larger set of computer hardware and software, which is referred to as the host. The host performs all of the familiar computer tasks: maintaining a user front-end interface, accessing mass storage devices, providing large scale data management, and supporting high-level languages. When the need arises for digital

**Chart 1.** Typical Digital Signal Processing Applications
(*Courtesy Texas Instruments, Inc.*)

| General-Purpose DSP | Graphics/Imaging | Instrumentation |
|---|---|---|
| Digital filtering | 3D rotation | Spectrum analysis |
| Convolution | Robot vision | Function generation |
| Correlation | Image transmission/compression | Pattern matching |
| Hilbert transforms | Pattern recognition | Seismic processing |
| Fast Fourier transforms | Image enhancement | Transient analysis |
| Adaptive filtering | Homomorphic processing | Digital filtering |
| Windowing | Workstations | Phase-locked loops |
| Waveform generation | Animation | |

| Voice/Speech | Control | Military |
|---|---|---|
| Voice mail | Disk control | Secure communications |
| Speech vocoding | Servo control | Radar processing |
| Speech recognition | Robot control | Sonar processing |
| Speaker verification | Laser printer control | Image processing |
| Speech enhancement | Engine control | Navigation |
| Speech synthesis | Motor control | Missile guidance |
| Text-to-speech | | Radio-frequency modems |

| Telecommunications | | Automotive |
|---|---|---|
| Echo cancellation | Fax | Engine control |
| ADPCM transcoders | Cellular telephones | Vibration analysis |
| Digital PBXs | Speaker phones | Antilock brakes |
| Line repeaters | Digital speech interpolation (DSI) | Adaptive ride control |
| Channel multiplexing | X.25 packet switching | Global positioning |
| 1200 to 19,200 bps modems | Video conferencing | Navigation |
| Adaptive equalizers | Spread spectrum communications | Voice commands |
| DTMF encoding/decoding | | Digital radio |
| Data encryption | | Cellular telephones |

| Consumer | Industrial | Medical |
|---|---|---|
| Radar detectors | Robotics | Hearing aids |
| Power tools | Numeric control | Patient monitoring |
| Digital audio/TV | Security access | Ultrasound equipment |
| Music synthesizers | Power line monitors | Diagnostic tools |
| Toys and games | | Prosthetics |
| Solid-state answering machines | | Fetal monitors |

signal processing, the host either routes the digital samples to the DSP chip or sets up a direct connection to the mass storage device.

Digital signal processing chips are not general-purpose computational devices and usually require the services of support hardware and software, such as a microcomputer or mainframe computer. However, some applications using DSP chips consist only of the DSP hardware, its resident control and algorithm software, and a minimal amount of support and interface hardware. Consider the example of automobile antilock brakes. A sensor on the brake generates a voltage or current which changes over time in proportion to the amount of pressure being applied (a signal). This signal is sampled and converted to digital form by an A/D converter and fed directly to the input of the DSP hardware. Based on its programmed algorithm, the DSP software changes the series of numbers representing brake pressure and directly outputs a new signal in digital form. This is converted to an analog signal by a D/A converter, and this signal can be used to control actuators which reduce or increase the brake pressure.

In this situation, the DSP chip might be referred to as an embedded controller. An embedded controller is a processor which is integrally built into a piece of equipment and is dedicated to a single task. The control and algorithm software usually exists in read-only memory (ROM) and is never changed during the life of the equipment. This is in contrast to a DSP chip with a microcomputer host. The microcomputer can easily load new algorithms into the DSP chip, changing the processor's function, for example, from a digital low-pass filter to a seismic vibration analyzer.

Digital signal processors physically take the form of very large scale integrated (VLSI) circuit chips, with packages about 1 inch square and 0.25 inch thick. There are usually upwards of fifty connection pins along the edges, and the chips are often mounted on cards containing support circuitry, such as clock generators, power supplies, A/D converters, D/A converters, and interface circuitry for connection to the host computer's data and memory busses. The ALU, the CPU, and the high-speed signal and program memories are all normally contained in the DSP integrated-circuit chip.

Designing hardware and writing software for these DSP chips requires a thorough knowledge of both the DSP chip itself as well as the host computer. Because of the optimization required for high-speed numeric processing, and the resulting streamlined signal and control paths, the software commands used to implement the processing algorithms are very hardware-specific. Furthermore, DSP chips made by different manufacturers all have their own unique collection of support hardware.

## Basic Architecture

A simple, generic DSP hardware structure is shown in Figure 1. The architecture is similar to that of many general-purpose microprocessors, which also place the functional blocks around a centrally located data bus. The data bus is used to carry software instructions, data, memory addresses, and I/O information between the blocks.

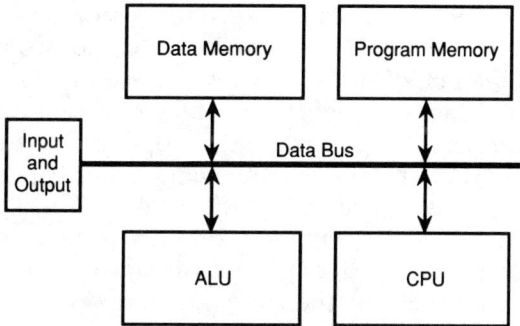

**Figure 1.** Simplified digital signal processor architecture.

The CPU (sometimes called the program controller or the program decoder) is responsible for executing the control and algorithm software. The software is stored as a series of single-step instructions in the program memory. These instructions are read, or fetched, by the CPU one at a time, at a rate determined by an external clock generator. The time interval between instruction fetches is called an instruction cycle, or sometimes just a clock cycle. Common operating speeds for CPUs found in DSP chips are on the order of ten million instructions per second (MIPS). This corresponds to a clock generator rate of about 40 MHz, which is much faster than the average desktop PC's clock rate of 10 to 20 MHz.

Although the CPU can execute ten million or more instructions per second, the program memory does not contain nearly that number of single-step instructions. Because digital signal processing centers on repeated, similar calculations, the CPU actually executes the same small list of instructions over and over again. When the end of the list is reached, the CPU jumps back to the beginning of the list and starts over. The housekeeping needed to determine which instruction to execute next, and when to repeat a group of instructions, is determined by the specifics of programming the control flow. A description of the actual instructions the CPU can fetch and execute is given in Chapter 14.

Most of the instructions executed by the CPU take the form of specific commands directed at the ALU. The ALU is a number-crunching workhorse, designed to perform the key DSP operations of addition, subtraction, and multiplication in a single instruction cycle. The phrase "in a single instruction cycle" is very important and distinguishes a DSP chip from a general-purpose microprocessor. Typical ALUs are capable of only one mathematical operation: addition. Performing subtraction is accomplished by adding the negative of the number. An extra instruction cycle is needed to generate the negative of the number. Multiplication is performed by adding a number to itself over and over. Since each addition requires a separate command from the CPU, each addition also occupies a separate and additional instruction cycle. Typical

ALUs contain electronic circuits which are only capable of performing addition. (Math or floating-point co-processors are often used to augment microprocessor ALUs that are limited in computational ability.)

The ALU in a digital signal processor, however, contains specialized hardware that can multiply, subtract, and add two numbers in a single instruction cycle. Much of the speed advantage in DSPs is gained in the ALU. The ALU usually includes temporary storage spaces for the operands. These storage spaces, called registers, hold the operands as they are moved from the data bus to the ALU. The result of the mathematical operation on the operands is stored in a special register called the accumulator. The operands and the results are moved to and from the ALU on the data bus. The width of the data bus corresponds to the word length, or sample size. The larger the data path, the larger the word size, and the larger are numbers that can be represented by a single word or sample. The data path is exactly as wide as the sample word. In DSP applications, however, the size of the data path takes on additional importance. The dynamic range of the signals that the DSP chip can accommodate is directly determined by the word size. An 8-bit sample word can represent enough integers to allow a dynamic range of 48 dB, while a 16-bit sample word supports a 96 dB dynamic range.

The standard bus width in most microprocessors is 16 bits. This, however, is often inadequate for DSP applications, which may be called on to add two or more 16-bit numbers together. This result of adding two 16-bit operands is a 17-bit result, and the size increases proportionately as subsequent calculations are carried out. General-purpose microprocessors solve this problem by representing large numbers with 32-bit words that are the concatenation of two 16-bit words. The data path size is still fixed at 16 bits, and many more instructions must be executed in order to carry out mathematical operations in a piecemeal fashion, dealing with the two parts of each number separately. Another option is to represent numbers in a form resembling scientific notation. A number is broken into its basic value (the mantissa) and the power of ten by which it is multiplied (the exponent). This format is referred to as floating-point representation, and differs from integer, or fixed-point, representation. Floating-point operations require additional instructions to be executed, since both the mantissa and the exponent must be shuttled on the data bus. ALUs designed to operate with floating-point number representation are relatively slower than fixed-point chips.

Obviously, these forms of large number representation are not in keeping with the DSP goal of carrying out all basic mathematical operations in a single instruction cycle. The solution is to increase the width of the data bus beyond the expected size of the sample word. A DSP chip designed to process 16-bit sample words may have a 24-bit–wide data path. The extra bits are not used for more dynamic range on the input signal, but rather to allow the processor to deal with numeric results that exceed 16-bit representation. Such overflows, as they are called, would introduce error into the computation, the algorithm, and the signal itself if the extra bits were not available.

Since the registers in the ALU take their operands directly from the data bus, they too are often 8 to 16 bits wider than the sample word size. In turn, the accumulator can be even wider than the registers. A DSP chip designed to process 16-bit sample words might likely have a 24-bit data path, 24-bit ALU registers, and a 56-bit accumulator. The oversized accumulator takes into account an operation called multiply and accumulate (MAC), which is the multiplication of two numbers and the addition of the result to a running total. Multiply and accumulate operations are at the core of many DSP algorithm implementations, and a 56-bit accumulator allows the single-step execution of a 16-bit MAC without overflow.

The program memory, which stores the software run by the CPU, is a high-speed random-access memory (RAM), usually smaller than 1K words. This is very much smaller than the 500K to several megabyte RAM memories common in modern microcomputers. DSP programs are small, fast, and repetitive, and generally do not require large amounts of program memory.

Software is loaded into the program memory through the program data bus. Typically, the host computer stores several programs in its larger memory and loads the particular software into the DSP chip as needed. For example, the software algorithm needed to implement a low-pass filter is different from that needed to perform notch or band-pass filtering. Depending on the type of filter called for, the host can download the appropriate program into the DSP chip and instruct it to begin execution. Another option is to store the software in a read-only memory (ROM) and transfer it to the program memory each time the DSP chip is powered up or reset. For example, a DSP chip acting as an embedded controller in a mobile telephone system will always run the same specialized software, which can be easily stored in a small ROM included in the DSP chip's support hardware.

A very small ROM, often called a bootstrap ROM, stores approximately 32 instructions that are executed each time the DSP chip is powered up. These bootstrap instructions prepare the DSP chip for operation and can either load the program memory from a nearby ROM and then turn execution over to the CPU or can wait for a program to come from the host to the program data bus.

Some newer DSP chips contain program memories and CPU hardware which let the program software be modified on the fly, as it is running. This allows software for similar algorithms to be overlaid in the same memory space without reloading the entire program. Furthermore, since most DSP algorithms require storage of and reference to numeric coefficients, dynamic software modification provides a method of changing such parameters as filter coefficients without changing the actual filter algorithm. Numeric coefficients can be stored as part of the program rather than in data memory to increase execution speed.

The data memory is also a high-speed RAM. It is used to store blocks of calculation results, tables of coefficients, waveform tables, and intermediate results of fast Fourier transforms (FFTs), for example. Typical data memory size is well under 1K words, and usually between 256 and 512 words. Applications requiring larger amounts of memory rely on off-chip expansion memories, which can be as large as 64K words, but these are used infrequently. Most DSP applications are throughput-intensive, not storage-intensive.

As an example, consider the use of a DSP chip as a disk drive controller. A disk drive controller translates generic requests for storage access into specific instructions to seek out a location on the disk and read or write the data stored at that location. The task is to move the disk head to the desired location as fast as possible, with minimal overshoot and error. The DSP data memory, however, may need to store disk locations as they are generated, or hold motion compensation values for the current target location. The DSP data memory provides a small, convenient data storage space that can be accessed without the additional input and output instructions that would be required to access an off-chip memory.

Most random-access memories are single-port; that is, a specific location can either be read from or written to only from one point. This arrangement serves traditional microprocessor needs very well, but some DSP architectures incorporate what are known as dual-port memories. These memories provide access to a single memory location through two separate input and output paths. A simple model for a single-port memory might be a mail box with a door at one end. First, the mail carrier opens the door, puts the mail in the box, and then shuts the door. Next, the recipient opens the door, removes the mail and closes the door again. Now, imagine that the mail carrier is delivering several hundred letters a minute for an entire day. It would be quicker and more convenient to have a door at both ends, so that delivery and removal of the mail would not interfere with each other. This is the model for dual-port memories, which allow the DSP data bus, as well as the input and output paths to the host computer, to have access to the data memory.

Dual-port memories can be used to solve the problems that arise when a slower host computer is transferring information to a DSP chip. If the host is transferring sample words one at a time to the DSP chip, then the DSP chip's speed is limited by the host's speed. This defeats the intended purpose of the DSP chip. However, if the host can transfer a block of 256 sample words at a time, then the host has a longer interval while the DSP chip processes all 256 values before it must have another block ready. The host is given access to one side of the dual-port memory, and the DSP chip is given access to the other. The host uses a direct memory access (DMA) instruction to quickly move a block of values to the dual-port memory, and the DSP chip steps through them one at a time from the other port. This sort of arrangement is common in applications where the input to the DSP chip is not from a nearby analog-to-digital (A/D) converter or other continuous stream of digital information, but rather from a device that can group its sample words in blocks.

Other methods of input and output are memory mapping and use of dedicated interface controllers. Memory mapping is a common technique that replaces a storage location in memory with a connection to an external device. The CPU or the ALU can input or output values by reading from or writing to an imaginary memory location that actually passes through to an external device such as an A/D converter, D/A converter, or host component.

Besides memory mapping, a DSP chip may have provisions for dedicated input and output interfacing in the form of parallel and serial interface ports.

Typical architectures include at least one parallel port that is the same width as the data bus. This port is used for transfer of sample words to and from a host device or from A/D converters and D/A converters. Also typically included are one or more serial ports, which are often used in communication applications. As an example, a DSP chip used in an office fax machine might receive much of its information in a serial fashion from a modem attached to a telephone line.

Finally, the input and output section typically includes several connection pins used for handshaking and interrupt requests. Handshaking refers to the information passed between two devices which communicates that a device has or has not successfully received information, or that it is or is not ready for the next item. Handshaking is needed when two devices that are not locked to a common clock are passing data back and forth. In a different method, interrupt requests are used to signal the receiving device that data is ready. When an interrupt request is issued, the receiving device temporarily suspends its current task and reads the incoming data. Having a slower host issue an interrupt request frees a faster DSP chip from continuously checking the input ports for fresh data.

Handshaking and interrupt considerations are an important, and often difficult, part of programming and interfacing DSP chips in both host and embedded controller environments. This topic is discussed further in the specific chip examples which follow and in Chapter 14.

# DSP Chip Architectures

The examples that follow are descriptions and comparisons of several commercially available DSP chips. The functional blocks, unique features and possible applications for each are highlighted, with the intention of providing an overview of current DSP chip architectures.

## Motorola DSP56001

One of the most widely used DSP chips is the DSP56001 56-bit general-purpose digital signal processor, manufactured by Motorola, Inc. A block diagram of this 88-pin, VLSI chip is shown in Figure 2. The DSP56001, more commonly called the 56001, has gained popularity because of its high computational speed and extensive on-chip resources and is an improved version of the earlier DSP56000 chip. The 56001 is found in audio, sound synthesis, communication, robotics, and image processing applications, and it may be the most common DSP chip used as a co-processor with microcomputers and minicomputers.

The 56001 is unique in that it contains not only a data bus, but also a program data bus and an address bus. This configuration is known as a Harvard Architecture. It allows three separate operations—moving data and operands

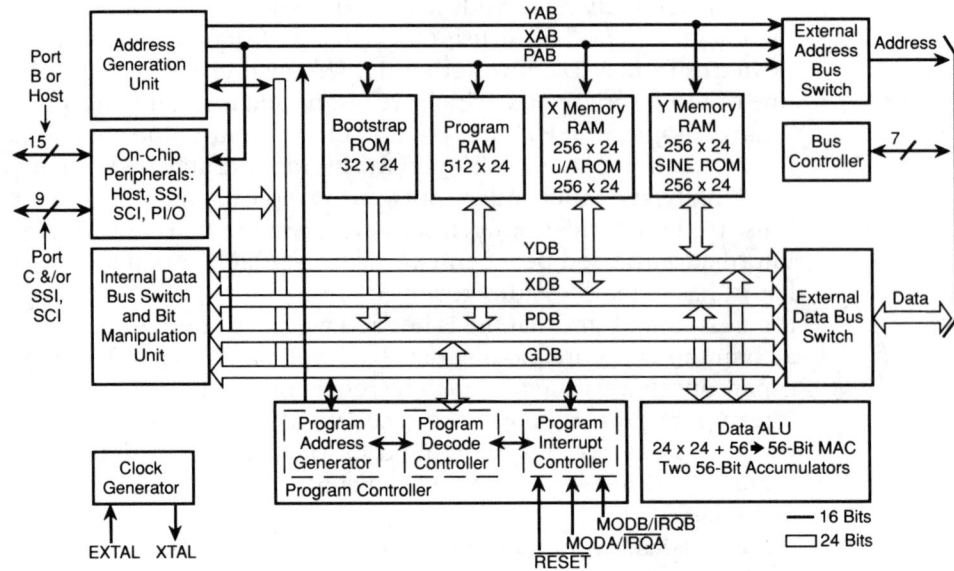

**Figure 2.** Block diagram of the Motorola DSP56001. *(Courtesy Motorola)*

from memory to the ALU, fetching software instructions from program memory, and sending addresses to the memory hardware—to take place simultaneously, without interfering with each other. This parallelism, often found only in minicomputers and supercomputers, offers tremendous speed advantages. Accordingly, the 56001 has a clock speed of 27 MHz and can execute 10.25 MIPS (million instructions per second).

The 56001 program controller, or CPU, is responsible for fetching software instructions from program memory and decoding them into hardwired commands for the ALU and other units. The program controller is what actually runs the software. The program controller also handles interrupt requests and generates the memory addresses needed to fetch from memory the next instruction to be executed.

The program controller contains a unique loop counter, which increments a hardware counter rather than adding one to a running counter total. This eliminates the need to store a software counter value as well as the extra steps needed to increment the counter each time a loop begins or ends. Because most DSP algorithms call for extensive repetition and looping, this can be a tremendous time saver both in terms of execution speed and programming effort. This is a good example of the kind of streamlining that DSP chips rely on for their speed and efficiency. A hardware counter might reduce the flexibility of a more general-purpose microprocessor designed to accommodate many kinds of looping (including conditional looping), but it tailors the 56001 program controller to the cyclic functions called for in DSP applications.

Software instructions are stored and fetched from the 56001 program memory, which can store up to 512 words. A small bootstrap memory is provided which can execute a small, preset software program upon power-up or reset. This small program usually initiates the load of a larger program into the program memory.

The data ALU contains four 24-bit input registers, which hold operands, and two 56-bit accumulators, which hold the results. The 24-bit size is designed to avoid a loss of precision when dealing with A/D converters that offer 16-bit resolution. Furthermore, the 56-bit accumulator provides enough range so that intermediate results on the 24-bit operands do not cause overflow errors. The 56001 can perform multiplication, addition, subtraction, division, multiplication, normalization, and bit shifting, each in one instruction cycle.

The data ALU contains hardware circuitry which enables it to multiply two 24-bit operands and add or subtract the result to a running total, in just one instruction cycle. As mentioned before, this operation is called a multiply and accumulate (MAC). The 56001 employs fixed-point processing.

The 56001 divides its RAM data memory into two regions, called X memory and Y memory. Many DSP applications generate or process data which lends itself to this sort of subdivision. Graphics applications identify points in an image plane using Cartesian coordinate axes, FFT and digital filtering algorithms call for coefficients and data, and many computations use real and imaginary number sets. Both the X and Y data memories store 256 24-bit words.

In addition to the X and Y RAM memories, the 56001 contains two additional read-only memories. One ROM contains a table of sinewave values, and the other contains $\mu$-law and $A$-law expansion tables. This avoids the need to compute these often-used values when programming DSP algorithms. Again, while this would be an overly specific feature in a general-purpose microprocessor, it is valuable in a DSP chip.

The 56001 uses four bidirectional, 24-bit data busses: one for transfer between the X memory and the ALU, another identical bus for the Y memory, one for transfer of instructions from program memory to the CPU, and one for transfer of information to and from the I/O ports. Provisions are also made for transfer from one bus to another. There are also three address busses, which carry addresses from the address generation unit to the various memories. The address generation unit translates the CPU's request for access to a particular memory location into the hardwired commands passed through to the actual memory hardware.

The input and output interface ports of the 56001 are among its more advanced and extensively implemented features. Using these ports, the 56001 can be configured with a host computer, or in serial or parallel arrangements with other DSP chips or peripherals. The 56001 I/O structure consists specifically of two ports, two interrupt pins, and a serial interface controller, all of which are memory mapped. The first port, port $A$, consists of a 24-bit bus designed to interface with external memories and other DSPs. It is a synchronous interface, which requires a common clock signal to exist between it and the device to which it is attached. The port $A$ interface has the capability of programming the timing of this bus through software control, which increases the number of compatible external devices. The second port, port $B$, is also referred to as the host interface. It is an 8-bit, bidirectional bus with seven additional control lines and is designed for direct connection to a host processor. Separate buffers are provided for incoming and outgoing data, and flags are provided for polled (keep checking until new data arrives) or interrupt driven (do something else until a "data ready" signal arrives) data transfer. The host interface allows the 56001 to appear as a memory mapped peripheral to the host computer.

Also part of the 56001 I/O section is the on-chip serial communication interface controller, which occupies a portion of port C. This interface uses three additional dedicated pins as data transmit, data receive and clock lines. The serial communication interface supports many different communication protocols, and can achieve transmission speeds of up to 2.5 Mb/s. The baud rate, word size, synchronization method, and handshaking considerations of this interface are all programmable, making it a very flexible I/O tool. The remainder of Port C is taken up by 9 pins, each of which can be programmed as either a general-purpose I/O pin or as a pin under direct software control. The direction, input or output, is also programmable. These pins might be used for sending simple on/off signals to external devices that do not need to share data with the DSP chip.

Finally, there are two general-purpose interrupt request pins. It is possible to program the 56001 to suspend execution of the current software and run a special portion of the program, called an interrupt service routine, whenever an interrupt signal is received. While each of the three interface ports use several interrupt pins, these two provide a means to service interrupts not tied directly to communications.

With its extensive communication and memory facilities, the 56001 is very well-suited for applications in a host environment. Evidence of this lies in its inclusion in a large number of digital audio expansion cards made for personal computers, and the fact that it is a standard co-processor in the NeXT computer system. A host system can easily load new software into the 56001, configuring it to run any number of DSP application programs. Moreover, its extensive on-chip resources allow it to run autonomously once configured.

Because of its complexity, however, the 56001 may also be considered too cumbersome for some applications, especially those calling for an embedded controller. Software written for a DSP chip is not isolated from the hardware, as is the case with software written for mainframes or personal computers (indeed, the purpose of general-purpose microprocessors and high-level languages is to insulate the programmer from the details of computer hardware). Software dealing with multiple I/O channels, interrupts, handshaking, dynamic memory allocation, and assembly language style software can be difficult to write and debug. In some cases, the resources of the 56001 might be more complex, time-consuming, and costly than the application warrants. The difference could be considered similar to the difference between using a desktop publishing page-layout program and simple word processor; sometimes simplicity is sufficient.

Credit must be given to Motorola, however, for providing many boilerplate routines, as well as extensive simulation and debugging software. The 56001 is also cost-effective in most medium and large scale applications. In all, the advantages of the 56001 far outweigh the disadvantages, and industry enthusiasm and opportunities for it continue to grow.

## Texas Instruments TMS320

Another group of very widely used DSP chips is the Texas Instruments TMS320 family. This 68-pin VLSI chip is well known for its compact architecture, which yields efficiency and high numerical throughput. Introduced in 1983, the TMS family was one of the first commercially successful DSP products. Several specific chips belong to the TMS320 family, the most recent of which is the TMS32025, and Texas Instruments offers versions of the chip with different memory and speed options to provide different cost versus performance tradeoffs.

The TMS32025 can execute at 10 MIPS, and the fastest TMS320 version, the TMS320C25-50, can execute at 12.5 MIPS. This corresponds to a clock cycle time of 80 ns. Like other DSP chips, the TMS family uses NMOS and CMOS fabrication technologies to obtain these fast operation speeds in a small, single-chip package.

A simplified block diagram of the TMS320 processor is shown in Figure 3. There are both program and data busses; memory addresses are sent over the program bus. The bus width is 16 bits, and the accumulator in the ALU is 32 bits wide.

The TMS320 CPU has access to 256 words of on-chip program RAM. In addition, the TMS320 features a 4K-word, erasable programmable read-only memory (EPROM). This EPROM can be loaded at the factory with user supplied program software. These memory configurations suggest that the TMS family is best intended for applications that do not require a host from which to download software, and they also call for software which will remain the same throughout the life of the DSP chip.

The TMS320 CPU also contains a hardware loop counter, which permits the rapid repetition of a single instruction step. For example, a single software instruction to add the contents of two memory locations can be repeated a certain number of times without any programming overhead. However, the counter cannot keep track of the number of times a group of instructions has been executed. The TMS hardware loop counter is nonetheless useful for successive multiplies, block moves, I/O transfers, and memory reads and writes.

The TMS320 can perform most of the necessary arithmetic operations, including a 16-bit by 16-bit multiplication, in one instruction cycle. However, it does not support single-cycle MACs. The ALU accepts two operands, one of which comes from the data RAM or the hardware multiplier, and one of which is always the output of the accumulator. This wraparound configuration is ideal for many DSP algorithms, which add results to a running total. The ALU also

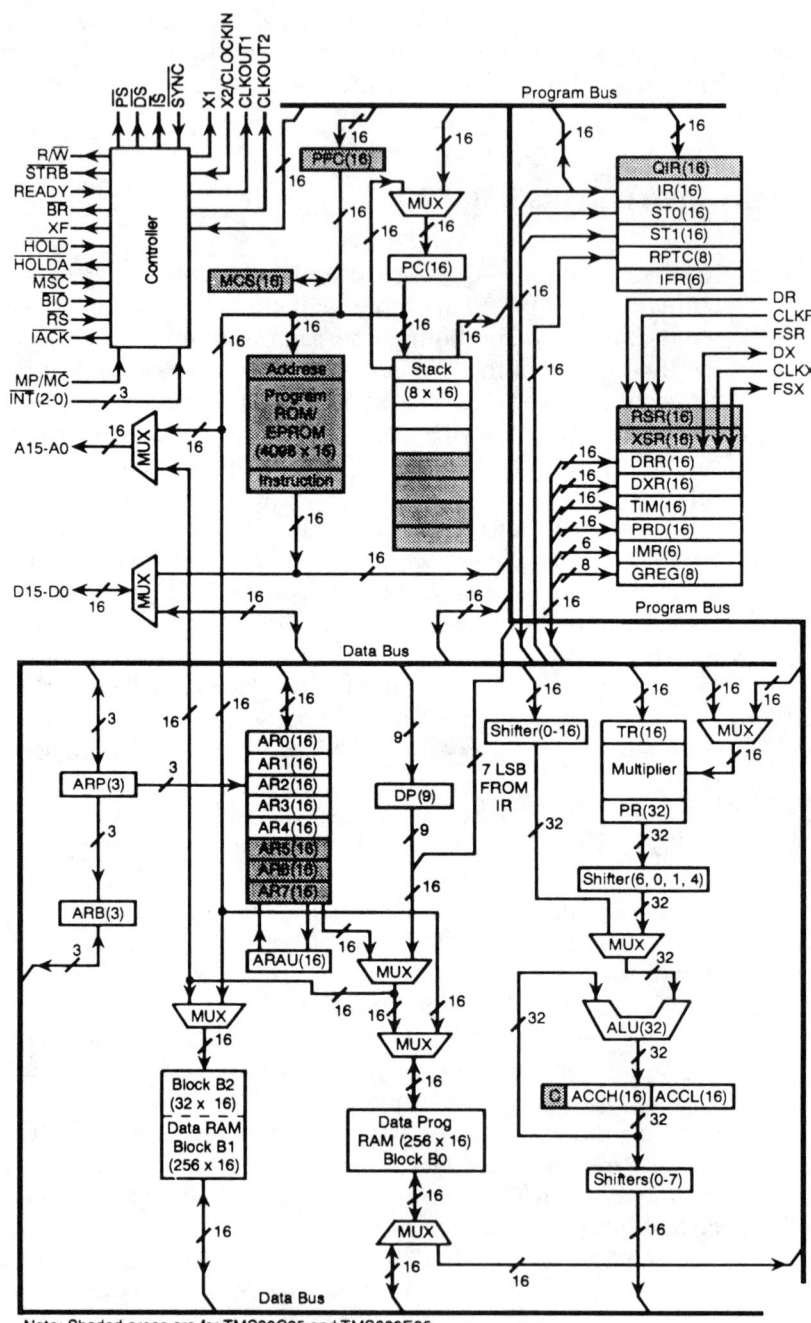

**Figure 3.** Block diagram of the Texas Instruments TMS32020.
*(Courtesy Texas Instruments, Inc.)*

supports floating-point operations when more than 96 dB (16 bits) of dynamic range is needed. The floating-point operations, however, are slower than the fixed-point, two's complement integer operations.

There are 288 words of data RAM, and the 256 words of program RAM may also be configured as data RAM. This brings the total on-chip memory RAM to 544 words. The TMS320 has the addressing capability to access up to 64K words of additional off-chip memory. Assuming the memory hardware is chosen to be sufficiently fast, there is no speed penalty for using this off-chip memory.

Input and output to the TMS320 are accomplished through a 16-bit input port and a 16-bit output port, both of which interface directly with the data bus. These ports are memory mapped, and a number of clock and status lines are provided to synchronize transfer. The direction of the ports is not programmable. There are also two general-purpose pins that can be directly controlled through software and may be used as flags or for simple device control. Finally, there is a serial port and three external interrupt pins. The serial port provides communication with serial A/D converters, modems and codecs, as well as other DSPs.

The Texas Instruments TMS320 family of DSP chips is very functional and efficient. While it may seem to lack some of the features of the Motorola 56000 family, it compensates with high speed and easy interfacing. From a programming point of view, TMS320 DSP algorithms are quickly implemented in software with a minimum of control software or glue logic. From a hardware and cost standpoint, the TMS320 is well suited to any number of military, industrial, and telecommunications applications, especially those which call for an embedded controller.

## AT&T WE-DSP16A and WE-DSP32C

Two DSP chips manufactured by AT&T are the WE-DSP16A and the WE-DSP32C. The DSP16A is 16-bit fixed-point chip, while the DSP32C is a 32-bit floating-point model. A block diagram of the DSP16A is shown in Figure 4. It contains only one data bus, which shares the functions of memory data and address transfer. The program memory data enters via an external bus. Some performance benchmarks include a 16-bit bus width, a 36-bit wide accumulator, and a 33 ns instruction cycle.

The DSP16A has no on-chip program RAM. Program memory consists of a 4K-word on-chip ROM, and addressing access is provided for up to 60K words of external ROM or RAM. Also provided is a special on-chip RAM called the instruction cache. In computer architectures, a cache is a small memory with a very fast access time used to hold often-needed data. The DSP16A cache stores up to 15 words of software instructions, and these can be looped up to 127 times with no software overhead. This is similar to the function of a hardware loop counter.

The DSP16A has an ALU (which AT&T refers to as the data arithmetic unit, or DAU) similar to other ALUs. It can perform a 16-bit by 16-bit multiplication

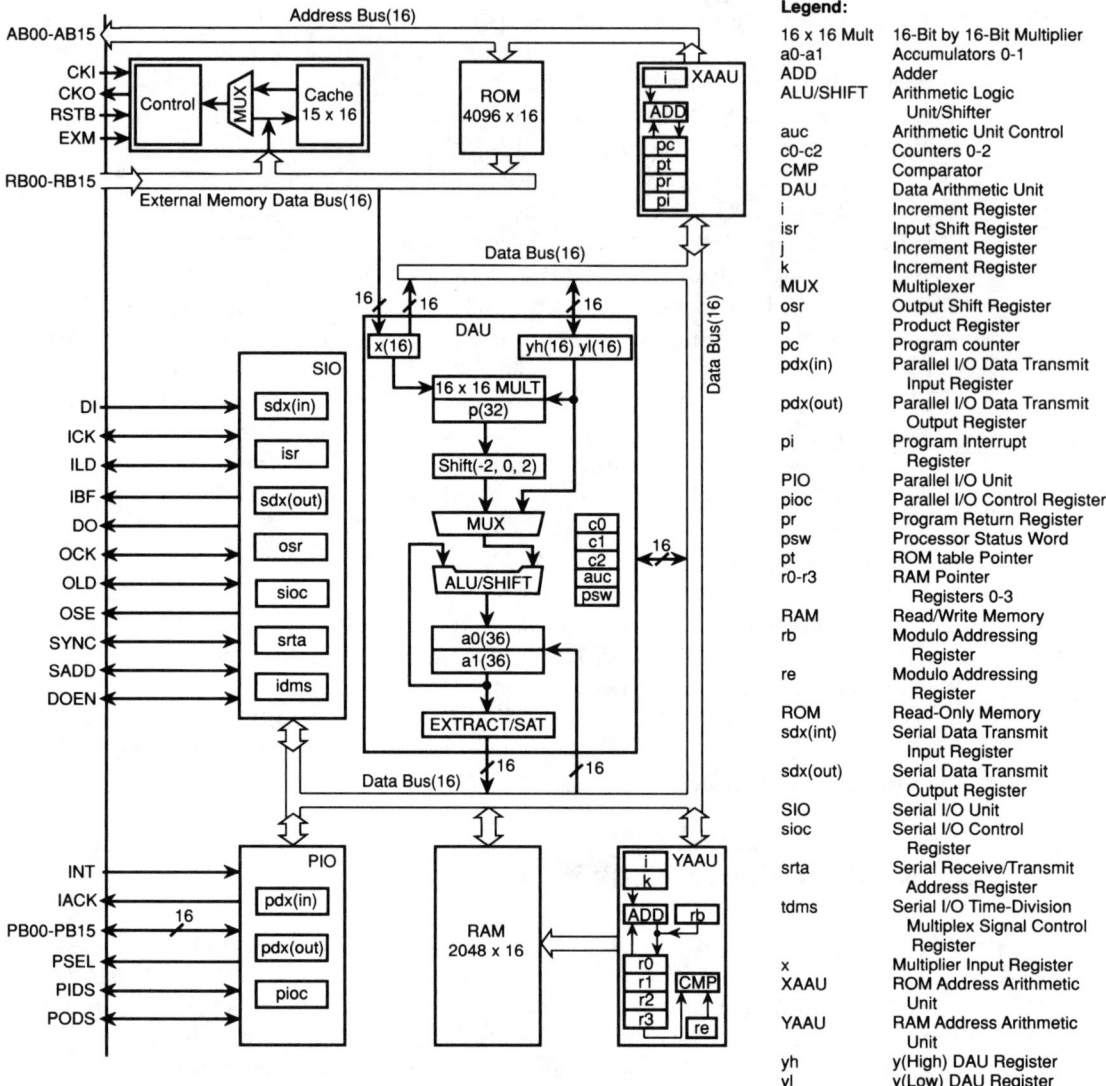

**Figure 4.** Block diagram of the AT&T WE-DSP16A. *(Courtesy AT&T)*

in one instruction cycle, and has two 36-bit accumulators. These accumulators can read or write to memory directly. A sizable 2K words of on-chip data RAM are provided. Communication to external devices uses either a serial or a parallel port. The serial port is a full-duplex, fully buffered, synchronous/asynchronous port capable of speeds up to 10 Mb/s. The parallel port is a simple 16-bit, bidirectional gateway to the data bus.

The DSP16A is similar to many math co-processors that are available for personal computer microprocessors. Its primary design function is as a high-speed arithmetic processor, as opposed to a device that operates in a more self-contained manner. The DSP16A might work best in devices that already have extensive hardware input and output facilities but need one or more high-speed arithmetic units. An example could be a digital video effects (DVE) device, similar to the type used to flip and rotate a picture within another picture. For speed, the DVE implements many of its functions in hardware, but it may use a device like the DSP16A to perform the repeated Cartesian axis calculations needed to compute the new picture location.

The DSP16A could implement other DSP algorithms, such as filtering and FFTs; however, its architecture makes it better suited as an arithmetic co-processor than as a stand-alone DSP processor. The DSP16A is also likely to be found working in parallel with one or more other DSP16As, each serving a separate division of a larger group of hardware or software.

The WE-DSP32C, also manufactured by AT&T, is a 32-bit floating-point DSP chip. Its block diagram is shown in Figure 5. The architecture is optimized for floating-point operations throughout, and all floating-point instructions are executed in a single instruction cycle. DSP32C speeds can reach up to 12.5 MIPS. Floating-point processing, high speed, and addressing access to a large amount of on- and off-chip memory make the DSP32C a powerful device.

The CPU (called the control arithmetic unit or CAU) features a large set of registers, much like those found in the ALU. The more registers in the CPU, the less often values have to be stored in memory. The DSP32C has 22 registers, compared to 6 in the Motorola 56001, and 8 in the Texas Instruments TMS320. These registers are available for general-purpose storage, and do not include program counters, address registers, etc. Nor is a stack considered the same as a general-purpose register.

The DSP32C offers many flexible memory configurations and management schemes. Either 1K words of RAM and 4 words of ROM, or alternatively a total of 1.5K words of RAM, may be made available as on-chip memory. The external memory interface can address 16 MB of off-chip memory. The DSP32C can partition this memory in eight different modes, and accesses can be made without regard to memory type or physical location. Access is equally fast for both 32-bit floating-point numbers and 8-, 16-, 24-, or 32-bit integer words.

The ALU block contains a hardware floating-point multiplier and adder, and four 40-bit accumulators. The multiplier and adder can work in parallel to perform up to 12.5 million MACs per second. One operand of the ALU always comes from the multiplier and adder combination, and the other can come from I/O or memory. There are also provisions for conversions between 8-, 16-, and 24-bit two's complement integer representation, IEEE floating-point format, and the resident 32-bit representation (8-bit exponent, 24-bit mantissa). As found in the Motorola 56001, the DSP32C CPU and ALU work in parallel for increased instruction and computation throughput.

The serial I/O unit can be programmed for many different configurations and synchronization formats, and it also performs serial-to-parallel and parallel-to-serial conversions on data words ranging from 8 to 32 bits wide. The parallel interface unit is based on a 16-bit–wide bus that can transfer entire

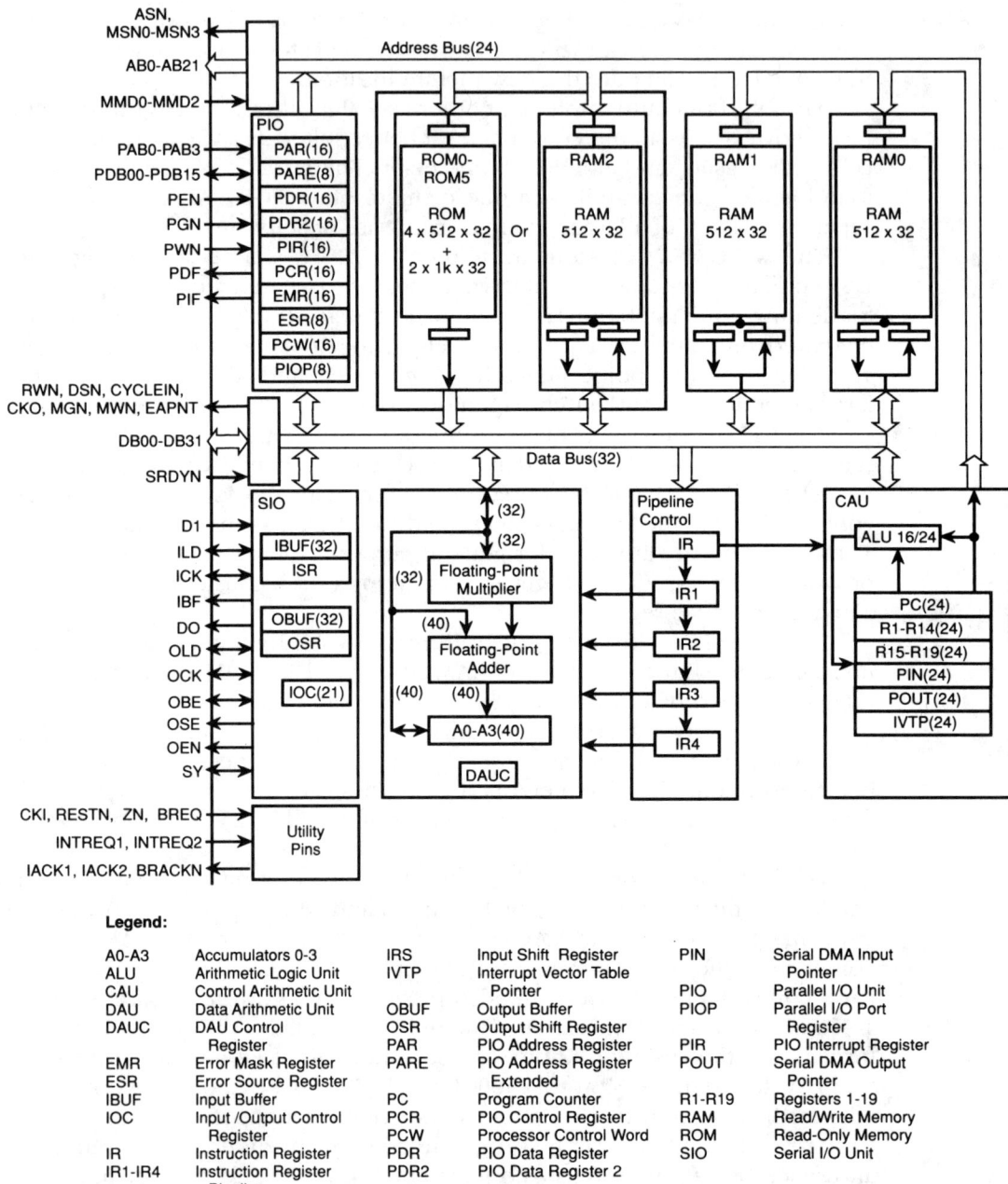

**Figure 5.** Block diagram of the AT&T WE-DSP32C. *(Courtesy AT&T)*

blocks of data from an external device into DSP32C memory without interrupting the software currently running. This direct memory access (DMA) transfer is a feature often reserved for much larger processors. The DSP32C provides two pins for external interrupt signals and can handle up to two million interrupts per second. Furthermore, the chip itself can generate interrupt signals when certain input and output buffers are full or empty.

The DSP32C is in a class of advanced DSP chips along with the Motorola 56001. However, the AT&T WE-DSP32C is set apart by its floating-point architecture, as well as its flexible access to a large set of memories. The DSP32C is a performance leader in the areas of high computational capacity and large scale memory management.

## Motorola DSP96002

The DSP96002, called the Media Engine by Motorola, is a 223-pin, floating-point DSP chip designed specifically for graphics and audio applications. A block diagram of its architecture is shown in Figure 6. The 96002 is designed to keep pace with the multimedia needs of increasingly sophisticated microcomputers and workstations, many of which include extensive graphics and audio capabilities. The brand name, Media Engine, was given to the 96002 because its architecture and computational functions are specifically designed for color graphics and multichannel audio applications. In addition, the 96002 can act as a co-processor to modern microprocessor chips such as the Intel 80486 or the Motorola 68040. Some of its applications are listed below:

- High-resolution color graphics
- Video and photo processing
- Medical imaging
- Digital audio processing
- Three-dimensional graphics and animation
- Robot vision
- Large-storage (gigabytes) disk drive
- Aircraft and ship navigation systems
- Military hardware

Functionally, the 96002 is essentially a 32-bit floating-point version of the 24-bit Motorola 56001. The major advances, in addition to the floating-point capability, are increased speed, a dual-port input and output structure, and the distinction of being the first DSP chip to comply with the IEEE standards for floating-point math. Most mainframe and supercomputer systems use the IEEE format, and floating-point DSP chips that do not comply with the IEEE standards must convert between proprietary formats and the IEEE format before data can be exchanged; this can result in a loss of numeric accuracy as well as time lost during the conversion process.

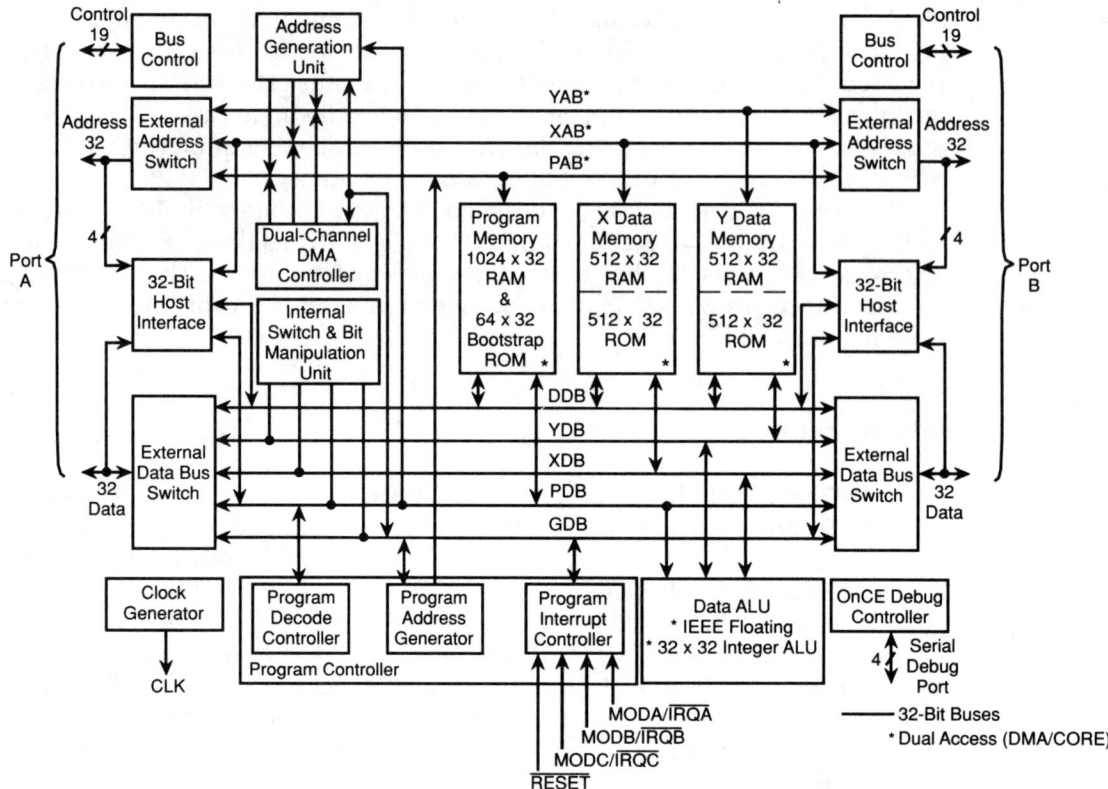

**Figure 6.** Block diagram of the Motorola DSP96002. *(Courtesy Motorola, Inc.)*

The 96002 has a clock rate of 33 MHz, which yields speeds of over 16 MIPS. However, the actual instruction throughput peaks at 50 million floating-point operations per second (MFLOPS) since the 96002's instruction format allows several floating-point operations to be carried out with a single instruction. Speed is also increased through the use of five data busses, three address busses, and a program instruction bus, all of which are 32 bits wide. The ALU features hardware units for 32-bit floating-point addition, subtraction, and multiplication, and the ALU is optimized for software division and square-root operations. There are ten 96-bit registers in the ALU, which can also be used as thirty 32-bit registers.

The feature most specifically intended for graphics and audio applications may be the 96002's input and output provisions. Because applications such as image processing can involve very large amounts of data (a high-resolution, digitized color image of a typical 8½- by 11-inch page can require over 50 MB of data), the transfer of that signal to and from the DSP chip can easily become the speed bottleneck in a DSP system. The 96002 has two identical input and output ports, each of which provides a 32-bit path to the data busses, a 32-bit path to the address busses, as well as 19 handshaking and control pins. These ports are both bidirectional and can operate independently of each other even while computation continues on-chip, allowing, for example, one port to control the transfer of video data while the other transfers audio data.

The two ports may also be configured for direct connection to additional 96002 chips without the need for any additional interface circuitry. In this configuration, one 96002 can directly send and receive data, and even software instructions, to other 96002 processors, and complex parallel arrays of DSP chips can be created. Parallel and array processors such as this are appearing more and more frequently in modern computer architectures and are needed for many image processing and artificial intelligence applications.

The architecture and processing power of the Motorola 96002 distinguishes it from general-purpose DSP chips, and its advanced and specialized features point out the increased use of DSP hardware by the computer and electronics industries. Without doubt, both general-purpose DSP chips and specialized processors such as the 96002 will continue to find their way into more and more applications and products.

# References

AT&T Microelectronics, *WE-DSP16A and WE-DSP32C Data Sheets and Applications Notes*, AT&T Microelectronics.

Motorola, Inc., *DSP56001 and DSP56000 Data Sheets and Applications Notes*, Motorola Microprocessor Products Group.

Motorola, Inc., *DSP96002 IEEE Floating-Point Dual-Port Processor User's Manual*, Motorola Microprocessor Products Group.

Texas Instruments, Inc., *Digital Signal Processing Applications with the TMS320 Family: Theory, Algorithms, and Implementations*, Texas Instruments, 1990.

Texas Instruments, Inc., *Second Generation TMS320 User's Guide*, Texas Instruments, 1989.

*Chapter 14*

# Digital Signal Processing: Programming and Interfacing

**Matt Booty**

## Introduction

As with any computer, a digital signal processing chip cannot function without software. The majority of software developed for DSP chips is written in assembly language; however, there is a small but increasing number of high-level and object-oriented language compilers available for some DSP chips. Each DSP chip has its own native assembly language, and, accordingly, software development for a DSP chip is similar to assembly language programming on general-purpose microprocessors. In a DSP chip, however, there is an increased focus on execution speed and throughput, on input and output, and also on the DSP chip's particular hardware architecture.

Many engineers and software programmers recoil in horror when they learn that most DSP chip programming is done in assembly language, which has traditionally been a meticulous and frustrating task. Even though some high-level language compilers are available, most DSP applications still involve assembly language programming, and there are several reasons for this. The most critical reason is the requisite need in DSP applications for fast execution speed and high signal and data throughput, and assembly language is almost always faster than C, Pascal, LISP, BASIC, or FORTRAN. Compilers for these languages have the difficult task of translating the easy to read, sentencelike source code written by the user into executable object code that can be loaded into the processor's program memory; the result is not always the most efficient or the fastest. The workings of computer language compilers and the reason why their output is sometimes less than optimal is beyond the scope of this chapter, but it is generally accepted that highly efficient or speed-critical applications should be written in assembly language, and this is particularly true in DSP applications.

Fortunately, the amount of assembly language programming required is often limited. DSP chip software rarely maintains any sort of user front end programming. Most DSP chips operate in a host environment, and user prompts, menus, file organization, and other tasks are handled by a PC or minicomputer. The user front end program can be a large part of the total software and is best written in a high-level language. In comparison, the amount of DSP software is usually very small. Finally, assembly language is used because it provides a more direct access to the specific hardware of the DSP chip. A comparison of assembly and high-level languages is summarized in Chart 1.

**Chart 1.** Comparison of Assembly and High-Level Languages

| Assembly Language | High-Level Languages |
|---|---|
| **Advantages** | |
| • Executes faster | • Convenient sentencelike description of tasks |
| • Direct access to specialized hardware | • Independent of particular computer hardware |
| • Better control over input and output | • Relatively faster development time |
| • More efficient code | • Easier to document and debug |
| • Requires less memory | • Well suited to large programs |
| **Disadvantages** | |
| • Difficult to remember instructions | • Slower execution |
| • Tasks must be explicitly detailed | • Language oriented to algebraic tasks |
| • Confusing to read and debug | • No direct access to specialized hardware |
| • Very hardware specific | • Code sometimes inefficient |
| • Time consuming to develop | • Large memory use for overhead |

This chapter will focus on the specifics of software development and assembly language programming for digital signal processing chips. Programmers not familiar with assembly language will find DSP chip programming easier if they first explore programming of a more simple microprocessor, such as the Motorola 6809.

# Programming Models

DSP programming begins with the chip's programming model, or conceptual arrangement of the basic functional blocks as they appear to the software programmer. The programming model reflects, but is not the same as, the overall chip architecture. The ALU input and accumulator registers, the address generation registers, the program controller, and the program and data memories are grouped according to the parts that the programmer can access and change.

The programming model for the Motorola DSP56001 chip is shown in Figure 1. From the programmer's point of view, the 56001 consists of three units operating in parallel. The ALU, the address generation unit, and the CPU or program controller all operate simultaneously during a single instruction cycle. The ALU features four 24-bit general-purpose registers that can be concate-

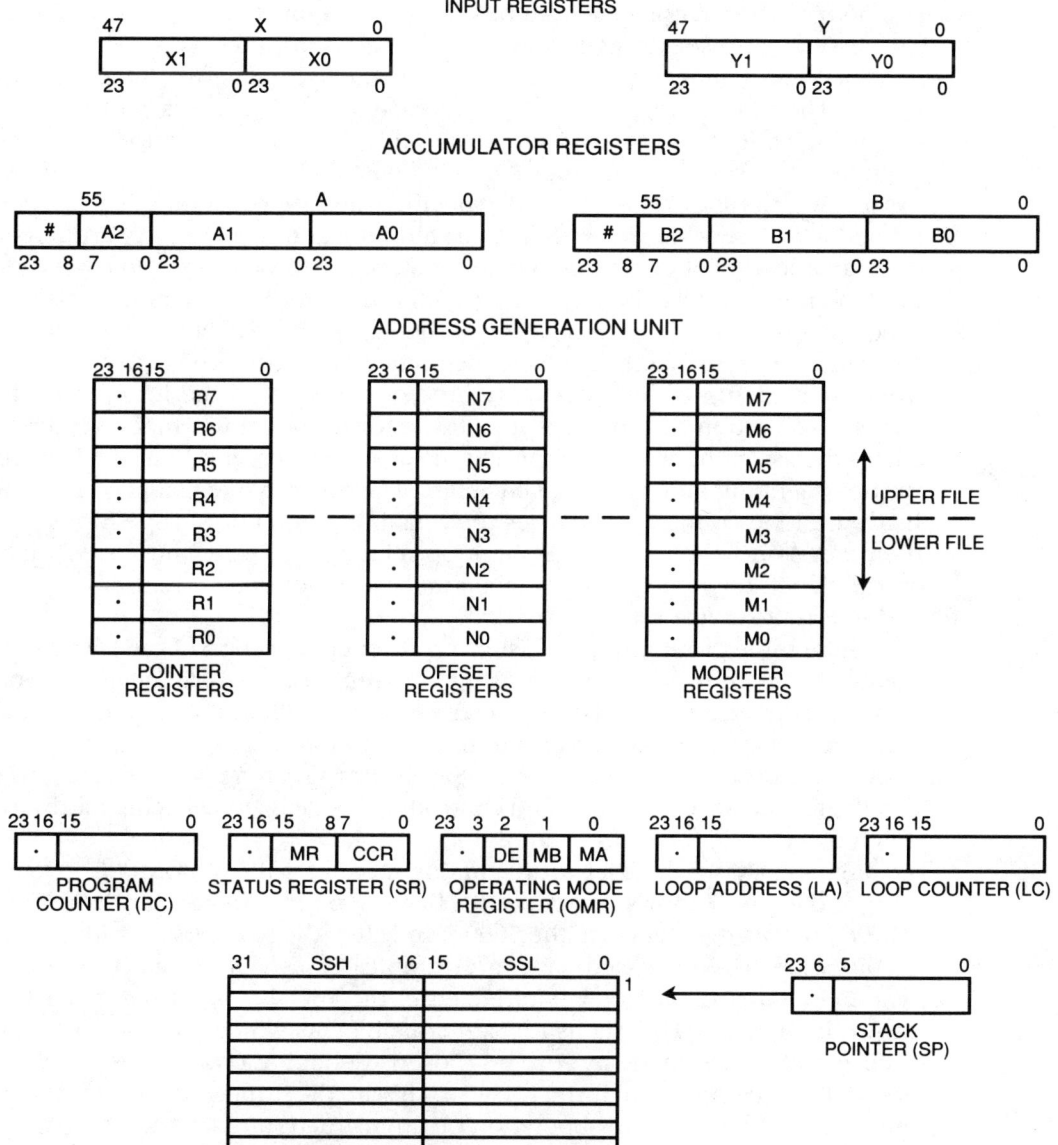

**Figure 1.** Programming model of the Motorola DSP56001. *(Courtesy Motorola)*

nated into two 48-bit registers. These registers are the storage space between the data busses and the arithmetic and MAC units, and they also allow new operands to be loaded for the next instruction while the contents are used by the current instruction. The values in the registers can also be read back onto the data bus. There are six accumulator registers, which are arranged to form two 56-bit accumulators, each 48 bits long with 8 bits for overflow, and they can serve as both source and destination operands for the ALU.

The address generation unit is a set of three registers used for memory access. The pointer registers contain the actual 16-bit addresses of memory locations, while the offset registers store values which are used to increment, decrement, or step through memory locations. Finally, each pointer register has its own modifier register, which specifies the type of arithmetic to be performed for address calculations. Linear, modulo, and reverse-carry arithmetic are available; the latter two are useful for stepping through lookup tables and implementing circular buffers. The pointer and offset registers may also be used to store general 16-bit data, since their contents only affect memory access when a specific memory access instruction is executed. The program controller is made up of five registers, three of which store the location of the current instruction, the processor status, and the operating mode. Figure 2 shows the assignments of each bit in the status register, and Figure 3 lists the various operating modes of the DSP56001. The other two registers control the hardware loop counter; one stores the remaining number of repetitions, and the other stores the location of the loop. Finally, there are the system stack and stack pointer, which store variables and processor status when subroutines and interrupt service routines are executed.

These three groupings of registers represent the portion of the 56001 chip that the programmer can control using the chip's assembly language instructions, and they are also the programmer's gateway to all input and output hardware. Although the 56001's programming model is very similar to some general-purpose microprocessors, DSP chip programmers must remember that the registers are accessing and controlling specialized DSP chip hardware, much of which operates in parallel.

Figure 4 shows the user programming model for the Motorola DSP96002 chip, and Figure 5 shows the chip's two floating-point formats. The 96002 contains more data registers than the 56001, and all of the registers are either 32 or 96 bits wide. The increased hardware complexity, as well as the overhead of floating-point computations, is reflected in the 96002's 32-bit status register, shown in Figure 6. This status register contains bits which are set to 1 to flag such conditions as divide by zero, ALU overflow, and ALU operands which are not in IEEE floating-point format. These flags are the primary method for keeping track of the DSP chip's operation on an instruction-by-instruction basis.

Programming models for other DSP chips, such as the Texas Instruments TMS320 or the AT&T DSP32, are similar. However, there are major differences in the registers and formats used to control chip hardware. Whereas data registers store values, and status registers report and store conditions, control registers dictate hardware action and operating modes, and these are usually very

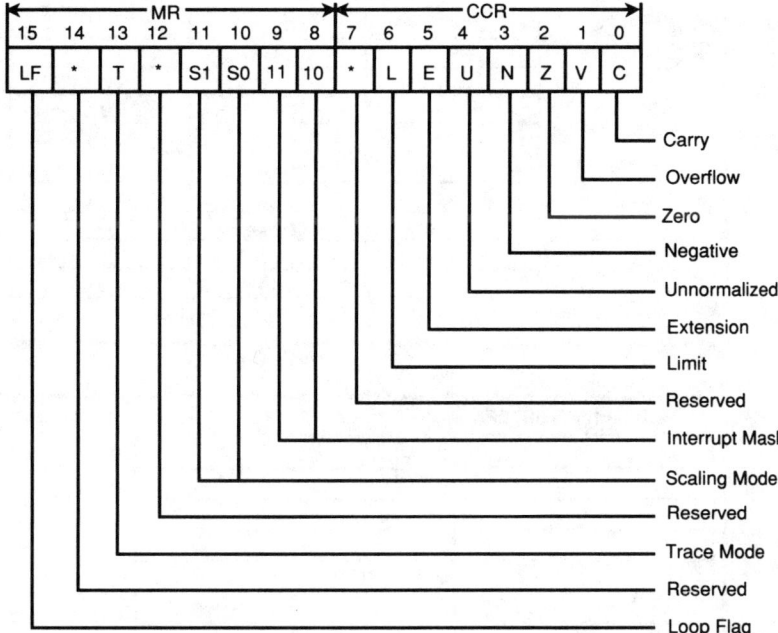

**Figure 2.** Status register bit assignments for the Motorola DSP56001. *(Courtesy Motorola)*

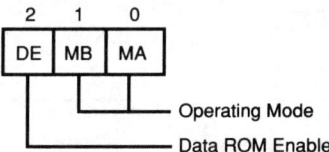

*(a) Operating mode register format.*

| Operating Mode | MB | MA | Description |
|---|---|---|---|
| 0 | 0 | 0 | PRAM enabled, reset at $0000 (internal) |
| 1 | 0 | 1 | Special bootstrap mode, after PRAM loading mode 2 is automatically selected |
| 2 | 1 | 0 | PRAM enabled, reset at $E000 (external) |
| 3 | 1 | 1 | PRAM disabled, reset at $0000 (external) |

*(b) Operating mode summary.*

**Figure 3.** Operating mode register format and summary for the Motorola DSP56001. *(Courtesy Motorola)*

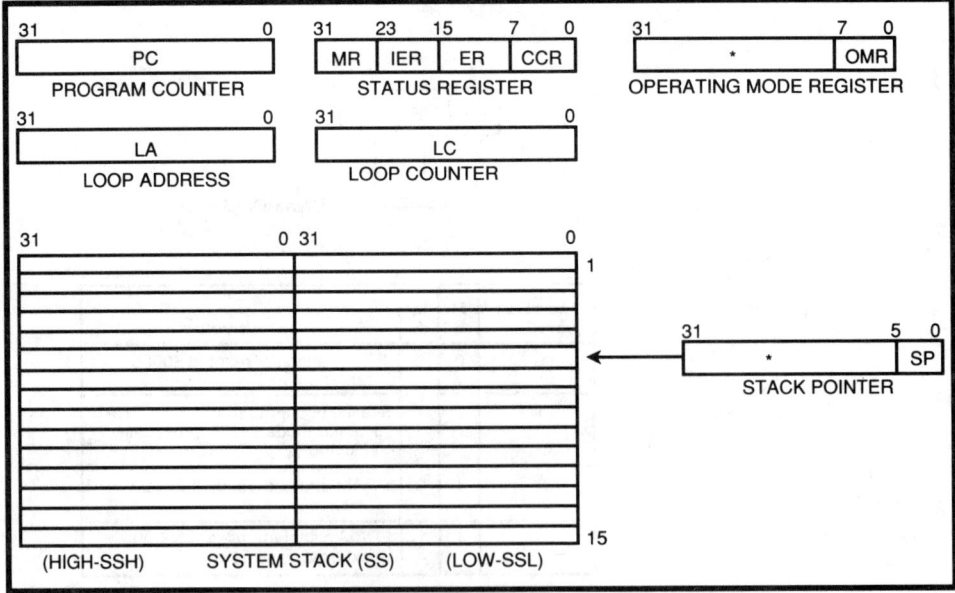

**Figure 4.** Programming model of the Motorola DSP96002.
*(Courtesy Motorola)*

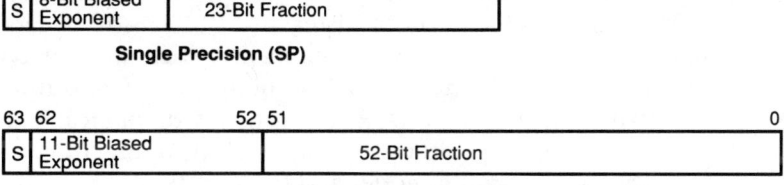

**Figure 5.** Floating-point arithmetic formats on the Motorola DSP96002. *(Courtesy Motorola)*

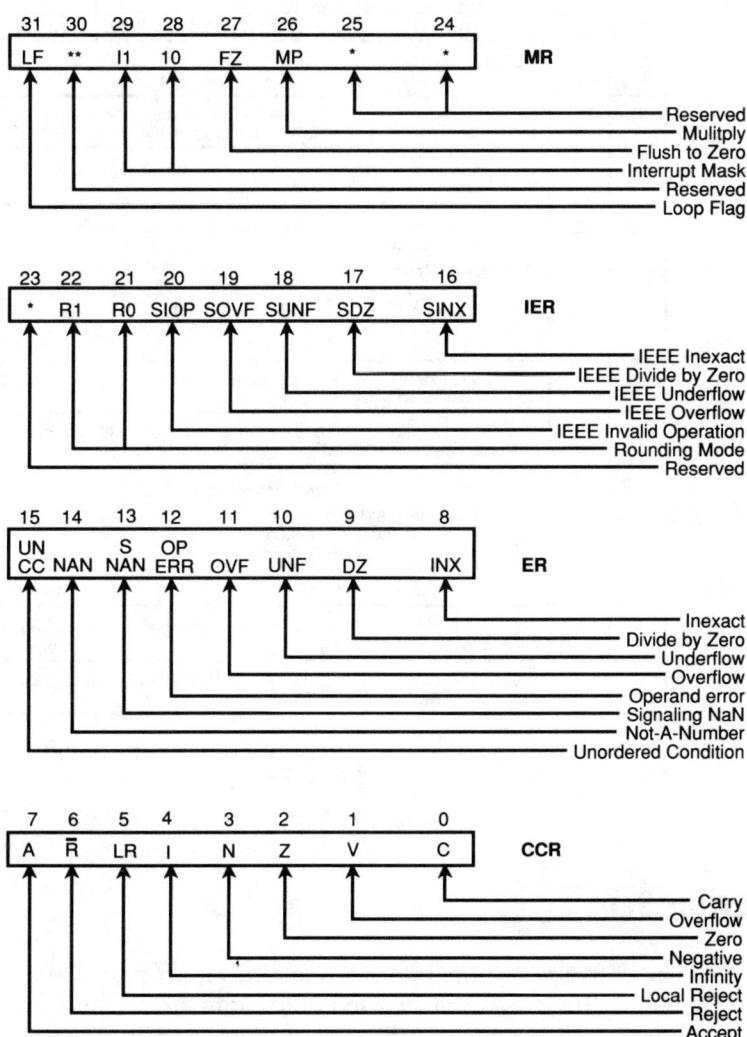

**Figure 6.** Status register bit assignments for the Motorola DSP96002. *(Courtesy Motorola)*

chip-specific. For example, Figure 7 shows the bit functions of the 21-bit input/output control register of the AT&T WE DSP32 chip, and Figure 8 shows an even more specialized register, the DSP32's 4-bit data arithmetic control unit register. For comparison, Table 1 gives the names and descriptions of several TMS320 control registers. The number, size and function of the control and status registers on a DSP chip can vary dramatically, and a DSP chip programmer's first step, after a tour of the chip architecture, is to become familiar with the programming model and the registers.

| Bit | 20 | 19 | 18 | 17 | 16 | 15-13 | 12 | 11 | 10 | 9 | 8 | 7 | 6 | 5 | 4 | 3 | 2 | 1 | 0 |
|---|---|---|---|---|---|---|---|---|---|---|---|---|---|---|---|---|---|---|---|
| Field | DSZ | O24 | CKI | OUT | IN | DMA | SAN | OLEN | | AOL | AOC | ILEN | | AIL | AIC | SLEN | | BC | ASY |

| Bit(s) | Field | Description |
|---|---|---|
| 0 | ASY | If 0, SY is external. If 1, SY is internal. When generated internally, SY = {ICK, OCK} / {256, 512, 1024}, based on IOC[BC] and IOC[SLEN]. |
| 1 | BC | If 0, ICK is used to derive the internal load and SY signals. If 1, OCK is used to derive the internal load and SY signals. |
| 3,2 | SLEN | These bits select the frequency ratio of the on-chip load signal to the on-chip SY signal. The possible ratios are:<br>**Bit 3  Bit 2  Ratio**<br>0     0     32<br>0     1     8<br>1     0     16<br>1     1     32 |
| 4 | AIC | If 0, ICK is external. If 1, ICK is generated internally with a frequency of CKI/8 or CKI/24, based on IOC[CKI]. |
| 5 | AIL | If 0, ILD is external. If 1, ILD is generated internally with a frequency of ICK/32 or OCK/32, based on IOC[BC]. |
| 7,6 | ILEN | These bits specify the length of the serial input data.<br>**Bit 7  Bit 6  Input Length**<br>0     0     32 bits (prior to ILD)<br>0     1     8 bits (after ILD)<br>1     0     16 bits (after ILD)<br>1     1     32 bits (after ILD) |
| 8 | AOC | If 0, OCK is external. If 1, OCK is internally generated with a frequency of CKI/8 or CKI/24, based on IOC[CKI]. |
| 9 | AOL | If 0, OLD is external. If 1, OLD is internally generated with a frequency of ICK/32 or OCK/32, based on IOC[BC]. |

**Figure 7.** Input and output control register bit assignments for the AT&T WE DSP32. *(Courtesy AT&T)*

# Instruction Sets

The software commands that make up a program come from the DSP chip's instruction set. Because some early chips designed for DSP and other high-speed applications could only execute a limited variety of instructions, they were sometimes referred to as reduced instruction set computers (RISC). These processors trade programming flexibility for execution speed by limit-

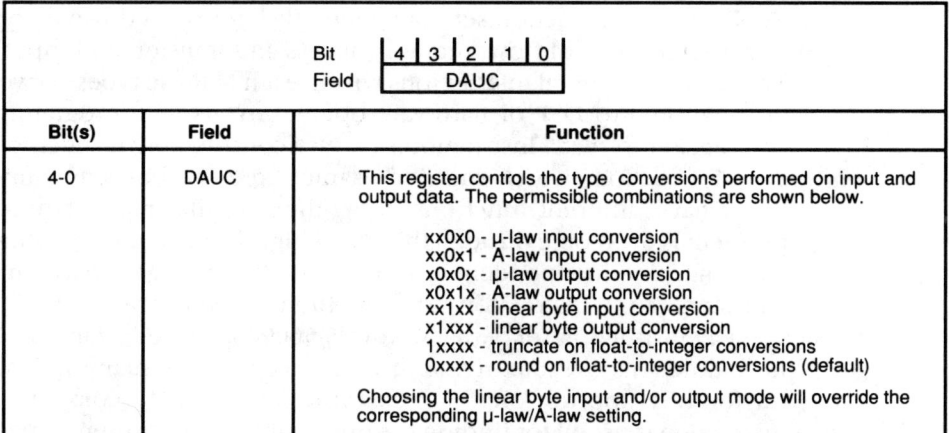

**Figure 8.** Data arithmetic control unit register bit assignments for the AT&T WE DSP32. *(Courtesy AT&T)*

ing the available number of instructions and optimizing the hardware to execute those that are available. Modern DSP chips, however, typically have instruction sets comparable in size to those of general-purpose microprocessors and are not usually considered RISC processors. A number of DSP chip instructions are tailored to specific chip hardware, but the overall quantity and functional variety of modern DSP chip instruction sets have moved well beyond early RISC processors, mostly because of rapid advances in VLSI chip fabrication technology.

**Table 1.** Control Registers of the Texas Instruments TMS320 Family. *(Courtesy Texas Instruments)*

| Register | Symbol | Function |
|---|---|---|
| Global Memory Allocation Register | GREG(7–0) | An 8-bit memory mapped register for allocating the size of the global memory space |
| Instruction Register | IR(15–0) | A 16-bit register used to store the currently executing instruction |
| Interrupt Flag Register | IFR(5–0) | A 6-bit flag register used to latch the <u>active</u>-low external user interrupts INT(2–0) and the internal interrupts XINT/RINT (serial port transmit/receive) and TINT (timer) interrupts. The IFR is not accessible through hardware |
| Interrupt Mask Register | IMR(5–0) | A 6-bit memory register used to mask interrupts |
| Auxiliary Pointer Register | ARP(2–0) | A 3-bit register used to select one or five/eight auxillary registers |

DSP chip instruction sets can generally be broken down by type: arithmetic, program control flow, memory access and transfer, and input and output. There will be general instructions within each of these types, as well as instructions tailored to DSP or hardware operations. As with programming models and register formats, instruction sets vary considerably from chip to chip, and even instructions which share the same three- or four-letter mnemonic will likely have different functions depending on the manufacturer. While this problem is overcome through the use of high-level languages which use a standard, machine-independent syntax, assembly language programming requires a mastery of the instruction set's mnemonics themselves.

The instruction set of the Motorola 56001 is listed in Table 2. The arithmetic instructions execute in one instruction cycle, and many are especially useful for DSP applications. For example, the CMPM (compare magnitude) instruction is useful for finding the minimum and maximum values in a table of floating-point data, and the NORM (normalize accumulator iteration) instruction performs a 1-bit normalization on floating-point numbers. One of the 56001's most powerful instructions is the MACR instruction, which is a signed, floating-point multiply-and-accumulate with rounding. Rounding is especially important in some aspects of digital filtering. Also valuable in certain FFT routines are the ADD and SUB instructions when followed by an L or R for a shift left or right, which multiplies or divides the accumulator value by two before adding or subtracting. Many of the 56001 instructions can also perform parallel data moves and prefetches in addition to the task indicated by the mnemonic.

**Table 2.** Instruction Set of the Motorola DSP56001 (*Courtesy Motorola*)

| Mnemonic | Arithmetic Instructions<br>Description |
|---|---|
| ABS | Absolute value |
| ADC | Add long with carry |
| ADD | Addition |
| ADDL | Shift left and add |
| ADDR | Shift right and add |
| ASL | Arithmetic shift left |
| ASR | Arithmetic shift right |
| CLR | Clear an operand |
| CMP | Compare |
| CMPM | Compare magnitude |
| DIV* | Divide iteration |
| MAC | Signed multiply-accumulate |
| MACR | Signed multiply-accumulate and round |
| MPY | Signed multiply |
| MPYR | Signed multiply and round |

**Table 2.** (cont.)

| Mnemonic | Description |
|---|---|
| **Arithmetic Instructions** | |
| NEG | Negate accumulator |
| NORM* | Normalize |
| RND | Round |
| SBC | Subtract long with carry |
| SUB | Subtract |
| SUBL | Shift right and subtract |
| Tcc* | Transfer conditionally |
| TFR | Transfer data ALU register |
| TST | Test an operand |
| **Logical Instructions** | |
| AND | Logical AND |
| ANDI* | AND immediate to control register |
| EOR | Logical exclusive OR |
| LSL | Logical shift left |
| LSR | Logical shift right |
| NOT | Logical complement |
| OR | Logical inclusive OR |
| ORI* | OR immediate to control register |
| ROL | Rotate left |
| ROR | Rotate right |
| **Bit Manipulation Instructions** | |
| BCLR | Bit test and clear |
| BSET | Bit test and set |
| BCHG | Bit test and change |
| BTST | Bit test on memory and registers |
| **Loop Instructions** | |
| DO | Start hardware loop |
| ENDDO | Exit from hardware loop |

**Table 2.** *(cont.)*

| Mnemonic | Move Instructions<br>Description |
|---|---|
| LUA | Load updated address |
| MOVE | Move data register |
| MOVEC | Move control register |
| MOVEM | Move program memory |
| MOVEP | Move peripheral data |

| Mnemonic | Program Control Instructions<br>Description |
|---|---|
| Il | Illegal instruction |
| Jcc | Jump conditionally |
| JMP | Jump |
| JCLR | Jump if bit clear |
| JSET | Jump if bit set |
| JScc | Jump to subroutine conditionally |
| JSR | Jump to subroutine |
| JSCLR | Jump to subroutine if bit clear |
| JSSET | Jump to subroutine if bit set |
| NOP | No operation |
| REP | Repeat next instruction |
| RESET | Reset on-chip peripheral devices |
| RTI | Return from interrupt |
| RTS | Return from subroutine |
| STOP | Stop processing (low-power standby) |
| SWI | Software interrupt |
| WAIT | Wait for interrupt (low-power standby) |

*These instructions do not allow parallel data moves.

The DO and ENDDO instructions set up the hardware loop counter. The DO command performs a number of housekeeping operations, such as initializing the number of loops, and the ENDDO instruction is used to break out of a loop prematurely. A less flexible instruction is the REP instruction, which allows the repetition of a single instruction a certain number of times and cannot be interrupted. Also of note are the STOP and WAIT instructions, which put the 56001 into a low-power standby mode until an interrupt is received. STOP and WAIT might be useful in a battery-powered or embedded controller application, such as a mobile telephone, where the processor could spend a large amount of time waiting for an incoming signal.

The much larger instruction set of the Motorola 96002 is listed in Table 3.

## Chapter 14: Digital Signal Processing: Programming and Interfacing

It contains 135 instruction mnemonics: 38 floating-point, 30 fixed-point, 13 logical, 4 bit manipulation, 4 loop, 9 move, and 35 program control instructions. The 96-bit floating-point instructions are those beginning with the letter F, and DSP specific instructions, such as FCMPG (graphics compare with trivial accept/reject flags), are also notable. Data is moved to and from memories and registers using the various MOVE commands, and access to external input and output ports is gained through the MOVEP (move peripheral) command.

**Table 3.** Instruction Set of the Motorola DSP96002 (*Courtesy Motorola*)

| Mnemonic | Floating-Point Arithmetic Instructions Description |
|---|---|
| FABS.S | Absolute value (single precision) |
| FABS.X | Absolute value (extended single precision) |
| FADD.S | Add (single precision) |
| FADD.X | Add (extended single precision) |
| FADDSUB.S | Add and subtract (single precision) |
| FADDSUB.X | Add and subtract (extended single precision) |
| FCLR | Clear |
| FCMP | Compare |
| FCMPG | Graphics compare with trivial accept/reject flags |
| FCMPM | Compare magnitude |
| FCOPYS.S | Copy sign (single precision) |
| FCOPY.X | Copy sign (extended single precision) |
| FGETMAN | Get mantissa |
| FINT | Convert to floating-point integer |
| FLOAT.S | Signed integer to SP floating-point conversion |
| FLOAT.X | Signed integer to SEP floating-point conversion |
| FLOATU.S | Unsigned integer to SP floating-point conversion |
| FLOATU.X | Unsigned integer to SEP floating-point conversion |
| FLOOR | Convert to floating-point integer with round to minus infinity |
| FMPY.S | Multiply (single precision) |
| FMPY.X | Multiply (extended single precision) |
| FMPY//FADD.S | Multiply and add (single precision) |
| FMPY//FADD.X | Multiply and add (extended single precision) |
| FMPY//FADDSUB.S | Multiply, add, and subtract (single precision) |
| FMPY//FADDSUB.X | Multiply, add, and subtract (extended single precision) |
| FMPY//FSUB.S | Multiply and subtract (single precision) |
| FMPY//FSUB.X | Multiply and subtract (extended single precision) |
| FNEG.S | Change sign (single precision) |
| FNEG.X | Change sign (extended single precision) |
| FSCALE.S | Scale a floating-point operation (single precision) |

**Table 3.** *(cont.)*

| Mnemonic | Floating-Point Arithmetic Instructions<br>Description |
|---|---|
| FSCALE.X | Scale a floating-point operation (extended single precision) |
| FSEEDD.S | Seed for reciprocal approximation |
| FSEEDR.X | Seed for square root reciprocal approximation |
| FSUB.S | Subtract (single precision) |
| FSUB.X | Subtract (extended single precision) |
| FTFR.S | Transfer floating-point register (single precision) |
| FTDR.X | Transfer floating-point register (extended single precision) |
| FTST | Test a floating-point operand |

| Mnemonic | Fixed-Point Arithmetic Operations<br>Description |
|---|---|
| ABS | Absolute value |
| ADD | Add |
| ADDC | Add with carry |
| ASL | Arithmetic shift left |
| ASR | Arithmetic shift right |
| CLR | Clear |
| CMP | Compare |
| CMPG | Graphics compare with trivial accept/reject flags |
| DEC | Decrement by one |
| EXT | Sign extend 16-bit to 32-bit |
| EXTB | Sign extend 8-bit to 32-bit |
| GETEXP | Get exponent |
| INC | Increment by one |
| INT | Floating-point to integer conversion |
| INTRZ | Floating-point to integer conversion, round to zero |
| INTU | Floating-point to unsigned integer conversion |
| INTURZ | Floating-point to unsigned integer conversion, round to zero |
| JOIN | Join two 16-bit integers |
| JOINB | Join two 8-bit integers |
| MPYS | Signed multiply |
| MPYU | Unsigned multiply |
| NEG | Negate |
| NEGC | Negate with carry |
| SETW | Set a long word operand |
| SPLIT | Extract a 16-bit integer |
| SPLITB | Extract an 8-bit integer |
| SUB | Subtract |

**Table 3.** *(cont.)*

### Fixed-Point Arithmetic Operations

| Mnemonic | Description |
|---|---|
| SUBC | Subtract with carry |
| TFR | Transfer data ALU register |
| TST | Test |

### Logical Instructions

| Mnemonic | Description |
|---|---|
| AND | Logical AND |
| ANDC | Logical AND with complement |
| ANDI | Logical AND immediate with control register* |
| BFIND | Finding leading set bit |
| EOR | Logical exclusive OR |
| LSL | Logical shift left |
| LSR | Logical shift right |
| NOT | Logical complement |
| OR | Logical inclusive OR |
| ORC | Logical inclusive OR with complement |
| ORI | Logical inclusive OR immediate with control register* |
| ROL | Rotate left |
| ROR | Rotate right |

### Bit Manipulation Instructions (Read-Modify-Write)

| Mnemonic | Description |
|---|---|
| BCHG | Bit test and change |
| BCLR | Bit test and clear |
| BSET | Bit test and set |
| BTEST | Bit test |

### Move Instructions

| Mnemonic | Description |
|---|---|
| LEA | Load effective address |
| LRA | Load PC relative address |
| MOVE | Move data register(s) |
| MOVEC | Move control register |
| MOVEI | Move immediate short data |
| MOVEM | Move program memory |
| MOVEP | Move peripheral |

**Table 3.** *(cont.)*

| Mnemonic | Description |
|---|---|
| **Move Instructions** | |
| MOVES | Move absolute short address |
| MOVETA | Move data register(s) and test address |
| **Loop Control Instructions** | |
| DO | Start hardware loop |
| DOR | Start PC relative loop |
| ENDDO | End current hardware loop |
| REP | Repeat next instruction |
| **Program Control Instructions** | |
| Bcc | Branch on condition cc |
| BRA | Branch always |
| BRCLR | Branch if bit clear |
| BRSET | Branch if bit set |
| BScc | Branch to subroutine if condition cc |
| BSCLR | Branch to subroutine if bit clear |
| BSR | Branch to subroutine |
| BSSET | Branch to subroutine if bit set |
| DEBUG | Enter debug mode |
| FBcc | Branch on floating-point condition cc |
| FBScc | Branch to subroutine on floating-point condition cc |
| FFcc | Data ALU operation on floating-point condition cc without update |
| FFcc.U | Data ALU operation on floating-point condition cc with update |
| FJcc | Jump on floating-point condition cc |
| FJScc | Jump to subroutine on floating-point condition cc |
| FTRAPcc | Execute software interrupt on floating-point condition cc |
| IFcc | Data ALU operation on fixed-point condition cc without update |
| IFcc.U | Data ALU operation on fixed-point condition cc with update |
| ILLEGAL | Illegal instruction interrupt |
| Jcc | Jump on fixed-point condition cc |
| JCLR | Jump if bit clear |

## Table 3.  (cont.)

### Program Control Instructions

| Mnemonic | Description |
|---|---|
| JMP | Jump |
| JScc | Jump to subroutine on condition cc |
| JSCLR | Jump to subroutine if bit clear |
| JSET | Jump if bit set |
| JSR | Jump to subroutine |
| JSSET | Jump to subroutine if bit set |
| NOP | No operation |
| RESET | Reset on-chip peripherals |
| RTI | Return from interrupt |
| RTR | Return from subroutine and restore status register |
| RTS | Return from subroutine |
| STOP | Stop processing (low-power standby) |
| TRAPcc | Execute software interrupt on fixed-point condition cc |
| WAIT | Wait for interrupt (low-power standby) |

*These instructions do not allow parallel data moves.

The instruction set of the Texas Instruments TMS320 chip is listed in Table 4. It contains many instructions particularly useful in DSP applications. The various MAC and MPY instructions perform multiply-and-accumulate operations, the SQRS instruction squares a value and subtracts the previous product, and the ZALR instruction can load the accumulator with a rounded value. The RPT instructions repeat a single instruction, similar to the Motorola DO and REP instructions. The block move commands, BLKD and BLKP, as well as the table read and write commands, TBLR and TBLW, are designed to move large blocks of ordered data and to access dual-port memories. These types of instructions find frequent use in FFTs, digital filtering, and other recursive or multitap DSP operations.

A short program using the TMS320 instruction set is shown in Figure 9. This program implements a digital filter, and the structure and flow is common to most filter programs written in DSP chip assembly language. First, a variable and an I/O port are established to receive the incoming digital signal, along with working memory and register space. Then each section of the filtering algorithm is repeated a number of times according to the size and type of filter the program is implementing. The results of the computations, the filtered signal, are collected into a single register and then transferred to an output port. The program then jumps back up to the input section and continues filtering with the next input value.

**Table 4.** Instruction Set of the Texas Instruments TMS320 Family
(*Courtesy Texas Instruments*)

**Accumulator Memory Reference Instructions**

| Mnemonic | Description |
| --- | --- |
| ABS | Absolute value of accumulator |
| ADD | Add to accumulator with shift |
| ADDC* | Add to accumulator with carry |
| ADDH | Add to high accumulator |
| ADDK* | Add to accumulator short immediate |
| ADDS | Add to low accumulator with sign extension suppressed |
| ADDT† | Add to accumulator with shift specified by T register |
| ADLK† | Add to accumulator long immediate with shift |
| AND | AND with accumulator |
| ANDK† | AND immediate with accumulator with shift |
| CMPL† | Complement accumulator |
| LAC | Load accumulator with shift |
| LACK | Load accumulator with short immediate |
| LACT† | Load accumulator with shift specified by T register |
| LALK† | Load accumulator long immediate with shift |
| NEG† | Negate accumulator |
| NORM† | Normalize contents of accumulator |
| OR | OR with accumulator |
| ORK† | OR immediate with accumulator with shift |
| ROL* | Rotate accumulator left |
| ROR* | Rotate accumulator right |
| SACH | Store high accumulator with shift |
| SACL | Store low accumulator with shift |
| SBLK† | Subtract from accumulator long immediate with shift |
| SFL† | Shift accumulator left |
| SFR† | Shift accumulator right |
| SUB | Subtract from accumulator with shift |
| SUBB* | Subtract from accumulator with borrow |
| SUBC | Conditional subtract |
| SUBH | Subtract from high accumulator |
| SUBK* | Subtract from accumulator short immediate |
| SUBS | Subtract from low accumulator with sign extension suppressed |
| SUBT† | Subtract from accumulator with shift specified by T register |
| XOR | Exclusive OR with accumulator |
| XORK† | Exclusive OR immediate with accumulator with shift |
| ZAC | Zero accumulator |
| ZALH | Zero low accumulator and load high accumulator |
| ZALR* | Zero low accumulator and load high accumulator with rounding |

**Table 4.** *(cont.)*

### Accumulator Memory Reference Instructions

| Mnemonic | Description |
|---|---|
| ZALS | Zero accumulator and load low accumulator with sign extension suppressed |

### Auxiliary Registers and Data Page Pointer Instructions

| Mnemonic | Description |
|---|---|
| ADRK† | Add to auxiliary register short immediate |
| CMPR* | Compare auxiliary register with auxiliary register AR0 |
| LAR | Load auxiliary register |
| LARK | Load auxiliary register short immediate |
| LARP | Load auxiliary register pointer |
| LDP | Load data memory page pointer |
| LDPK | Load data memory page pointer immediate |
| LRLK* | Load auxiliary register long immediate |
| MAR | Modify auxiliary register |
| SAR | Store auxiliary register |
| SBRK* | Subtract from auxiliary register short immediate |

### T Register, P Register, and Multiply Instructions

| Mnemonic | Description |
|---|---|
| APAC | Add P register to accumulator |
| LPH* | Load high P register |
| LT | Load T register |
| LTA | Load T register and accumulate previous product |
| LTD | Load T register, accumulate previous product, and move data |
| LTP* | Load T register and store P register in accumulator |
| LTS* | Load T register and subtract previous product |
| MAC* | Multiply and accumulate |
| MACD* | Multiply and accumulate with data move |
| MPY | Multiply (with T register, store product in P register) |
| MPYA† | Multiply and accumulate previous product |
| MPYK | Multiply immediate |
| MPYS† | Multiply and subtract previous product |
| MPYU† | Multiply unsigned |
| PAC | Load accumulator with P register |
| SPAC | Subtract P register from accumulator |
| SPH† | Store high P register |
| SPL† | Store low P register |
| SPM* | Set P register output shift mode |

**Table 4.** *(cont.)*

### T Register, P Register, and Multiply Instructions

| Mnemonic | Description |
|---|---|
| SQRA* | Square and accumulate |
| SQRS* | Square and subtract previous product |

### Branch/Call Instructions

| Mnemonic | Description |
|---|---|
| B | Branch unconditionally |
| BACC* | Branch to address specified by accumulator |
| BANZ | Branch on auxiliary register not zero |
| BBNZ* | Branch if TC bit ≠ 0 |
| BBZ* | Branch if TC bit = 0 |
| BC† | Branch on carry |
| BGEZ | Branch if accumulator ≥ 0 |
| BGZ | Branch if accumulator > 0 |
| BIOZ | Branch on I/O status = 0 |
| BLEZ | Branch if accumulator ≤ 0 |
| BLZ | Branch if accumulator < 0 |
| BNC† | Branch on no carry |
| BNV* | Branch if no overflow |
| BNZ | Branch if accumulator ≠ 0 |
| BV | Branch on overflow |
| BZ | Branch if accumulator = 0 |
| CALA | Call subroutine indirect |
| CALL | Call subroutine |
| RET | Return from subroutine |
| TRAP* | Software interrupt |

### I/O and Data Memory Operations

| Mnemonic | Description |
|---|---|
| BLKD* | Block move from data memory to data memory |
| BLKP* | Block move from program memory to data memory |
| DMOV | Data move in data memory |
| FORT* | Format serial port registers |
| IN | Input data from port |
| OUT | Output data to port |
| RFSM† | Reset serial port frame synchronization mode |
| RTXM* | Reset serial port transmit mode |
| RXF* | Reset external flag |
| SFSM† | Set serial port frame synchronization mode |

**Table 4.**  *(cont.)*

### I/O and Data Memory Operations

| Mnemonic | Description |
|---|---|
| STXM† | Set serial port transmit mode |
| SXF* | Set external flag |
| TBLR | Table read |
| TBLW | Table write |

### Control Instructions

| Mnemonic | Description |
|---|---|
| BIT* | Test bit |
| BITT* | Test bit specified by T register |
| CNFD* | Configure block as data memory |
| CNFP* | Configure block as program memory |
| DINT | Disable interrupt |
| EINT | Enable interrupt |
| IDLE* | Idle until interrupt |
| LST | Load status register ST0 |
| LST1* | Load status register ST1 |
| NOP | No operation |
| POP | Pop top of stack to low accumulator |
| POPD* | Pop top of stack to data memory |
| PSHD* | Push data memory value onto stack |
| PUSH | Push low accumulator onto stack |
| RC† | Reset carry bit |
| RHM† | Reset hold mode |
| ROVM | Reset overflow mode |
| RPT* | Repeat instruction as specified by data memory value |
| RPTK* | Repeat instruction as specified by immediate value |
| RSXM* | Reset sign-extension mode |
| RTC† | Reset test/control flag |
| SC† | Set carry bit |
| SHM† | Set hold mode |
| SOVM | Set overflow mode |
| SST | Store status register ST0 |
| SST1* | Store status register ST1 |
| SSXM* | Set sign-extension mode |
| STC† | Set test/control flag |

*This instruction is specific to the TMS320C2x instruction set.
†This instruction is specific to the TMS320C25/E25 instruction set.

```
*       THIS SECTION OF CODE IMPLEMENTS A SECOND-ORDER DIRECT-FORM II IIR FILTER
*       d(n) = x(n) + d(n-1)a  + d(n-2)a
*                           1          2
*       y(n) = d(n)b + d(n-1)b + d(n-2)b
*                  0         1         2
*
NEXT            IN      XN,PA2          *   NEW INPUT VALUE XN
*
                LAC     XN
                MPYK    0               *   CLEAR P REGISTER
*
                LARP    AR1
                LRLK    AR1,>03FF
                CNFP                    *   USE BLOCK B0 AS PROGRAM AREA
*
*       d(n) = x(n) + d(n-1)a + d(n-2)a
*                           1         2
*
                RPTK    1               *   REPEAT 2 TIMES
                MACD    >FF00,*+
*
                APAPC
                SACH    DN,1            *   d(n)
*
*       y(n) = d(n)b + d(n-1)b + d(n-10)b
*                  0         1          2
*
                ZAC
                MFYK    0               *   CLEAR P REGISTER
*
                MPY     >FF02
*
                RPTK    1
                MACD    >FF03,*-
*
                APAC
                SACH    YN,1            *   SAVE FILTERED OUTPUT
*
                OUT     YN,PA2          *   YN IS THE OUTPUT OF THE FILTER
                B       NEXT
```

**Figure 9.** A TMS32020 assembly language program.
*(Courtesy Texas Instruments)*

The AT&T WE DSP32 has a unique format for its instruction set, which is listed in Table 5. In contrast to a traditional assembly language format which uses an instruction mnemonic followed by one or more operands, the DSP32 uses a sentencelike format that resembles C language syntax. Instructions consist of small sentences containing operands and qualifiers strung together with standard arithmetic and relational operators. Mnemonics are replaced with more descriptive identifiers such as if, goto, ieee, dsp, float, call, and return, etc. Although the DSP32 instruction set may seem confusing at first, it becomes more readable once actual register names and memory locations are inserted, as in the short program shown in Figure 10. The upper portion of this program initializes input, output, and address variables and sets up a call to the subroutine which implements the filter. The first half of the filter code accesses the variables by calculating a series of offsets from the address passed to the subroutine. The second half performs the calculations needed to implement the filter and, finally, the filter output is returned to the main program. The lengthy comments included in this program are a good example of the documentation that should accompany all assembly language programs.

```
main:       .
            .           /* beginning of main program */
            .
        call fir (r14)          /* routine call */
        nop                     /* latent instruction */
int     N                       /* arguments */
int     &in, &out
int     &h[N-1], &firS
.align 4                        /* set proper alignment */

            .       /* Next instr. executed after routine call */
            .
            .       /* End of main program. fir routine appended. */
/*
*   fir routine -- The following code implements the FIR equation
*
*       y(n) = h[N-1]x[n-(N-1)]+h[N-2]x[n-(N-2)] +...+h[1]x[n-1]+h[0]x[n]
*
*   The routine tests a repeat instruction before the accumulation
*   of the next partial product. This minimizes memory requirement
*   at the expense of execution speed.
*
*       Routine code            = 60 bytes
*       Filter coefficients     = 4N bytes
*       State variables         = 4(N-1) bytes
*       No. instructions executed = 11 + 2*N (including routine call)
*
*   A description of the arguments follows:
*
*       N           Length of the FIR filter (N > 2)
*       &in         Memory location storing the input value
*       &out        Memory location for the output value
*       &h[N-1]     Memory location of the (N-1)th filter coefficient
*                   where the coefficients are arranged in reversed
*                   order and stored in consecutive memory locations
*       &firS       Starting memory location for the static
*                   variables
*/
fir:    r1 = *r14++             /* r1 = length of the FIR filter */
        r5 = *r14++             /* r5 points to input */
        r6 = *r14++             /* r6 points ot output */
        r2 = *r14++             /* r2 points to coeficients */
        r3 = *r14++             /* r3 points to static variable */
        r1 = r1 + (1-4)         /* r1 is used as loop counter */
        r4 = r3                 /* r4 points to static variable */
        a1 = *r4++ * *r2++              /* a1 = h[N-1]x[n-(N-1)] */

firA:   if (r1-- >=0) goto firA
        a1 = a1 + (*r3++ = *r4++) * *r2++

        a0 = a1 + (*r3 = *r5) * *r2
        *r6 = a0 = a0

        r1 = *r14++             /* dummy load instruction */
        return (r14)
        nop                     /* end of fir routine */
```

**Figure 10.** A WE DSP32 assembly language program. *(Courtesy AT&T)*

Opinions on the approach of the DSP32 instruction set are mixed; some programmers feel that assembly language is made even more confusing by attempts to change the traditional mnemonic and operand format, and others are confused by what appears to be C language at first glance, but is actually assembly language. Still others applaud its highly readable format.

**Table 5.** Instruction Set of the AT&T WE DSP32 (*Courtesy AT&T*)

## CA Data Move Instructions

| Instruction | CAU Flags Affected | Format | Description |
|---|---|---|---|
| rD = N | NZ00 | 6c | 16-bit immediate load |
| rDe = M | — | 8b | 24-bit immediate load |
| {ioc†,dauc} = VALUE | — | 5a | 5- or 21-bit immediate load |
| {MEM,*N,obuf,piop} = {rSh,rSl} | — | 7 | MEM, *N, piop, and obuf are 8 bits |
| {MEM,*N,obuf,pdr,pdr2,pir,pcw} = rS,pcsh | — | 7 | MEM, *N, and obuf are 16 bits |
| {MEM,*N,obuf} = rSe,psche | — | 7 | MEM, *N, and obuf are 24 bits |
| {rDh,rDl} = {MEM,*N,ibuf,piop} | NZ00 | 7 | MEM, *N, piop, and ibuf are 8 bits |
| rD = {MEM,*N,ibuf,pdr,pdr2,pir,pcw} | NZ00 | 7 | MEM, *N, and ibuf are 16 bits |
| rDe = {MEM,*N,ibuf} | NZ00 | 7 | MEM, *N, and ibuf are 24 bits |
| MEM = {ibufl,piop} | — | 7 | 8-bit transfer |
| MEM = {ibufl,pdr,pdr2,pir,pcw}‡ | — | 7 | 16-bit transfer |
| MEM = {ibufe,pdre}‡ | — | 7 | 32-bit transfer |
| {obufl,piop} = MEM | — | 7 | 8-bit transfer |
| {obuf,pdr,pdr2,pir,pcw} = MEM | — | 7 | 16-bit transfer |
| {obufe,pdre} = MEM | — | 7 | 32-bit transfer |

## CA Control Insructions

| Instruction | Flags Affected | Instruction Format | Description |
|---|---|---|---|
| if (CA COND) goto {rH, N, rH+N,} | None | 0 | Conditional branch |
| if (rM -->= 0) goto {rH, N, rH+N,} | None | 3a | Conditional branch |
| if (DA COND) goto {rH, N, rH+N,} | None | 0 | Conditional branch |
| if (IO COND) goto {rH, N, rH+N,} | None | 0 | Conditional branch |
| call{rH, N, M, rH+N,}(rM) | None | 4 | Call subroutine |
| return(rM) | None | 0 | Return from subroutine |
| ireturn | None | 0 | Retrun from interrupt |
| do J,{K,rH} | None | 3b, 3c | Do next J + 1 instructions K + 1 (or rH + 1) times.§ J = 0, 1, 2, . . . , 31. K = rH = 0, 1, 2, . . . , 2047 |
| goto{rH, N, M, rH+N,} | None | 0 | Unconditional branch |
| nop | None | 0 | No operation |

## Table 5. (cont.)

### CA Arithmetic/Logic Instructions

| Instruction | CAU Flags Affected | Instruction Format | Description |
|---|---|---|---|
| rD[e] = rH+N | nvzc | 5a, 5b | Three-operand add with 16-bit sign-extended immediate |
| rD[e] = rS1+rS2 | NZVC | 6a, 6b | Triadic add |
| rD[e] = rD+rS | NZVC | 6a, 6b | Diadic add |
| rD[e] = rS1−rS2 | NZVC | 6a, 6b | Triadic left subtract |
| rD[e] = rS2−rS1 | NZVC | 6a, 6b | Triadic right subtract |
| rD[e] = rD−{N, rS} | NZVC | 6a, 6b, 6c, 6d | Right subtract |
| rD[e]−{N, rS} | NZVC | 6a, 6b, 6c, 6d | Compare |
| rD[e] = {N, rS}−rD | NZVC | 6a, 6b, 6c, 6d | Left subtract |
| rD[e] = rD&{N, rS} | NZ00 | 6a, 6b, 6c, 6d | AND |
| rD[e] = rD1&rS2 | NZ00 | 6a, 6b | Triadic AND |
| rD[e] &{N, rS} | NZ00 | 6a, 6b, 6c, 6d | Bit test |
| rD[e] = rD | {N, rS} | NZ00 | 6a, 6b, 6c, 6d | OR |
| rD[e] = rS1 | rS2 | NZ00 | 6a, 6b | Triadic OR |
| rD[e] = rD^{N, rS} | NZ00 | 6a, 6b, 6c, 6d | XOR |
| rD[e] = rS1^rS2 | NZ00 | 6a, 6b | Triadic XOR |
| rD[e] = rS/2 | NZ0C | 6a, 6b | Arithmetic right shift |
| rD[e] = rS>>1 | 0Z0C | 6a, 6b | Logical right shift |
| rD[e] = rS>>>1 | NZ0C | 6a, 6b | Rotate right through carry |
| rD[e] = rS<<<1 | NZVC | 6a, 6b | Rotate left through carry |
| rD[e] = −rS | NZVC | 6a, 6b | Negate |
| rD[e] = rS*2 | NZVC | 6a, 6b | Arithmetic left shift |
| rD[e] = rD#{N, rS} | NZ0C | 6a, 6b, 6c, 6d | Diadic carry reverse add |
| rD[e] = rS1#rS2 | NZ0C | 6a, 6b | Triadic carry reverse add |
| rD[e] = rD&~{N, rS} | NZVC | 6a, 6b, 6c, 6d | Diadic AND with complement |
| rD[e] = rS1&~rS2 | NZVC | 6a, 6b | Triadic AND with complement |
| rD[e] = rS | NZVC | 6a, 6b | Assignment |
| rD[e] = rS{+,−} 1 | NZVC | 6a, 6b | Increment/decrement |

### Data Multiply/Accumulate Instructions

| Instruction | DAU Flags Affected | Description |
|---|---|---|
| [Z=] aN = [−]aM {+,−}Y*X | NZVU | The product of the X and Y fields is added to or subtracted from the accumulator aM, and the result is stored in accumulator aN. The result can also be output according to the Z field |
| aN = [−]aM {+,−} (Z=Y)*X | NZVU | The Y field operand is output according to the Z field. The product of the X and Y fields is added to accumulator aM, and the sum is stored in accumulator aN |
| [Z=] aN = [−]Y {+,−} aM*X | NZVU | The product of the X field and accumulator aM is added to or subtracted from the Y field. The result is placed in accumulator aN and can also be output according to the Z field |

## Table 5. (cont.)

### Data Multiply/Accumulate Instructions

| Instruction | DAU Flags Affected | Description |
|---|---|---|
| [Z=] aN = [−]Y∗X | NZVU | The product of the X and Y fields is added to or subtracted from zero. The result is stored in accumulator aN and can also be output according to the Z field |
| aN = [−](Z=Y)∗X | NZVU | The value of the Y field is output according to the Z field. The product of the Y and X fields is stored in accumulator aN |
| [Z=] aN = [−]Y {+,−}X | NZVU | The sum or difference of the Y and X fields is stored in accumulator aN, and the result can also be output according to the Z field. Note: X is a multiplier input |
| [Z=] aN = [−]Y | NZVU | The value of the Y field is placed in accumulator aN and can also be output according to the Z field |
| aN = [−](Z=Y) {+,−} X | NZVU | The sum or difference of the Y and X fields is stored in accumulator aN, and Y can also be output according to the Z field |

### DA Special Functions

| Instruction | DAU Flags Affected | Description |
|---|---|---|
| [Z=] aN = ic(Y) | NZ00 | Input conversion, $\mu$-law, $A$-law, 8-bit linear to float |
| [Z=] aN = oc(Y) | — | Output conversion, float to $\mu$-law, $A$-law, 8-bit linear |
| [Z=] aN = float(Y) | NZ00 | 16-bit integer to float |
| [Z=] aN = float(24) | NZ00 | 24-bit integer to float |
| [Z=] aN = int(Y) | — | Float to 16-bit integer (round or truncate, DAUC[4]) |
| [Z=] aN = int24(Y) | — | Float to 24-bit integer (round or truncate, DAUC[4]) |
| [Z=] aN = round(Y) | NZVU | Round float(40) to float(32) |
| [Z=] aN = ifalt(Y) | — | If(aN<0) then [Z=] aN=Y else [Z=] aN |
| [Z=] aN = ifaeq(Y) | — | If(aN=0) then [Z=] aN=Y else [Z=] aN |
| [Z=] aN = ifagt(Y) | — | If(aN>0) then [Z=] aN=Y else [Z=] aN |
| [Z=] aN = dsp(Y) | NZVU | IEEE to DSP32 format conversion |
| [Z=] aN = ieee(Y) | — | DSP32 to IEEE format conversion |
| [Z=] aN = seed(Y) | NZ0U | 32-bit to 32-bit reciprocal program seed |

†ioc = VALUE may not be used in an interrupt routine.
‡MEM = {pde,pdr2,pir,pcw,pdre} cannot be used in the presence of PIO DMA.
§The **do** instruction and the instructions it encompasses are not interruptible for K times. For the last iteration, interrupts are enabled. Further, the **do** instruction cannot be used in an interrupt routine.

# Implementing Algorithms with Software

Most software designers agree that it is often best to separate the algorithm design for a program from the actual software coding; this is especially true with respect to DSP chip programming. The algorithms at the heart of applications such as digital filtering, speech recognition, and signal synthesis are heavily rooted in theory and mathematics, and both the DSP algorithm and a tactic for implementing it as a computer program should be worked out well before beginning any assembly language programming. In fact, it is not uncommon for the two tasks, algorithm design and programming, to be divided between specialists in each area.

As an example, the general equation describing one type of digital filter may be expressed as

$$y(n) = \sum_{i=0}^{M} a_i x(n-i) - \sum_{i=1}^{N} b_i y(n-i).$$

This nomenclature is described in detail in Chapter 10. Briefly, the input to the filter is the series of discrete samples held in $x(n)$, and the output consists of $y(n)$. The frequency response of the filter is determined by the filter coefficients held in $a_i$ and $b_i$. As denoted by the summation of the $x(n-i)$ terms, the filter response is determined not only by the present input value, but also by the past $M$ input values. The actual number of past input values used in the calculation is called the filter length. Because the filter also depends on past output values, the $y(n-i)$ terms, it is said to have feedback or to be recursive. If the $b_i$ terms are all set to zero, then the recursion is eliminated, and the equation becomes

$$y(n) = \sum_{i=0}^{M} a_i x(n-i).$$

This equation describes a finite impulse response (FIR) filter, a common type of digital filter that can be designed to have good phase response and, because it lacks the feedback terms, is inherently stable. A block diagram of the FIR filter structure is shown in Figure 11, and this form helps to visualize how the filter could be implemented in software. The input to the DSP chip will be the $x(n)$ terms and will consist of a continuous series of discrete sample values passed from an A/D converter or some other peripheral device. These values will be stored in memory, and each will be multiplied by its corresponding coefficient. The results of these multiplications will be accumulated, and this single value will be output as the $y(n)$ term to a D/A converter or other device. When the next input value is received, each value stored in memory will be shifted one location, with the value in location $M$ being dropped since it is no longer needed.

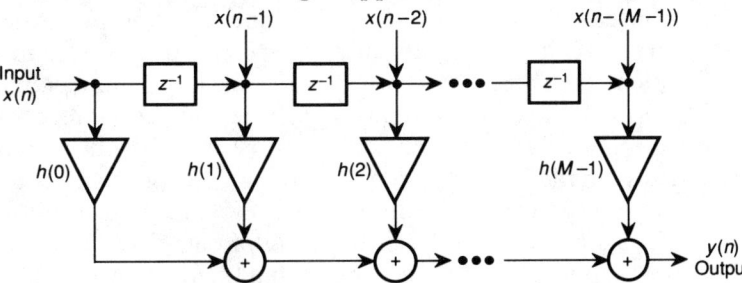

**Figure 11.** FIR digital filter block diagram.

One of the first details to consider in this implementation is the allocation of sufficient memory to store the input values and the coefficients. FIR filters require long filter lengths to realize steep cutoffs, and lengths as long as 1024 are not uncommon. A filter of this size would need 1024 words of memory for past input values and 1024 words of memory for coefficients. Furthermore, the input memory must be organized so that the last $M$ values can be easily accessed with a minimum of shifting and reorganizing; this is often accomplished using circular buffer schemes. Input values are stored in data RAM, while the filter coefficients can be stored in RAM or ROM, depending on whether or not the filter coefficients, and therefore its response, will be changed. Storage spaces for the input value, the final output value, and several intermediate calculations are also required. While the input values and coefficients are stored sequentially in memory, the intermediate calculations are almost always stored in ALU registers to minimize the number of memory access instructions. While these memory considerations might be trivial on larger computer systems, they are a fundamental concern when working with DSP chips that may only have 2K words of on-chip data RAM, and when trying to reduce the number of instructions needed to step through 1024 multiplications and additions.

Another software concern is the quantization which takes place in the basic FIR filter equation when using fixed-point, finite-precision arithmetic. In designing a digital filter one must determine the set of coefficients which yields the desired frequency response. When these coefficients are represented in fixed-point arithmetic, quantization will take place, and the filter response will be affected. Figure 12 shows the effects of coefficient quantization on an FIR filter that has been implemented using 6-bit, fixed-point arithmetic. While 6-bit quantization is severe, quantization effects become noticeable when less than 12 bits are used, and the effects worsen as greater demands are placed on the filter cutoff slope. A similar situation exists with the arithmetic used to perform the multiply-and-accumulate instructions needed to multiply the input values by the coefficients and maintain the running sum. The results of multiplications are often rounded or truncated so they will fit in the accumulator register of the DSP chip's ALU. Furthermore, the final sum, which is the output of the filter, may require truncation so it does not exceed the bit width of the final accumulator, register, or output device.

The effects of coefficient quantization can cause undesirable filter responses, while the effects of arithmetic rounding, overflow, and truncation during addition and multiplication can cause noise and nonlinearities. More serious is the fact that even if the level of the input signal is reduced to avoid overflow, the arithmetic noise will remain the same, reducing the filter's signal-to-noise ratio. If the filter gain is reduced, the coefficients become relatively smaller, which intensifies the effects of quantization. For these reasons, the DSP chip programmer must pay particular attention to the arithmetic scheme as well as digital dithering of data used in an FIR filter implementation, especially when dealing with a fixed-point DSP chip. While wider word widths and floating-point arithmetic lessen many of these effects, they also

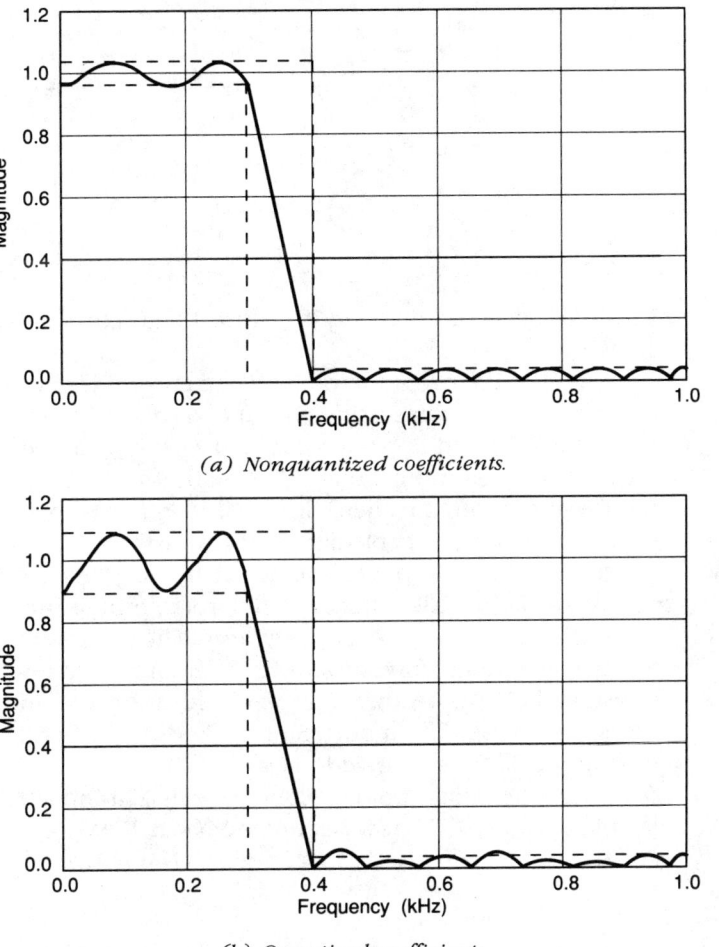

**Figure 12.** Effects of coefficient quantization on FIR filter design.
(*Courtesy Atlanta Signal Processors, Inc.*)

require a greater amount of instruction and computational overhead, which slows execution time and increases memory requirements.

As a practical example, a length-five FIR filter is shown in block diagram form in Figure 13. The difference equation that describes it is

$$y(n) = x(n-4)h(4) + x(n-3)h(3) + x(n-2)h(2) + x(n-1)h(1) + x(n)h(0).$$

The first step in designing the filter is to allocate storage for the input values and the coefficients. The five input values, the coefficients, and the output value will be stored in memory, while the running sum will be stored in the accumulator. An additional register will be used for intermediate calculations.

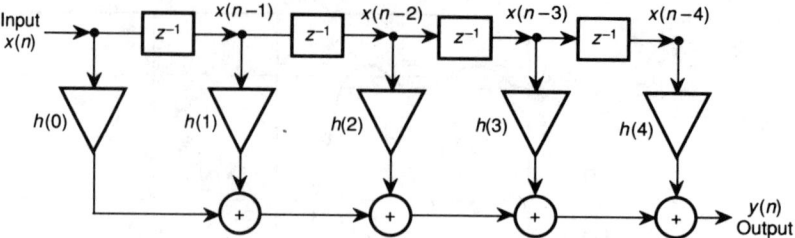

**Figure 13.** Length-five FIR digital filter.

This scheme is summarized in Figure 14, and the TMS320 code for the filter is shown in Figure 15.

Step 1 in the program transfers the current input value from an I/O port to the memory location storing XN, and zeros the accumulator in preparation for calculating the running sum. Step 2 uses the LT instruction to load the T register with the past input value XNM4, and the MPY instruction then multiplies the T register with the coefficient stored in H4. Step 3 multiplies XNM3 and H3 in the same way, except that the LTD instruction also adds the result of the multiplication in Step 2 to the accumulator. This process of multiplying and adding the last result to the accumulator continues until Step 6. Step 7 adds the last multiplication to the accumulator. The accumulator now contains the filtered output value corresponding to the input value received in Step 1. Step 8 transfers the output value to memory location YN, and then memory location YN is sent to the output port. Step 9 jumps the program back to Step 1, and the next input value is received.

When the next input is received, each past input value must be shifted one location. That is, XN now becomes XNM1, XNM1 becomes XNM2, XNM2 becomes XNM3, and XNM3 becomes XNM4. Note that XNM4 is no longer

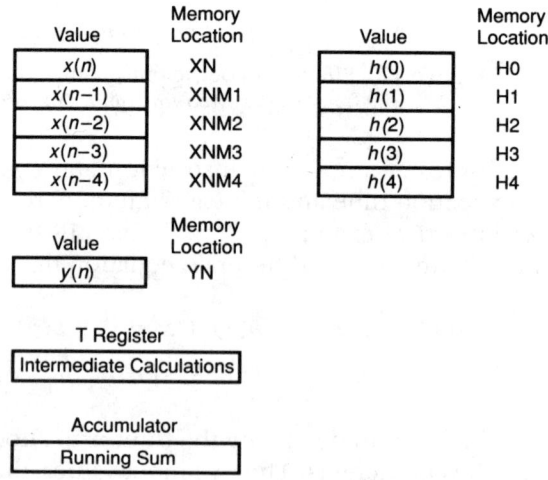

**Figure 14.** Memory allocation scheme for implementing a length-five FIR digital filter.

```
        * THIS SECTION OF CODE IMPLEMENTS THE FOLLOWING EQUATION:           *
        * x(n-4)h(4) + x(n-3)h(3) + x(n-2)h(2) + x(n-1)h(1) + x(n)h(0) = y/n *
        *
STEP 1 ┌ NXTPT   IN XN,PA2        * GET THE NEW INPUT VALUE XN FROM PORT PA0 *
       │ *
       └         ZAC               * ZERO THE ACCUMULATOR *
         *
STEP 2 ┌         LT XNM4           * x(n-4)h(4) *
       └         MPY H4
         *
STEP 3 ┌         LTD XNM3          * x(n-4)h(4) + x(n-3)h(3) *
       └         MPY H3
         *
STEP 4 ┌         LTD XNM2          * SIMILAR TO THE PREVIOUS STEPS *
       └         MPY H2
         *
STEP 5 ┌         LTD XNM1
       └         MPY H0
         *
STEP 6 ┌         LTD XN
       └         MPY H0
         *
STEP 7 ┌         APAC              * ADD THE RESULTS OF THE LAST MULTPLY TO *
       └                           * THE ACCUMULATOR *
         *
STEP 8 ┌ *       SACH YN,1         * STORE THE RESULT IN YN *
       └         OUT YN,PA2        * OUTPUT THE RESPONSE TO PORT PA1 *
         *
STEP 9 ┌         B NXTPT           * GO GET THE NEXT POINT *
```

**Figure 15.** The TMS320 assembly language program for implementing a length-five FIR digital filter. *(Courtesy Texas Instruments)*

needed. It would be possible to perform these shifts at the end of each filter pass, or between Steps 8 and 9 in the example. However, the unique LTD instruction of the TMS320 shifts each value to the next higher address after it transfers it to the T register. Therefore, by the end of Step 8, each value has already been shifted to its correct position. The LTD instruction, which loads a register with a value, adds the result of the last multiply to the accumulator, and then shifts the value to the next memory location, is a good example of the specialized instructions found in DSP chips and the benefits of parallel execution they offer to a programmer.

In addition to the code described above, some programming would be needed to initialize the input and output ports, and to store the numeric values of the filter coefficients in memory locations H4 through H0. Furthermore, one or more steps are required to either synchronize the input and output steps with the external I/O hardware, or to synchronize each pass with an on- or off-chip clock signal. This synchronization or clock signal will directly determine the sampling rate of the filter, which cannot exceed the time it takes Steps 1 through 9 to execute.

This FIR filter example uses straightline code which performs separate multiply-and-accumulate instructions for each of the five taps of the filter. For longer filters, this would create impracticably long programs, and a looped instruction which steps through the past input values would likely be used in conjunction with a slightly altered memory storage scheme. However, because the TMS320 has no provisions for looping a group of instructions without setting up a software loop counter, the straightline version will execute faster. Design decisions such as this, which concern memory size, execution speed, and selection of the available instructions, are some of the most challenging aspects of DSP chip programming.

# Interfacing

Once a DSP algorithm and the implementing software have been developed, the next step is often to design the hardware interface and program the software for input and output. There are three basic categories of input and output for a DSP chip: memory, interrupts and control, and data. The hardware for these interfaces usually consists of the I/O device itself, such as an A/D or D/A converter, and some amount of glue logic, which refers to the additional chips and electronic components that condition and control the electrical signals passed to and from the I/O chips to the DSP chip. The software used to access these external devices usually consists of a memory move or read/write instruction, or a specific instruction designed to access a dedicated I/O port. Peripherals are accessed with memory read/write instructions when the device is memory mapped to create an address on the bus that reads or writes through to a device rather than to a RAM or ROM location. Peripherals interfaced to dedicated I/O ports are accessed with instructions such as the IN (input data from port) instruction on the TMS320, or the MOVEP (move peripheral data) instruction on the DSP56001.

Two basic types of DSP chip memory interfaces are used for adding additional program or data RAM and ROM, or EPROM and other preprogrammed memory devices. Direct memory expansion adds an amount of memory that fits within the bit width of the address bus. For example, if the address bus is only 4 bits wide, then sixteen address locations (words) can be directly accessed, and if the bus is 16 bits wide, then 64K words of memory can be accessed. Direct memory expansion for a DSP chip often consists of a combination of program and data RAMs and ROMs, each of which occupies a portion of the total address space. For example, a 16K-word address space could be divided into two 2K-word ROMs, a 4K-word data RAM, and an 8K-word program RAM.

A circuit for adding 2K words of program RAM and 2K words of program PROM to the Texas Instruments TMS320 chip is shown in Figure 16. The RAM chips are U2 and U3 (Advanced Micro Devices Am9128-70) and the PROM chips are U4 and U5 (Texas Instruments TBP28S166), and both provide 2K words of 8-bit–wide memory. The memory chips are mapped to the first 4K words of program memory on the TMS320 chip, beginning with location 000, using two buffer chips and a small selection of TTL gates which decode the 16-bit address to select the correct chip. Designing such an interface requires a knowledge of the memory control signals output by the DSP chip as well as an understanding of memory timing diagrams, which detail the correlation between the numerous electrical signals controlling address generation, read/write selection, memory chip output, and timing. While these hardware design considerations are usually not the programmer's concern, the programmer does need to be aware of the read and write times accommodated by the memory chip. Some interface and memory chip combinations can read or write to memory in one instruction cycle, while others may require the use of prefetching or software WAIT instructions to buffer a longer memory access time.

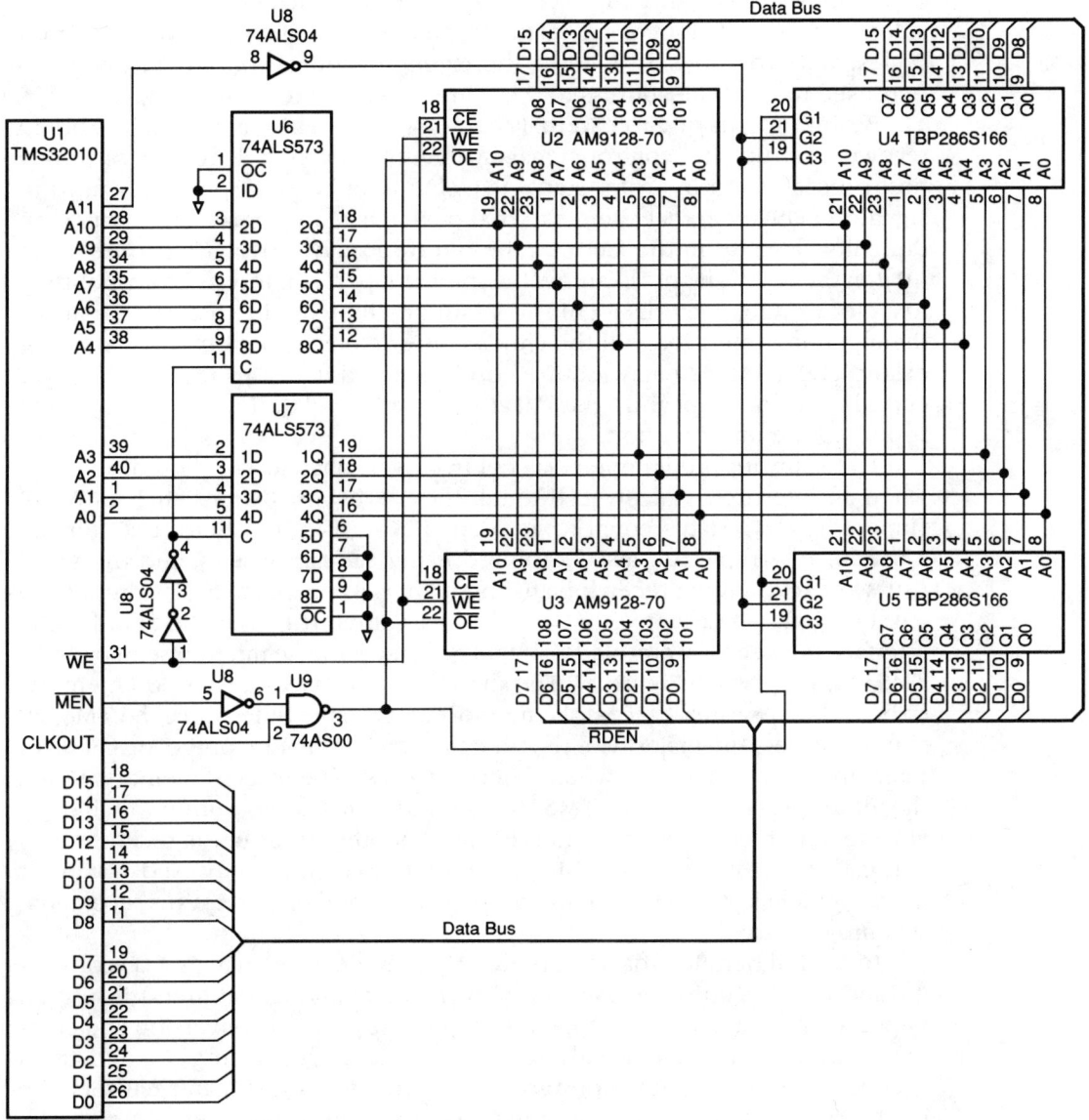

**Figure 16.** Direct memory expansion circuit for the TMS320. *(Courtesy Texas Instruments)*

If more memory is needed than the DSP chip can directly access, an expanded memory can be added using bank switching techniques. With bank switching, addresses are calculated using modulo arithmetic to decode an address and switch in one of many memory banks. For example, a 64K-word memory could be implemented as four 16K-word banks; when an address referencing location 60,032 is received, the fourth bank is switched in. A reference to location 1,280 would cause the first bank to be switched in, and so on. Because this decoding is performed by TTL circuitry, bank-switched memory expansions often require more complicated glue logic circuits. Software considerations include longer access times, on the order of two to five instruction cycles, and provisions for loading the extended address onto the bus or I/O port used to send the address to the memory hardware.

Besides interfacing to external memory, a DSP chip must also be able to receive interrupt and control signals from either external hardware or from a host processor. An interrupt signal is a simple pulse that is typically used to signal the DSP chip that some external device is ready to exchange data. The difficulty arises from the fact that the interrupt may arrive at any time, even though the chip can only examine the interrupt port at intervals corresponding to a clock cycle. Some DSP chips contain on-chip circuitry that synchronizes the incoming interrupt with the chip's master clock, while others do not. For example, the TMS320 may require edge-triggered flip-flops (SN7474) to synchronize the interrupt and to hold it at a constant level during the time the chip is polling the interrupt port.

The software programmer has control over what action should be taken when an interrupt is received through the use of instructions such as BIOZ (branch on I/O status equal to zero) or IDLE (idle until interrupt) for the TMS320. Interrupts can also be polled using status registers, and the usual course of action upon receiving any interrupt is to branch to an interrupt service subroutine. An example of the usefulness of interrupts can be found in the interface between a Motorola MC68000 general-purpose microprocessor and a TMS320, as shown in Figure 17. The shared memory is not a dual-port memory, so both devices cannot access the memory at the same time. Since the TMS320 chip has access the majority of the time, it is most efficient for the MC68000 to interrupt the TMS320 only when it needs access to the shared memory. When the interrupt is received, the TMS320 could branch to a subroutine which stops any memory reads or writes and idles until another interrupt is received, signaling that the MC68000 is finished. This method is commonly used to transfer tables or blocks of data from a host to a DSP chip, and often requires substantial off-chip circuitry.

Input and output of data to other peripheral devices, such as A/D converters and UARTs (universal asynchronous receiver and transmitters) is also an important part of DSP chip programming. These types of conversion and communications devices are usually accessed by reading or writing to a dedicated I/O port. Because of the wide variety of device types and manufacturers, the hardware can often include many analog and digital components. Once again, this emphasizes the hardware-oriented aspect of DSP chip programming.

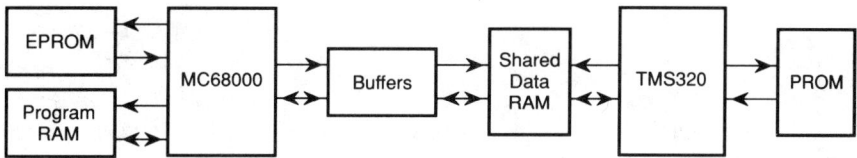

**Figure 17.** Block diagram of TMS320 and Motorola 68000 interface.
*(Courtesy Texas Instruments)*

Figure 18 shows the interface of a linear, 12-bit, analog-to-digital converter to the TMS320 chip. The design uses a TL072 operational amplifier configured as an anti-aliasing filter, an Analog Devices sample and hold (AD585) and A/D (ADADC84) chip set, and flip-flops and other TTL logic for generating a data ready signal. The A/D hardware is accessed using the TMS320 IN and OUT instructions, and the sampling frequency is controlled from the DSP chip. A counter is loaded with a value used to subdivide the chip's master clock down to the desired sampling frequency, and an interrupt is sent to the A/D chip once each time the counter reaches zero. (This connection can be seen in Figure 18 as the connection between pin 21, CONVERT, on the ADADC84, and the XF pin on the TMS320.) The exact sampling frequency is determined by the value the programmer loads into the counter. The upper limit for the sampling frequency depends on the TMS320 chip clock frequency (typically 40 MHz), the instruction cycle length, and the number of instructions to be executed between each successive sample conversion.

A linear 10-bit D/A interface is shown in Figure 19. An Analog Devices ADDAC100 digital to analog converter chip is used in conjunction with a 10-bit flip-flop buffer, as well as circuitry to generate a conversion-done signal. One TL072 op amp is used as a current-to-voltage converter, and another implements a second-order, low-pass, reconstruction filter. Software access to the D/A involves writing to Port 1 of the TMS320 and waiting until a conversion-done signal is received from the interface. In some cases, the number of instructions needed to prepare the next data may provide sufficient delay to eliminate need for any wait or interrupt instructions.

In many situations, A/D and D/A circuitry may be combined onto a single chip called an AIC (analog interface circuit). Analog interface circuits are often tailored to a specific DSP chip and use serial data exchange to reduce the interface hardware to just a few data, clock, and handshaking lines, without any glue logic components. One type of AIC is a codec (en*co*der and *deco*der) which performs nonlinear, pulse-code-modulated A/D and D/A conversion, as well as anti-aliasing and reconstruction filtering. A nonlinear codec converts an analog input into a companded sample value, which the DSP chip software must convert to linear form. The $\mu$-law and $A$-law expansion tables stored in ROM on the Motorola DSP56001 can be used for this purpose. The DSP chip also sends a companded value to the D/A section of the codec over the serial port, which is then output in linear analog form. The interface of a Texas Instruments TCM29C16 codec to a TMS320 is shown

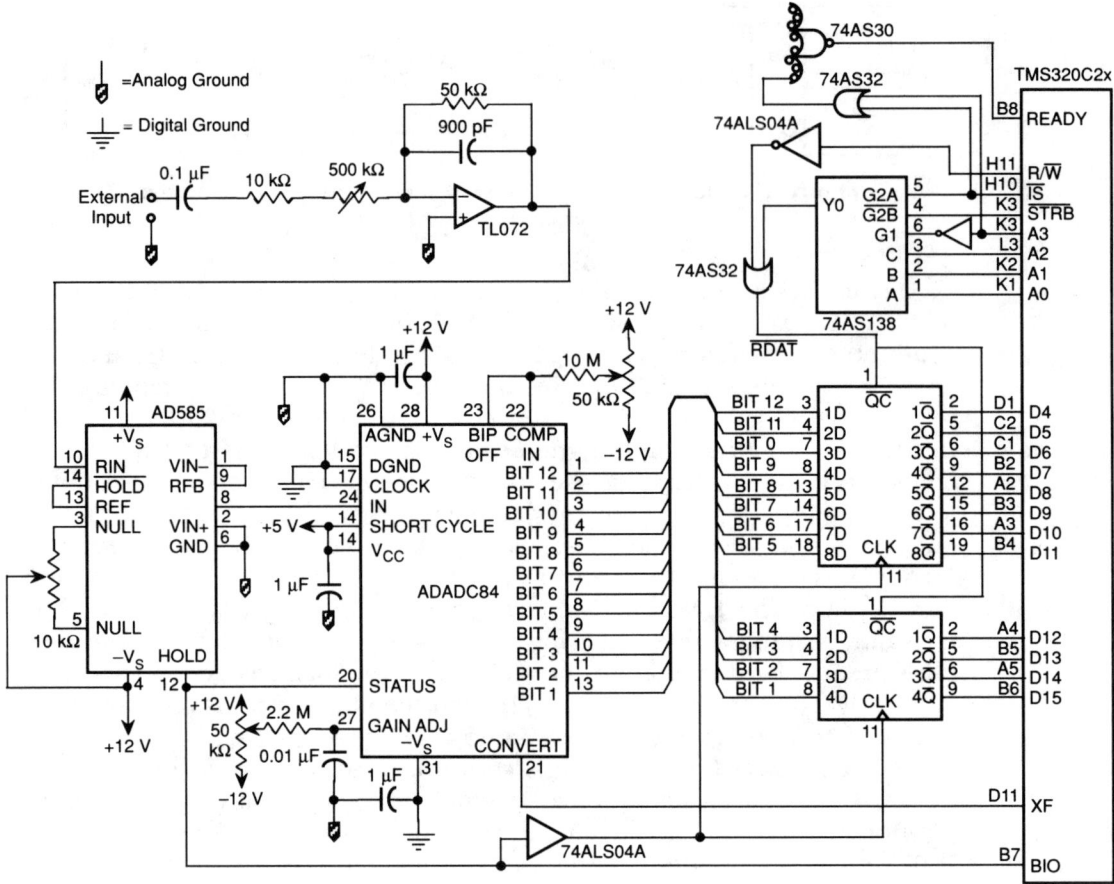

**Figure 18.** Analog to digital converter interface for the TMS320.
*(Courtesy Texas Instruments)*

in Figure 20. The serial port connections are through DR, DX, PCMOUT, and PCMIN, and the additional circuitry divides the master clock to generate the sampling frequency clock.

In summary, the interfacing of expansion memory and I/O hardware to a DSP chip involves three basic tasks. First, the hardware interface between the DSP chip and the peripheral device is designed using a combination of analog components, TTL devices, and specialty VLSI chips. Next, communication with the device is established by isolating the software instructions which control and access the DSP chip's parallel or serial ports. Finally, the program software is written with careful consideration given to timing and synchronization between the DSP chip and the peripheral device. Interfacing is one of the more difficult DSP programming tasks, but once a software routine successfully interfaces with a peripheral, it can be used repeatedly in many different applications. Furthermore, software and schematics are available in the public

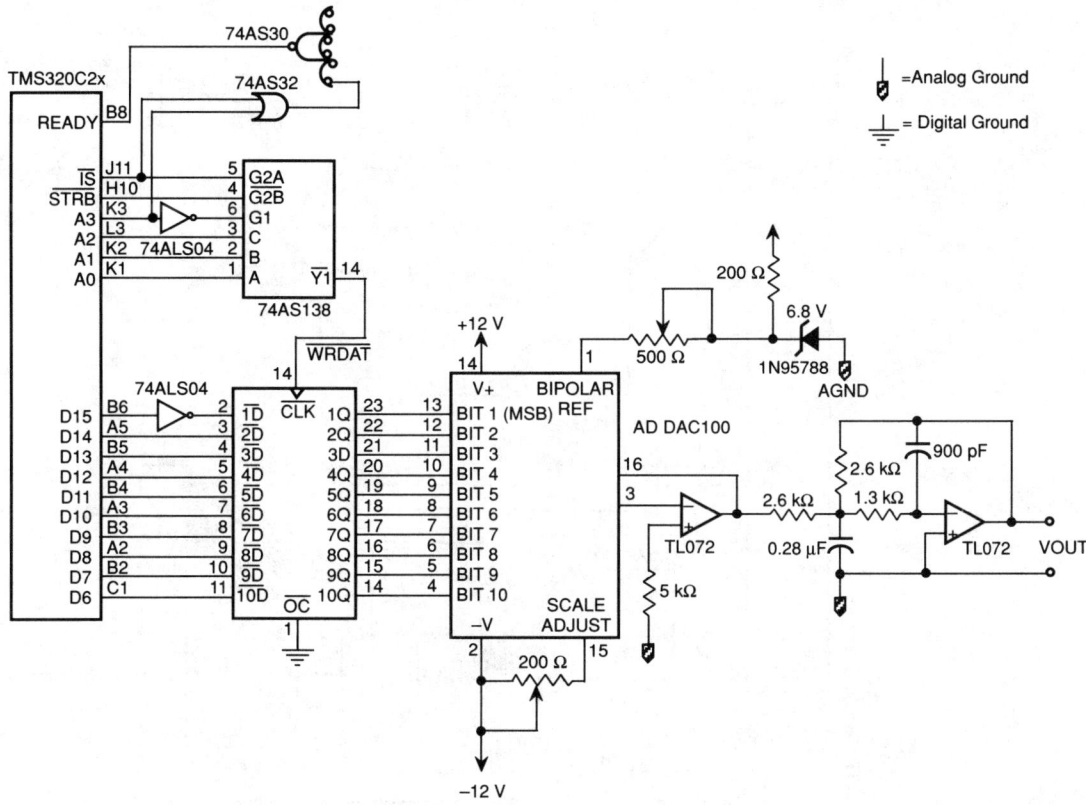

**Figure 19.** Digital to analog converter interface for the TMS320. *(Courtesy Texas Instruments)*

domain from manufacturers and third party developers for most DSP chips and general-purpose devices such as A/Ds, D/As, codecs, and UARTs.

## Assemblers, Compilers, and Development Packages

Developing software for any system, whether it be a digital signal processing chip, a general-purpose microprocessor, or even a mainframe computer, requires a number of supporting software and hardware tools. A system of these tools and their interrelationships are shown in Figure 21. A user supplied text editing program is used to create and modify the source code. An assembler program translates assembly language source code into executable object code. A compiler program is used to translate high-level languages into object code. Any digital signal processing chip will have its own unique assembler and compiler program to match its architecture—these are usually developed and supplied by the manufacturer. In addition, software support includes provisions for maintaining and linking together libraries of often used routines,

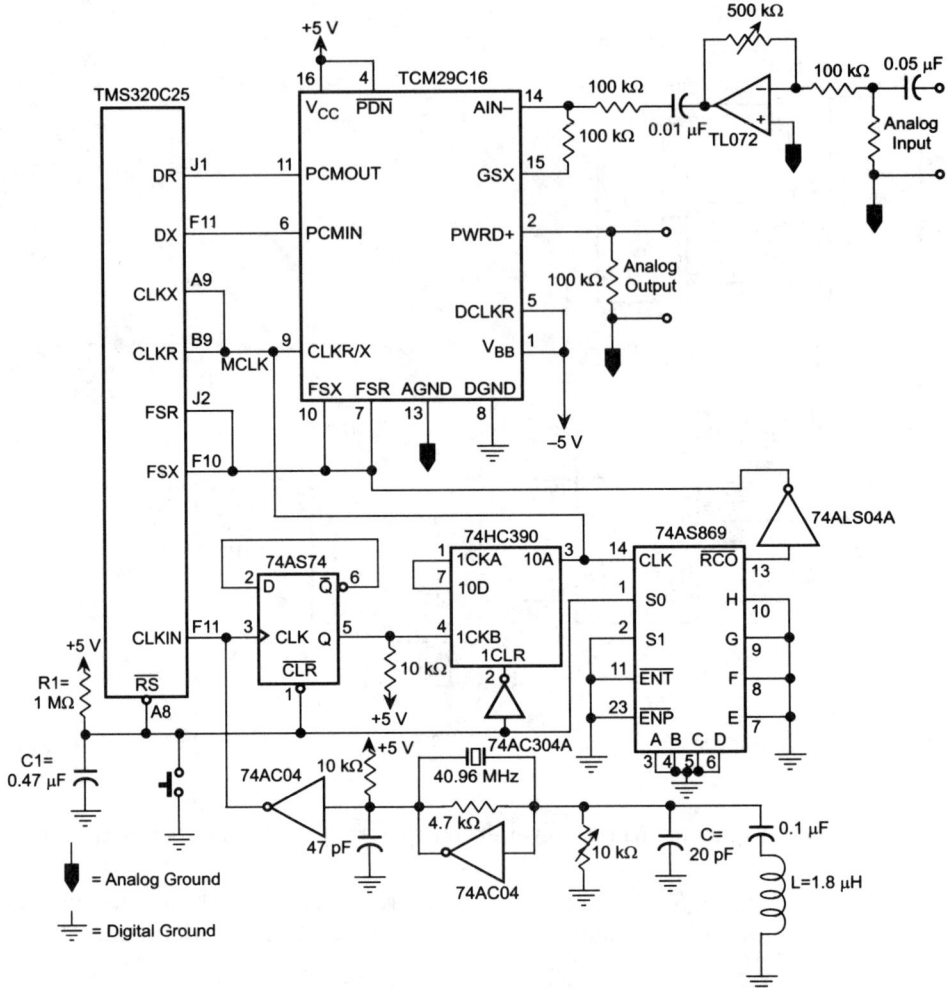

**Figure 20.** The codec interface for the TMS320. *(Courtesy Texas Instruments)*

simulator programs which emulate in software the hardware functions of the DSP chip, and debugging utilities for analyzing code and viewing memory location contents. These software development programs run on any number of microcomputers.

For example, Motorola markets an assembler, a linker and a simulator for its DSP56001, all of which are available for the IBM PC, the Macintosh II, the SUN-3, and the VAX. The Motorola 56001 assembler, like most others, is a macro assembler, which allows the substitution of macros for often used blocks of source code. It can generate absolute object code to be directly loaded into the 56001, or relocatable code to be combined with other modules by the linker. The 56001 assembler includes programming aids tailored to chip architecture features such as the X and Y data memories, wave table ROMs, and

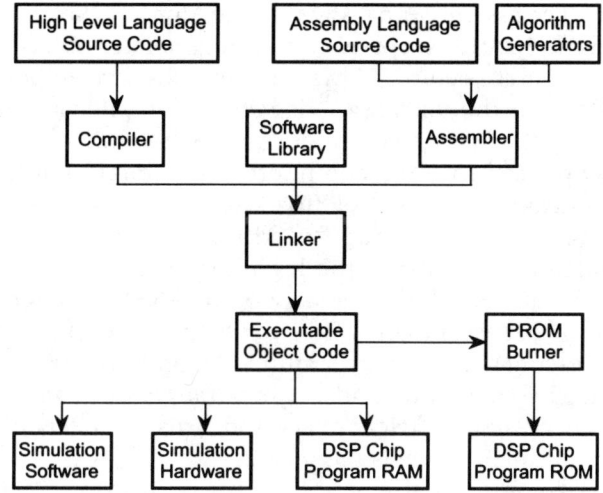

**Figure 21.** Digital signal processing software development tools.

parallel execution abilities. There are functions for common transcendental math computations such as sine, cosine, logarithm and square root, and there are assembler instructions for defining circular buffers and data conversion. During assembly, error checking provides a report of any violations which may be specific to the digital signal processing environment, such as improper setup of hardware controlled loops, poorly ordered instructions, invalid addresses, etc. Another feature helpful in DSP chip programming is the available count of instruction cycles and memory usage.

The linker combines multiple object modules translated by the assembler with library routines and adjusts address references so the code can run in a specific location in program memory. The linker also creates standardized files which are used to form ROM program memories.

The 56001 simulator program emulates almost all of the functions of the 56001 chip, including code execution, internal and external memory spaces, and interrupts and interfaces. The simulator loads assembled and linked executable object code into a simulated 56001 memory map, and instructions are executed a step at a time or continuously. During execution, register and memory location contents can be displayed and changed. Tools such as the 56001 simulator program are valuable to digital signal processing software developers because they provide a measurement of often critical code execution times and spare the software designer from time consuming and difficult debugging on the chip hardware itself.

Motorola provides some of the C language source code and object code modules for its 56001 simulator program, allowing designers to customize the hardware environment they need to simulate in software. This is helpful when the intended application involves controlling an external device or, more commonly, when complicated communications must be established with a peripheral.

The assembler package for the Motorola 96002 is nearly identical with that for the 56001, except, of course, it generates object code executable within the 96002 architecture. However, there are two important tools available exclusively for the 96002: a C language compiler and an on-chip simulator. The 96002 C language compiler supports the full ANSI definition of the language, as opposed to some compilers which might implement only a subset of the language. It allows IEEE floating-point constructs and can integrate in-line assembly language code for critical routines. Also included is an optimizer, which helps translate the high-level C language into assembly language in a way that best takes advantage of the 96002 architecture. This is different from the 56001 assembler which, in contrast, merely provides commands for easily accessing the hardware features. An optimizer analyzes the intent of the C language source code and may rearrange and combine instructions to achieve faster or more efficient execution. This is especially critical in a highly parallel environment such as the 96002.

An additional example is an optimizing C compiler for the Motorola 96002. The package includes an optimizing C compiler, a Motorola-compatible macro assembler, and a debugging program. It is designed to run on the IBM PC. The compiler features optimization techniques including instruction scheduling and coalescing, which reduce execution time by ensuring that the instruction pipeline is full whenever possible. It examines C loop constructs and determines when to use the hardware loop mechanism, and it analyzes the lifetimes and frequencies of all variables within a program, determining the best strategy for using the large register set. The InterTools compiler is available from Intermetrics, Inc.

Advances in compiler and optimizer design, combined with the increasing sophistication of DSP chips, have made these tools practical on such highly specialized processors. Previously, a high-level language compiler for a DSP chip would have been impractical due to the sometimes cumbersome code generated and its resulting slow execution time. (Even on larger computer systems, speed-critical routines are often written in assembly language). However, optimizers, and the increased number of registers and ALU functions on chips such as the Motorola 96002 make compilers a new and useful tool for DSP software development. Furthermore, as future DSP chips acquire more of the functions of general-purpose microprocessors, while retaining their speed and throughput advantages, many of the software tools currently reserved for microcomputer applications will certainly find their way into DSP software development.

For example, the Motorola 96002 DSP chip includes a hardware unit called OnCE, for *on*-chip *e*mulation. The OnCE hardware, an integrated part of the overall 96002 on-chip architecture, can freeze instruction execution, single step, and trace through instructions, and read data from the various busses and peripheral ports. There is also a history buffer which lists the last five instructions executed. Accessed through a dedicated serial port, OnCE is a powerful tool for debugging software which has been loaded into a working DSP system, for performing production diagnostics, and for in-the-field diagnosis and repair.

The software support programs offered by AT&T for their family of digital signal processing chips, the WE-DSP16 and the WE-DSP32, run on microcomputers under the UNIX, MS DOS, or VMS operating systems. Included in the software development packages are a macro assembler, linker, library and utility programs, and a software simulator. The simulator delivers I/O timing and execution reports and access and display of register and memory contents.

Software support for the Texas Instruments TMS320 family of DSP chips is extensive. A macro assembler and linker, software simulator, and C language compiler are all available. The Texas Instruments assembler package and simulator run under MS DOS, VAX-VMS, and SUN-3 UNIX operating systems and include the standard features found in other packages.

In addition to these, the Digital Filter Design Package (DFDP) is a useful and innovative tool for developing TMS320 software. It is an interactive, menu-driven program which runs on the IBM PC and aids in the design of filters according to Kaiser and Parks-McClellan FIR and Butterworth, Chebychev, and elliptic IIR algorithms. The user supplies a piecewise-linear filter specification in terms of response amplitude versus frequency and then specifies which digital filter type should be employed to approximate the analog filter specification. The magnitude and impulse responses, phase, and pole/zero maps of both the linear analog and digital filters can be displayed, and the program will perform iterations and prompts for modifications until the filter specification is met as closely as possible with the selected algorithm. Finally, the program uses the filter coefficients to generate fully commented, ready to assemble source code which implements the filter on the TMS320 family of DSP chips. (Code generators for other DSP chips are also available.) Translating digital filter coefficients and algorithms into software is a common DSP programming task, and as other DSP applications become more common, software such as DFDP will likely be employed to speed up the implementation of DSP algorithms in executable code. The Digital Filter Design Package is available from Atlanta Signal Processors, Inc.

Because software development for digital signal processing chips is dependent on the specific chip architecture, as well as the hardware configuration of the final application, many DSP chip manufacturers and third party companies offer hardware development packages to be used in conjunction with software development tools. These hardware packages range from systems that download program data from a PC into a DSP chip, to plug-in cards for a PC containing a DSP chip and all its needed support, A/D, D/A, and I/O hardware.

The Texas Instruments Software Development System (SWDS) is a hardware package that includes a plug-in circuit board for the IBM PC. It contains a TMS320 chip, a clock source and program and data memories, a cable for connecting each pin on the TMS320 chip to the target application, and a cable for connecting the card to the Texas Instruments Analog Interface Board (AIB). Essentially, SWDS provides a way to insert a functional TMS chip into a hardware application, while retaining control over the chip through the PC, including monitoring and accessing chip I/O. The SWDS includes the Texas

Instruments assembler/linker package and performs many of the same features as the TMS320 software simulator, with the advantage that the simulation and debugging software are working with a chip that is connected to the actual application, rather than with a software model of the application.

Working in conjunction with SWDS, or as a stand-alone unit, is the Analog Interface Board, which is shown in a typical configuration in Figure 22. The AIB is shipped with empty sockets, which may accommodate a TMS320 chip or the SWDS interface cable. Contained on the board are a 16-bit A/D converter with sample-and-hold circuit and anti-aliasing filter, a 16-bit D/A converter with low-pass filter, a function generator for on-board generation of square, sinusoidal, and triangular waves, and input and output audio amplifiers. Also included are a clock source, a codec, and a 16-bit digital I/O port. The AIB operates as a hardware platform for the development of digital filters and other digital audio processing applications, especially those that concentrate on a general algorithm that eventually will be ported into several target systems. In those cases, the AIB is useful because it provides the designer with generic hardware to get analog signals into and out of the digital signal processing development system.

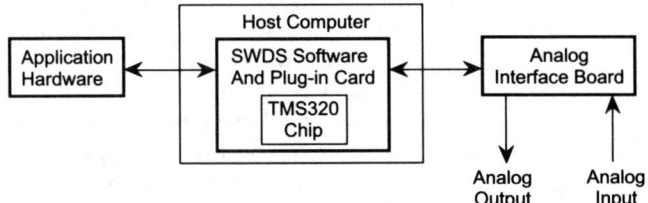

**Figure 22.** Texas Instruments software development system (SWDS) and analog interface board (AIB). *(Courtesy Texas Instruments)*

Texas Instruments also offers a system called XDS, a sophisticated hardware emulator for the TMS320 chip family. XDS is a self-contained hardware device that combines the TMS320 chip with debugging and simulation functions implemented in hardware for speed and efficiency. It can operate in one of four modes: stand-alone, with a host computer, PC mode, and as one of many multiprocessors. In stand-alone or host computer mode, user communication with XDS is made via a terminal. While in PC mode, the PC can act as both the terminal and the host computer. XDS is similar to SWDS except that it can trace code execution, set operation breakpoints, and update registers and memory, all in real time and at full execution speed. It can also reverse assemble the object code in the chip's program memory back into assembly language instructions. Up to nine XDS emulators can be daisy chained together to emulate multiprocessor configurations. A typical XDS system configuration is shown in Figure 23 and would likely be used to debug and analyze the performance of complicated DSP multiprocessor and device control systems.

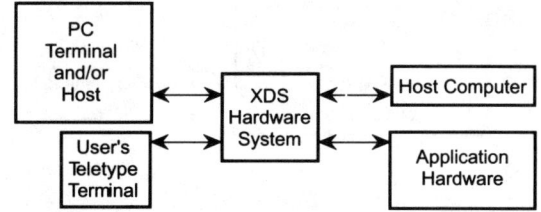

**Figure 23.** Texas Instruments XDS emulation and development system. *(Courtesy Texas Instruments)*

Hardware support for the Motorola 56001 chip is available as the 56001 Application Development System (ADS) and is shown in Figure 24. ADS works in conjunction with Motorola support software and provides the software developer with the basic interconnection of the DSP56001 chip and the PC development platform. An MS-DOS–based program handles control and communication with the 56001 chip, and a 96-pin connector accommodates A/Ds, D/As, and other user supplied peripherals. A similar system for the 96002 chip is available for IBM PC, Macintosh II, and SUN-3 platforms. It takes advantage of the 96002's on-chip emulation hardware, as shown in Figure 25.

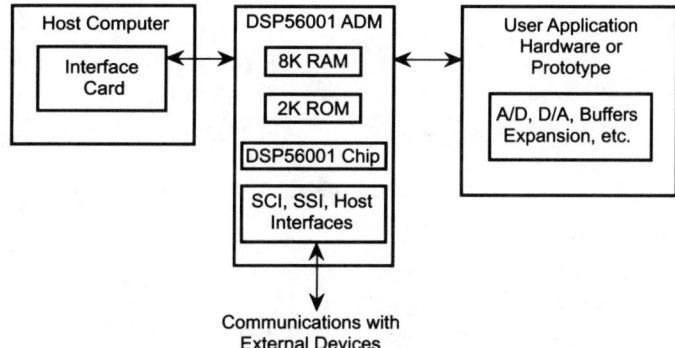

**Figure 24.** Motorola application development system (ADS) for the DSP56001. *(Courtesy Motorola)*

AT&T hardware support for the WE-DSP16 chip consists of a stand-alone board called the DEMO16, shown in Figure 26. This board contains a WE-DSP16 chip, two codecs for analog I/O, an RS-232 port for communications with a host computer, and four EPROM sockets which can hold up to 16K words of user supplied program memory. The DEMO16 package also includes several demonstration algorithms and software, including an echo canceler, an adaptive differential pulse-code (ADPCM) coder, and a multichannel, dual-tone, multifrequency (DTMF) receiver.

In addition to hardware and software support, Motorola, Texas Instruments, and AT&T all offer electronic bulletin board services. Technical question and answer forums, application notes and documentation, and public domain software and algorithms are available on these dial-up bulletin boards.

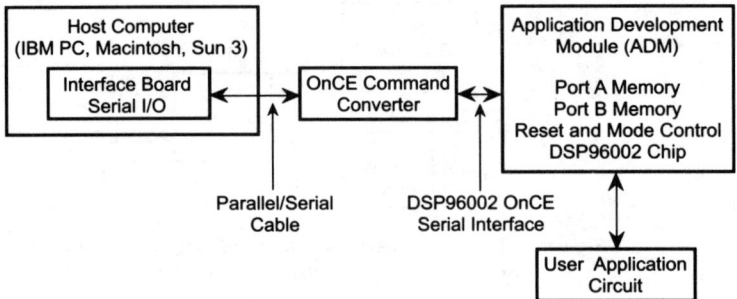

*(a) Application development.*

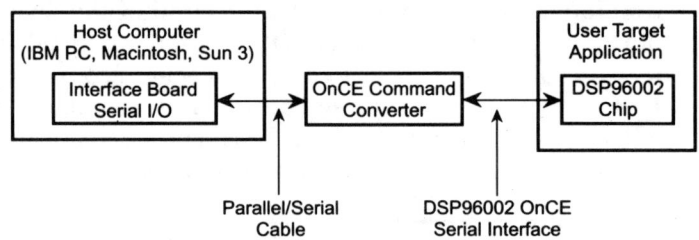

*(b) Target circuit emulation.*

**Figure 25.** Hardware development tools for the Motorola DSP96002.
*(Courtesy Motorola)*

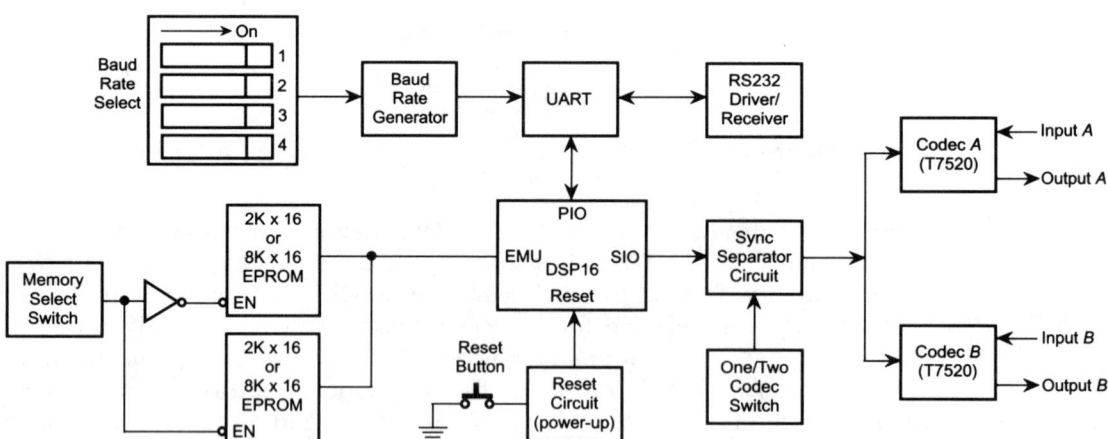

**Figure 26.** AT&T DEMO16 development board for the WE-DSP16.
*(Courtesy AT&T)*

# References

AT&T Microelectronics, WE-DSP16A and WE-DSP32c Data Sheets and Applications Notes, AT&T Microelectronics.

DFDP Manual and Program Disk, Atlanta Signal Processors, June, 1987.

*Digital Signal Processing Applications with the TMS320 Family, Theory, Algorithms and Implementations*, vol. 1-3. Texas Instruments, 1989.

*Filter Design and Analysis Program*, Momentum Data Systems, Inc.

*InterTools for the 96002*, Intermetrics, May, 1990.

Jackson, L. B., *Digital Filters and Signal Processing*, Kluwer Academic Publishers, 1986.

Leventhal, L. A., *6809 Assembly Language Programming*, Osborne, McGraw-Hill Book Co., 1981.

Morris, R., *Digital Signal Processing Software*, Carleton University, 1983.

Motorola, Inc., DSP56001 and DSP56000 Data Sheets and Applications Notes, Motorola Microprocessor Products Group.

Motorola, Inc., *DSP96002 IEEE Floating-Point Dual-Port Processor User's Manual*, Motorola Microprocessor Products Group.

Parks, T. W., and C. S. Burrus, *Digital Filter Design*, John Wiley & Sons, 1987.

"Programs for Digital Signal Processing," *IEEE Transactions*, ASSP DSP Committee, IEEE Press, 1979.

*Second Generation TMS320 User's Guide*, Texas Instruments, 1989.

Texas Instruments, Inc., *Second Generation TMS320 User's Guide*, Texas Instruments, 1989.

*WE-DSP32-SL Support Software Library Data Sheet and WE-DSP16-SL Support Software Library Data Sheet*, AT&T, June, 1987.

*Chapter 15*

# New Markets in Digital Audio

**Daniel C. Mikat**

## Introduction

In the preceding chapters, we examined the state of digital audio systems and related technologies. By contemplating the growth of electrical and computer technology in the recent past it is clear that more significant and advanced developments have and will come about almost before we are able to digest the current technologies. Mankind's view of technology and social change in general is often clouded by attitudes of complacency or efforts to preserve the past. In this chapter we shall attempt to relate social attitudes and industrial progression to engineering of the future and, in particular, the role of advanced audio systems in a mushrooming technological environment. We will see how the unique properties of the digital audio media lend themselves to collaboration with various other technologies, yielding integrated fields which hold vast promise and perhaps even fortune for many.

Many fear that technology is advancing without prudent supervision and will become a controller of society instead of a convenience. For example, people are confused and misled by the notion of a thinking computer. The mere mention of automation and robotics provokes concern over unemployment. In today's society, technology holds a stigma of manipulation. In reality the opposite of these views would be more appropriate. As we shall see, technology might just be the saving grace of civilization. As the world's resources are depleted and the demands of mankind become more urgent, it becomes all too clear that industrial, social, and technological change likewise becomes more urgent. By understanding how these changes will occur, the transition into the next wave of civilization can be smoothed. We would be well advised to avoid holding onto the past and instead look to the promise that the future holds for technologists, and the power that they will exercise in the "Third Wave."

# The Third Wave

In 1980, Alvin Toffler authored *The Third Wave*, in which he described society as changing in "waves." Toffler explains that in the history of humankind on earth, there have been two complete waves of social change. Before the first wave of change, mankind existed as scavengers. Cave dwellers banded together to hunt and protect themselves. With the advent of the First Wave, the agricultural revolution, mankind learned to cooperate in order to cultivate and tame the land, domesticate animals, and organize societies. People lived in small, self-perpetuating societies where they provided for their own needs by working together. It was found that if everyone did one job, the productivity of the society would increase. If town $A$ produced crops and town $B$ raised livestock, they found that they could trade with each other to achieve a higher yield than if both societies were totally self-sufficient. This phase lasted for about 3000 years until the dawn of the industrial revolution.

As society moved into the industrial period, the Second Wave, people began to work for others. They commuted to a workplace, performed one simple specialized task, and received payment with which they could purchase their temporal necessities. Societies which moved most quickly into this wave found prosperity and improved lifestyle and standard of living. This industrial phase is currently ending after approximately 300 years.

The primary emphasis of industry during the second wave has been "produce, produce, produce." Industry took from the earth in order to make a profit. Only recently has humankind become aware of the limitations in this pursuit and the self-destruction that is assured by unlimited energy expenditure and indiscriminate harvesting of the earth's resources. A new attitude toward and from industry is now imminent, and new technology can help bring absolution from past crimes.

We are now entering the Third Wave of civilization, known by some as the technological revolution, characterized by large scale automation and super-specialization. Four industries will form the backbone of the third wave, at least two of which relate directly to our study of digital audio technology. Let's examine each of these four areas before we explore the impact of the third wave on audio engineering and its contribution to an advanced society.

One of these backbone industries, and perhaps the most clearly related to digital systems, is the industry of electronics and computers. In the past thirty years there has been as much advancement in electronic technology as has occurred in the whole of history. The use of computers in all areas of industry and the private sector has grown tremendously. *Computerworld* magazine stated that "If the auto industry had done what the computer industry has done in the last 30 years, a Rolls-Royce would cost $2.50 and get two million miles to the gallon." The electronics industry has advanced similarly, especially in the field of entertainment. Engineers have developed new music and video systems with better fidelity at a lower cost. Television and telecommunication systems are being revamped with satellite transmission and fiber optics. As an example of the rapidity of change that the entertainment field experiences, consider the impact of the compact disc medium. Within a few years of its introduction, nearly every household that actively purchases recorded music has purchased CD players, CD libraries, and peripheral gear. Most record companies have phased out the production of vinyl LPs altogether. It seems that technologies are changing faster than current ones can be absorbed.

The "thinking computer" has already proved very valuable to industry in the form of expert systems. Computerized control of automation systems is affecting many major industries, and advanced electronic devices appear in automobiles, offices, factories, libraries, supermarkets, and every room in the house. It is clear that people are becoming more and more dependent on their electronic devices just to complete their daily routine. This dramatic change in lifestyle serves as an example illustrating the volatile times that the world is now experiencing and will continue to experience throughout the duration of the third wave.

The second area of forthcoming industrial value in the third wave society is in the area of space manufacturing and industry. Few people realize the incredible importance of this field. Such fields as space energy generation, highly controlled antigravity manufacturing, medical drug and instrument handling and construction, alloy development, super high-temperature and super low-temperature manufacturing, and space communities are no longer extremist's dreams and fantasy movie plots, but real notions which, if not already being implemented, are well into planning stages for serious pursuit.

The third area of development lies in the utilization of natural resources. For example, economists are beginning to realize the importance of oceanic industry. The use of the oceans in industry is the third area which will undoubtedly see incredible advancement in the future. Methods to alleviate a part of the world hunger problem are being explored in the form of oceanic farming. Means have been established to economically raise fish for harvest without indiscriminate strip-mining of the seas.

The use of natural energy in the form of wind, thermal currents, and tides can provide nondestructive means of utilizing energy sources. As oceanic energy and food production become more economical, their uses will evolve and demonstrate an incredible resource which is often overlooked. As the third wave unfolds it will become more and more clear that a key to success in all industry lies in the solution to current inefficiencies.

The fourth major field of development in the third wave is biotechnology. This field involves biomedical applications, bioindustry, and genetics, all of which are becoming more viable with each passing year. Biomedical advancements bring hearing to the deaf, vision to the blind, and cures for the incurable. Many of these applications are still crude but hint toward new treatment of serious health concerns in the future. Bioindustry uses biological matter to produce useful by-products such as oil or electrochemical energy.

Knowledge of genetics doubles every two years according to Toffler, and the field is ripe for business. The notion of genetic alteration of people causes great concern and rightly so. The idea of creating a super race of genetically altered individuals to perform certain tasks more appropriately is not a far-fetched idea but a real possibility given the advancements in genetics. In a less controversial example, parents could be tested for possible genetic problems in offspring before pregnancy, and problems could be corrected before the development of the fetus. Genetic engineering could help to prevent and treat diseases which now are incurable. Medical drug companies are currently spending fortunes in researching the possibility of treating disease and dysfunction genetically. The real value or liability of this incredible field is unforeseeable. However, it must be addressed with great care to maintain control of its powers.

Although these four areas are clearly not the only ones which will advance, they are undoubtedly ones which may solve important problems that society faces upon entering this new phase of civilization. The most important notion of industry appears to be in creating new efficiencies of production and in integrating various emerging technologies. Becoming aware of these trends and moving with them is the only hope for success in the confusing times to come.

# A Perfect Audio System

With respect to digital audio, several difficult problems must be addressed before an acoustically "perfect" result can be achieved. Development of new technologies, applications, and media will be required. A serious problem which has faced digital audio engineers from the introduction of digital systems has been that of noise. Due to the nature of quantization, the introduction of noise and distortion into a signal is inevitable. The use of dither to correct this distortion has in effect changed the correlated distortion into an uncorrelated, static noise floor. This has vastly improved the sound quality by changing the nature of the existing noise. In Chapter 2 the nature of dither was discussed, and Chapter 12 demonstrated how noise shaping circuits can further reduce the noise floor associated with the introduction of dither. Further improvements in high-order noise shaping technology to control the effects of dither are certain to have a great influence on the design of new digital audio systems.

A tremendous inefficiency in current digital audio systems lies in the handling of the enormous amount of data required to produce even a modest duration of program material. Music is perhaps 90% redundant but in linear systems the redundant data is treated in the same regard as valuable and important information. It seems reasonable that, instead of being a hindrance, this redundancy can be used to our advantage. Through the use of data compression methods such as pattern recognition a more efficient manner of encoding audio data could be devised.

A pattern recognition element would analyze a block of data to find a pattern which repeats itself more than ten times. It would then encode that pattern with a label such as PATTERN 'A.' Every time that pattern occurs throughout the block of data, instead of recording the entire pattern it could merely code a cue to REPEAT 'A', which would take considerably less space on the recorded medium.

For a pattern encoded system to be implemented without any audible artifacts, study must first be employed to determine the greatest variance from a pattern which cannot be audibly perceived. This will determine the number of cycles a pattern can be repeated within a given block of material. An exact repetition of a sequence actually occurs quite infrequently because of the dynamic properties of music. To overcome this, the pattern recognition element might employ the command REPEAT 'A' EXCEPT ... and then describe the manner of variance from the pattern labeled 'A.' The listing of variances would require vastly less space on the recorded medium than is required to describe the entire block of data. The efficient use of pattern recognition could decrease the amount of data required to deliver a given length of program material by 40% to 70%.

The use of pattern recognition schemes could also be very useful in concealment of errors. In any medium where large amounts of data are handled with high density, some errors are inevitable. With error detection and concealment strategies, most errors are corrected or concealed and become inaudible. With large errors however, in which more than 5 ms of data is lost, interpolation is of no value and a muting of the invalid data can become audible. It is with large blocks of missing data where pattern recognition is most beneficial. If a section involving 30 ms is lost and a pattern is established in the prior 5 ms, that pattern could be repeated until new valid data resumes. By examining patterns which follow the lost data, an even better, less audible concealment might be possible by adjusting the prior pattern to more closely match the following one. Within the important field of data compression, the study and implementation of pattern recognition has received relatively little attention. As the problems of efficiency in storage and transmission become more urgent, however, the value of such schemes will become self-evident. As the third wave unfolds, pattern recognition and other means of data compression will become increasingly important.

Chapter 5 demonstrated the efficiency of optical-fiber transmission. This will be a field of continued growth and development throughout the third wave. Almost all major cities in the United States and abroad are converting their telephone transmission paths from copper wire to optical-fiber transmission lines. These pathways transmit data many times more efficiently than their copper predecessors. Photons travel virtually resistance-free through optical fiber while electron motion is considerably restrained within copper pathways. The manufacture of optical fiber is also far more efficient than copper (by about three orders of magnitude).

Telephone companies, financial institutions, commercial retailers, schools, and practically every business imaginable all rely on information transmission and retrieval in order to function properly. With virtually trillions of dollars worth of industry relying so heavily on data transmission, advances in digital transmission, optical technology, and satellite link-up engineering are paramount in the furthering of industry throughout the next phase of civilization. It is clear that data transmission will receive much attention in the coming years.

The importance of digital signal processing (DSP) is now recognized in many areas of technology as noted in Chapters 10 through 14. It is important not only in ambience processing and equalization for entertainment systems; it will also be applied in a variety of technologies to achieve impressive results, particularly in the field of bioelectrical medical applications. A few of these bioelectrical techniques are discussed below in the context of digital audio applications. The speed limitation in many DSP applications seems to be the greatest hindrance in the science to date. Complicated multichannel filtering, for example, is difficult to do in real time. The acceleration of such functions using pipelining strategies and parallel processors will undoubtedly be the thrust of study in the near future.

With respect to storage, the compact disc shows great promise as the primary consumer digital medium for many years to come. Its durability makes it appropriate for the abuse which is ensured by consumer usage. It offers digital fidelity and nearly infinite longevity. Its next phase will take the form of recordable discs available on a widespread basis to the consumer. As more and more recordable/erasable optical disc systems are made available, prices are certain to come down and consumers will have access to optical disc recording capabilities in their own home.

Digital audio tape (DAT) is finding widespread use, especially for on-location recording. It is of great benefit to film and video production companies to have digital-quality recording with full SMPTE time code capabilities in a shoulder pack. The DAT system offers an affordable way for many production companies to go digital. The DAT system is not without its problems, however. The nature of magnetic tape exhibits problems of tape stretching and breaking. Tape-cut editing is not possible because of the helical scan recorded tracks. Given these problems, the DAT medium still provides a superb signal-to-noise ratio and versatility surpassing that of analog media at a comparable cost.

The most inefficient aspect of any mechanical storage system is its transport mechanism. Handling data electronically consumes very little electrical energy opposed to delivering data mechanically to a pickup element. For example, rotating the CD and controlling the laser carriage uses about 100 times the energy of handling the data after pickup. This inefficiency will surely lead design engineers to explore the use of portable memory modules that do not require physical motion to be read. The use of a memory card as a medium is one possible solution to the problem of efficiency. A programmable logic array (PLA) network on a memory card 2 inches by 4 inches in surface area can hold as much digital audio information as a CD. A memory card would allow data to be electronically read into a cache memory via a complex array of decoders for random access. Memory cards will become important in areas where energy efficiency is most crucial. They would be best appreciated in applications requiring battery power supplies such as portable audio units.

As the focus of technology in the third wave becomes clear, efficiency, compactness, versatility, and longevity will all become key factors. Digital audio lends itself to all four of these categories. The success of new media and encoding/decoding systems in the marketplace will depend on their development as related to all four factors.

# New Digital Audio Markets

The key to success in industry in the third wave will lie in the ability to integrate various technologies to deliver new, efficient, nondestructive means of production. Digital audio and signal processing developments lend themselves directly to many efficient applications when used with other existing technologies.

The entertainment industry, for example, has been the focus of digital audio since its introduction in the early 1970s. The main intent of digital audio systems has been to improve fidelity, increase dynamic range, and to extend the longevity of the medium. All of these goals have been accomplished, leading us to new applications which are integrated with other technologies to further increase flexibility and improve transmission quality.

Already being implemented are DSP units found in automobile stereo systems and television sound systems. These ambience processors produce surprisingly effective results. Stereo speakers at the sides of a television can deliver effective stereo imaging via a set of digital delay registers. These delays introduced in a complex manner between the two speakers effectively trick the brain into hearing the sound from a location removed from either speaker. This "aural holography" is an evolving and important notion in future entertainment systems.

Automobile sound systems comprise a large portion of the consumer audio market. These systems are increasing an area of study, research, and development in many audio companies. The car has become the universal place for listening to music following the widespread influx of televisions in the home. Even though the market indicates more emphasis on music in the car than the home, industry has been slow to tackle the unique problems of audio in such a small, moving space. Lately, however several companies have shown great success in the marketing of automobile elitist systems. These systems employ non–shock-sensitive transport systems, automatic equalization schemes, digital delay lines to achieve phase coherence for the driver, and ambience processors that realistically mimic the ambience of a variety of preset listening environments. The resulting sound quality is clearly improved to the point that many consumers will spend extraordinary sums of money for their installation. The development of advanced digital audio systems for automobiles will clearly be a promising venue for audio companies in the coming years.

In professional applications, digital recording systems and workstations have been used for several years and demonstrate important qualities which promise their use throughout the third wave. The digital workstation provides a single musician/engineer with the tools to create complete works which even the most aurally astute cannot delineate from a group of acoustical musicians. To be a truly flexible musician or producer one must also be an informal engineer. With the high cost of using professional musicians for recording, a single, well-paid, workstation musician can achieve comparable results for a fraction of the cost. Such workstations must provide sampling at a variety of user adjustable sampling rates, complete editing capabilities, sequencing and sequence list editing, DSP functions, and SMPTE triggering among many other tools.

In the multibillion dollar telecommunications industry, the fields of computer engineering, fiber optics, lasers, radio transmittal, satellite technology, and data compression/multiplexing are all integrated with digital audio engineering to handle the multitude of messages sent around the world. Digital audio is inherently appropriate for transmittal because of its robust nature and great versatility in processing and transport. Digital transmission was discussed in Chapters 8 and 9 as it deals with modulation codes, broadcasting, and the field of data compression.

## High-Tech Education

Education has demonstrated very little significant advancement in theory or practice in the past century. Although computers and other high-tech subjects are being introduced to children of a very young age, the methods of teaching have remained basically unchanged. With illiteracy on the rise in America and abroad, it is clear that some change is needed in teaching to combat the problems that this illiteracy causes. The challenge is probably one of sociology rather than engineering, but a high-tech learning system might spur motivation in some people and help to alleviate some of the problem. If the means of education is made interesting to students, results will not be far behind.

In the early 1970s "learning centers" cropped up across the country, advertising computer-based learning systems to teach children school subjects via a computer terminal. The problem with the systems was that software interfaces were not user-friendly and were especially inappropriate for young children. Students usually developed better skills and knowledge base with a teacher or tutor and so the computer learning center concept was short-lived. The idea, however, was actually a very good one and might show great promise if implemented with the help of child psychologists and sociologists, who may have a better understanding of positive teaching strategies and motivation than do programming engineers.

Some universities have introduced lecture courses on video tape where students sit in a large class room and watch an authority lecture on his or her subject of expertise. The method has been used to a great extent as a teaching aid for many years. The problem with its use as a primary means of teaching lies first in the quality of presentation. If video or film could be presented with better visual resolution along with high-quality audio, the overall message would be enhanced. The use of laser discs with digital audio and video would ensure both high quality and a long lasting data medium. The problem, however, lies in the lack of teacher/student interaction. This interaction has always been a means of directly stimulating motivation. Colleges and universities have found that if recorded lecture series are presented, discussion sections should also be required in which the topics presented could be reemphasized or clarified to whatever extent is necessary. A positive side to prerecorded course lectures lies in the student's ability to replay any lecture which was difficult to understand or that the student missed. It would allow students to learn a subject from an absolute authority who can also effectively deliver the subject matter. Technology currently supports these new wave learning systems. The problem lies in their implementation. It is certain that as new means are tried, new forms of education will evolve. It seems reasonable that with such technology available, the value of high-tech education will become more and more obvious and its implementation shall become necessary.

CD-ROM and CD-I will prove to be an increasingly important medium base in the coming third-wave society because of its extraordinary cost and space efficiencies. Information retrieval systems are already being used in many professions. The medical field, for example, uses databases on symptom/disease association, drug and medicine information on uses and side-effects, patient histories, and other administrative records. The optical disc is an extremely efficient means of storing such data. The initial cost of the system is high but is very efficient in medium cost, size, and longevity. In the long run, these systems save considerable cost and space. As recordable CD systems are made more available and affordable, far more data storage will be done on disc in the digital domain. Since one compact disc can hold the same amount of information as an entire encyclopedia set, the advantage in space regained becomes clear.

The purchase of "books on disc" in both text and audio form will likely become another useful implementation of digital systems in the near future. Current technology allows for such storage of these text files and transfer of data, but few titles have been successfully marketed. The storage is simple using CD-I or an audio compact disc. Speech can be stored on a compact disc much more efficiently than music because of the lesser dynamic range and frequency band required to deliver suitable fidelity. The purchase of text in either audio or text form could be individually done on a home PC with a modem. The consumer could call the company to purchase a file, which is then downloaded to the consumer's home system. Since the primary motivation in business and industry in the third wave will be efficiency, it will be reasonable to conclude that this efficiency motive will be applied to education, learning systems, lecture series, and in large data/text files.

# Biomedical Applications

Biomedical engineers are fast learning the value of digital engineering in the treatment of many physiologic dysfunctions. Perhaps the most dramatic application, receiving much attention in the biomedical community is the treatment of deafness. For certain types of dysfunction, hearing aids can provide great enhancement of whatever hearing ability is still intact. They have become considerably smaller and less obtrusive, while offering increased intelligibility through the use of frequency selective filtering schemes. For patients who suffer from complete bilateral deafness, a hearing aid would be of no benefit, but new advances in digital signal processing lend themselves to several other important treatments.

The three major areas of treatment are in tactile stimulation, cochlear implantation and direct electrical stimulation of the auditory cortex.

To give cues regarding speech or environmental stimuli to the deaf using tactile stimulation, an acoustic signal is filtered into several bands each of which stimulate a vibrating transducer placed on the skin. The vibrations delivered by the transducer contact are felt tactilely by the patient. In this way the patient can sense when an acoustic signal is present and at what loudness. The sense of loudness is delivered by a greater displacement of the transducer element on the skin. By filtering the signal into bands, the patient can determine by the location of excitation which frequency band is acoustically active at any given time.

Tactile stimulation can deliver some speech intelligibility in very motivated patients. Some problems faced by these patients involve difficulty in keeping the contacts adequately affixed to the skin, inconvenience of carrying a speech processing unit, and the discomfort of the contact array, which is generally worn across the stomach. The primary drawback of a tactile stimulator lies in the fact that the patient still cannot hear anything. The stimulator delivers cues to a subject but restores no hearing.

The cochlear implant utilizes digital audio technology to actually restore hearing to the deaf. An array of electrodes are surgically implanted into the human cochlea. Each electrode carries information from one narrow band of frequencies. When acoustic energy occurs in a particular band, the electrode delivers an electrical stimulus to the region along the basilar membrane which corresponds to that band of frequencies. When the electrical stimulus enters the organ of Corti, a sensation of hearing is delivered. An acoustic signal enters a microphone (placed at the patient's ear canal) and is sent to the signal processor, where it is filtered into the appropriate number of frequency bands. Each of these bands control a single electrode in the array. High frequencies stimulate the region closest to the oval window, while low frequencies are stimulated at the apex of the cochlea.

The signal processor holds the key to the success of the system, and it is in the signal processor where the DSP application is used. To deliver good intelligibility to the patient, at least eight frequency channels are needed. The current state of the art device employs a 22-channel system. These frequency bands require brickwall filters to avoid crossover between channels. Theoretically, with perhaps 100 frequency channels, near-normal hearing could be restored. The DSP filtering described in Chapter 10 delivers this in real time.

The cochlear implant delivers good intelligibility of speech for many patients. Unfortunately, however, it cannot restore complete hearing and the device is only available to patients with sensorineural hearing loss in which the auditory nerve is still intact.

Research is currently under way of attempt direct stimulation of the auditory cortex in the brain. The theory is similar to that of the cochlear implant except that the stimulating electrodes are affixed to the appropriate frequency location at the auditory cortex. This means of stimulation would be useful in bypassing yet other kinds of deafness and may be used more extensively as the science progresses. The problem with this manner of stimulation is the difficulty in accurately mapping appropriate stimulation points at the cortex. It takes a great deal of testing to accurately predict what response will be felt with a given stimulation.

Related to the field of auditory prosthesis are other neurological treatments for muscular and spinal dysfunction. The electrical stimulation of muscles has been used in sports medicine and in physical therapy for about 20 years. By sending a step-pulse electrical current through a muscle, it is possible to cause a contraction or relaxation regardless of what neurological signal is being sent from the brain. This technology, in addition to recent study in neurological coding and transfer, has led biomedical engineers toward the idea of spinal injury bypass. If a clear understanding of the relationships between neurological impulses and muscular excitation is understood, it would be possible to receive messages from the undamaged portion of the spinal cord and deliver a mimic of the signal directly to the muscles involved. To achieve this result, a precise mapping of the nerve fiber functions must be obtained and their firing must predictably cause the same desired response. Once an accurate mapping is completed, nerve firing combinations can be decoded in a large database and instructions then forwarded to the appropriate muscle groups. For this to be done in real time, superfast DSP decoding and encoding systems are required. This manner of addressing a very serious health issue is still very experimental and has met with very limited success, but as knowledge and understanding of neurological functions are made more clear, these theories will find realistic applications.

## Conclusion

As this new wave of civilization unfolds, new applications of high-performance digital systems will become apparent. The mushrooming of technology in the past 20 years is an indicator of some of the advances to come and the importance that integrated technologies will hold throughout the next phase of humanity. The first wave of civilization lasted about 3000 years, the second is now ending after only 300 years. The third wave as described by Toffler will likely last only about 30 years. With knowledge, efficiency, and technology changing so quickly, it is clear that the only way to maintain position is to be a leader of change. Digital audio in particular illustrates this change in society and technology. It demonstrates society's demands for improvements in efficiency, compactness, versatility, and longevity. As digital audio continues to progress and become integrated with other important sciences, designers and users alike will lead the industry in its pursuit of these varied approaches and applications.

## References

Toffler, A., *The Third Wave*, Morrow, 1980.

# Index

## Symbols

1-inch type C video, 228, 230-231
18-bit D/A conversion architecture, 63-76
4:1 digital audio data compression system, 261
8mm format, 221-228
   audio conversion process, 223-225
   audio-only mode, 226
   automatic track following (AFT) signals, 225
   multitrack mode, 226-227
   physical properties, 221
   track format, 222-223
   video performance, 228

## A

A/D (analog-to-digital) conversion, 76-81
   low-bit, 403-412
      chip, 413-414
      with D/A, 417-419
   multilevel quantizer, 414-417
absolute linearity error, 57
absorption, 85
   fiber optics, 127
   induced transitions, 90
   radiation field energy density change, 92
active medium, 88
adaptive delta modulation (ADM), 256
adaptive delta pulse code modulation (ADPCM), 256-257
additive white Gaussian noise (AWGN), 14
AES/EBU output, baseband signals, 270
AIC (analog interface circuit), 479
aliasing, 12
ambience processing, 8-9
amplification for lasers, 92-93, 97-99
amplitude quanta values, 22
amplitude shift keying (ASK), 273
analog dither, 47
analog interface circuit (AIC), 479
analog-to-digital (A/D) conversion, 76-81
angle of acceptance, 125-126
angle of incidence (refraction), 122-124
antennas
   isotropic, 277
   parabolic, 278, 280
   satellite design, 280
APD (avalanche) photodetectors, 144
aperture error, 32
areal effect, 4
arithmetic data compression coding, 251-253

AT&T WE-DSP16A DSP chip, 437-441
   DEMO16 board, 487
   software support packages, 485
AT&T WE-DSP32C DSP chip, 437-441
   instruction set, 466-470
   programming models, 452
   software support packages, 485
attenuation (optical fibers), 162
audio
   fiber optic applications, 158-164
   laser diode applications, 114-116
   processing, D1 (component digital video) format, 211-212
audiology, spatial, 8-9
auditory
   capabilities, 1-18
   habituation, 9, 16
   physiology, 1-6
automatic muting, 18
automatic track following (AFT) signals, 225
AWGN (additive white Gaussian noise), 14

## B

band-pass interference filter as demultiplexer, 151
bandwidth (fiber optics), 162
base-line wander, 26
baseband signals
   AES/EBU output, 270
   delta modulation (DM), 270
   digital broadcasting studios, 269-270
   modulation, 270
   pulse code modulation (PCM), 270
   pulse width modulation (PWM), 270
   super high frequency (SHF), 270
basilar membrane, 5
BBC 14/10 digital audio data compression scheme, 259-260
Bessel filters, 336
bilinear transform, 335-336
binary coding, 22
binary phase shift keying (BPSK), 275
bipolar D/A converters, 66-67
biquadratic realization of digital filters, 332

bit coding in pulse modulation, 27
bit error probability, 28
bit error rate (BER), 159, 174
bit rate reduction (BRR) encoder, 257
bit-switching D/A converters, 64-66
brickwall characteristic, 37
Burrus LED (light-emitting diode), 138
burst errors, 173
butt-type connectors, 157-158
Butterworth filters, 336

## C

cable DAB systems, 289-290
cables
   *see also* optical cables
   optical, 128, 130-136
   plenum, 132
   ribbon, 132
   snakes, 158
   submarine, 132-133
canonic realization of digital filters, 332
CATV audio tuners, 290
Cauer filters, 336
CD *see* compact discs
CD-I (compact discs-interactive) digital audio data compression format, 260-261
CD-WO (compact-disc, write-once) format, 188-193
   applications, 192-193
   audio applications, 195
   constant linear velocity (CLV), 191
   disc structures, 192, 195
channel coding, D1 (component digital video) format, 213
Chebyshev filters, 336
CIRC (cross-interleaved Reed-Solomon code), 170, 172-174
Claude I. Shannon, theories of data compression, 244-245
CMOS (complementary metal-oxide semiconductor), 68
cochlea *see* inner ear
coherence (lasers), 99
coherent demodulation, 273
color spectrum, 121-122

community concert hall example, 159-164
  bit error rate (BER), 159
compact disc player, 180
compact discs, 496
  bit error rate (BER), 174
  burst errors, 173
  constant linear velocity (CLV), 169
  cross-interleaved Reed-Solomon code (CIRC), 170, 172-174
  D/A (digital-to-analog) converter, 188
  data format, 169-170
  decoding, 186-187
  digital filter, 187-188
  eight-to-fourteen modulation (EFM), 170, 178-179
  electrical systems, 186-188
  encoding, 172-179
    sampling, 172
    subcode word, 174-178
  International Standard Recording Code (ISRC), 178
  low-pass filter, 188
  medium, 169
  messages, 168
  random errors, 173
  structure, 171-172
  Universal Product Code (UPC), 177
companded predictive delta modulation (CPDM), 256
complementary metal-oxide semiconductor (CMOS), 68
component digital video *see* D1
computer application write-once optical recording systems, 193-195
  constant angular velocity (CAV), 194
  constant linear velocity (CLV), 193-194
connectors
  butt-type, 157-158
  fiber optic losses, 163
  fiber optics, 156-158
  lens-type, 158
constant angular velocity (CAV)
  computer application write-once optical recording systems, 194
  magneto-optical recording (MOR), 199
constant linear velocity (CLV), 169

CD-WO (compact discs, write-once), 191
computer application write-once optical recording systems, 193-194
magneto-optical recording (MOR), 199
continuous transforms, 302
continuous-wave (CWS) lasers, 102-103
conversion
  digital-to-analog (D/A), 55-56, 59
  multibit, 55-81
converters
  bipolar D/A, 66-67
  bit-switching D/A, 64-66
  dual D/A with bias, 73-76
  dual-channel A/D, 77-81
  dynamic element matching (DEM), 56
  integrating D/A, 56
  linear 18-bit D/A, 67-68
  linearity, 58
  multiple-bias D/A, 68-71
  ROM-compensated D/A, 71-73
  sigma-delta, 407-408
convolution (digital systems), 298-300
cooling systems (transmitters), 142
copper cable, problems associated with, 119
couplers
  optical, 147-148
  transmitters, 142
CRCC (cyclic redundancy check code), 176
critical angle (refraction), 122-123
critical sampling, 35
cross-interleaved Reed-Solomon code (CIRC), 170, 172-174
cyclic redundancy check code (CRCC), 176

# D

D/A (digital-to-analog) conversion, 55-56
  low-bit with A/D, 417-419
  quasi-fourth-order noise shaping, 401-403
  second-order noise shaping, 390-394
  third-order noise shaping, 394-400

D1 (component digital video) format, 203-213
   audio processing, 211-212
   cassette shell specifications, 204
   channel coding, 213
   magnetic tape specifications, 204
   track pattern, 205-209
   transport, 205-209
   video processing, 210-211
D2 (composite digital video) format, 214-220
   cassette shell, 214
   magnetic tape specifications, 214
   processing, 218-220
   track pattern, 215-218
   transport, 215-218
DAT 16/12 digital audio data compression scheme, 259-260
data compression, 241-264
   adaptation, 248-249
   block vs. nonblock compression, 247
   digital audio systems, 258-264
   distortion measure, 245-246
   distortion rate theory, 245
   distortion-rate function (DRF) minimization, 246
   information theory, 243-245
   masking, 259
   measurement, 242
   noiseless vs. noisy coding, 246-247
   optimal performance theoretically attainable (OPTA), 246
   practices, 246-249
   prediction vs. interpolation, 248
   quantization, 247
   rate-distortion function (RDF) minimization, 246
   systems, 258-264
      4:1, 261
      BBC 14/10, 259-260
      CD-I (compact disc-interactive), 260-261
      DAT 16/12 scheme, 259-260
      DCC (digital compact cassette), 263-264
      designs, 259
      DVI (digital video interactive), 260-261
      Eureka, 262-263
      PASC (precision adaptive subband coding), 263-264
      Video 8 10/8 scheme, 259-260
   techniques, 249-258
      adaptive delta modulation (ADM), 256
      adaptive delta pulse code modulation (ADPCM), 256-257
      arithmetic coding, 251-253
      delta modulation (DM), 255-256
      differential pulse code modulation (DPCM), 254-255
      digital filters, 258
      Huffman coding, 249-251
      Shannon-Fano technique, 251
      transformations, 257-258
      Ziv-Lempel coding, 253-254
   theories, 242-246
data formats, compact discs, 169-170
DCC (digital compact cassette) digital audio data compression systems, 263-264
decibels, 7, 13
decimation, 364
delta function, 297
delta modulation (DM), 255-256, 378-379
   baseband signals, 270
DEM (dynamic element matching) converter, 56
demodulation, 273
demultiplexers
   band-pass interference filter as, 151
   fiber optics, 149-151
   prism grating, 150-151
   slanted rod, 150
   wavelength division multiplexing (WDM), 149-151
differential linearity error, 57
differential pulse code modulation (DPCM), 254-255
diffraction, 122

# Index

digital audio
    automobile sound systems, 497
    biomedical applications, 499-501
    compact disc, 496
    digital audio tape (DAT), 496
    digital workstations, 497
    education, 498-499
    entertainment industry, 497
    fiber optics, 495
    film, 203, 231-237
    new markets, 491-501
    pattern recognition element, 494-495
    satellite broadcasting, 267-290
    The Third Wave, 492-494
    video, 203-231
digital audio broadcasting (DAB), 267
digital audio design
    dynamic range, 10, 13-14
    filtering scheme, 12-13
    frequency response, 10
    limitations, 10-14
    loudness, 13
    minimum sound pressure level, 13
    sampling rate, 12
digital audio systems, quantization error, 16
digital audio tape (DAT), 168, 496
digital broadcasting studios, 269
    baseband signals, 269-270
digital dither, 48
digital filters, 258, 313-343
    *see also* filters
    classification, 319-324
    components, 319
    design techniques, 333-343
        computer-aided FIR filter design, 340-342
        equiripple FIR filters, 343
        FIR filters with arbitrary responses, 336, 339
        frequency sampling, 342-343
        general analog-to-digital transformations, 334-336
        windowing, 339-340
    errors, 329-331
    examples, 324-329
    pole patterns, 315
    stability, 316-318
    system response, 315
    types of realizations, 331-332
        biquadratic, 332
        canonic, 332
        direct form I, 331-332
        direct form II, 332
    zero patterns, 315
digital frequencies, 312-313
digital mixing consoles, 365-368
digital signal processing (DSP), 423-488, 495-496
    allocating memory, 472
    applications, 347-371, 425
    architectures, 423-443
    arithmetic logic unit (ALU), 424, 427-429
    as disk drive controller, 430
    assemblers, 481-488
    chip architecture, 431
        AT&T WE-DSP16A, 437-441
        AT&T WE-DSP32C, 437-441
        Motorola DSP56001, 431-435
        Motorola DSP96002, 441-443
        Texas Instruments TMS320, 435-437
    compilers, 481-488
    consumer audio, 359-362
    CPU (central processing unit), 427-428
    dedicated interface controllers, 430-431
    development packages, 481-488
    dual-port memory, 430
    early restoration processes, 369-370
    handshaking, 431
    I/O to other peripheral devices, 478
    implementing algorithms with software, 470-476
    instruction sets, 452-470
    interfacing, 476-481
    interrupt signals, 478-479
    length-five FIR filter, 473-476
    memory mapping, 430
    programming languages, 445-446
    programming models, 446-452
    quantization, 472
    restoration workstation, 370-371

standard bus width, 428
tapeless studio, 368-369
theory, 293-343
digital signals, 293-313
   digital frequencies, 312-313
   discrete Fourier transform (DFT), 304-306
   fast Fourier transform (FFT), 306-307
   Fourier transform, 304
   frequency domain, 301
   Laplace transform, 302-304
   one-sided, 295
   spectra, 301-302
   time domain, 301
   transforms, 302-312
   two-sided, 295
   unit circle, 312-313
   z-transform, 307-312
digital soundfield processing, 348-359
   car, 355-359
   home, 350-355
digital systems, 293-313
   convolution, 298-300
   discrete, 295-296
   homogeneity, 296
   impulse response, 297
   superposition, 296
   time-variant, 296-297
digital transmission processing, 271-276
   modulation, 272-276
   multiplexing, 271-272
digital-to-analog (D/A) conversion, 55-56, 59
direct broadcast satellite (DBS) system, 268, 283-288
   Australia, 286
   Europe, 286-288
   Japan, 284-286
direct current modulation, 110
direct form I realization (digital filters), 331-332
direct form II realization (digital filters), 332
directional optical couplers, 147
discrete digital systems, 295-296
discrete Fourier transform (DFT), 304-306

discrete transforms, 302
dispersion, 126-127
   intermodal, 126
   material, 126
   waveguide, 126-127
dispersion-shifted single mode optical fibers, 130
distortion, 14-17
   measure, 245
   zero-cross, 60-63
distortion rate theory of data compression, 245-246
distortion-rate function (DRF) minimization, 246
dithering, 16, 47-50
   analog, 47
   average noise power, 49-50
   digital, 48
   Gaussian, 47
   high-frequency, 48
   noise shaping, 388-390
   rectangular probability density function (RPDF), 47-48
   triangular probability density function (TPDF), 49
double-heterojunction (DH) laser diode structure, 107, 110
dual D/A converter with bias, 73-76
dual-channel A/D converters, 77-81
DVI (digital video interactive) digital audio data compression format, 260-261
dye polymer recording, 200
dynamic element matching (DEM) converter, 56
dynamic range of digital audio design, 10, 13-14

## E

ear canal, 2
eardrum, 2-4
edge-emitter double-heterojunction LED (light-emitting diode), 140-141
eight-to-fourteen modulation (EFM), 170, 178-179
elastic mechanical optical splicing, 152-153

electrical systems of compact discs,
  186-188
    D/A (digital-to-analog) converter, 188
    decoding, 186-187
    digital filters, 187-188
    low-pass filters, 188
electro-optic switches, 144-146
electromagnetic waves, 122
elliptic filters, 336
embedded controllers, 426
envelope demodulation, 273
equalization
    sample-and-hold circuits, 40
    soundfields, 348
erasable optical recording systems,
  196-201
    applications, 201
    disc drives, 201
    dye polymer recording, 200
    magneto-optical recording (MOR),
      197-200
    phase change recording, 200
etched-well LED (light-emitting diode),
  138-140
Eureka digital audio data compression
  systems, 262-263
Eureka/147 terrestrial DAB system,
  288-289
Eustachian tube, 2

## F

fast Fourier transform (FFT), 306-307
fiber optics, 119-164
    absorption, 127
    advantages, 119-120
    angle of acceptance, 125-126
    audio applications, 158-164
    community concert hall example,
      159-164
    connectors, 156-158
      losses, 163
    demultiplexers, 149-151
    diffraction, 122
    digital audio, 495
    disadvantages, 120
    dispersion, 126-127
    electromagnetic waves, 122
    intermodal dispersion, 126
    intramodal dispersion, 126
    manufacturing optical cable, 133-136
    multiplexers, 149-151
    numerical aperture (NA), 125-126
    optical
      cables, 128, 130-136
      couplers, 147-148
      fibers, 128-130, 136-137
      properties, 121-128
      spectrum, 121-122
      splicing, 152-156
    polarization, 127-128
    principles, 120-128
    Rayleigh scattering law, 127
    receivers, 143-144
    reflection, 122-124
    refraction, 122-124
    scattering, 127
    switches, 144-146
    transmitters, 137
      cooling systems, 142
      couplers, 142
      laser diode, 141
      light-emitting diode (LED), 137-141
      photodetector, 142
film
    digital audio, 203, 231-237
      fluorescent layer system, 234
      imaging dye technique, 234-237
    digital sound design, 231-233
      bandwidth, 232-233
      durability and high-speed
        duplication, 233
      dynamic range, 233
      number of channels, 232
filters
    *see also* digital filters
    Bessel, 336
    Butterworth, 336
    Cauer, 336
    Chebyshev, 336
    digital audio design, 12-13
    elliptic, 336

finite impulse response (FIR), 321-322, 324
   ideal low-pass, 34
   infinite impulse response (IIR), 319, 322, 324
   low-pass, 37-39
   nonrecursive, 37, 320-321
   oversampling, 42
   recursive, 37, 319
finite impulse response (FIR) filter, 321-322, 324
floating-point converter, 16
fluorescent layer system film digital audio, 234
format conversion ratio, 25
Fourier transform, 304
   infinite impulse response, 39
   sampling theory, 33-34
   time function, 29-31
frequency division multiplexing (FDM), 272
frequency domain of digital signals, 301
frequency response, 10
frequency shift keying (FSK), 274
frequency spectrum of sampled signals, 32-33
fused optical splicing, 155-156
   local injection and detection (LID), 156
   profile alignment systems (PAS), 156

## G

GaAs laser, 107
GaAsAl laser, 107
gain saturation, 93-95
Gaussian dither, 47
Gibbs phenomena, 339
   low-pass filtering, 39
graded-index multimode optical fibers, 129-130

## H

Hartley-Shannon law, 25
Harvard Architecture, 431
HDLC (high-density linear converter), 398-400

high-frequency dither, 48
homojunction laser diode structure, 107
Huffman coding, 249-251
   disadvantages, 250-251
human auditory capabilities, 1-18
   error correction, 17-18
   limitations in digital audio design, 10-14
   noise and distortion, 14-17
   psychoacoustical foundations, 6-8
   psychoacoustical phenomena, 9
   spatial audiology, 8-9

## I

imaging dye technique film digital audio, 234-237
   application, 236-237
impulse function, 297
impulse response in digital systems, 297
in-band noise power, 46
index of refraction, 122
induced transitions
   absorption, 90
   stimulated emission, 90-91
infinite impulse response (IIR) filters, 319, 322, 324
   Fourier transform, 39
information theory of data compression, 243-245
   entropy of source theory, 244
   entropy of words theory, 245
initial delay gap (IDG), 349
inner ear, 1, 4-5
instruction sets, digital signal processing (DSP), 452-470
integrating D/A converter, 56
intensity, 7-8
interference intersymbol, 36
intermodal dispersion, 126
International Standard Recording Code (ISRC), 178
interpolation, 18
   data compression, 248
intersymbol interference, 36
inverse Fourier transform, 304
   sampling theory, 34

inverse Laplace transform, 303
inverse z-transform, 307
isotropic antenna, 277

## J-L

jitter, 26
Laplace transform, 302-304
laser diode, 104-116
    audio applications, 114-116
    driver circuitry, 113-114
    fiber optic transmitters, 141
    maintenance, 111-113
    manufacturing, 111
    modulation, 110-111
    operation, 104-107
    reliability, 111-113
    structure, 107-110
        double-heterojunction (DH), 107, 110
        homojunction, 107
        single-heterojunction, 107
        single-heterostructure (SH), 107
        stripe-geometry, 110
lasers
    amplification, 92-93, 97-99
    applications, 86-87
    characteristics, 99-104
    coherence, 99
    continuous-wave (CW), 102-103
    cooling requirements, 104
    decay population density rate of change, 94
    directionality of beam, 100
    efficiency, 102
    elements, 88-89
    GaAs, 107
    GaAsAl, 107
    gain coefficient, 96-97
    gain saturation, 93-95
    history, 85-87
    induced transitions, 90-91
        population density rate of change, 94
    lifetime, 102-103
    lineshape, 91
    longitudinal modes, 96, 101
    medium, 88
    mode structure, 101
    monochromatic light, 99
    optical cavity, 88
    oscillation, 95-99
    physics, 87-88
    polarity of beam, 100
    population density rate of change, 92-93
    power, 101
    pulsed, 102-103
    pumping source, 88
    radiation field
        absorption energy density change, 92
        energy density rate of change, 92
        gain of energy density, 96
        population density rate of change, 93
        stimulated emission energy density change, 92
    semiconductor *see* laser diode
    spontaneous emission, 89-90
        radiation field energy density change, 93
    technology, 85-116
    theory, 89-91
    transversal modes, 101
    transverse electrostatic mode (TEM), 101
least significant bit (LSB), 58-59
LED (light-emitting diode), 105
    Burrus, 138
    edge-emitter double-heterojunction, 140-141
    etched-well, 138-140
    fiber optic transmitters, 137-141
    planar-surface, 138
    structure, 138-141
lens-type connectors, 158
light-atom interactions, 85-86
light-emitting diode (LED), fiber optic transmitters, 137-141
linear 18-bit D/A converters, 67-68
linearity, 58
lineshape (lasers), 91

local injection and detection (LID) fused optical splicing, 156
localization, 8
longitudinal mode of laser cavity, 96
loose tube
    mechanical optical splicing, 153
    optical cables, 131-132
loudness
    digital audio design, 13
    intensity, 7-8
    soundfields, 348
loudspeaker crossover networks, 362-364
    decimation, 364
low-bit conversion, 375-377
    A/D, 403, 405-412
    A/D-D/A converter, 417-419
    converter chip, 413-414
low-pass filtering, 37-39
    compact discs, 188
    cutoff frequency, 37
    Gibbs phenomenon, 39
    ideal, 34
LSB (least significant bit), 58-59

## M

magnetic recording systems, 167-168
magneto-optical recording (MOR), 197-200
    constant angular velocity (CAV), 199
    constant linear velocity (CLV), 199
    Kerr effect, 197
masers, 86
MASH system, 394-398
masking, 15-16, 259
material dispersion, 126
MCVD (modified chemical vapor deposition), 135
mean square quantizing error, 45
meatus *see* ear canal
mechanical optical splicing, 152-155
    elastic, 152-153
    loose tube, 153
    ribbon, 153
    rotary, 155
    silicon chip array, 153-155
    V-groove, 152

mechanical recording systems, 167
mechanical switches, 144
microbending, 124-125
middle ear, 1-4
modal dispersion *see* intermodal dispersion
modified chemical vapor deposition (MCVD), 135
modulation
    amplitude shift keying (ASK), 273
    baseband signals, 270
    delta, 378-379
    digital transmission processing, 272-276
    frequency shift keying (FSK), 274
    laser diode, 110
    phase shift keying (PSK), 275
    quadrature amplitude (QAM), 275
    sigma-delta (SDM), 377-380
monochromatic light (lasers), 99
most significant bit (MSB), 58-59
Motorola DSP56001 DSP chip, 431-435
    56001 Application Development System (ADS), 487
    assembler, 482-483
    Harvard Architecture, 431
    instruction set, 454-456
    linker, 483
    programming model, 446-448
    simulator program, 483
Motorola DSP96002 DSP chip, 441-443
    assembler, 484
    instruction set, 456-461
    OnCE (on-chip emulation), 484
    optimizing C compiler, 484
    programming model, 448
MSB (most significant bit), 58-59
multibit conversion, 55-81
    18-bit D/A architecture, 63-76
    analog-to-digital (A/D), 76-81
    digital-to-analog (D/A), 55-56
    nonlinearity, 56-60
    zero-cross distortion, 60-63
multilevel quantizer A/D converter, 414-417

## Index

multimode optical fibers, 128-130
   graded-index, 129-130
   step-index, 129-130
multiple-bias D/A converters, 68-71
multiplexing
   digital transmission processing, 271-272
   fiber optics, 149-151
   frequency division (FDM), 272
   time division (TDM), 272

### N

noise, 14-17
   requantization, 43
noise shaping, 16-17, 375, 380-419
   dithering, 388-390
   HDLC (high-density linear converter), 398-400
   idle patterns, 388
   MASH system, 394-398
   quasi-fourth-order in D/A conversion, 401-403
   second-order in D/A conversion, 390-394
   third-order in D/A conversion, 394-401
   thresholding, 388
   VANS (Victor advanced noise shaping), 401-403
noncoherent demodulation, 273
nonlinearity, 57-60
nonmonotonicity, 57
nonrecursive filters, 37, 320-321
NRZ full binary pulse type, 27-28
numerical aperture (NA), 125-126
Nyquist frequency, 33, 35
Nyquist sampling theorem, 33-36

### O

optical cables, 128, 130-136
   design, 130-133
   loose tube, 131-132
   manufacturing, 133-136
      vapor phase deposition method, 133-134
      modified chemical vapor deposition (MCVD), 135
      optical time domain reflectometer (OTDR), 137
      outside vapor deposition (OVD), 134
      plasma chemical vapor deposition (PCVD), 135-136
   plenum, 132
   ribbon, 132
   submarine, 132-133
   tight buffer, 131
   tight tube, 131
   vapor axial deposition (VAD), 135
optical couplers, 147-148
   directional, 147
   star, 147-148
optical discs, 167-201
   encoding synchronization word, 179
optical fibers, 119, 128-130
   attenuation, 162
   bandwidth, 162
   buffer, 128
   characteristics, 130
   cladding, 128
   core, 128
   failures of, 136-137
   fracturing, 136
   maintenance, 137
   microbending, 124-125
   multimode, 128-130
   single mode, 128-130
optical pickup *see* pickup
optical properties
   diffraction, 122
   reflection, 122-124
   refraction, 122-124
   spectrum, 121-122
   total internal reflection, 124-125
optical recording systems, 168-201
   erasable, 196-201
   WORM (write-once read-many), 188
   write-once, 189-196
optical spectrum, 121-122
optical splicing, 152-156
   fused, 155-156
   mechanical, 152-155

optical systems, 180-186
focusing, 183-184
optical pickup, 180-185
pickup control, 186
tracking, 184
optical time domain reflectometer (OTDR), 137, 155
optimal performance theoretically attainable (OPTA) data compression, 246
organ of Corti, 5
orthogonal transformations, 258
oscillation (lasers), 95-99
outer ear, 1-2
outside vapor deposition (OVD), 134
oval window, 3-5
OVD (outside vapor deposition), 134
oversampling, 42-44
    filters, 42
    ratio, 43

## P

P subcode word, 175-176
PAM (pulse amplitude modulation), 22, 55
parabolic antennas, 278-280
PASC (precision adaptive subband coding) digital audio data compression systems, 263-264
pattern recognition circuitry, 18
PCM (pulse code modulation), 22-25
PCVD (plasma chemical vapor deposition), 135-136
PDF (probability density function), 47
phase change recording, 200
phase shift keying (PSK), 275
photodetectors
    APD (avalanche), 144
    PIN (positive-intrinsic-negative), 143
    transmitters, 142
photonic switches *see* electro-optic switches
photons, 88
pickup, 180-183
    control, 186
    polarization beam splitter (PBS), 181
    quarter-wave plate (QWP), 181
    single-beam, 184-185
PIN (positive-intrinsic-negative) photodetectors, 143
pitch, 6-7
pitch-place theory, 6
planar-surface LED (light-emitting diode), 138
plasma chemical vapor deposition (PCVD), 135-136
plenum cables, 132
polarization, 127-128
polarization beam splitter (PBS), 181
power systems in satellite design, 279-280
PPM (pulse position modulation), 21-22
prediction (data compression), 248
prism grating demultiplexers, 150-151
probability density function (PDF), 47
profile alignment systems (PAS) fused optical splicing, 156
psychoacoustics, 6-8
    auditory habituation, 9
    intensity, 7-8
    localization, 8
    other phenomena, 9
    pitch, 6-7
    pure tones, 7
    selective discrimination, 9
    timbre, 7
    tones, 7
pulse amplitude modulation (PAM), 22, 55
pulse code modulation (PCM), 22-25
    baseband signals, 270
    conversion, 376
pulse modulation, 21-27
    base-line wander, 26
    bit coding, 27-29
    jitter, 26
    measuring deterioration, 26
    practical limitations, 25-27
    theoretical performance, 23-25
pulse position modulation (PPM), 21-22
pulse width modulation (PWM), 16, 21-22
    baseband signals, 270
pulsed lasers, 102-103

# Index

pure tones, 7
PWM (pulse width modulation), 16, 21-22

## Q

Q subcode word, 176-178
quadrature amplitude modulation (QAM), 275
quadrature phase shift keying (QPSK), 275
quanta values, 22
quantization, 44-47
    data compression, 247
    digital audio systems error, 16
    digital signal processing (DSP), 472
    errors, 44-46
    in-band noise power, 46
    mean square error, 45
    scalar, 247
    uniform, 45
    vector, 247
quantum mechanics, 87-88
quarter-wave plate (QWP), 181
quasi-fourth-order noise shaping in D/A conversion, 401-403

## R

random errors, 173
rate-distortion function (RDF) minimization, 246
Rayleigh scattering law, 127
receivers
    antenna gain, 283
    APD (avalanche) photodetectors, 144
    fiber optics, 143-144
    noise temperature, 283
    PIN (positive-intrinsic-negative) photodetectors, 143
    satellite, 281-283
recording systems, 167-168
    magnetic, 167-168
    mechanical, 167
    optical, 168
rectangular probability density function (RPDF), 47-48
recursive filters, 37, 319

reduced instruction set computers (RISC), 452
reflection, 122-124
    total, 122
    total internal, 124-125
reflective star optical couplers, 147-148
refraction, 122-124
    angle of incidence, 122-124
    critical angle, 122-123
    index of, 122
region of convergence (ROC), 310-312
requantization noise, 43
reverberation in soundfields, 348
ribbon cables, 132
ribbon mechanical optical splicing, 153
RISC (reduced instruction set computers), 452
ROM-compensated D/A converters, 71-73
rotary mechanical optical splicing, 155
round window, 3
RPDF (rectangular probability density function), 47-48
RZ half binary pulse type, 27-28

## S

sample-and-hold circuits, 39-41
    equalization, 40
sampled signals
    defining nature, 32
    frequency spectrum, 32-33
sampling
    compact discs, 172
    critical, 35
    function, 33
    rate in digital audio design, 12
sampling systems
    bit coding, 28-29
    brickwall characteristic, 37
    dither, 47-50
    frequency spectrum, 33
    low-pass filtering, 37-39
    Nyquist sampling theorem, 34-36
    oversampling, 42-44
    pulse modulation, 21-27
    quantization, 44-47

sample-and-hold circuits, 39-41
time sampling, 29-32
sampling theory, 33-36
   Fourier transform, 33-34
   inverse Fourier transform, 34
satellite broadcasting, 267-290
   digital studio, 269
   digital transmission processing, 271-276
   direct broadcast satellite (DBS) system, 283-288
   receivers, 281-283
   satellite
      design, 278-281
         transmitters, 276-278
   terrestrial DAB systems, 288-289
satellite design, 278-281
   antennas, 280
   power systems, 279-280
   stabilizing subsystem, 279
   station-keeping subsystem, 279
   transponders, 280-281
satellite
   receivers, 281-283
   transmitters, 276-278
scalar quantization, 247
scattering, 127
second-order noise shaping in D/A conversion, 390-394
selective discrimination, 9
semiconductor laser *see* laser diode
series transforms, 302
Shannon-Fano technique, 251
sigma-delta converters, 407-408, 413-414
sigma-delta modulation (SDM), 377-380
sign and magnitude binary notation, 22
signals, defining nature of sampled, 32
silicon chip array mechanical optical splicing, 153-155
$(\sin x)/x$ function, 33, 36, 40
sinc function, 30, 34
single mode optical fibers, 128-130
   dispersion-shifted, 130
   step-index, 130

single-beam pickup, 184-185
single-heterojunction laser diode structure, 107
single-heterostructure (SH) laser diode structure, 107
slanted rod demultiplexers, 150
snakes, 158
Snell's law, 123-124
sound
   collection by outer ear, 2
   error correction, 17-18
soundfields, 348-349
   equalization, 348
   extension, 348
   loudness, 348
   reverberation, 348
spatial audiology, 8-9
spontaneous emission, 85, 89-90
   radiation field energy density change, 93
   transition rate, 89-90
stabilizing subsystem, 279
star optical couplers, 147-148
   reflective, 147-148
   transmissive, 147-148
station-keeping subsystem, 279
step-index
   multimode optical fibers, 129-130
   single mode optical fibers, 130
stimulated emission, 86
   induced transitions, 90-91
   radiation field energy density change, 92
stripe-geometry laser diode structure, 110
subcode word, 174-178
   P, 175-176
   Q, 176-178
submarine cables, 132-133
super high frequency (SHF), 270
switches
   electro-optic, 144-146
   fiber optics, 144-146
   mechanical, 144

synchronization word, 179
synchronous demodulation, 273

## T

tapeless studio, 368-369
TEM (transverse electrostatic mode), 101
temporary threshold shift (TTS), 7
terrestrial DAB systems, 288-289
    cable DAB systems, 289-290
    CATV audio tuners, 290
    Eureka/147, 288-289
Texas Instruments TMS320 DSP chip, 435-437
    Analog Interface Board, 486
    Digital Filter Design Package (DFDP), 485
    instruction set, 461-466
    software support, 485
    Texas Instruments Software Development System (SWDS), 485-486
    XDS hardware emulator, 486
third-order noise shaping in D/A conversion, 394-401
threshold of hearing, 10, 12-13
tight buffer optical cables, 131
tight tube optical cables, 131
timbre (psychoacoustics), 7
time division multiplexing (TDM), 272
time domain, 301
time function of Fourier transform, 29-31
time sampling, 29-32
time-variant digital systems, 296-297
tones, 7
total internal reflection, 124-125
total reflection, 122
TPDF (triangular probability density function), 48-49
tracking in optical systems, 184
transforms, 257-258
    bilinear, 335-336
    continuous, 302

digital signals, 302-312
discrete, 302
discrete Fourier (DFT), 304-306
fast Fourier (FFT), 306-307
Fourier, 304
inverse Fourier, 304
inverse Laplace, 303
inverse z, 307
Laplace, 302-304
orthogonal, 258
series, 302
z, 307-312
transmissive star optical couplers, 147-148
transmitters
    cooling systems, 142
    couplers, 142
    fiber optics, 137-142
    photodetector, 142
    satellite, 276-278
transponders in satellite design, 280-281
transverse electrostatic mode (TEM), 101
triangular probability density function (TPDF), 48-49
TTS (temporary threshold shift), 7
two's complement binary notation, 22
tympanic membrane *see* eardrum

## U

uniform quantization, 45
unit circle, 312-313
Universal Product Code (UPC), 177

## V

V-groove mechanical optical splicing, 152
VAD (vapor axial deposition), 135
VANS (Victor advanced noise shaping), 401-403
vapor axial deposition (VAD), 135
vapor phase deposition optical cable manufacturing, 133-134
vector quantization, 247

video (digital audio), 203-231
    1-inch type C video, 228-231
    8mm video, 221-228
    D1 (component digital video) format, 203-213
    D2 (composite digital video) format, 214-220
Video 8 10/8 digital audio data compression scheme, 259-260

## W

waveguide dispersion, 126-127
wavelength dispersive devices, 149-151
wavelength division multiplexing (WDM), 149-151
wavelength selective devices, 151
write-once optical recording systems, 189-196
    CD-WO (compact-disc, write-once format), 188-193
    computer application, 193-195
    recording methods, 189-190

## Z

z-transform, 307-312
    region of convergence (ROC), 310-312
zero-cross distortion, 60-63
Ziv-Lempel coding, 253-254